SHAPING THE DAY

Shaping the Day

A History of Timekeeping in England and Wales 1300–1800

PAUL GLENNIE
AND
NIGEL THRIFT

UNIVERSITY PRESS

OXFORD
UNIVERSITY PRESS

Great Clarendon Street, Oxford, OX2 6DP,
United Kingdom

Oxford University Press is a department of the University of Oxford.
It furthers the University's objective of excellence in research, scholarship,
and education by publishing worldwide. Oxford is a registered trade mark of
Oxford University Press in the UK and in certain other countries

© Paul Glennie and Nigel Thrift 2009

The moral rights of the authors have been asserted

First Edition published in 2009

All rights reserved. No part of this publication may be reproduced, stored in
a retrieval system, or transmitted, in any form or by any means, without the
prior permission in writing of Oxford University Press, or as expressly permitted
by law, by licence or under terms agreed with the appropriate reprographics
rights organization. Enquiries concerning reproduction outside the scope of the
above should be sent to the Rights Department, Oxford University Press, at the
address above

You must not circulate this work in any other form
and you must impose this same condition on any acquirer

Published in the United States of America by Oxford University Press
198 Madison Avenue, New York, NY 10016, United States of America

British Library Cataloguing in Publication Data
Data available

Library of Congress Cataloging in Publication Data
Data available

ISBN 978–0–19–960512–5

Contents

List of Figures	vi
List of Plates	ix
List of Tables	x
Acknowledgements	xiii
1. Introduction: The Measured Heart	1
2. Clocks, Clock Times, and Social Change	22
3. 'Not Everyone Occupies the Same Now': Reconceptualizing Clock Time	65
4. Clock Times in Medieval and Early Modern Bristol	100
5. The Provision of Clock Time in Pre-Modern England	135
6. Clock Times in Everyday Lives	181
7. Precision in Everyday Lives	244
8. 'Posted Within Shot of the Grave': Seafaring Times	279
9. The Pursuit of Precision	329
10. Clocks from Nowhere? John Harrison in Context	359
11. Some Concluding Remarks	407
Bibliography	419
Index	451

List of Figures

1.1	Galileo's rolling ball experiment to time acceleration under gravity	3
1.2	Places from where the Galilean moons of Jupiter had been observed, 1610–11	6
2.1	The core components of mechanical timekeepers	30
2.2	Diagram of verge and foliot mechanism	32
2.3	Human involvement in the running of early turret clocks	33
2.4	Diagram of pendulum regulator and going train	34
2.5	Human involvement in early pendulum-regulated clocks, c.1670	36
2.6	Three dimensions of time discipline	45
3.1	Locations of the earliest documented English clocks	76
4.1	Locations of clockmakers and watchmakers in Bristol in the 1770s	111
4.2	Clock- and watchmaking apprenticeships begin in Bristol, 1640–1819	113
4.3	Communications as time-cues in late eighteenth-century Bristol	118
4.4	Hackney coach stands in late eighteenth-century Bristol	120
4.5	Increases in clock/watch ownership by gross wealth of inventories	125
5.1	Clocks on civic buildings in early modern England	146
5.2	Annual spending on the church clock at (Bishops) Stortford, Hertfordshire, 1470–1560	149
5.3	English parishes with surviving churchwardens' accounts, 1400–1700	150
5.4	Numbers of parishes with surviving churchwardens' accounts	151
5.5	The uneven geographical survival of churchwardens' accounts for rural and non-market parishes	152
5.6	Parishes with surviving accounts, by recorded adults in 1676	153
5.7	Proportion of documented parishes with church clocks, overall and for different settlement types	154
5.8	Trends in church clocks for a selection of parish cohorts according to their earliest surviving accounts	155
5.9	Percentage of parishes whose churchwardens' accounts refer to church clocks	156

List of Figures

5.10	The frequency of church clocks by recorded adults in a parish in 1676	158
5.11	Clocks recorded in the Visitation of the Archdeaconry of Bedford, 1708	159
5.12	Trends in various means of signalling clock times	160
5.13	Spatial variations in inventoried consumer goods, 1660–1730	165
5.14	Testators' ownership of clocks by probated wealth, East Sussex 1710–39	169
5.15	Locations of clockmakers active before 1730 for whom one or more clocks survive	173
6.1	Locations mentioned in this chapter	183
6.2	Components of researching clock time practices	184
6.3	Clock time references by Samuel Pepys and his contemporaries	197
7.1	Specific places for which almanacs were published	248
7.2	Progressive improvements in the accuracy of mechanical clocks	251
7.3	Places from where durations of totality reported to Edmund Halley in 1715	259
7.4	Considerations in the demand for precision time-use	262
7.5	Late-sixteenth century English post-towns	267
8.1	Time-space arrangement of the naval watch system	316
9.1	The shrinkage of France, following the 1693 Picard-Cassini mapping of France with longitudes based on Jovian eclipses	337
9.2	Inaccuracies of latitudinal and longitudinal positions in maps of England by Saxton (1580s) and Adams (1670s)	338
9.3	Progressive reduction of longitude errors for selected major ports in Spanish and Portuguese territories	340
9.4	Table of predicted positions of Jovian satellites, the first in England, by the Astronomer Royal, John Flamsteed, 1683	343
9.5	Astronomical precision-timing as a network of interdependent precisions	349
9.6	Recording sheet for lunar distance observations, and calculations of longitude based on them	352
9.7	The marine chronometer as a network of minimal human interference	354
9.8	Interacting communities of practice in clockmaking	356
10.1	Locations of Barton-on-Humber and Barrow-upon-Humber	361
10.2	Clockmakers in northern England with extant work	367
10.3	Genealogy of John Harrison	379
10.4	Centres of horse-breeding and racing before 1750	385

Every effort has been made to contact copyright holders. The publisher will be pleased to rectify any omission in subsequent impressions.

List of Plates

1. A clockmaker at work, *c.*1350–75
2. Birth dates and times of Robert Hill's children, 1517–21
3. A sketch by Hans Holbein for his portrait of the family of Sir Thomas More
4. Mechanism of a turret clock by Leonard Tennant of London, *c.*1600–20
5. Churchwardens' recorded spending on parish church clocks, 1540–1700
6. A London clock for export markets, *c.*1700
7. Cottage interior, illustrating plebeian clock ownership
8. Fine time-telling from single-hand clocks
9. Fine time-telling from a sundial
10. John and Joseph Harrison's long-case clock of 1728
11. Illustration from Jeremy Thacker's *The Longitude Examin'd*, 1714

Every effort has been made to contact copyright holders. The publisher will be pleased to rectify any omission in subsequent impressions.

List of Tables

2.1	Chronology of technological progress in clocks	35
2.2	Conventional dimensions of clock time and signalling	38
2.3	Chronology of some major changes in social-disciplinary technologies	43
4.1	Regular timetabled coach departures from Bristol, 1775	106
4.2	Clock- and watchmaking apprenticeships begun in Bristol, 1640–1819	112
4.3	An advertisement to recover a stolen watch	114
4.4	Known Bristol clockmakers, 1750–1800	115
4.5	Types of evidence available on Bristol clockmakers before 1800	126
4.6	Evidence for church clocks in early modern Bristol	127
5.1	Documented public clocks in medieval European cities	139
5.2	Consumer goods in English probate inventories, 1670s–1720s	164
5.3	Valuations of clocks made by probate appraisers: East Sussex, 1730–7	170
5.4	The clock sales of Samuel Roberts, 1755–74	172
5.5	Rooms in which clocks were located in selected inventory collections	174
6.1	Daily routine of Cecily, Duchess of York, *c.*1490	185
6.2	Authors of the early modern diaries and memoranda analysed	195
6.3	Uses of clock times in early modern English diaries and memoranda	199
6.4	Part of Richard Hill's records of his children's birth times, 1510s and 1520s	208
6.5	Dorchester deponents using clock time to locate narrative within day or night	215
6.6	Ending of letter from Thomas Betson to Katharine Stonor, June 1474	216
6.7	Some uses of clock times in Henry Best's 'Farming Book', 1642	218
6.8	Time-items imported through London by native and Hanseatic merchants, 1567–8	221
6.9	Entries on a page of the Huntingdon postmaster's book, late August 1585	222
6.10	Feoffees' instructions to the master of Solihull Free School, 1669	227
6.11	Teaching hours at boys' grammar schools	228

List of Tables

7.1	Contexts for precise clock times in five almanacs for 1642	247
7.2	Precision and its uses in early modern English diaries and memoranda	263
7.3	Selective precision in Samuel Jeake's diary entries, 1670–2	270
8.1	The will of Leonard Tele of Leigh, mariner, 20 April 1570	295
8.2	Inventory of first Frobisher voyage in search of the North West Passage, 1553	298
8.3	An early ship's navigator's instruments, *c*.1631	298
8.4	Instruments sold by John Seller at his house in Wapping or his city shop by the Royal Exchange, second half of seventeenth century	299
8.5	Instruments sold by Thomas Heath and Tycho Wing from their shop near Exeter Exchange in the Strand, 1750 and 1765	299
8.6	Instruments sold by Benjamin Cole from his shop in Fleet Street, 1768	300
8.7	List of instruments issued by the younger George Adams from his shop in Fleet Street, 1789	301
8.8	Timekeeping equipment on three naval ships *c*.1660	304
9.1	Required elements for two methods of determining longitude	341
10.1	Occupational profile of Barton and Barrow testators with probate inventories	372
10.2	Probate inventory of John Harrison's father, Henry Harrison	374
10.3	Churchwardens' payments for work on the church clock of St Mary's, Barton-upon-Humber	381
10.4	Locations of male Thacker baptisms, 1650–1730	394
11.1	Chronology of some major changes in English clock-time practices	408

Acknowledgements

A book like this has necessarily called on the skills and concerns of many interested parties. It may not constitute its own community of practice but it has certainly intersected with the members of many different communities of practice as it has been written. In particular, we want to thank the following for their support, encouragement, and pointers to new sources of information: Trevor Barnes, Richard Britnell, Adrian Evans, Derek Gregory, Carl Griffin, Paul Griffiths, Peter Haggett, Brian Harrison, the late Les Hepple, Chris Humphrey, Dave Kilham, Bruno Latour, Anne Laurence, Davis Ley, Victor Morgan, Simon Naylor, the late Brian Outhwaite, Mark Overton, Robert Poole, Paul Slack, John Styles, Mark Taylor, Lynda Thrift, Anne de Windt, Edwin de Windt, Andy Wood, and Tony Wrigley. We thank also audiences at conferences and seminars in the University of Bristol; Cambridge University; Dartington Hall; the University of Lancaster; the Departments of Urban History and English Local History, University of Leicester; the Institute of Historical Research, London; the London School of Economics; Oxford University; the University of Warwick; the University of British Columbia, Vancouver; and various of the annual or bi-annual conferences of the Institute of British Geographers, the Social Science History Association of America, the Social History Society, and ASSET. We thank the National Endowment for the Humanities for their support of, and Garrett Sullivan and Dan Beaver for inviting, our participation in the July 2002 Faculty Workshop on Society and Space in the Past at Pennsylvania State University, and other participants for their comments.

Generations of researchers have held the expertise and patience of staff at record offices and archives in high regard, and we are no exception. For their helpfulness, professionalism and dedication, which have contributed in no small way to the work in this book, we are very grateful to the archivists and other staff of: the Bedfordshire and Luton Archives and Records Service, Bedford; the Berkshire Record Office, Reading; the Borthwick Institute for Historical Research, York; the Bristol Record Office, Bristol; the City of Bristol Reference Library archives section, Bristol; the Centre for Buckinghamshire Studies, Aylesbury; the Cambridgeshire County Record Office, Cambridge; the Cambridgeshire County Record Office, Huntingdon; the Cheshire and Chester Archives and Local Studies Centre, Chester; the Cornwall Record Office, Truro; the Cumbria Record Office, Carlisle; the Cumbria Record Office, Kendal; the Devon Record Office, Exeter; the North Devon Record Office, Barnstaple; the Devon Local Studies Library, Exeter; the Dorset History Centre, Dorchester; the East Riding of Yorkshire Archives Service, Beverley; the Essex Record Office, Chelmsford; Gloucestershire Archives, Gloucester; the Hampshire

Record Office, Winchester; the Herefordshire Record Office, Hereford; the Hertfordshire Archives and Local Studies Service, Hertford; the Medway Archives and Local Studies Centre, Rochester; the Lancashire Record Office, Preston; the Record Office for Leicestershire, Leicester and Rutland, Wigston Magna; Lincolnshire Archives, Lincoln; the Guildhall Library, London; the London Metropolitan Archives, London; the Norfolk Record Office, Norwich; the Northamptonshire Record Office, Northampton; Nottinghamshire Archives, Nottingham; the Oxfordshire Record Office, Oxford; Reading Central Library, Reading; Shropshire Archives, Shrewsbury; the Somerset Archive and Record Service, Taunton; the Staffordshire County Records Office, Stafford; the William Salt Library, Stafford; the Suffolk Record Office, Ipswich; the Suffolk Record Office, Bury St. Edmunds; the Suffolk Record Office, Lowestoft; the Surrey History Centre, Woking; the East Sussex Record Office, Lewes; the West Sussex Record Office, Chichester; the Warwickshire County Record Office, Warwick; the Wiltshire and Swindon Record Office, Trowbridge; the Worcestershire Record Office, Worcester; the Worcestershire Library and History Service, Worcester; the North Yorkshire County Record Office, Northallerton; the former Surrey History Centre, Kingston-upon-Thames; the former Centre for Kentish Studies, Maidstone.

The archival work underlying the book involves several hundred days over some fifteen years. Many friends provided convivial hospitality during archive visits, including Patrick and Vicky Boggon, Carolyn Brodie, Chris and Judith Carey, David Gillard, Peter Glennie, Jessica and John Haslam, Kate and Rob Jones, Paul Truman, and the Master and Fellows of Jesus College, Cambridge. The book would have taken significantly longer without two periods of study leave, for which Paul Glennie thanks the University of Bristol. He is also grateful to Dr Paul Dorman and Mr M Campbell, formerly of the Bristol Royal Infirmary, Dr A Lavalle and others at Dean Lane Family Practice, and Professor N Scolding, Denise Owen, Carole Copstake, and other staff at Frenchay Hospital's Neurology department for their care and professionalism since January 1992, and to the University of Bristol for help with specialist software and computing equipment.

We thank Tom Holland, for his help with Latin translations and identifying sources of Latin quotes; Simon Godden, for his cartographic skills; Carmel Thomasson, for checking references in churchwardens' accounts at the County Durham Record Office; Emma Jones for clarifying the identity of a manuscript in the Bodleian Library; William Rupp for general research assistance and help with the proofs, George and Annie Coates for discussion of children's development of time-telling skills; and Alastair Glennie and Zinya Coates for their participation in a small experiment.

1
Introduction: The Measured Heart

1.1 GALILEO: THE MEASURED HEART

We want to introduce this book by way of a story told of Galileo Galilei by his first biographer (and former pupil) Vincenzio Viviani. Some time before 1600 (Viviani says 1583, but this is disputed) Galileo became intrigued by the movement of pendulums, his interest arising from watching oil lamps slowly swinging back and forth from the ceiling of the cathedral in Pisa. In trying to formulate laws of motion to explain the rhythms of their swing, Galileo performed a series of experiments. For example, he compared how the timing of pendulum swings varied with the length of the pendulum; the weight of the 'bob' at its lower end; the height from which the pendulum started swinging; the time for which it had been swinging; and so on. How did different combinations of circumstances, he asked, affect the length of the swing, and (as he supposed) the deceleration after the initial push that set them swinging?

Some modern biographies regard the tale of the cathedral oil lamps as either apocryphal or invented, as simply a device through which a contemporary could highlight how apparently mundane observations led Galileo to pose hitherto unasked questions. Others ignore it, moving straight to the conclusions—some of them incorrect—that Galileo drew from the results of the 'proper' experiments without conveying much of a sense of how provisional and open these experiments actually were (Garfinkel, 2002).

But both reactions gloss over a simple, yet vital, question. Since precise mechanical clocks did not yet exist, how could Galileo time the movements of the cathedral's oil lamps, of the swinging pendulums in his experiments, or indeed the acceleration of falling weights, rolling balls, or all the other objects he was interested in? Of course, by 1600 there had been mechanical clocks for some 300 years, as well as older timing devices like water clocks and sand-glasses, but none of these devices permitted accurate and repeated measurement of times as required by Galileo, ranging between a fraction of a second and a part of a minute. Indeed, the development of precise mechanical clocks became possible only with the elaboration of Galileo's work on pendulums by the Dutch mathematician Christian Huygens, in the late 1650s. The situation thus presents a classic circular problem: how were the movements of pendulums to be timed if the possibility of sufficiently accurate clocks would itself depend on knowing the result of accurately timing pendulums?

According to Viviani, unable to rely on mechanics, Galileo's solution was to turn to human biology. He timed the movements of pendulums using his own heartbeat as the timing device, comparing the number of oscillations completed with the heartbeats counted. In a striking inversion of what we take for granted today, he timed the machine with his pulse, not the other way round. Of course this was a short-term improvisation: physicians had long been aware that heart rates varied with health and activity, and that Galileo would need to seek standard or constant circumstances to achieve a steady pulse.[1]

Using his pulse to measure the time, Galileo found—somewhat contrary to common-sense expectations—that pendulums were close to being isochronous: that is, for a given pendulum, the period of its swing did not vary significantly with the length of the arc through which it swung. A metre pendulum hung from a point at, say, shoulder height, that was released at shoulder height would move rapidly as it dropped and begin oscillating to and fro. But its movement slowed as it swung to successively lower heights. It swung with the same period, the same number of times in a given time interval, as when it swung from a starting position only a few centimetres away from its resting perpendicular.[2]

His own heart was not the only example of Galileo using an embodied technique to measure time. Another, documented in Galileo's own notebooks, was the use of song. Singing served a specific purpose, where Galileo wanted to establish a sequence of equally short time intervals in experiments on rates of acceleration. But not just any song would do. It needed to possess the key attribute of a strong, regular rhythm, so that the gaps between beats identified equal periods of time. The set-up of a key series of experiments is shown in Figure 1.1.

The experiment was designed to reduce friction between the ball, and the surface down which it rolled, to a minimum. The catgut strung across the groove caused an audible 'click' when the ball ran over it, and Galileo adjusted the spacing between the catgut lengths so that the clicks coincided with the beats of the song. When these coincided precisely, the lengths between the gut-strings were measured, the increasing distances between them indexing the acceleration of the ball.

To historians of the 'progress of technology', such quaint and obviously inaccurate practices go to demonstrate the magnitude of the ensuing technological changes leading to modern precision timekeeping. By comparison, medieval or sixteenth-century practices appear crude, primitive, even haphazard. It is a near-miracle that any 'proper' scientific work could be generated at all. However, we read Galileo's use of his heartbeat and song as timing devices rather differently, emphasizing that 'accuracy' has to be assessed in the context in which the work took place. In context, his techniques were strikingly accurate, and his practices in

Introduction: The Measured Heart 3

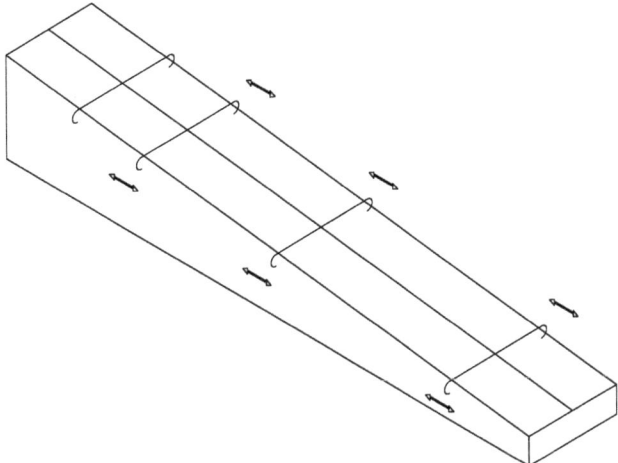

Fig. 1.1. Galileo's rolling ball experiment to time acceleration under gravity.
Without accurate clocks and watches Galileo used other means to measure times and intervals. These means included his own pulse, and also the apparatus shown here. Singing a strongly rhythmic folksong, Galileo adjusted the position of wires across the groove down which a brass ball rolled, such that the clicks of ball on wire coincided with the beat of the song. The pattern of increasing distances between successive clicks was a basis of his writing on acceleration.

pursuit of precision were neither primitive nor irrational but showed considerable sophistication (that is the display of expertise, skill, and patience) in the treatment of diverse materials, the performance and repetition of measurements, and the making of calculations. Nor were they necessarily atypical. One of Galileo's fiercest adversaries, Riccioli, used strikingly similar practices in an attempt to demonstrate that Galileo's rule of free fall was wrong.[3]

These stories are interesting not only because of their content, but also because of their subject. Galileo is—or rather has become—one of the key figures in popular narratives of the development of science, and of modern civilization. But, just as in other cases, such as those of Isaac Newton or Charles Darwin, seizing upon Galileo as a proto-modern, living somehow 'ahead of his time', has resulted in a somewhat partial picture of what Galileo himself thought he was doing. Just as with Newton and Darwin, Galileo did not operate, speculate, and experiment in an entirely 'modern' cultural mindset. He held many ideas that today appear to sit oddly together, just as to modern eyes Newton's work in 'scientific' alchemy and astrology still often appears as hard to reconcile with his 'modern' work on mathematics and mechanics. Until comparatively recently, the instinct of most historians of science was to suppress these apparent aberrations in thinking, in favour of emphasizing the modernity of the great figures in the development of European science. But in recent years, it is no exaggeration to say that the reconciliation and integration of these apparently contradictory facets of people such as Galileo and Newton has become a central issue for many biographers, and, in turn, illustrates an aspiration to get away from categories like 'modern' and 'pre-modern' which hold history hostage to preformed accounts whose understanding is too dependent upon contemporary cultural understandings (e.g. Latour, 1993).

In other words, with the benefit of hindsight, too many of Galileo's practices have frequently been labelled 'modern'. Such an ascription is *post hoc* term, and hardly an appropriate description of what Galileo was doing; his were not specifically 'modern' practices at the time he carried them out. It would be similarly inappropriate to talk of Galilean practices as 'hybrids' of pre-modern and modern practices. However neatly they may now appear to slot into those categories as part of a narrative of 'technological progress' (the sorts of narratives that have dominated twentieth-century—and for that matter nineteenth-century—histories of science), such narratives are histories of instrumental techniques, not histories of *practices*. That distinction is of major significance in this book.

Particular practices were responses—not necessarily fully articulated—to particular situations and questions. They require 'judging' in terms of those situations and questions, rather than as a commentary on how their performance, or effectivity, compares with that of later techniques or instruments whose situations and questions were particular to their later, often very different, contexts. From a present-day standpoint, many such practices may well appear strange, but this should not obscure the fact that they often worked with considerable accuracy *in their own terms*. To the people 'performing' these practices, in other words, they often appeared to be quite adequate, especially since notions of 'adequacy' themselves depend, at least in part, on the degree to which accuracy *is able to be regarded as achievable*.

Applying these strictures to Galileo specifically, we would stress that his efforts to achieve accurate timing of very short periods were not motivated by some general concern with measurement accuracy for its own sake. Rather, his was a novel approach in so far as his ideas of what constituted 'achievable accuracy' differed from those of virtually all his predecessors and contemporaries.

1.2 GALILEO AND MODERNITY

This juxtaposition of supposedly 'modern' and 'non-modern' traits becomes still clearer when we turn to another Galilean contribution to scientific interest in time, namely his proposal to use precise astronomical determination of time as the basis for accurate navigation, following especially from his observations of the moons of Jupiter in late 1610 and early 1611. That project illustrates a further set of points that are central to the content of this book and which we will briefly allude to here as a means of demonstrating what our chief concerns are here.

The first point, and probably the most familiar to modern readers, is Galileo's search for *precision*, using a really precise 'natural' indicator of time taken from the physical universe. In particular, there was the case of his efforts to use the positions of the four satellites of Jupiter that he discovered in 1610 as exact time-signals visible to observers on earth, efforts that both illustrated that Galileo was able to hold a notion of natural super-precision and also his ability to immediately fix on and disseminate major practical applications (particularly in survey and navigation).

The second point concerns Galileo's interaction with his peers. Here was someone who was located within several different *communities of practice* whose knowledges significantly affected (and were affected by) his conduct. Galileo was not a detached and isolated scientist, in contact only with other scientists as he/they pursued related scientific and technical goals. Anything but, in fact. At least four of Galileo's community memberships come to mind: membership of court; membership of the astronomical community; membership of the church (and especially the contact this gave him with Jesuit astronomers[4]), and membership of numerous networks of personal, political, and professional contacts. These various communities of practice linked larger or smaller groups of people over larger or smaller 'reaches' of geographical space, but their functioning was hugely variable. Take the example of the astronomical community, nowadays usually the first to come to mind in connection with Galileo. In some respects this group of (almost all) men was extremely specialized, dependent on patronage, independent means, teaching positions within the few universities in sixteenth-century Europe, or some combination of the three. New instruments, such as telescopes, were still rare, and much work comprised the re-processing of old, even antique, observational data. The network was small in size, and spread over large distances. Much communication was through publications, or via letters or messages, carried by travellers, often churchmen. Speeds of travel, communication, and publication were generally slow, but it was nevertheless possible on occasion for knowledge to travel quite rapidly and widely among members of this small community. For example, by the end of 1610 not only had news of Galileo's telescopic observations of Jupiter's first four moons in February that year circulated among European astronomers, but similar observations had been made in several other, widely separated, places, and those observations had in turn circulated in the astronomical community (Figure 1.2).

Such rapid diffusion of information and replication of conduct is clearly impressive. However, it was not always so. Dissemination of knowledge could be much more hesitant. Again, Galileo supplies an instance, this time countervailing: he seems to have been unaware of the hybrid geo-heliocentric model of the solar

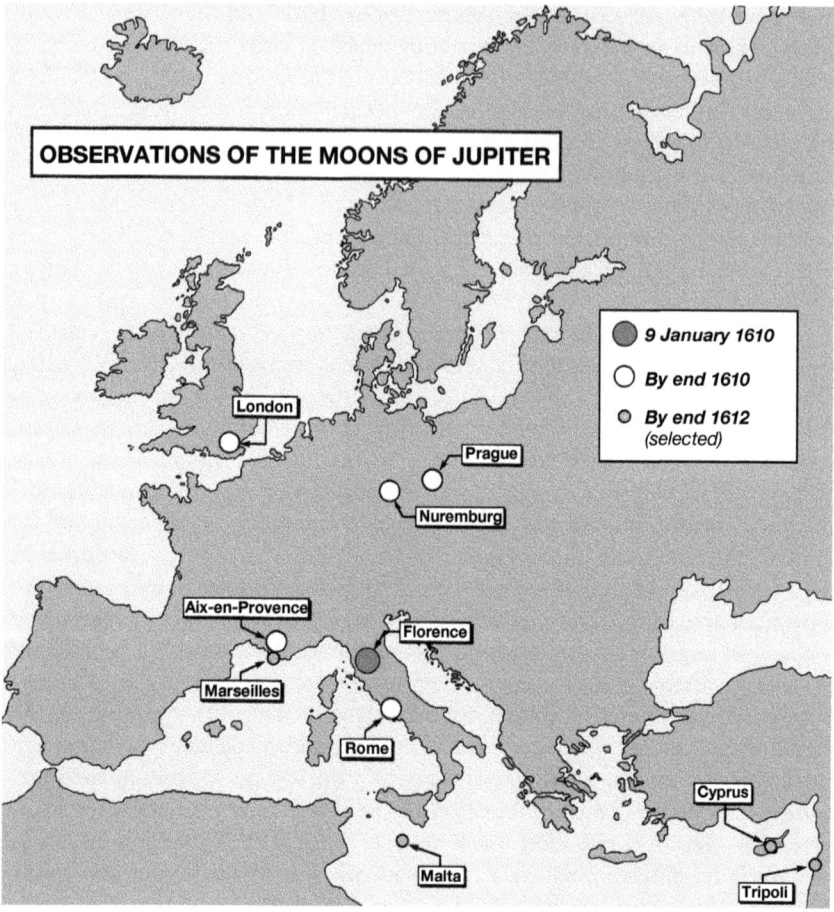

Fig. 1.2. Places from where the Galilean moons of Jupiter had been observed, 1610–11. Little more than a year after Galileo's announcement of Jupiter's moons, his observations had been repeated at several European sites. Though the community of telescope astronomers was very small, it is clear that news could circulate relatively rapidly within specialized networks.

system proposed by Tycho Brahe. So ideas did not always spread very far, or very fast, even when present in a specialized network.

One reason for this was that both the communities themselves, and the knowledges and practices which formed part of them, had pronounced geographies. They were *distributed*, not general features of the social world of sixteenth-century Europe. At least three facets of these geographies can be distinguished here. First, there was the spatial location of communities generating questions such as: where were these communities relatively dense and sparse? how were their connections sustained? Second, there was the extent to which different communities overlapped or intersected. For example, to what extent did information or ideas move from Galileo as part of an astronomical community to, or through, other communities of which he was part (whether willingly or no)? Third, we need to reiterate emphatically that communities of practices did not just happen, and they did not function automatically. That depended on the shaping of particular interests by their participants out of what was sometimes a maze of particular priorities and motives. This texture of interests shaped whether particular information did spread, how fast, and how widely.

The third point is concerned with the deployment of *trust*. Highlighting the significance of interests within and among communities of practice leads us to note the significance of relationships of trust and authority. Scientific knowledge was decidedly non-neutral. What counted as information was at times critically dependent on the trust vested in its source by others in the relevant communities. Thus Galileo's empirical skills were generally regarded as good but not necessarily perfect. Riccioli, for instance, believed that Galileo had fabricated, or at least presented in improved or idealized form, certain of his observations and claimed experimental results. Skills, instruments, and results all depended on trust, and trust-building was not something that could be taken for granted as information or knowledge circulated through space and time. The reception of information or ideas could be a very active process of questioning and reinterpretation, as Gingerich shows for Copernicus's *De Revolutionibus*, through studying the annotations made by readers on the books themselves (Gingerich, 2006).

One last point needs to be made. Galileo's attitudes towards time and timing were grounded within particular *wider senses of space and time*, the relationship between them, and the ways in which time and space were registered (that is both the dimensions of space and time that could be recognized, and the sensory facets thereof). There is still considerable debate about what these wider senses of space and time might have been and how they might have intersected with early scientific work.

Of course, as an historical individual Galileo was in many ways far removed from the specific developments in clock time in England that we examine here,[5] but his struggles to measure and understand time dramatically illustrate many of the major themes that recur throughout the book. The example of Galileo, in other words, provides a thematic link to our histories of clock times in England. Specifically, notions of timing as embodied practices recur in Chapters 3, 6, and 8; precision timekeeping is the chief subject of Chapter 9, and is also prominent in Chapter 10; discussions of intersecting communities of temporal practice appear in Chapters 4, 7, and 9; while the importance of trust, and especially temporal specificity as an indicator of the reliability of testimony, arises in Chapters 6 and 7.

Having raised many of the issues that will occupy us in this book through the example of Galileo, let us now set out more exactly what our concerns are by addressing the issue of clocks and clock times head on.

1.3 WHAT THIS BOOK IS ABOUT

This is a book about a phenomenon—clock time—that we take as a simple given in our world. It is one of the cultural equivalents of breathing: a continuous performance of clocks and other timekeeping devices, what are now global standards like Greenwich Mean Time, institutions enforcing time-discipline like workplaces, schools, and media timetables, small phrases and sayings, and the reinforcement of our own comings and goings, which produces a background to our lives that we take as read. Even when—very infrequently—we challenge this foundation, we tend to do so via hackneyed responses that only confirm the deeply ordinary power of clock time, thereby keeping us in check rather than providing us with any great insights.

Writing about the history of something so ordinary, trying to make its ordinariness remarkable while also retaining exactly its power of making things mundane, is difficult. But, as we hope to show, it is not therefore impossible. This is especially the case if, as we have striven hard to do, we restrict the subject of our research to clock time and do not, as many authors, move off into a shotgun exploration of all kinds of other dimensions of time, trying somehow to summarize every facet of temporal experience (senses of past, present and future, memory and forgetting, circadian rhythms, other temporal frameworks like the calendar, and so on) through history as though they somehow added up.

This book takes what we hope is an original look at histories of clock time(s), especially in England. Its originality is to be found in both its substantive-cum-empirical and its theoretical-cum/conceptual dimensions. On the histories of clocks and related devices in England (and indeed in western Europe in general), we depart substantially from standard horological works. This divergence arises in part from the sorts of clocks with which we are concerned: most obviously, where the latter have been largely preoccupied with private timekeepers (domestic clocks and pocket watches), we place much more emphasis on how public clocks were key components in creating a general grasp of several facets of clock time among the population of late medieval and early modern England.

As important to the book's originality are differences in how 'clock time' is conceived, defined, and investigated. This book is distinctive from much other writing on clock time in its treatment of clock time as a complex of practices, rather than as an object in itself. It is perhaps easiest to begin by spelling out what we are trying to get away from, the sort of technological triumphalism laid out by authors like David Landes:

Introduction: The Measured Heart

I would not want simply to say that time measurement and the mechanical clock made the modern world and gave the West primacy over the rest. That they did.

But the clock in turn was part of a larger open, competitive Western attitude toward knowledge, science and exploration. Nothing like this attitude was to be found elsewhere. Attitude and theme came together, and we have all been the beneficiaries, including those civilizations and societies that are now learning and catching up.

Vive l'heure! Et vive l'horloge!   (Landes, 2003: 26)

Landes and many other authors view clock time as an object in the world, but our view is quite different. Put bluntly, there is no such *thing* as clock time. Rather, clock time comprises a number of *concepts, devices,* and *practices* which have meant different things at different times and places, and even in any one place have not had a single unitary meaning. Our history of clock time, therefore, is not a triumphal narrative of the relentless development of 'modern time' (though long-run improvements in timekeeping clearly cannot be denied). No, our history of clock times is much more tentative and comprises the stories of many different types of practice—engaged in by larger or smaller groups of people—that have changed at highly variable rates, not always in the direction of the present day, and which have often interacted with and influenced one another. In the course of these interactions, certain practices that were originally highly specialized have spread to become much more general, while the incidences of some other practices have shrunk, in some cases to the point of disappearance, or have become confined to particular highly specialized communities.

We seek to show some of these waxings, wanings, and interactions in order to demonstrate how contemporary practices, so easy to take for granted as 'natural' and unproblematic, have complex histories. It is emphatically *not* the case that 'clock time' was created by accurate pendulum-based timekeeping from the late seventeenth century (or by industrial work discipline in the late eighteenth century), and then has changed little since. Many different clock times have been formed over time and space, by many different groups of people; groups which can be conceived of as social networks or communities. All these various clock times have been formed in the execution of practices and understandings of timing among various temporal communities. The clock times of different temporal communities may have much or little in common with those of other such communities, even those located close in space and time. At a particular time, certain elements of clock times may be shared in common across many different temporal communities, and can thus appear as an autonomous technological feature of an abstract clock time in their own right. Certainly much horological history has been written in such terms, whereas we want to avoid this reification of features of practices into a generalized or universal object called 'clock time'.

If readers' first reactions are that such an approach is unnecessarily complex since the nature and meaning of clock time are 'obvious', we would reply that such reactions demonstrate just how engrained a particular view of clock time has become. In all kinds of ways, clock time is commonly taken for granted both as a feature of everyday modern western life and as something that is defined, unproblematically, through the clock as a technological object, whether mechanical, electrical, or atomic. Yet what is so 'natural' about there being twenty-four hours in a day? Or about counting hours from one to twelve twice a day? Or about the idea that all hours are the same length? Or about representing the passage of time on a dial? These are more than hypothetical questions since, as we shall see, however much all these practices are taken for granted today, earlier European societies have done things differently and, in some cases, very differently indeed.

A corollary of the frequent taking for granted of present-day notions of clock time as somehow always present is a striking lack of sensitivity to the very different registers and dimensions of clock times in the past. Many illustrations of this insensitivity could be given, but here we will cite just four, from the diverse fields of English literature, art history, and history of science. The first illustration concerns references to times of day made by the fourteenth-century poet William Langland, in his *The Vision of Piers the Plowman*. The date of the poem's original composition is uncertain, but was prior to 1362; several versions exist from that date shortly afterwards. Then, as now, several versions were in circulation, at the time driven mainly by various additions or improvements for particular medieval audiences, or more recently through literary scholars' attempts to recover a version close to the 'original' composition. With dozens of medieval manuscript versions extant, *Piers Plowman* was clearly among the most widely circulated of medieval texts in English, along with such works as *The Canterbury Tales* and *The Travels of Sir John Mandeville*.

In book 6 of *Piers Plowman*, entitled 'Piers sets the world to work', Piers hires several men to assist in agricultural tasks, and in the course of this various disputes arise about attitudes to work, including timekeeping. Langland wrote of work beginning 'at hye prime', prime in this context being the service at the first hour of the monastic day. Since the counting of hours in the monastic day began at dawn, ecclesiastical hours fell at varying 'modern' clock times according to the time of year: the hour of prime did not simply equate to a fixed time 'o'clock'. Nevertheless, J.F. Goodridge's translation for the Penguin Classics series, unproblematically translates 'At hye prime' as 'Then, at nine o'clock in the morning'. Silently, without any explanation that this has been done, Goodridge treats two times that were reckoned under fundamentally different systems, as equivalent to each other. There is no discussion of the assumptions on which this conversion rests, or a note to the effect that the two expressions are incommensurate: instead, the older form is treated as a clock time, one just lacking the 'correct' clock terminology. This drastically changes the sense of the phrase, but the translator seems not to notice that this is what he has done.

Our second illustration nicely exemplifies the context-specific character of 'reading' visual representations, and comes from the eminent biographer Peter Ackroyd's recent *The Life of Thomas More* (1998). Ackroyd uses the sketch for Hans Holbein's the Younger's now-destroyed portrait of Thomas More and his family (Plate 3) as a basis for discussing family relationships. Perhaps the single most striking material object in the sketch, and in the adapted version of the picture later commissioned from Rowland Lockey by Thomas More's grandson in c.1593 was a clock. Given the interest in self-presentation on the part of More, like other courtiers of Henry VIII who sat for portraits, and given Holbein's interest in the representation of scientific objects in his painting, the clock is highly unlikely to be so prominent just by chance. That the clock was actually situated as depicted would seem unlikely, since the clock would have been fixed to the wall behind, inhibiting the movement of the curtain or creating a gap in it that would negate the curtain as a draught excluder.

Even if the precise significance of the clock remains elusive, it is hard to doubt that it was significant, as Ackroyd recognizes in discussing it. Yet he reads it highly anachronistically, misplacing it by nearly a century and a half:

On the wall, above the seated figure of Thomas, there hangs a weight-driven clock and pendulum. It would have possessed only an hour hand. (Ackroyd, 1998: 246)

The anachronism here is Ackroyd's reference to a pendulum, an innovation that did not appear in European clockmaking until the late 1650s. The two weights are clearly visible and a hand is hinted at in the sketch, and shown in the painting. But this clock no more had a pendulum than a liquid crystal display. Such a slip is significantly out of character for Ackroyd, whose wonderfully vivid and informed depictions of London life in this and other biographies have been widely praised. So it is revealing that he should take the nature of 'an old clock' so much for granted as to commit such a solecism—revealing, that is, of what is now taken for granted.

Our third example comes from Susan Woodford's *Looking at Pictures*, published in 1983. Ironically for a book whose central theme is the importance of reflective observation of what is being viewed, here again we find a writer making anachronistic presumptions about the clocks in a picture. The picture in question is the version of Jan Steen's *The Dissolute Household*, sometimes known as *Beware of Luxury*, in the collection in the Wellington Museum at Apsley House in London.[6] The exact painting date is unknown, but the consensus is c.1665–c.1668. Like many seventeenth-century Dutch genre paintings, Steen's picture has a strong moral message, through its counter-position of sober and irresponsible living. Woodford's interpretive discussion centres on the moral that:

the disgraceful behaviour of their elders is obviously setting a bad example for the children and even has an adverse effect on the servants; the dog, similarly, is about to succumb to his animal appetites; and the monkey is literally 'wasting time'. (Woodford, 1983: 31)

Time is an integral part of the message and effect:

> The clock by the door in the picture shows it is five minutes to five and a late afternoon light seeps in through the mullioned window high up on the left. . . . the monkey perched on the bed canopy playing with the weights of the clock. Perhaps it isn't five to five after all! What does time matter in such a household? (Woodford, 1983: 31)

However, that the clock shows five to five is based on the assumption that this mid-1660s painting depicts a clock with an hour hand and a minute hand. But domestic clocks with two hands were virtually unknown prior to the application of the pendulum to clock regulation, and the first such clocks were made in London in 1658. Our reading of this picture is that Steen has painted a single-handed clock. Clocks shown in other Dutch paintings from this period are all single-handed. Woodford's reading of the time shown is heavily constrained by the presumption that the clock has two hands, pointing to a time at which the minute hand is exactly opposite the hour hand, but on our reading the clock could as easily be pointing to just before half past eleven in the morning, as just before five o'clock in the evening.

Finally, Peter Galison's recent book on the context from which early twentieth-century theories of relativity arose reproduces the conventional account, this time in considering late nineteenth century recognition of problems of time-coordination:

> time awareness had become acute. Before the nineteenth century, clocks normally did not even have minute hands. . . . the late nineteenth century public wanted their second adjusted, [although] astronomers had long grown used to far greater precision. (Galison, 2003: 95)

Yet if the last of these claims is true, the second is highly debatable (it is far from clear that the public took much notice of seconds), and the first is patently wrong.[7] But, as part of a familiar-seeming story, it is reproduced uncritically as a historical canon.

These four examples each illustrate the ways in which even expert commentators can lack a sense of the specificities of clock time(s) in the past. We take their infelicities to suggest a more general need for clarification of what was involved in the clocks and the clock times of the past, a clarification which we seek to provide in this book.

Our contention is that concepts of clock time need to be approached as historically constituted. That is, clock time does not have an intrinsic meaning: it has meanings (plural) that have developed over several centuries, taking many different forms over this time which, even if now obsolete, were certainly significant in their day. We make this argument by examining the long-run histories of clock time and of mechanical timekeeping, which make clear the unstable ways in which clock time has been defined, and how contemporary senses of clock time include many components, each with their own histories.

Works on time all too often start out with a ritual quotation from Saint Augustine about the difficulty of defining time. Many then move on to a consideration of the different contours of time to be found in different societies and finally end up with a diatribe on modern western clock time, a form of life which often seems to be blamed for all the evils and iniquities currently besetting the world.

This book is different in four ways. First, we do not want to invest the passage of time with mystical qualities: for us, time is a resolutely material and mundane set of procedures and practices of aggregation. Second, we do not believe that the passage of time takes on mysteriously different qualities in different cultures, which is not to say that different cultures may not be collectively minded to believe this. As Gell tellingly puts it:

[there is no] struggle between different kinds of time, but the struggle between different classification systems and the real world. All classification systems have their difficulties, which stem from tendencies of real world objects to fail to conform to the criteria laid down for them. When Lévi-Strauss writes . . . of societies with an attitude toward history which denies, not that history has taken place, but simply that it has not made any difference to them, he surely gets to the heart of the matter. The 'conflict', is between this attitude, faith in the ability of a certain set of event-and-process classifications to embrace all foreseeable events, and the unfortunate tendency of real events not to occur normatively. The clash is between classifications and reality, not between irreconcilable versions of reality.   (Gell, 1992: 53)

Third, we want to count time as a historically variable process. So what is called clock time continually absorbs what were once regarded as contingencies, incorporates new sources of variation, and restructures itself in the process of doing so (Lefebvre, 2003). Fourth, we see no reason to blame clock time for all the ills of the world. Indeed, it is possible to argue that the gradual putting into place of those procedures and practices of aggregation that for convenience we call 'clock time' has been as much a liberatory as an oppressive force. It has allowed as much it has disciplined. New entities, capacities, and experiences have become possible which did not exist before and there is no reason to believe that all of these have been negative.[8]

In particular, in contradistinction to much writing in this area, we are convinced of the need for theoretically informed empirical work that can make clear what it is that is at stake. Too often those who have written on clock time seem convinced that they know what it is before they start writing about it, whereas for us one of the main problems has been sorting out all the different ways in which what we unproblematically call 'clock time' manifests itself. In particular, we have tried to avoid making excessively quick connections between the larger philosophical, historiographical, and anthropological questions and our own work, not so much because we are not interested in these questions but because we are wary of allowing these investigations—which often include all manner of presuppositions—from dominating our enquiries and so preventing us from seeing all manner of new things.

That said, it is important to make as clear as possible the chief historical protocols that inform this work and have allowed it to breathe. This first chapter is therefore an exposition of our general approach and, simultaneously, a kind of settling of theoretical accounts. But it is important to stress that our intention is not to weigh down this book with an enormous theoretical infrastructure: throughout its writing our concern has been to leave a space open for the various sources to speak, not least because by their very nature these sources have often proved fugitive and determinedly partial.

Put baldly, our approach rests on four principles. The first of these is that no singular history of clock time exists. The second is that it is important to understand that practices of clock time were (and are) remarkably diverse: they cannot be tied down to one abstract frame. The third is that it is important to take devices seriously and not just as adjuncts of something called the social. The fourth is that clock time is therefore something which is determinedly ordinary and, as a result, goes largely unremarked upon: it lines 'the routines and rhythms of everyday life' (Harootunian, 2000: 21). In what follows, we will expand on each of these points. Subsequently, we will show how they produce a methodological challenge, both on the level of immediate interrogation of the historical record and on the level of more general historical principle, which this book intends to take up.

1.4 THE STRUCTURE OF THE BOOK

The organization of the book is intended to combine several strands of research into clock time. The next two chapters respectively situate the book in the context of earlier debates on the history of (clock) time in England and Wales, and set out a new conceptual framework for thinking about clock time. Thereafter, five chapters deal with different substantive facets of histories of clock time in England and Wales, deliberately seeking to offer views from several different perspectives, and at a variety of scales, of clock times in the lives of individuals or in particular places, or within particular temporal communities. There is a brief, concluding chapter.

Chapters 2 and 3 set out what might be termed our historical terms of reference, the theoretical and methodological rules of engagement which we hope will guide us to the heart of clock time. Non-specialists might want to pass parts of it by. But we hope not because what we most want to achieve in these chapters is some sense of the historical protocols that need to be followed if we are to take the ordinary into account. This is, of course, hardly a new ambition in history—it was one of the objectives of the *Annales* School (Revel and Hunt, 1995). It preoccupied a good number of ethnologists as they sought out folk traditions. More recently, the history of practices and objects has become a key focus of, for example, the history of everyday consumption (Roche, 2000) and histories of science and technology (Mokyr, 2002; Pickstone, 2000).

Early sections of Chapter 2 outline the two major prevailing historical accounts of clock time. One is based on drawing direct relationships between the technical development of clocks and clock time. Whilst no one would want to deny that there must be technical relationships, we contend that, in a classical case of technological determinism, these have been drawn as tight and unproblematic when they are nothing of the kind. This section also allows us to provide the reader with a brief history of the technical development of clocks which can act as a point of reference through the book. The other prime historical account draws direct relationships between socio-economic changes and clock time, as proposed in E.P. Thompson's massively influential paper, 'Time, work-discipline and industrial capitalism', which first appeared in 1967. The paper raised and attempted to answer a string of questions about the role(s) played by clock time in the modernization of western European (and especially British) society. Attaching particular importance to the period from the late eighteenth century, it is without doubt one of the most influential historical papers of the late twentieth century, being influential not just among historians but within many other disciplines as well (as its prominence in citation indices across the humanities and social sciences makes clear).

Chapter 2 also discusses several later responses to Thompson's paper, from various disciplines. Many dealt with limited parts of Thompson's argument, and all tended to regard their contribution as making small modifications to Thompson's thesis, rather than indicating that a fundamental reworking was needed. This was certainly Thompson's own view, for a quarter-century on, he reasserted his original position, writing that 'while interesting new work has been done on the question of time, none of it seemed to call for any major revision of my article' (1991: vii). We argue that, on the contrary, such major revision is both possible and desirable on several grounds. In part, Thompson's paper has been left behind by extensive work in horological history and the history of consumption, which has transformed our understanding of the density and types of clocks over a long period before the late eighteenth century. More than this, we argue for a quite different conceptualization of clock time to that used by Thompson, one which builds on both anthropological and historical material, and which focuses on clock time as formed by communities of practice, not as an abstract object.

Chapter 2 then turns to two increasingly prevalent narratives of 'the triumph of clock time'. Both 'the world as production line' and 'the accelerating world' are underlain by Thompsonian framings of clock times as disciplinary, and as especially related to industrial capitalism, financial flows, and communications. The remaining sections of Chapter 2 consider methodology, and relevant historical models.

Chapter 3 attempts to set out the important elements of an alternative approach to histories of clock time. It is important to stress from the outset, though, that we are not seeking to simply substitute some different material and/or emphases within a basically similar structure to that of the standard models. Rather, we disavow the idea that an alternative grand theory of clock time and western modernization is even appropriate. In Chapter 3 we therefore go about synthesizing an account of long-run change in a quite different way, by concentrating on a theory of time telling and keeping practices. The chapter builds through three sections to our own account of the history of clock time, an account sharply different from the mainly negative histories of clock time that we set out in Chapter 2.

The first section develops the historical principles running through our account of plural histories of clock times: that keen attention must be paid to practices; that devices must be taken seriously, and clock time is part of ordinary everyday life, not an alien intruder. These principles figure in new conceptions of times that have emerged over the last thirty years, which are at odds with, and present a radical challenge to, standard histories of clock time. The following section sets out how new accounts of clock time might be generated against this new background. Drawing on work in anthropology, human geography, and the sociology of science, it builds an alternative account of clock time as a series of elastic but rigorous practices which have become general but not necessarily ascendant. Resting on the arguments set out beforehand, the final major section then offers a new account of clock time, an account developed out of a parallel process of both empirical study and theoretical evolution, which acts as the foundation of this book.

The first of the substantive studies, in Chapter 4, concerns the provision and uses of clock time in late medieval and early modern Bristol. We begin with a case-study of a particular place because this format enables us to juxtapose many of the theorising issues, raised in Chapters 2 and 3, to the further empirical themes of Chapters 5 to 10. The approach of Chapter 4 is retrogressive: beginning with an examination of clock times in Bristol in the second half of the eighteenth century, we then work backwards towards the high Middle Ages. By doing this we are able both to present a wide range of evidence about the uses of clock time, and also to show very clearly the much more limited nature of the evidence for earlier periods. This is an advantage because the restricted documentation available for medieval and early modern societies has sometimes led to assumptions that senses of clock time were somehow correspondingly narrow. The short answer to this is that frequently we are simply unable to say what, exactly, was the significance of clock time for different people in the city. If this was ever spelled out (which in many cases we doubt), almost all relevant documentary evidence has failed to survive. It is frustrating to conclude 'we do not know' to many questions, but at least that is an honest answer. Too often in the past the absence of evidence about a practice has been taken as evidence that the practice was absent. Hence, explicit attention to the (un)availability of types of evidence from which familiarity with, and use of, clock times can be shown is always necessary. Bristol contained no one thing called clock time, but many different sorts of practices, devices, and skills involving clock *times*. Chapter 4 then concludes with a set of questions raised by looking at clock times among Bristolians; questions which are each the subject of a later chapter.

Obviously Bristol—a major provincial city, a port of wide regional, and in some respects national and international importance—was not the whole of England and Wales. Chapter 5 therefore explores the extent to which Chapter 4's conclusions about the dense infrastructure of public and private clocks, and about the embedded familiarity of clock times in everyday life, held for other parts of England and Wales. The main themes here are the numbers and distributions of public clocks, especially church clocks (based on the first systematic analysis for this purpose of parish accounts across England); how these were used for signalling times of day, and for what purposes; and the ownership of domestic clocks and watches prior to the late eighteenth century period that lies at the heart of Thompson's account. We build on the very large literature on horological collecting and connoisseurship, but are critical of its relative neglect of public clocks, and of the period before the so-called 'horological revolution' of the late seventeenth century. Our central argument is that the density of time-signalling was much greater than usually recognized.

In Chapter 6 we explore what people did with clock times, and the extent to which the clocks and signals discussed in Chapter 5 entered the fabric and conduct of everyday lives. We analyse the degree to which time signals were commonly used resources for the conduct of everyday life, and to what ends. This discussion uses a wide range of documentary and literary sources—diaries, letters, (auto)biographical pieces, court depositions, and more—to trace the availability, observation, and uses of clock times, addressing the obstacle created by the rarity of recording of plebeian writing or utterances. We emphasize how readily the signalling of particular times for quite specialized disciplinary, trading, or religious purposes was adopted as a much more general resource for the organization of daily life.

Chapter 7 is likewise concerned with everyday lives, focusing more tightly on (relatively) precise uses of clock times. This discussion uses a wide range of documentary and literary sources to trace the availability, observation, and uses of relatively precise time indications or measurements. We pay particular attention to the circumstances within which people used precise clock times. The thrust of this chapter is that concepts of precision were not simply dependent on technologies of precise timekeeping, but that their applications were fashioned from whatever temporal cues were to hand in everyday environments. The use of clock times, both precise and generalized, consistently demonstrate the degree to which uses of precision were highly selective. It follows from this that even when people commonly eschewed precision in many areas of everyday life, this did not mean that they were incapable of doing so (any more than casual or imprecise use of clock times today indicates that people could not use clock times precisely when they deemed it appropriate).

In Chapter 8 we examine in detail temporal communities among seafarers. Oceanic navigation has recently received widespread attention through Dava Sobel's bestselling *Longitude*, but we are concerned with far more than attempts to establish longitudinal positions using very accurate timekeeping. Longitude was just one—and in many respects a rather late—navigational use of clock time. Navigators had many other uses for clock time, and these involved specialized skills, training, instruments, and everyday practices within shipboard life. But clock times were not, indeed could not be, reliant on the presence of clocks. Rather, via other devices and practices, clock times shaped many facets of shipboard life. Once these diverse clock times and the practical rhythms they induced are included, shipboard life embraced not just a narrow community of 'theoretical' navigators, but the much broader community of practice of seafarers at large. Moreover, with a considerable throughput of men through the merchant marine and navy, especially in wartime, the wider influence of seafaring (and other military) timekeeping practices could be considerable.

Chapter 9 discusses the history of very precise timekeeping, using the topic to highlight the development of, and interactions among, several communities of practice for whom precision was a concern. This especially involved technical and scientific communities connected with astronomy and navigational techniques and instruments. However, it would be wrong to treat precision as solely the concern of small, mathematically-skilled communities. It was, after all, essential that precise navigational methods and techniques could be carried out at sea, not just in the observatory, so the practical use of precise navigation required that skills be accessible to at least some seamen and officers. As Chapter 9 shows, important extensions of precision were less to do with theoretical precision, 'in the abstract', than with making precision a practical proposition, realizable 'in the everyday shipboard world'.

Chapter 10 brings arguments developed through the book, especially in Chapters 5 and 6 (on the early modern *zeitgeist* of clock times) and Chapter 9 (on technical and practical precision) to bear on the principal character of Sobel's *Longitude*, John Harrison. Harrison is characterized by Sobel as an outsider, both personally and geographically, originating from an area outwith communities of clockmaking experience and expertise but, paradoxically, able to solve 'the longitude problem' because of his freedom from the constraints of orthodox thinking about timekeeping. In part based on new research findings, we re-appraise Harrison's background and experience, and suggest that his massive contribution to chronometry did not come 'from nowhere'. Rather, it needs to be seen as an outgrowth of wider processes.

Finally, the concluding chapter, Chapter 11, outlines some of the major questions which our research has thrown up but which remain largely unanswered. In particular, we point to four areas of research which we believe provide keys to neglected but important aspects of the history of clock time, namely the way in which tinkering and other forms of experimentation provided much of the dynamic of innovation, the importance of maintenance and repair, the role of watches, and the role of schools and other educational institutions in teaching clock time.

1.5 CONCLUSION

What we hope to provide in this book is a well-theorized history of clock time and timing in England and Wales between 1300 and 1800. We have spent very large amounts of time, more than twelve years, in order to reach a well-specified understanding of the temporal understandings in the period we consider. But we would not want to put readers off. This is not, we hope, a formidably theoretical book which simply substitutes one theoretical orrery for another: at all points we have tried to let the historical record (such as it is) speak. Indeed, this is a crucial element of the book, since we believe that the history of time in England and Wales has too often been the subject of a rush to theoretical judgement based on theoretical frameworks which are heavy-handed or simply inappropriate. In other words, if we let the minutiae of historical research make their way in to the record, they can often stimulate different—and exciting—historical accounts.

Indeed, we are quite clear that we do *not* want to provide some overarching theoretical framework that can become a new standard—much to the distress of some of our colleagues who want a kind of certainty. And there are good reasons for this cautious stance. First, we do not believe, given the current state of the evidence, that such a definitive account is possible. Indeed, our book is very much intended as a spur to further research. Second, we are not sure that such an account is even necessary. Perhaps we should see history instead as something which is constantly provisional—as it was for those who lived it. Third, and in turn, such a stance allows us to make room for those constantly improvised acts of 'performance' which go to make up so much of what we consider. The galvanizing heroic leap into the unknown of the kind ascribed to Galileo or Harrison is a rarity: more often than not it is an emphasis on modest increments in practice that counts in the history of clock time and timing. Such an emphasis, we believe, may ultimately be more interesting, more revealing, and more likely to give us back some notion of the temporal skills of (extra)ordinary people.

NOTES

1. Once the temporally stable behaviour of pendulums was known, physicians augmented their longstanding qualitative use of raised or lowered pulse as a symptom of fevers or other conditions, with measuring the pulse as a diagnostic tool.

2. Half a century later, the Dutchman Christian Huygens would show that a freely swinging pendulum is precisely isochronous if, rather than swinging to and fro along a line (when seen from above), it swung through a cycloid (resembling a shallow figure of eight). Huygens's development of ways to force cycloidal oscillation on a pendulum immediately enabled the construction of much more precise clocks than hitherto.

3. Riccioli made the first good measurements of the constant free fall. He did so expecting to defeat Galileo's rule. He began his experiments in 1640, when teaching philosophy at the Jesuit College in Bologna. Building on Galileo's observation of the regularity of pendulum beats, Riccioli used a chain and a weight as a clock. But how to find the precise number of seconds in each beat of the pendulum? Riccioli's answer, which raised his determination of the constant of (gravitational) acceleration to the 'first sustained attempt at an experimental measurement', was to choose a pendulum of such a length that its bob took exactly one second to make one swing. He proposed to find this convenient length by experiment.

   Riccioli and Grimaldi chose a pendulum $3'4''$ long, Roman measure, set it going, pushed it when it grew languid, and counted, for six hours, by astronomical measure, as it swung back and forth, 21,706 times. That came close to the number desired: $24 \times 60 \times 60 = 21,600$ seconds. But it did not satisfy Riccioli. He tried again, this time for an entire 24 hours, enlisting more of his brethren including Grimaldi: the result, 87,998 swings against the desired 86,400. Riccioli lengthened the pendulum to $3'4.2''$ and, with the same team: this time they got 86,999. That was close enough for them, but not for him. Going in the wrong direction, he shortened to $3'2.67''$ and, with only Grimaldi and one staunch counter to keep the vigil with him, observed on three different nights, 3,212 swings for the time between the meriodional crossings of the stars Spica and Arcturus. He should have had 3,192. He estimated that the length required was $3'3.27''$, which—such is the confidence of faith—he accepted without trying. It was a good choice, only a little further out than his initial one, as it implies a value of 955 cm/sec^2 for the constant of gravity.

   Armed with this information, a smaller, faster pendulum calibrated by it, and balls of wood and lead, and accompanied by a chorus of musical brethren to complete their clock, Riccioli and Grimaldi repaired to the Torre degla Asirelli. The musical brethren chanted 'do', 're', 'mi', etc. as the pendulum beat so that Riccioli only needed to keep track of blocks of eight, rather than individual, swings. As everyone had expected, Galileo was disproved. The lead ball always hit the ground before the wooden one when they fell from the same height. (Heilbron, 1999: 180–1)

4. It is worth referring to the work of Heilbron here, concerning the widespread high regard for Jesuit observation and measurement, which in turn was important as a source for Flamsteed, and hence the navigational communities discussed in Chapter 9.

5. Although we return to his work in relation to very specific points in Chapters 7 and 8.

6. There is another version at the Metropolitan Museum of Art in New York, but the composition is more tightly framed than in the Apsley House picture, omitting much of the right-hand side of the painting, including the clock.

7. The citation for the first part of this claim is Dohrn-van Rossum's *History of the Hour* (1996), though this chiefly concerns medieval public clocks, rather than seventeenth- or eighteenth-century timepieces (which were overwhelmingly domestic clocks and pocket watches).
8. We subscribe to Stiegler's (1998) idea that technologies are as likely to temporalize as they are to flatten time out.

2
Clocks, Clock Times, and Social Change

2.1 INTRODUCTION

In this chapter we examine the main ways in which the history of clock time has been told, from horological history centred on clocks and other timekeeping technology, social/historical accounts of time and timing as dimensions of social discipline, to conceptualizations of time as a facet of cultural modernity. Each of these narratives has something important to say, and each is a useful resource in thinking about time and social life, about time and social change, and about time and everyday life in the long run. But each also has limitations. While our approach engages with all three, it does so critically, and this chapter also sets out some of the ways in which that critical engagement has led to the approach taken in the later chapters.

Section 2.2 asks the deceptively simple question, 'what is clock time?' and examines some of what is taken for granted (thereby becoming hidden from us), making the answer to seem as obvious as it does. Section 2.3 then asks 'so what did clocks do?', introducing the history of horological technology, which we reconceptualize as a network of devices and actors, rather than simply a machine. Section 2.4 turns to disciplinary conceptualizations of clock times, in which the clock is best understood as an invention to enhance or intensify social discipline. We focus on E.P. Thompson's analysis of clock time as part and parcel of the increasingly time-sensitive work-discipline required in early factories, easily the most influential example of such an approach, which we then argue is both too narrow, and too negative. Nonetheless, a broadly Thompsonian narrative has been a significant influence on the prevailing standard folk models of clock time, inscribed in numerous books and articles. We consider two broad groups of accounts in section 2.5, respectively focused on 'the world as a production line' and 'the accelerating world'. These narratives tell of a world gradually taken up and taken in by clock time, in a chiefly negative history of reasoning and rationalization.

Clock time makes its way into the historical record in diverse ways which show that what we regard as history or the historical record cannot be taken for granted. In particular, one feature of (broadly) disciplinary narratives is that they draw on evidence largely generated by the disciplines themselves, rather than the kinds of temporal practices central to our argument. Sections 2.6 and 2.7 therefore set out to discuss the evidence and methods used as we have attempted to access everyday practices and perceptions, facing both the intrinsic problem of getting at things thoroughly taken for granted, and the particular problem of doing so for people now long dead.

2.2 WHAT IS CLOCK TIME?

The answer to this question may seem so obvious as to scarcely merit the time it takes to reply. Clock time, we might say, is the reckoning of the time of day/night in hours, minutes, and seconds, which we see on clock faces in our homes or cars or computer screens, or on our wristwatches, or on a host of other locations nowadays, shown either by hands on a dial, or digitally. Hours are counted in two sets of twelve, from 12 noon and 12 midnight, and all the hours are of the same length. What more needs to be said about this constant whisper of activity?

In truth, quite a lot more. The characterization just outlined summarizes certain generally accepted features of current everyday clock time; of what might—given that it stems from several centuries' experience of mechanical clocks—be termed 'mature' clock time. However, that summary hardly suffices as a *general* description of clock time. We tend to experience its elements as components of an entire package, encompassing conventions about the units into which time is divided, the ways in which time is measured, counted, and signalled. As we shall show, these components have their own complex histories. Neither current practices nor changes in clock time over time can be considered to be 'natural'. There is little that was inevitable about how practices regarding clock time have changed. Although our twentieth-century, everyday formulations of clock time may now seem entirely obvious and workaday, all their elements are nothing more or less than *social conventions*, conventions that are durable (having survived while one-time rivals disappeared), and that have come to be very widely held. Looking back to periods when alternative ways of defining clock time still coexisted alongside those that survived to be taken for granted nowadays, the components of clock time are much more clearly recognizable as these conventions.

In replying to the question 'what is clock time?' we will briefly elaborate four main themes. First, we highlight the interdependence of concepts of 'clock time' and 'natural time'. Second, we emphasize the public dimensions of clock-timekeeping in late medieval and early modern England (and much of continental Europe), a dimension underplayed in the horological literature. Third, we emphasize the critical importance of an historical transition in thinking about everyday temporality, from thinking and reckoning in 'unequal hours' to 'equal hours', a transition decisively accelerated—though not caused—by the late medieval availability of mechanical timekeepers. Finally, we point to the sheer diversity of ways in which clock time has been counted, a topic with a complex history which has been almost entirely lost in contemporary daily life.

(i) Clock Time in Relation to 'Natural Time'

Both everyday and intellectual notions of clock time have been commonly constructed in opposition to implicit concepts of some 'natural time'. Indeed, characterizations of clock time invariably start by differentiating 'clock time' from 'natural time'. Historians have often constructed narratives around the replacement of the latter by the former (most notably E.P. Thompson, whose work we discuss in Chapter 1). Historians not explicitly focused on time have often built dichotomies of temporality into accounts of long-run change. Raymond Williams's *The Country and the City* provides a common example of this tendency, presenting a rural natural time (simple, slow, nostalgia-soaked) superseded by an urban clock time (complex, quick, unsentimental). In the last decade or so, however, much more attention has been paid to 'natural' times, and this has shaped our view that important elements of clock times arise from their derivation in relation to the prevailing notions of natural time in particular historical contexts. Therefore, unpacking the widespread idea of 'natural time' is a prior issue that lies at the heart of the meanings of clock time.

(ii) The Publicness of Clock Time

In late medieval and early modern Europe, a central dimension of timekeeping was its 'publicness': clock time was essentially, and explicitly, a public concept. Clock time revolved around public devices and public practices, rather than being something that was kept privately. Yet this central public dimension of timekeeping is almost entirely absent from a horological literature overwhelmingly focused on surviving private timepieces. This preoccupation is unsurprising, not only because household clocks and pocket watches dominate the surviving timekeeping artefacts available for collection and study, but also because they accounted for most of the massive increase in production of timepieces from the late seventeenth century 'horological revolution' onwards. It has proved easy, even natural, for horologists to reconstruct the general timekeeping environment from these objects, rather than to research it from documentary sources in connection with timekeepers that no longer survive, especially where these appear from the current vantage point as—at best—an evolutionary cul-de-sac in the evolution of clock design.

However, for several centuries after the first development of mechanical timekeepers, watches and household clocks were *not* the forms in which mechanical timekeeping was usually encountered. Before the later seventeenth century, and for parts of the population a good deal later, the 'typical' clock was a much larger machine, located in a (quasi-)public building such as a parish church, a market hall, or a town hall. Domestic clocks were, until the later seventeenth century, essentially derivative from these large public clocks. That the latter have been largely written out of horological histories has obscured several important features of earlier timekeeping, and distorted general histories of clock time by accepting particular facets of mature clock time as 'natural', universal, even as inevitable.

(iii) Defining 'the Hour'

Just as means of measuring the 'natural' time of day long pre-date the appearance of mechanical clocks, so too do systems of dividing the day into units such as 'hours', which date back to at least Assyrian and Babylonian times, several centuries BCE (Macey, 1994; Lippincott, 1999). Two different definitions of 'hour' circulated over several centuries. First, and the more widespread in medieval Europe, there was the idea that the hour was a fixed proportion of the daylight hours. This definition was relatively easy to keep track of using sundials during the day and somewhat similar devices, sometimes known as nocturnals, during the night traced the apparent rotation of specific stars or constellations around the pole star, Polaris. The term *unequal hours* highlights the fact that these hours were not of constant length. At any given time of the year, the twelve hours of the day and those of the night, differed in length, except when it happened that day and night were exactly the same length, and sunrise to sunset was the same length of time as sunset to sunrise. Moreover, the total daylight and night-time periods varied with the seasons, so midsummer daytime hours were very much longer than midwinter daytime hours.

The second definition of an hour involved *equal hours*, defined as one twenty-fourth of the period between successive noons. This definition was much more difficult to measure from the sun or stars, but had long been used by certain specialized groups. For example, before equal hours could be measured, astronomers and astrologers (not that these were distinguishable activities) used equal hours in recordings of observations, and in calculating the appearances of eclipses, planetary conjunctions, star risings, and similar phenomena. Thus Islamic astronomers, and perhaps Babylonian astronomers a thousand years or so earlier, had already used units of 1/360th of the 24-hour day-night period—four minutes—in giving precise times for eclipse events. This four-minute unit appears specific to astronomers in that area, and serves to illustrate how the co-existence of unequal and equal hour reckonings was entwined with substantial local variations in the schemes and terminologies used to define and name particular hours, and various shorter periods.

The two co-existing definitions of 'the hour' rested on quite different conceptions of 'natural time', and on quite different approaches to timekeeping. Unequal hours relied on cosmological observation of fixed points of the day (sunrise, noon, and sunset), and sub-divided the intermediate periods into set proportions, indicated by angular shifts in shadows or stellar alignments. 'Natural' here took the form of (something like) 'directly observable from the cosmos'. In the case of equal hours, on the other hand, the meaning of 'natural' was different, (something like) 'proceeding at a constant rate in the cosmos'. However, it was impossible to keep track of equal hours directly from observable celestial movements. Reckoning in equal hours relied on being able to measure, or to calculate from observations and tables, uniform-length divisions of noon-to-noon periods. So equal hours could not be, as it were, directly revealed from the universe itself, in an equivalent way to the monitoring possible with sundials and similar devices. Their measurement rested on attempts to construct devices that operated at an exactly constant rate, principally water clocks and, more crudely, sand glasses of various types.[1]

Thus, even before the quantification of the 'equation of time', and its rapid diffusion in the seventeenth century, clocks were taking timekeeping further away from the cosmos as the direct source of time. The 'hours' indicated by clocks were a set accumulation of the movement of a mechanical device, which ultimately had to be consistent with celestial movements, but which proceeded independently of them except through being set so as to maintain the conjunction between clock and cosmos. Clocks shifted from being purely proxies or intermediaries for what a sundial would show, were it not cloudy, to *being themselves the source of times* to which causal powers could be ascribed.

The emergence of mechanical clocks must be set, therefore, within an earlier usage of both unequal and equal hours. The units (such as 'an hour') or points (such as five o'clock) central to clock time were already both imagined and calculated. This point needs to be remembered alongside Dohrn-van Rossum's argument that mechanical clocks were important, above all, for firmly establishing equal hour reckoning, so long the subordinate hour-reckoning system. While transition from unequal to equal hours was not instantaneous, and for a time the two hour-reckoning systems were used alongside one another, the shifting balance towards equal hours is unmistakable. The rapid spread of mechanical clocks in late medieval Europe went alongside a relegation of unequal hour reckoning to the status of the derivative system. But although the switch towards thinking and reckoning in 'equal hours' was decisively accelerated by the late medieval availability of mechanical timekeepers, it was not technologically determined. Neither did the spread of mechanical clocks necessarily entail unchanging equal hours. 'Nuremburg hours' involved adjusting both the number of hours in day and night (ranging between 16 daytime and 8 night hours in mid-summer, and 8 daytime hours and 16 night hours around mid-winter) and the exact length of day hours and night hours as the seasons changed (Dohrn-van Rossum, 1996: 115).

To reiterate: the spread of clocks, the use of clocks to keep equal hour time, and the use of time markers to structure daily life were all intertwined. We cannot see one of these as determining the others, whichever way we point the causal arrow. Although equal hour clock time became used for very many purposes, it was often taking over existing—and essentially similar—functions of unequal hour time. Equal hours appear in the documentary record through changes in people's ways of performing and describing actions, not because the actions themselves were entirely new.

(iv) Diverse Ways of Counting and Marking Time

Here, too, the history of timekeeping and time-counting has often been written as a 'genealogy of the present', attaching little significance to practices in the past unless they appear directly antecedent to practices today. We therefore briefly outline some 'failed' approaches to measuring, counting, and signalling clock time, in order to show how current practices are not the only ones that there might have been.

The earliest mechanical clocks made use of the several various different hour-reckoning systems that co-existed in late medieval Europe. The ancient division of the day–night cycle into twenty four hours was more or less standard, but there was considerable variety in other respects, such as the pattern of hour-numbering, and when the count should begin. There was no initial standardization around the now-familiar twice-daily striking from one to twelve, anchored to noon and midnight. During the fourteenth and fifteenth centuries several striking cycles were tried in different places (Table 2.1). Religious feasts[2] often comprised a night–day period of twenty-four hours from sunset, and some early mechanical clocks were used to count equal hours from sunset or sunrise rather than noon (Dohrn-van Rossum, 1996: 114). The sun might reach its highest point at around four o'clock, or eight o'clock, depending on the time of year. Such systems were eventually superseded by having midday anchored at noon, while the hours of both sunrise and sunset varied, but this process took many decades.

The practice of numbering hours from one to twenty-four starting half an hour after sunset, first evident in some northern Italian cities around 1300, persisted in various parts of Italy, Bohemia, Silesia, and Poland into the seventeenth and eighteenth centuries (Dohrn-van Rossum, 1996: 114–15), notwithstanding both the heavy wear and tear on bells, hammers, and ropes caused by 300 strikes every day, and the inconvenience of the frequent adjustments of the start of the count that were necessary as daylight lengthened and shortened through the year. To counter the first difficulty, shorter counting cycles were adopted more or less widely, not just the 'twice twelve hours' familiar today, but also schemes involving four times six hours, or three times eight hours (Dohrn-van Rossum, 1996: 113–17). The shift to calibrating cycles of counting at noon (even if the formal numbering might begin at midnight) stabilized the hitherto irregular intervals created by starting hour-counting from sunset, but here too exceptions (as at Nuremburg) survived. Given that the survival of documentary evidence is very sparse, it is striking that it includes some discussions of the relative merits of different systems (Dohrn-van Rossum, 1996: 119). It is at all events clear that standardized hour-counting emerged from a landscape of earlier practices that displayed considerable variety.

In the long run, changes in hour-counting produced convergence among the diverse systems encountered in late medieval Europe, and in the metrics from which public signals were derived. A purely national focus on England and Wales, within which medieval time-counting conventions seem largely homogenous, would both obscure that diversity and blind us to the conventional nature of all hour-counting systems. Much of this material seems very unfamiliar and strange at first blush, having been altogether lost from daily life. But its apparent strangeness also serves to emphasise the extent to which we experience present-day practices as natural and inevitable. Many historical perceptions of what was necessary to, and required for, clock time to be a useful resource were very different from those now familiar to us. Modern, 'obviously right', elements of clocks and hour systems came about through extended processes of experimentation, selection, influence, and pure chance, rather than as inevitable consequences of some intrinsically superior 'design'.

2.3 SO WHAT DID CLOCKS DO? HOROLOGICAL HISTORIES OF CLOCK TIME

Questions about what clocks did or showed, or about how they worked, did not have fixed answers in medieval and early modern Europe. What clocks did varied greatly in different times and places. Standardized practices only slowly became general, and were contingent rather than inevitable. A recurring theme through this book is that neither ideas about clock time, nor counting and signalling schemes, nor imaginings of clock time, have been merely derivative of prevailing technologies. The particular forms taken have been contingent on societal practices and everyday sociality, as well as the familiar imperatives of technological and social disciplinary influences.

Two approaches to understanding the history of clock time have predominated. The first approach, which is often cited as though it were simple common sense, bases the history of clock time on technological advances (Table 2.1). This is hardly surprising. After all, the history of clocks is so often made homologous with the history of clock time that the two have become all but fungible. It would, of course, be foolish to claim that the changing character of timekeeping devices has had no influence on the history of clock time, not least because we want to argue below and elsewhere that devices have a degree of independence from the 'social' (though our point will be that the separation of the 'technical' from the 'social' is precisely the conceptual problem that needs to be surmounted).

But what we are firmly against is the kind of technological determinism to be found in the work of otherwise sophisticated writers like Landes (1983), Barnett (1998) and Cipolla (1978) which views clock time as an object in the world, rather than as a variable set of practices. In particular we are against the kind of work that views current usages of devices like clocks as if they were a foregone conclusion by ignoring all the embodied, situated, and concerted work that went into making these usages social facts, rather than others (Garfinkel, 2002).

However, none of this is to say that there is not considerable value in considering the history of timekeeping devices. There quite clearly is and therefore in this section we set out a history of mechanical clocks as a referent for the book as a whole. Our aims are twofold. First, we want to provide a brief but clear account of how various timekeeping *devices* have worked. Second, we likewise need to outline *chronologies* of clock time technology, covering both how time was measured using clocks, and how clock times were signalled within a public domain. Some specialized terminology is unavoidable here, but its use is important in specifying what was denoted by terms like 'clock' at different times, and in charting the changing relationships between clock technologies, on the one hand, and the ways in which clock time was defined, imagined and used, on the other.

Nowadays it seems natural to think of the alarm clock as derivative of clocks more generally. Historically, however, this reverses the relationship between alarm devices and true clocks. For water clocks and mechanical clocks alike, alarm devices were the underlying technology. Especially for waking monks for nocturnal prayer, the clockwork alarm was important in its own right. Though it may seem counter-intuitive to regard something resembling a clockwork kitchen-timer as a major technological breakthrough, the making of alarms was where many of the component technologies of timekeeping were developed. If clocks were mechanically more complex than earlier alarms, they were considerably simpler than another of their precursors, the planetarium (Price, 1959; North, 2005: 160–6). Planetariums showed solar, lunar, and planetary positions, and also fuelled desires to synchronize revolving clockwork with the apparently revolving heavens.

A mechanical clock performs a series of linked tasks (Figure 2.1), in all of which there have been dramatic changes since the first mechanical clocks appeared in *c.*1270 (Rawlings, 1993). The essential task of a timekeeper is to maintain a count of some regular event, laying down a 'grid' of divided time over everyday life. This task involves several components: a way of harnessing energy to power the device; a way to control the release of that energy over an extended period; a way of creating oscillatory movements within the device; a way to count those oscillations; a way to signal the accumulated total oscillations in terms of the time of day; and the means to speed up or slow down the whole task, to keep the timekeeper in accordance with cycles of daylight and night.

Our discussion is laid out around the five topics of *drive* (how was the movement of a clock powered?), *regulation* (how were key movements maintained at a constant rate?), *escapement* (how were movements transferred to a counting device?), *signalling* (how was the passage of time, or of moments such as 'nine o'clock', struck or shown?), and *coordination* (how were clocks kept in phase with

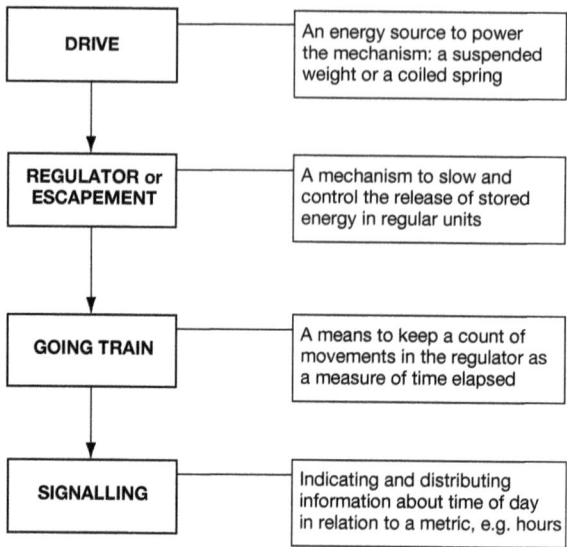

Fig. 2.1. The core components of mechanical timekeepers.
Though it seems natural to think of clocks or other mechanical timekeeper as singular machines, they synthesized a number of distinct functions. In early clocks these functions were much more distinct than they later became.

diurnal rhythms?). It is mechanically convenient, if slightly artificial, to discuss these topics in turn.

Any mechanical timekeeper requires an impulse, that is, something that generates a movement in the first place. Pre-mechanical attempts to measure time had generally made use of gravity, as in the case of water clocks that measured time through the emptying or filling of containers of water or other liquids. Early mechanical clocks and alarm devices also relied on gravity, using the systematically relieved fall of a weight to provide a continual impulse to machinery. The maintenance of the potential energy of the weight, literally through 'winding-up', was vital to all mechanical clocks throughout our period. The critical problem for medieval craftsmen lay not in the provision of impulsive power, but in how to restrain it in such a way that it could produce a regular movement through which a count of time could be kept. Thus, water clocks needed to maintain constant head pressure to stop the outflow of water slowing as water levels fell in the reservoir, and mechanical clocks needed to counteract the changing impulse exerted by the weight as it fell between windings. (We are concerned mainly with clocks here, but similar principles apply in the design of watches, where the impulse was provided by the uncoiling of a spring after it had been 'wound up', and which faced very similar problems of ensuring an even impulse as the spring unwound.) Recognition of this central problem is already clearly expressed by a number of late thirteenth century commentators: that craftsmen had made clockwork machinery, but they had yet to perfect a version which could consistently indicate temporal positions within the 24-hour day/night cycle (Dohrn-van Rossum, 1996: 89–96).

The novelty of mechanical timekeepers lay in three linked features. First, their timekeeping was detached from constant monitoring of celestial movements. In other words, clocks measured the passage of time in parallel with the cosmos, not directly from it, as was the case with sundials. Second, they indicated the time of day through 'fusion' rather than 'fission'. By this, we mean that mechanical clocks measured periods of time (e.g. hours) by accumulating a count of very many, very short periods (such as the movements of a regulator), rather than the division of longer periods (e.g. one-twelfth of daylight, as in unequal hours sundials, or one-twenty-fourth of noon-to-noon). Third, clock times occurred within a framework of equal hours rather than unequal hours. These three features of mechanical clocks, in conjunction with their relative reliability, and the accessibility of the time they kept, were important elements in the wider shifts in analytical thinking in Europe from the late thirteenth century onwards (Dohrn-van Rossum, 1996), though this has often been conceived in grossly schematic form (e.g. Crosby, 1998; Landes, 2003).

Early mechanical clocks were mostly large 'turret clocks' (Plate 2.1), rather than domestic clocks or watches (Baillie, 1951; Dohrn-van Rossum, 1996: 118–23). They ranged widely in form, technological details, and priorities. Various versions of most technological elements were devised and constructed, even in fundamentals like the escapement (e.g. North, 2005: 175–90). Various types of particular device elements were retained, or modified, if subsequently discarded. Such technological changes involved both clocks as material objects, and the relationships between material and human components of keeping time. The latter is of considerable importance for our overall argument.

Late thirteenth-century advances in clockmaking came from the better calculation and cutting of gear trains, on the one hand, and from more effective regulators, on the other. Of solutions to the problem of maintaining regular movements that were pursued in late thirteenth- and fourteenth-century Europe, the *verge and foliot* was easily the most common, and formed the 'standard' clock movement for more than three centuries (Figure 2.2). The aim of maintaining mechanical movement at a constant rate was approached through releasing the weight in small increments controlled by the movements of the verge and foliot. These used the pressure exerted by the weight to propel one another, at a pace determined by the size and positions of the weights on the foliot. The aim, therefore, was to produce even rotation of the pallet-restrained wheel, and those wheels (the 'going train') to which it was connected by gears.

More formally, early clocks comprised a regulator (the verge and foliot), a going train, and a striking train (Figure 2.2). The regulator was a mechanism able to produce a regular movement and usually comprised a verge and foliot.

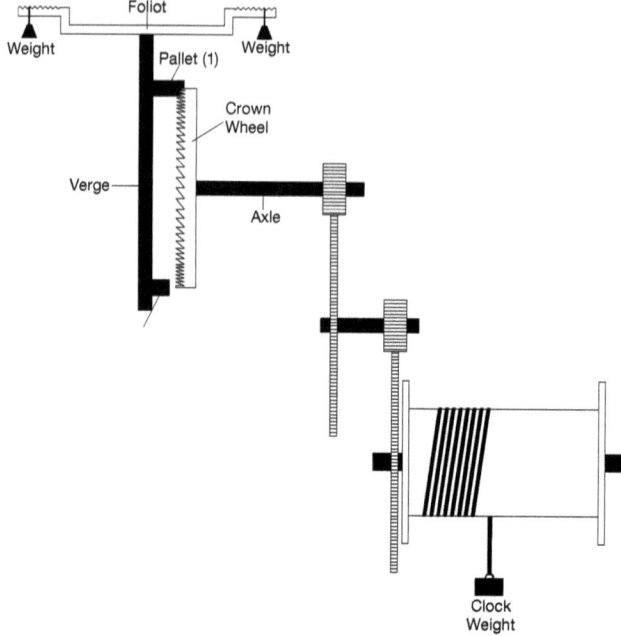

Fig. 2.2. Diagram of verge and foliot mechanism.
The verge and foliot act both to slow the descent of the clock-weight, and to produce a rotary motion. The large weights and forces in these clocks led to their large size, and to heavy wear on their components, one of several sources of inaccuracies in timekeeping.
Source: Rawlings (1993): 101.

This was a device powered by the fall of a suspended weight, which used the back and forth rotation of a vertical rod carrying a weighted bar to regulate the rotation of a horizontal wheel via gears. The going train was the system of gears and cogs that converted the oscillation of the regulator into the cumulative movement of a mechanism, enabling a count to be kept. The striking train was the apparatus that caused given numbers of movements to produce a signal, in early examples through the sounding of an alarm device set off by the release of a restraining device.

Since any such alarm device required a striking train as well as a going train, it was a comparatively simple matter to add a larger-scale mechanical linkage to a bell reserved for the striking of the clock. Direct, mechanical linkage of the clock mechanism to a clock bell or to a set of chimes appears already to have been the most common configuration of clock and indicator by the fifteenth century, whereas arrangements with a dial and rotating pointer were uncommon until rather later. The use of more than one pointer, through the introduction of minute (and later a second) hand, came later still.

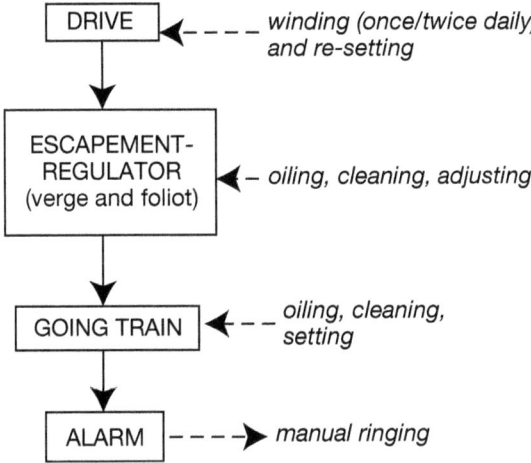

Fig. 2.3. Human involvement in the running of early turret clocks.
Human activity was integral to how early clocks worked; there were several tasks to be undertaken frequently in keeping the machinery running.

Even where hour-striking mechanisms had been mechanized, though, early turret clocks required very frequent human involvement to maintain their performance (Figure 2.3). Few could run for much more than 24 hours without the rewinding of their weight, and winding as often as every eight hours was common. To maintain accurate time, given that their going rates were at best approximate, they required setting with reference to an accurate sundial as often as possible, though noon was the only exact moment of direct solar time determination.

Some of the technical shortcomings of verge and foliot clocks were intrinsic to the design itself, others to problems in producing parts that could function as accurately and consistently in practice as they might in theory. Of the intrinsic issues, the most serious was that the verge and foliot had no natural period of its own, no inherent rhythm of oscillation. Oscillatory tendencies had to be designed into the wheel-work itself. The evidence of surviving verge and foliot clocks seems to indicate that oscillatory periods of two to three, occasionally even four, seconds were normal, amounting to some thirty to fifty thousand oscillations per day. It is stating the obvious to say that even very small inconsistencies in speed of oscillation could result in significant errors in timekeeping. Although an accuracy of 99 per cent might sound impressive, this equates to an error of nearly fifteen minutes a day.

Besides the lack of an intrinsic oscillatory period, verge and foliot clocks also suffered from the hand-produced nature of their components. For example, the teeth and gear wheels were filed by hand, and this made the even spacing of

Fig. 2.4. Diagram of pendulum regulator and going train.
The pendulum, especially in conjunction with the anchor escapement, was a much gentler and more consistent regulator, substantially reducing the mechanical stress on a clock's moving parts.

Source: Rawlings (1993): 101.

gear teeth very difficult to achieve. Whether this was a problem depended on the demands made upon the timekeeper. For example, if an hourly strike was produced by one complete revolution of a particular wheel, it might be of little short-term importance than the teeth of the wheel were unevenly spaced: just so long as the wheel rotated once an hour (although in the longer run this might have important consequences for friction, wear and slippage in other parts of the mechanism). In other cases, though, signals might be produced several times per rotation of a wheel, and here even spacing of teeth would be considerably more important. It might be one thing to produce sixty teeth on a wheel, quite another to produce a wheel with exactly one tooth per 6 degrees of arc.

The lack of a natural oscillatory period was remedied, and the friction relating to escapements was greatly reduced, through two critical developments, the *pendulum* and the *anchor escapement*. The adoption of these two innovations by late seventeenth-century clockmakers, along with the balance spring in pocket-watches, were key components of the *horological revolution* (e.g. Landes, 1983). They changed the performance, the costs and availability, and the potential uses of clock time in utterly fundamental ways, as temporal possibilities in every facet of life were transformed by the achievability of more accurate timing (Table 2.1).

Somewhat against his own expectations, in the late sixteenth century Galileo found that the period of swing of a pendulum depended on its length, and not on the height from which it was released, or the weight of its 'bob' (Drake, 1990). Subsequently, it was established by Huygens that this was true as long as the path of the pendulum was cycloidal (Yoder, 1988). In other words, the pendulum did not swing precisely to and fro but, when seen from above or below, followed a very shallow figure-of-eight path. The mathematics of cycloids were well understood, having been explored as a 'pure' research topic by

Table 2.1. Chronology of technological progress in clocks

1200		
	First mechanical clocks	
1300		
1400		
1500		
	First watches	
1600		
	'Horological revolution': pendulum clocks, balance spring, anchor escapement	
1700	More precise time-indication on many temporal devices	Expanding market for domestic timepieces
	Marine chronometer	
1800		
	Working-class ownership of clocks and watches	
1900		
	Electric clocks	
	Quartz watches, atomic timekeeping	

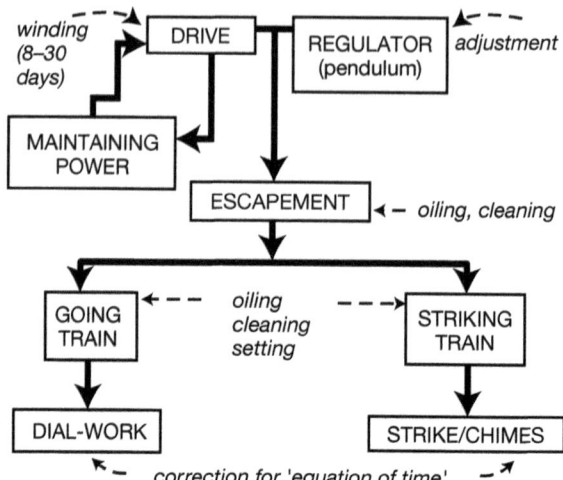

Fig. 2.5. Human involvement in early pendulum-regulated clocks, c.1670.
The necessity for intense human involvement in pendulum clocks was much less than for their predecessors.

ancient Greek mathematicians and their successors (an interesting early example of the practical dividends from abstract research). The anchor escapement was particularly well suited to transmitting the movements of a pendulum to the rotation of a wheel. Friction was much reduced, less lubrication was required, and the rotation of the wheel itself returned energy to the pendulum. The pendulum came into general use from the 1660s. Within decades it was standard even for clocks at the bottom of the market (Pryce and Davies, 1985).

After the horological revolution clocks generally kept better, if still approximate, time. In so doing they required significantly less, and less constant, human intervention. But there was very considerable variety, for several reasons. One reason was age. Clocks remained in use for long periods: they were no short-life throwaway consumer good. At any given moment therefore, the 'population' of English clocks included many whose needs for maintenance were those of previous generations of clocks. Another reason was scale. The turret clocks central to public time involved devices of considerable size and physical stresses that precluded the very delicate components involved in producing domestic clocks that required winding only every eight days, monthly, or even less often. Turret clocks continued to be wound, and maybe re-set, on a daily basis for very much longer yet. A third reason was the social range of the market for clocks. The combination of better timekeeping and less frequent winding came at a price. At the lower end of the market, thirty-hour clocks dominated: their relative robust and straightforward construction made for cheapness at the expense of accuracy.

A fourth reason was that the pattern of discrepancies between apparent solar time and mean clock time was only slowly being refined. It constituted, as it were, a 'moving target' for clockmakers, and so was problematic to 'build in' to clocks even where sufficiently precise gear wheels could be calculated and cut, especially when produced by hand-filing. In the long run, though, and not withstanding variations in the ages, scales, and costs of clocks, and imprecise knowledge of the solar time, the overall trajectory was clearly one of a diminishing frequency of human involvement in clocks' normal working, across the entire range of timepieces (Glennie and Thrift, 2002, 2005).

The maintenance of movement at a constant rate overlaps with issues of counting and signalling. The counting of clock time chiefly required gear wheels designed so as to rotate in relevant periods, such as one, three, twelve or twenty-four hours. In this way, the movement of particular parts of the clock related to equal subdivisions of the day, not solely to a given number of movements of the verge or pendulum.

Connecting oscillatory movements to a counting device to keep and display time could take various forms. For early clocks this was overwhelmingly through aural rather than visual displays of time, since bells provided the main means of marking the time. At its simplest, a clock could function merely as a rather spartan prompting device, sounding after a particular interval, or repeating at set intervals.

Alongside the progressive reduction of ongoing human intervention in clocks, and the standardization of hour-counting schemes, went a convergence towards standardized ways of displaying times, whether by ringing or on dials. The latter, especially, was heavily conventional, and took some centuries to settle on a single system (Table 2.2).

In large late medieval clocks, alarms were cues to servants or officials, who then manually rang a bell to signal the time for a particular service or other activity, and it seems likely that this was the context in which mechanical clocks (proper) were first constructed. The idea of more directly linking clocks to signalling apparatus rapidly became widespread. Where gearing was arranged to produce several strikes per complete revolution of a wheel, it was a relatively simple matter to modify the 'striking train' to ring successive signals in different ways (for example, to sound the appropriate number of strikes for each hour, from a wheel that rotated once every twenty-four hours).

It would be too hasty, however, to presume that medieval time signalling necessarily took the modern form of ringing the hours. Particular moments during the day were marked by ringing long before the invention of mechanical clocks, for both general purposes (such as the opening time of a city's market, or in advance of church services) or for specific groups of people (such as council meetings, or the commencement and conclusion of shoemakers' lunchbreaks). These were all practical rather than abstract signals: it was important that

Table 2.2. Conventional dimensions of clock time and signalling

Counting the hours
(a) How many hours in the day–night cycle?
    (usually 24)
(b) What defines an hour?
    (equal length, one-twelfth of daylight & of night, hybrid schemes)
(c) How are hours counted?
    (1 to 24, 1 to 12 twice, 1 to 6 four times, other schemes)
(d) From when does counting start?
    (sunrise, midnight, noon)

Sensory register of signalling times of day
    (by ringing, by visual indication on a dial, combined schemes)

The nature of visual indication on clock dials
(a) What moves?
    (the pointer, the dial-plate)
(b) How many revolutions in one 24-hour day?
    (one, two, other schemes)
(c) Position/orientation of numbers on the dial
    (12 at top, 12 at bottom, 12 at right-hand side)
(d) How many hands/pointers?
    (hour hand only, hour and minute hands, other schemes)
(e) Location of hands/pointers
    (on a separate dial for each hand; on a single dial)
(f) Rotation direction of the hand(s)
    (clockwide, anticlockwise)

key organizational moments were signalled on the appropriate days. What distinguished one signal from another were the location of the bell that was rung; and the pattern and style of striking—how many strikes? how close together? rung 'sharply', 'hard', 'softly' or in various other styles? The effect created by the co-existence of different groups with their own signalling requirements resulted, especially in large cities, in what has been aptly described as 'acoustic chaos'. With a bewildering array of different signals, not necessarily occurring in a fixed sequence from day to day, since they had their own promptings, it was certainly true that times were being signalled, but it cannot be said that this necessarily amounted to the signalling of clock time.

The attractions of signalling of clock time for its own sake were the potential simplification of these complex acoustic environments, and, usually at any rate, the placing of hour-ringing outside the control of potentially conflicting groups, such as employers and employees (Dohrn-van Rossum, 1996: 198). In other words, clock time became something more abstract, more neutral, than the earlier signalling embedded in particular social relations (Humphrey, 2001).

While bells provided the main means of signalling and clock dials were rare, three dimensions of the visual representation of time were important in medieval

England. First, some of the earliest large-scale clocks were fitted with large astronomical displays, like those surviving in several major cathedrals. These depicted the relative position of the sun, moon, and zodiac as (apparently) they revolved around the earth. Several did not indicate the time of day at all, and were entirely objects of celestial spectacle. Second, the striking of bells could itself be made into a spectacle, especially through the use of decorated 'jacks', figures who publicly struck the bell(s) with hammers to sound the hours. These were certainly present on a number of parish churches in the fifteenth century, for example. And third, the spread of equal hours timekeeping had important ramifications for the design of sundials, which could show 'clock time' if the dial was appropriately carved, though this was no simple matter.

Although the circumstances of early solutions are only imprecisely known, a rapid diffusion of mechanical clocks from the 1270s meant that clocks were scattered across England by the mid-fourteenth century (see Figure 5.1).

Part of the history of clock times, then, involved the establishment of conventions, both conceptual (as in defining 'an hour') and practical (for example, the senses to which time was signalled). Signalling of hours or shorter periods involved several variants on practices and conventions that we now regard as completely natural or obvious, thus showing the way in which clock time has over the centuries become something strongly associated with vision, whereas it had earlier been much more strongly associated with hearing. The change also highlights a long-run shift from public or quasi-public timekeeping to the private keeping of collective time.

It is important to appreciate that changes in the public indication of time have involved major sensory shifts, as well as technological change. To late twentieth-century perceptions, telling the time is a visual activity. We most frequently establish the time of day by looking at a clock or watch with either a dial, where the time is indicated either by the position of hour and minute hands (and perhaps a seconds hand as well), or a digital display, where the units of display may go down to tenths or even hundredths of a second. The time of day is something we read, then, with our eyes. The historical specificity of the practice needs to be emphasized: the prioritizing of the visual in indicating and apprehending the time is a considerably more recent development than is the use of public clocks and time signalling.

Of course, instruments such as sundials were a longstanding visual means of apprehending 'clock' time (at least during daylight hours and so long as the sun was shining), so we are not suggesting a complete switch between the senses in time-signalling and reading. But sundials did not readily signal the equal-hour divisions of the day that are fundamental to mechanical timekeepers. The very variable lengths of time for which the sun was above the horizon at different times of the year, and the varying angles and lengths of shadows cast, meant that—even assuming uninterrupted sunlight(!)—the indication of time measured in equal hours required sundials of highly sophisticated design and construction.

Most early medieval time-reckoning, especially that of the church, remained working with 'unequal hours', in which the hour was defined as the twelfth part of the day, or night. Thus at any given time of year day-time and night-time hours were of unequal lengths. The twelve hours of the day between sunrise and sunset were longer than the twelve night hours in the summer, and shorter in the winter. Only when the day and night were each exactly twelve hours long would the length of the daytime hours and the night-time hours be equal, and equivalent to the 'equal hour'.³ As we have already pointed out, it might well be argued that the adoption of equal hour time-reckoning (which had been confined to specialist groups such as astronomers), rather than reckoning in the unequal hours indicated by the sun, was more significant a change in medieval European timekeeping than any specific advances in timekeeping technology (Dohrn-van Rossum, 1996).

Timekeeping with daytime 'hours' that were longer than an (equal) hour in the summer, and shorter in the winter, was obviously problematic for mechanical timekeeping, an observation more appropriately put the other way round: mechanical timekeeping was comparatively unhelpful while time-reckoning used unequal hours whereas, notwithstanding the numerous imperfections of early clocks, it provided immediately applicable information within an equal-hours framework.

At all events, it is worth stressing just how widespread was the use of equal hours in England, as in much of western Europe, by the end of the Middle Ages. And it is worth stressing, too, that what later became the standard way(s) of signalling the time of day emerged over a long period, rather than as a finished whole.

A theme running throughout this book is that ideas about clock time, and imaginings of clock times, are not merely derivative of prevailing technologies. Typically, historical notions of clock time have been considerably more sophisticated than the abilities of clocks at those times to deliver on them. This is of immediate importance to this book, since the forms of clocks, and especially signalling, has been strongly shaped by prevailing notions of the possibilities of clock time, rather than the latter being a consequence of autonomously available technologies.

Hourly striking also attracted comment for its relative coarseness for those to whom shorter periods were relevant. (It is an interesting reflection how quickly these comments appear following the use of hour-striking, and highlights the latent utility of more frequent signalling.) More precise signalling could be attained in a number of ways, and several of these were pursued in the fifteenth and sixteenth centuries, even though the ostensible precision of such signalling was rendered spurious by the clocks' own mechanical limitations (modern reconstructions of verge-and-foliot clocks have found it difficult to obtain accuracies to greater than several minutes per day). Provision of more detailed signals could be aural, through more frequent ringing or chiming of the half and quarter hours. Particular hour signals could be accentuated by more sustained ringing of certain hours, or the playing of particular tunes by chiming mechanisms.

More refined signalling was not necessarily aural, though. It could also, or instead, be achieved visually through the use of dials. Dials were not, to reiterate, regarded as integral parts of early mechanical clocks—they took several centuries to become part of the definition of 'what a clock is'. Indeed, there are still cases from the second half of the nineteenth century in which parish churches commissioned clocks with chimes, but without a dial (see the case of Bedfordshire, Pickford, 1997). Even where clock dials incorporated only one hand, the subdivision of the dial's hour ring enabled the distinction of at least quarters of hours. (Note here that one-handed clocks most often had three marks and hence four subdivisions between numerals [though some had seven and eight respectively], whereas minute-hand clocks replace this with four marks and five subdivisions). The timing and frequency of different signalling devices, and their distribution, are discussed in Chapter 5. For the present, we want to stress that more refined signalling systems were not simply dependent on changes in clock mechanisms, such as the more accurate timekeeping enabled after the invention of the *isochronic pendulum*.

By the early eighteenth century, a dial (even if it possessed only one hand) had become integral to how people thought of 'a clock'. Clock time signalling, in other words, was being generally understood as something visual as well as auditory. And following quite closely behind the proliferation of dials was a sustained increase in sounding of smaller time intervals, such as quarter-striking clocks, that rang every 15 minutes rather than hourly.

A further impact on visual timekeeping came from the impact of clock time on sundials. Of course, sundials were of considerable importance for clock time, since the observation of local noon provided a (potentially) daily cue for clock-setting. (Again, note that it was not until the 1650s that it was realized that the interval between noon on successive days varied substantially through the year, so that an accurate clock by definition required frequent altering to maintain correspondence to the inconsistent apparent movements of the sun.) Only a sundial of considerable accuracy allowed the 'correction' of a clock at other times of day, so the spreads of both mechanical clocks and of the use of clock time, were powerful stimuli both to the construction and maintenance of more sundials, and to the provision of better sundials.

Attempts to provide signals more frequently than hourly, and the more sophisticated use of sundials, both illustrate ways in which groups of people expended considerable effort in maintaining clock timekeeping, and resorted to sometimes complicated systems of meeting the financial and logistical demands of keeping clocks working, and corrected to apparent solar time. In both cases this could involve them or their parishes in significant expenditure, and in turn this sometimes stimulated exceptional arrangements for raising the necessary money.

The key point here is that, in their signalling of the time, the clocks of medieval and early modern England were for a long time more aural than visual; more about making a noise than showing a sight. Even when dials and the visual indication of time became more commonplace, aural indications of time through bells and chimes still retained considerable importance. Telling the time then involved some hybrid of the aural and the visual: it is misleading to suppose a once-and-for-all transformation between 'aural' and exclusively 'visual' time.

The distinction between aural and visual indication of the hour potentially makes a great difference when clock time is considered as a set of practices. This is particularly the case when we consider the everyday apprehension of clock time, and the skills that people required in order to be able to 'tell the time'. On the face of it (no pun intended), fewer and less abstract skills were required by a bell-based system as opposed to a dial-based system. Signalling hours (and possibly their subdivision into quarters) arguably required only that listeners be able to count the strikes, after which their position in a sequence of hours was determined, so long as they comprehended the idea of days and nights split into twelve hours, from noon or midnight. On the other hand, interpretation of a clock face, especially where it featured both hour and minute hands, was an intrinsically more taxing task. Its abstract element was greater, and the minute hand indicates numbers other than those to which it appears to be pointing. Of course, visual time indication did possess significant advantages, particularly in its clearer depiction of time's continuousness, and of continuity in the measurement of time. A dial, whether with one hand or two, offered the continual (if not perpetual) presence of something observable in a way that periodically striking bells did not. But this advantage in continuity of time indication was at the cost of making telling the time a more difficult task. A moral we draw here is that when late twentieth century perspectives may regard telling the time as a more complicated procedure than it was several centuries ago, they may exaggerate the numerical and other skills required. If so, presumptions that few English people could tell the time in, say, 1600 are highly likely to be suspect.

Thus 'telling the time' comprised differing repertoires of skills and activities in different places, at different times, and with respect to different timekeeping devices. Telling the time was not always necessarily the comparatively complex task of reading that we now think of, which involves interpreting the position of two (or even three) hands on a clock face.

2.4 SOCIAL DISCIPLINARY APPROACHES TO CLOCK TIME

The second approach to understanding the history of clock time is the reverse of the first. It understands clocks and other timekeeping devices as simply social inventions: from this perspective, clocks are but 'mechanical movements of a

Table 2.3. Chronology of some major changes in social-disciplinary technologies

	Monastic 'church time'	
1200		
	'Merchants' time'	
1300		
1400		
1500	Protestant/Puritan valuing of time	
1600		Establishing global positions
		Land surveying
	Concept of 'mean time'	
		Oceanic navigation
1700		
		Factory work discipline
1800		
	National standard times	
	Standard international time zones	
1900		

particular type, employed by people for their own specific ends' (Elias, 1984/1992: 118). Thus clock time becomes a socio-symbolic invention concerned with more precisely regulating and coordinating the repetition of various social phenomena. In effect, clock time is a social custom (Table 2.3). This kind of thinking flows over into perhaps the most influential social determinist account of clock time—the work of E.P. Thompson—which finds the reason for its existence in new forms of social discipline.

E.P. Thompson's pioneering paper (Thompson, 1967), 'Time, work-discipline and industrial capitalism' demonstrates some of the advantages and disadvantages of such an approach. Published over thirty years ago, this extraordinarily influential paper is still a remarkable tribute to the power of grand historical synthesis and to Thompson's particular genius in that field. Thompson took what theoretical and empirical sources were available at the time and produced a powerful account of clock time as the vanguard of industrialism. Therein, of course, lies the problem. The account has been so powerful, its influence so immense, that it has, like the mythical Upas tree, smothered all beneath it. The paper has been taken as *the* standard account of the history of clock time, not just in Britain but elsewhere. More importantly perhaps, in terms of influence in the academic community at large, the paper has been taken by non-historians as *the* historical account: countless paragraphs or footnotes in more general works of social and cultural research cite Thompson's account as not just an intellectual baseline for further research but as something approaching the fact of the matter (e.g. Doane, 2002).

In examining time as a part of the changing culture of eighteenth-century English working people, Thompson sought to unpack the connections between the restructuring of industrial working habits and changes in people's inward notation of time. Thompson's chief target was the view that changes in time-discipline (greater synchronization of labour and more exact time-routines) were simple byproducts of new manufacturing techniques. On the contrary, argued Thompson, such changes involved much broader cultural changes: a transformation of the general work ethic and orientation to labour. These changes entailed the imposition and eventual internalization of a specific 'time orientation' to labour and life. New customs in common replaced the old.

Central to this transition was the replacement of 'task orientation' (the organization of time according to the necessity of performing particular tasks, with little attention paid to time in labour: 'the day's tasks... seem to disclose themselves, by the logic of need') by 'time orientation' (work organized by regular, coordinating time-disciplines). The progression was also one from natural, irregular, and humanely comprehensible time, blurring work and leisure (indeed, these are anachronistic concepts for task-oriented societies), to an 'unnatural' life tyrannized by the clock and timed labour. New time-disciplines were, initially, externally imposed through official timepieces, and systems of communicating time to the workforce and enforcing continuous work during the working day. But these disciplines became internally realized in quite new everyday time-senses among the labour force, and came to dominate society as a whole, not least through the school system. The process of internalization was greatly facilitated (but certainly not caused) by time ethics that had evolved from seventeenth-century Puritanism. Thompson dated this great transformation from the end of the eighteenth century when many people were acquiring access to precise clock time.

Indeed, a general diffusion of clocks and watches is occurring (as one would expect) at the exact moment when the industrial revolution demanded a greater synchronisation of labour. (Thompson, 1967: 69)

One striking consequence, and indication, of the powerful influence of Thompson's paper was that historians and others have subsequently tended to take its exploratory and pioneering proposals as given, rather than as a basis for sustained empirical research. Thompson's points were often adopted as an axiomatic framework for linking changes in time-sense to other dimensions of societal change (e.g. Thrift, 1981; Landes, 1983; Hopkins, 1982; Harvey, 1989). Widely anthologized (e.g. Flinn and Smout, 1974), the paper was, and is, a key text in the socialization of history students, rapidly gaining an authoritative status which belies the extent to which it was as much a paper of questions as a paper of answers.

2.4.1 Reconceptualizing time-disciplines

But time-discipline is actually a highly problematic category which consists of a number of dimensions of activity, not just one, and these dimensions

Clocks, Clock Times, and Social Change 45

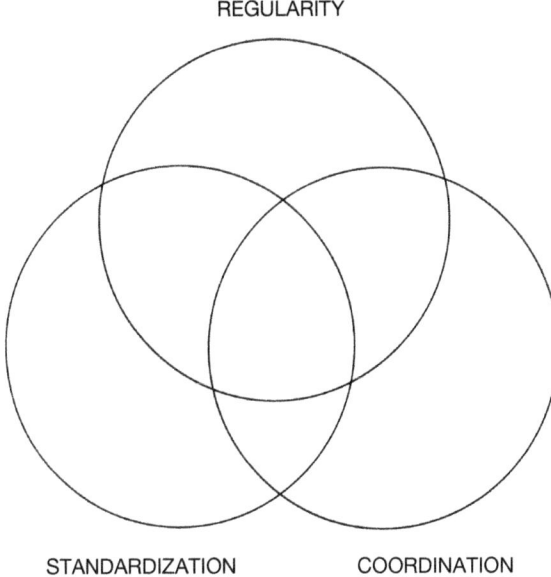

Fig. 2.6. Three dimensions of time discipline.
Time-discipline is not a singular thing, but can take varying forms depending on the importance attached to various disciplinary elements.

are conflated with one another at the writer's peril. Three dimensions seem particularly important, namely standardization, regularity and coordination. By *standardization* we mean the degree to which peoples' time-space paths are disciplined to be the same as one another's. By *regularity* we mean the degree to which peoples' time-space paths involve repetitive routine. By *coordination* we mean the degree to which peoples' time-space paths are disciplined to smoothly connect with one another's. These dimensions are shown schematically in Figure 2.6.

It is possible to depict these dimensions, and their variable inter-relationships in various ways, although none of these are unproblematic. Though a Venn diagram is not an ideal form to display each of the dimensions (Figure 2.6), other forms (for example, axes in three-dimensional space) raise difficulties that seem at least as serious. Here the representation of three dimensions of time-discipline allows the illustrative depiction of trajectories of industrial change. In addition, there are further dimensions of discipline which are not incorporated within these diagrams: for example, the degree to which patterns were enforced through coercive means, or adopted more or less voluntarily.

Nevertheless, such complications do not greatly affect our main point, which is simply that various permutations of these dimensions are possible whereas conceptions of time-discipline that follow Thompson's approach require all three to be part of a single disciplinary force. Fordist factory discipline does indeed involve high degrees of standardization and regularity and coordination. But other sorts of arrangements of production are not hard to find, in both industrial and other contexts, and could function efficiently.

For example, we can point to proto-industrial work (in which work was coordinated but not regular, and only in some respects standardized), or many agricultural tasks (in which work might be regular but is neither coordinated nor standardized to any great extent). In any case, the three-pronged intensity of Thompson's time-discipline did not need to await the factory, but characterized particular tasks such as harvesting (in which seasonal work was standardized, coordinated, and regular). In contemporary western society, so-called flexible production systems are very much less standardized or regularized than those of Fordist production, but all manner of coordinating mechanisms have become ever more finely honed.

Moving to sources of time-discipline outside productive activity, the commercial disciplines of exchange (which are much more important than Thompson allows) required coordination and to a lesser extent regularity for activities as different as getting produce to market or setting up bills of exchange (Thrift, 1996).[4] Again, monastic disciplines such as the Rule of Benedict stressed standardization and regularity, and only occasionally placed much weight on coordination as such (McCann, 1976).

In other words, Thompson's conception of time-discipline represents a particular and historically unusual combination of high degrees of standardization *and* regularity *and* coordination. Above all, early factories imposed a need for the schedules of many different people to be standardized as exact duplicates of one another. It is in the combination of all three facets of discipline that the factory appears distinctive, especially in reimposing a need for the schedules of many different people to be exact duplicates of one another. However, *contra* Thompson, the factory cannot be considered as some kind of end-point. Today's so-called 'flexible production' regimes, on which there is now a large literature, characteristically involve much more intense coordination, but sharply diminished requirements for standardization or regularity (Gertler, 1988). It is also worth underlining that work can be intense and morally compelling without work patterns necessarily being standardized, or coordinated, or regular. Much agricultural work, which is obviously crucial in our period, takes this form.

More than this, we want to emphasize that notions of standardization, regularity, and coordination are themselves culturally determined to a very considerable degree (Woodhouse, 1996; Zerubavel, 2003). 'Timing and timeliness are defined differently in various cultures and under different historical circumstances' (Hareven, 1991: 169), and studies around the world have shown how wrong it would be to assume that western conceptions about these notions are the only ones possible. For example, as the western language literature on Japanese society has mushroomed during the last few years, it has become increasingly obvious that both social and work practices in Japanese industry are very problematic for western frames of reference and analytical categories (Smith, 1986; Hareven, 1991). That contemporary developments are not a special case in this regard is evidenced by the parallel problems encountered in analyses of Japanese agriculture past and present, particularly in regard to rice cultivation (Shimada, 1994).

In summary, Thompson's notion of time-discipline is both too narrow, and too contextually specific. These propositions can be verified by reference to the historical record, the record of other cultures, and contemporary analytical developments. Work in all these areas suggests other means of time-discipline which conceive of time more flexibly—but which still demand very considerable rigour. Over time, the flexibility and rigour of work patterns has fluctuated in response to many factors. More generally, two important lessons may be drawn from this empirical work. On the one hand, there is the danger of 'simply assum[ing] connections between the way people work, or are suppose to work, and the way that they think' (Foster and Woolfson, 1989: 52). On the other hand, even in supposedly strict temporal regimes, it should be recognized that workers may act only minimally towards the script they are given, often never taking on some of its key premises, and in general finding ways of giving themselves room to manoeuvre (Birth, 1999; Rule, 1994). Thus work on time-discipline demands substantive research on actual temporal practices of the kind we endeavour to provide in this book, and not just the demonstration that explicit temporal frameworks and rules existed, since these may well have only been honoured in the breach or, more generally, interpreted in quite different ways by different parties.

2.5 'THE TRIUMPH OF CLOCK TIME': FEATURES OF A PERSUASIVE NARRATIVE

It is often argued, and even more often simply assumed, that clock time constitutes a singular entity. Such a typification usually rests on selecting one aspect of the practice of clock time as dominant. Marxist accounts often fall prey to this kind of reductionism but they are not alone in wanting to ascribe to clock time just one singular motivation. In certain senses, writers from very different traditions do exactly the same. For example, Heidegger argued throughout his works that clocks provide no insight into time, a view in part formed from the notion that clock time simply represents a process of destructively superficial objectification which therefore glosses over the unstable temporalities of being. Clock time represents the remorseless advance of science, blotting out the coalescent and emergent qualities of a more human time. Whatever the case, two approaches have been particularly dominant in explaining the history of clock time and we will now turn to each of these in more detail, using them both as exemplars and as a means of introducing important material.

A famous woodcut by Albrecht Dürer called *The Triumph of Time* shows time wreaking havoc on the human population, causing the wreckage of all hopes and leaving death presiding over an icy field. Dürer's bleak vision holds for much writing in which clock time's seemingly inevitable progress sucks all life from the world, leaving only a rationalized, dead husk: 'time is everything, man is nothing; he is, at the most, time's carcass' (Marx, 1976/1874: 127). In other words clock time is conceptualized as an almost entirely negative phenomenon, with this negative identification resting on three main features.

First, clock time is an oppressive monster which has produced a total social makeover, forcing societies around the world to model themselves on the clock. In Marx's version of this oft-told story, 'all economy ultimately reduces itself' to the 'economy of time' (Marx, 1973/1858: 173). In capitalist societies, everything is gradually reduced to commodities, which are mere moments of exchange; perishable, transient. Only money is an imperishable commodity and that is labour time objectified by means of a contract which extracts a terrible price from the worker (Marx, 1954/1887). Worse, the whole madcap dance is constantly speeding up as capital constantly tries to reduce its own circulation time:

while capital must on one side strive to tear down every barrier... to exchange, and conquer the whole earth for its market, it strives on the other side to annihilate this space with time, i.e. to reduce to a minimum the time spent in motion from one place to another. The more developed the capital, therefore, the more extensive the market over which it circulates, which forms the spatial orbit of its circulation, the more does it strive simultaneously for an even greater extension of the market and for greater annihilation of space by time.   (Marx, 1973/1858: 539)

A similar story was told by Oswald Spengler (1926), for whom:

the modern era is characterised by the restless striving of the Faustian soul and is inherently temporal.... eventually produc[ing] the pocket watch that accompanies the individual to remind him constantly of his temporal existence. The drama of Spengler's message is prepared by his emphasis on the importance of a sense of the future in the modern world. While the classical world bowed in 'submission to the moment', the modern world has an unsurpassably intense Will to the Future: Western culture glorifies hard work as 'an affirmation of Time and the future' and, with its meaning embodied in the future, it is particularly sensitive to the pessimistic vision that Spengler sketches.   (Kern, 1983: 105)

Later still, the spread of an 'ordinary', inauthentic conception of time is told by Heidegger as 'a fallen form of the existential modality of "within-time-ness"' (Osborne, 1995: 62), in which the inauthentic time of reckoning, of dating and of other means of time measurement—of the technology of calendars and clocks, in other words—covers up the character of temporality as a mode of *Dasein's* (Heidegger's term for his particular philosophical conception of human beings as questioning beings; literally 'there-being') existence by producing a conception of time as an endless and irreversible succession of instants which refuses to acknowledge Dasein's finitude and represents, 'thereby, a fleeing in the face of death' (Heidegger, 1967: 479). Clock time is therefore a fallen mode of time consciousness.[5]

Then, finally, it is the story of an imperial, colonial, and then postcolonial, 'time of the victors' (Serres, 1982) which has been imposed on the world at large (McLintock, 1996). This is 'the time of progress' which leaves those outside the vanguard of modernity to rot in a world that is supposedly timeless (McLintock, 1995; Spivak, 1991).

To summarize:

> The invention of the mechanical clock in the thirteenth century inaugurated a new representation of time. It . . . embodies a homogenising representation of time, offering a grid within which all events everywhere are commensurable because locatable within the same system of coordinates. Such a device makes possible the 'hour of equal length', granting us independence from the changing seasons and allowing us to be indifferent to the distinction between night and day. The representation afforded by the clock allows for the coordination of disparate activities on a global scale. Like the 'world-as-picture', the temporal representation effected by the clock is rooted in an interest. The clock is a piece of technology that is a means of disposing over, of mastering, time.
>
> [T]he clock ushers in a 'new time sense [that] finds one expression in a heightened awareness of mortality' (Thompson, 1967), . . . an ever-present reminder to us of time's relentless march, and is thus at the same time a spur to technological development. So we might say that technology both depends upon and is fundamentally challenged by the clock. In this way we can begin to see how the clock as a representation of time as linear, as irreversible, as the bearer of the irretrievable, is a key to the technological phenomenon. (Simpson, 1995: 23)

Second, clock time is unnatural. Though it is derived from 'nature', clock time has produced at best a shallow simulation, at worst a mockery, of nature. Clock time lacks the existential depth of natural times. It does not dwell in the world but is built over it; it is a mechanical cuckoo that has taken over the nest. As a result, people were pushed out of the gentle, rural rhythm of the seasons, and into a rapid-fire urban environment where they are bent to its will. Calculation becomes all, appreciation is as naught. Writers connecting the rise of a homogenizing (and mean-spirited) clock time to the decline of heterogeneous (and generous) nature are typified by Griffiths:

> The West's obsessive time measurement has gone . . . beyond the point of usefulness, and the clock of the present is not the realization of time, but its betrayal. Society begins to think in the forms it has structured for itself, linear, artificial, over-fragmented, modelling itself in the image of its machinery. Today's timekeeping pretends to itself it is describing time ever more accurately; what it really describes is modernity, a telling self-portrait. Modernity ascribes to time a driving, rigidly linear, impersonal, coercive and dominating character, overcrowded and overwound, which harries its victims, people. You hear the screech, time is running out—as if that were time's fault—but it is modern society itself which is overscheduled and domineering in its timings.
>
> . . .

The mono-time of Greenwich *mean* time offers no sense of time's generous variety, nor any intrinsic character or colour to time, it is an artificial construction and, more than anything, modernity's time—the global present—is increasingly standardized, increasingly the *same*... The very seasons impinge less on people in cities:... [that] ever more independent of nature, use the unspecific global rhythm of the global clock.

The forest is the symbolic opposite of the city... Time is everywhere in nature. In urbanized life, clocks are needed precisely because there is no other way of telling the time. But while nature knows a million varieties of time, the clock of modernity knows only one. The same one. Everywhere. (Griffiths, 1998: 12–15)

Third, clock time is omnipotent and omnicompetent: an adaptable, flexible monster making its way into every area of human life, producing all manner of time-based obsessions and perversions. This is the domain of the hard taskmaster, the hoarder, and the authoritarian minister of religion, but also of the sex addict and the serial killer; all those who measure desire by the clock (Barthes, 1976). Because of its refusal to adapt to human rhythms, clock time is seen as an intruder whose adaptations pervert natural time.

2.5.1 The World as Production Line

This sense of an inauthentic, unnatural, omnipotent and omnicompetent clock time is usually figured through various overarching narratives. Two of these narratives have proved particularly important (Duffy, 1992: 577–86; Burgess, 1988: 56–94, 199).[6] In the first of these, the world becomes a production line, in which time is split up into finer and finer mo(ve)ments, each of which can be described, analysed, and mastered.

The very form of time has been chopped up into uniform, repetitive clock time units. Work rhythms based on clock time are no longer context-dependent... there is no longer a way to take account of specific conditions and particular needs.

[Hence] the need for external normative and legislative protection which has to impose socially meaningful boundaries on the endlessly uniform strip of time units. Contemporary industrial work organisation has to have its rhythms artificially imposed on the standard metronomic beat of clock time. (Adam, 1990: 12)

Time as quality and abstract exchange value is no longer merely used, passed or filled, instead becoming an integral part of production, a commodity. Time-as-commodity, and people-as-clockwork-automata, become irreducible aspects of industrial societies, culminating in late nineteenth and early twentieth-century 'scientific management'. F.W. Taylor, the doyen of the movement, observed skilled workers to determine the exact series of elementary operations that made up their jobs. He selected the fastest series, timed each elementary operation with a stop watch to establish minimum times, and then reconstituted jobs as composites with the minimum times as standard. Taylor 'began to publicise his methods in 1895, stressing that workers complete jobs in the shortest possible times. The following year a Massachusetts builder, Sanford Thompson, devised a "watch book" with stop watches in the cover so that they could be operated without the worker's knowledge' (Kern, 1983: 116). People were transformed into 'modern industrial workers' in which time and the body reify one another through the medium of clock and stopwatch.

The clock, because it makes possible coordination, comparison, and so on, increases our control by enabling us to orchestrate practices and processes and to improve them along the lines of efficiency. F.W. Taylor's classic time-and-motion studies, undertaken at the turn of the century in the interest of heightening industrial efficiency and productivity would have been impossible without Taylor's dreaded stopwatch.... the autonomy of technology is circumscribed by technology's obedience to the clock. (Simpson, 1995: 22–3)

In the clockwork epoch of the commodity, and mass production, time becomes:

an infinite accumulation of equivalent intervals. It is irreversible time made abstract: each segment must demonstrate by the clock its purely quantitative equality with all other segments. This time manifests nothing in its effective reality aside from its *exchangeability*.... This is time devalued—the complete inverse of time as 'the sphere of human development'. (Debord, 1995, No. 147)

Recent variants on this narrative incorporate 'flexible production systems', which rely on minimizing inventory through just-in-time production and delivery systems, as symptomatic of a further powering-up of clock time's influence. Such systems entail even greater fractionation of time, through technologies offering immediacy of contact among suppliers and consumers; even greater pressures on individual temporal coordination (Sennett, 1998); and, in some accounts, an even greater push to consume (the result of new generations of never-sat(isfi)ed customers, always about to move on to the next commodity). In the most extreme accounts, this produces a

'real-time culture' in which 'culture will be founded on a new interrelation between companies and customers created by the capacity of fully integrated real time systems for acute sensing, for dialogue and for responsiveness, systems in touch with the marketplace twenty-four hours a day. And that is how managers will cope with the outstanding feature of the age of the real time consumer: the eventuality of anything'. (McKenna, 1997: 176)

2.5.2 The Accelerating World

The 'production-line world' narrative is often paralleled by another, expressing similar fascinations with, and fears of, clock time. This is that the everyday world is intensifying and speeding up; becoming ever more frantic, and producing a general shortage of time. This second narrative has a remarkable historical constancy. Popular books predicting an ever-faster world (Gleick, 1999) and forecasting 'the end of patience' (Shenk, 1999) are the latest in a long line of multi-stranded jeremiads across several centuries, centred on the pathological consequences of speed (Thrift, 1995, 1996). Current versions of this narrative emphasize a Whiggish progression of events in the history of time, especially emphasizing the establishment of 'world time' at the 1884 Washington International Meridian Conference, which made Greenwich the prime meridian, and several simultaneous technological innovations in transport and electronic telecommunications which allowed the world to become 'an order of simultaneity' (Kern, 1983).

> Once the linear system of time was set... acceleration could start in the form of motion making everything dynamic, which seemed to stop at nothing. In the *tourbillon social* which broke out with the industrial revolution and wrenched people out of their countless 'small worlds', out of the small towns and villages with self-contained social structures and forms of existence, and kindled in them the inevitable desire for growth,... acceleration became the experience of modernisation overshadowing and shaping everything else. The pace became more important than the destination: anyone who stands firm stands still; everything, above all time, becomes frantic motion: the new myth was speed. (Nowotny, 1994: 84)

Nowadays this narrative, which long predates Marx's characterization of the 'annihilation of space by time' (Thrift, 1995), has been enthusiastically adopted by contemporary commentators right across the ideological spectrum, from Marxist doom-sayers to corporate Panglossians. Thus David Harvey discusses a general shrinkage of space under pressure of time that he calls 'time-space compression' (Harvey, 1989). Combining earlier geographical ideas of 'time-space convergence' with work on changing structures within contemporary capitalism, and on the constitution of western human subjects, Harvey constructs an epochal history of western civilization premised on necessary relations between successive waves of time-space compression, the emergence of new modes of capitalist accumulation, and new cultural forms, especially in the domain of the image.

Paul Virilio is perhaps the best-known cartographer of this 'Great Acceleration'. For Virilio, as the chronicler of instantaneous time, Harvey only describes the initial effects of speed-up since he still presupposes a journey of departure-(physical)displacement-arrival. Now, however, electronic technology enables 'instant' transmission, eliminating journey time. A new 'generalized arrival' has occurred. Individuals can simultaneously be in two places, as 'alternative beings' or 'trajects' who are both transmitters and receivers. The result is a 'crisis of the temporal dimension of the present moment' as:

> one by one, the perceptive facilities of an individual's body are transferred to machines, or instruments that record images or sound,... that can replace absence of tactility over distance. A general use of telecommunications is on the verge of achieving permanent telesurveillance. What is becoming critical here is... [the] temporal dimension... the present itself. (Virilio, 1993: 4)

In turn, this instantaneous, 'intensive' time foreshadows a remarking of the world because in . . . territorial development, 'time' now counts more than 'space'. . . . [N]o longer . . . some chronological local time, . . . but universal world time, opposed not only to the local space of a region's organisation of land, but to the world space of a planet on the way to becoming homogenous. From the urbanisation of the real space of national geography to the urbanisation of the real time of international telecommunications, the 'world space' of geopolitics is gradually yielding its strategic proximity to the world time of a chronostrategic proximity without any delay and without any antipodes. (Virilio, 1997: 69)

Whichever narrative is told,[7] life in contemporary societies seems to resemble the time of the movie *Groundhog Day*: we are doomed to repeat even our most minute bodily movements, trapped in an eternal present. The clock holds dominion over time; time has become the clock.[8]

2.6 METHODOLOGY

Although fuller discussion awaits the next chapter, especially section 3.4, it will already be clear that our focus on clock times as everyday practices, rather than as technologies or disciplines, entails a rather different emphasis from the prevailing histories and narratives so far discussed. In particular, an emphasis on clock times as unreflective uses and meanings, which involve relationships with devices embedded in societal networks, making up the fabric of everyday life, raises questions of appropriate questions, appropriate evidence, and appropriate methods.

In this section we draw together some of the observations made in the previous sections by considering the methods used to produce the evidence laid out in this book. These methods are of four kinds. First, there is the arduous task of scouring the archives. By drawing on a number of sources, each of which has their own insights and distortions (Ginzburg, 1999) but each of which are able to provide productive accounts of the usage of clocks and other timekeeping devices, we have tried to provide some sense of the multivalent nature of what we fondly tend to regard as the simple outcome of 'telling the time'. We draw upon a wide range of archival sources, and try to provide a sense of the range and reliability of evidence that each source provides. Second, as we have already pointed out, we have tried to take devices seriously. We do not see devices such as clocks as passive tools. Rather, devices are active parts of events which have their own powers and have to be used skillfully to bring those powers forth. Devices are not only used in situations, they also define many of the contours of perception of what those situations entail, both by acting as extensions of the body and by providing a culturally sedimented interpretative frame. For example, a clock provides timekeeping capacities but it also assumes a particular interpretative frame which has been built up over many centuries. Thus, our method has been to stick closely to accounts of objects in use whilst, at the same time, attempting to outline the ways in which those usages could at various points in history, have been radically different. Often only the smallest deviations have cemented a particular set of conventions in place. Third, we have thought seriously about how we can identify extra-linguistic performative modes in what are chiefly linguistic sources. Many elements of clock time, from gaits to gestures, from various kinds of space to all manner of performances, can be reduced to language, but 'it is particular to them that one has to do without spoken language in order to trigger or control appropriate actions, attitudes or modes of behaviour through them' (Koselleck, 2002: 25). Picking up these modes in the historical record presents a particular range of problems in trying to trace the effects of things that go unmentioned or hardly mentioned but which are undoubtedly 'there' and undoubtedly have influence. Fourth, at various points in the book, we attempt to grasp the 'inside' of thought, in classical Collingwoodian style. Thus, we attempt to draw on sources in which individual human agents give accounts of time usages. However, we are cautious about the value of these kinds of accounts, not only because they very often tend to be the views of elites dealing in standard cultural narratives (for example, that rural labourers are temporal innocents) but also because they rather rarely give descriptions of the specifics of timekeeping practices and instead tend to fall back on stylized cultural metaphors about their efficacy of which we have our suspicions.

Our belief is that the real currency of 'consciousness' of clock time chiefly lies in a few micro-practices which are repeated very often, usually without any great degree of cognition, and which are therefore often utterly *taken for granted*. We want to end this section with an actual example of this taken-for-grantedness, which also incorporates each of the principles we have so far been pursuing—no singular history, the importance of a practice-based approach, the centrality of devices and the need to pursue the ordinary—but wrapped up in the practical methodological conundrums that we have had to negotiate.

However, the taken-for-grantedness of everyday life creates evidential problems. There are few topics more difficult to investigate historically than those ethnomethodological details that were so a much part of the mental world of people in the past that there was never, or only very rarely, any call to set them down explicitly. Certain facets of everyday times have proved very difficult for us either to follow through in the literatures, or to investigate ourselves. Many difficulties in these areas stem from one or both empirical uncertainty, and the fact that temporal behaviour and attitudes were frequently implicit rather than explicit and so do not enter the documentary historical record. It is, of course, difficult to demonstrate clearly that particular material was not recorded because it was 'too obvious'. Such an argument may smack all too easily of the cavalier dismissal of apparently negative evidence. And the principle of Ockham's razor might suggest that it would be unnecessary: would it not be more economical to suppose that something not mentioned was absent, rather than present but taken for granted? These are not arguments to be dismissed lightly: to counter them effectively it is necessary to show the taken-for-granted in operation.

We can demonstrate how we have often had to proceed through an increasingly well-known piece of writing dating from about 1560. This was an essay written by a man named Roger Martin who lived in the small Suffolk clothmaking town of Long Melford. His subject was his local parish church: *The Church of Long Melford as I have known it*. Of course, the timing of Roger Martin's writing was anything but accidental. The swings in institutionalized religious practices over the previous thirty or forty years had had massive effects on everyday lives among the laity, perhaps especially in East Anglia (Duffy, 1992: 478–501). And Roger Martin was, from an autobiographical standpoint, exceptionally well qualified to write on his topic. He had lived in the parish since his birth in about 1520, but his family context was more wide-ranging, with connections for example to London through involvement in a legal practice based at Lincoln's Inn (Duffy, 1992: 37–9, 496–7, 510). The Martin family at this time, as later, were among the most active in parish administration and in the array of formal and informal guilds that served a variety of formal and social functions. They were, not only on Roger Martin's account but from surviving parish and guild records, active and (presumably) enthusiastic participants in what some revisionist historians have termed 'popular Catholicism'. Roger Martin himself served as churchwarden on a number of occasions. In particular he was one of two churchwardens during the return to Catholic practices under Queen Mary in the mid-1550s, and their re-suppression following the accession of Queen Elizabeth I. It is evident from the slowness with which the Elizabethan suppression was put into practice in Long Melford that it did not command popular support among those responsible for its local implementation.

This context, within which Roger Martin wrote his memoir, largely explains his threefold focus: on the fabric and furniture of the church as central to the parish community; on the active involvement of parishioners in religious and parish-welfare activity; and on the senses and rhythms of time conveyed by (and embodied in) the festival calendar, which was again being dismantled and stripped of its significance at the end of the 1550s, as it had been a decade or so earlier under Edward VI.

Nearly four and a half centuries later, it is instructive to juxtapose Martin's narrative of the church in parish life with the account books of the churchwardens who were largely responsible for administering it. The accounts, some of them compiled and signed by Roger Martin himself, are, of course, dry and prosaic by comparison with the narrative. But for the most part it is inescapable that, while employing a different sort of description, they are describing 'the same thing'. Spending associated with the religious festivals; spending on repairing the church, bells and vestments; spending on the church hall and on parish-owned houses, all are clear. Not surprisingly, the upsurge of revisionist work on the Reformation and popular religion has been extremely interested in what Roger Martin had to say.

Our concern here, however, is with something that Roger Martin *did not* say. His narrative description prepares us well for reading the churchwardens' accounts, with the (literally) striking exception of the church clock. There is no question but that the striking of the clock would have been familiar to Roger Martin, for Martin had been the churchwarden who paid nine shillings to one 'Clowthe' for mending the clock in 1556, as part of substantial renovations to the clock, as well as making a variety of small payments for wire, oil, and [wooden] 'polys' relating to the clock and bells. This 'Cloughe' appears in the accounts through to a hiatus in the mid-1580s. The clock was already quite old; a man called Ynholde had been brought from Sudbury to repair it in 1549, the first full year of extant accounts.[9] Further, Clough was paid nearly three pounds for a new clock in 1581, the date of which strongly implies a much earlier date for the clock that was being replaced.[10]

Now we could argue that Martin left the clock out of his account of the church explicitly and deliberately. Clocks were not an explicit focus for church reformers from the 1530s through the 1560s (though it is an interesting question why not—if the motive for public clocks was largely related to church services, as has often been suggested). In this interpretation the clock was omitted as peripheral to what Martin wanted to say. The problem with this argument is that exactly the same could be argued with respect to many objects and practices that he *does* write about, that were not specifically liturgical or ceremonial. So why should the clock be any different?

Our contention, which we will back up many times over later in the book, is that Martin's omission of the clock from his account of church and parish life provides a specific instance of the 'taken-for-grantedness' of timekeeping. Of course, this is a difficult claim to argue in a short excursus like this one. But it is not impossible. Our first argument centres on the sheer numbers of church and other public clocks in sixteenth-century England, as will be discussed in Chapter 5. Church clocks were not ubiquitous, but they were common in places like Long Melford at this time. And they were used. The evidence of spending on repair shows that parishes were too concerned to have and maintain clocks for the idea that they were not thought to be important to be sustainable. And parishes used information about the time of day in too many ways for us to accept that they did not think it mattered. We do not think that the church clock was omitted from Martin's account because he thought it unimportant to the parish.

What Martin's description does seem to demonstrate is that public time, although maintained largely through church clocks, had become widely perceived as only related to the church through locational convenience. Martin's omission of the clock from a description of the church was not, however, an isolated example. Only a few years earlier, in the national survey of parish church property under Edward VI, churchwardens across the country had been asked to list objects in their church. There was no explicit checklist to which they were to respond, and the majority, like Martin a decade or so later, omitted clocks. This was not universal: clocks were occasionally listed in churchwardens' responses, but the great majority of clocks whose existence is attested to by churchwardens' accounts were not included in the parish inventories. Only around 10 per cent of the parishes whose surviving accounts reveal them to have had a working church clock thought to list the clock among the parish possessions. We are not seeking to argue here that clocks were a special case in this regard: a similar argument has recently been made for rings of church bells.[11] But the Edwardian church inventories support the argument that Martin's taking for granted of the church clock was a widespread phenomena in mid-sixteenth-century England, a taking for granted of that which simply did not need to be noted or explained.

This argument clearly has implications for the methodological and interpretative strategies used to explore the place of clock time in the taken-for-granted everyday. *We should not expect to find many explicit commentaries on the everyday use and value of clocks.* Rather, most of our evidence will be found in the remaining traces of the ways in which clocks and time signals were installed and maintained, and in quite incidental—often almost accidental—recording of the uses made of them.

2.7 HISTORY AND METHOD

In drawing this chapter to a close we want to touch upon some more general questions of historical method as they relate to time. Our intention has been to write a book whose form conforms to our own strictures about clock time. We do not believe that clock time consists of one thing. Rather, we envisage it to be a variable element of a set of networks of practice which are only partially connected. We have therefore chosen to employ an *episodic* form in this book, one which recognizes that all the different networks of practice we will do our best to intrude upon ran at different paces and covered different territorial expanses and, not surprisingly therefore, could have surprisingly different time senses.

In the process of constructing this episodic account (or rather accounts) of clock time we have therefore had to leave behind the simple and sometimes glib ideas of linear progression from untimed to timed cultures, via intermediate stages in which the untimed and the timed *coexisted*, with the inexorable undermining of the former as a singular modern 'industrial time consciousness' makes it way through society and across space.[12] This account has proved not just inaccurate but damaging over many years, not least because it is caught up in what Fabian (1983) calls 'allochrony', a map of time which places the spaces of western modernity at some kind of civilizational peak in time and the spaces that do not conform to these features as somehow 'back in time', as not modern. Fabian's identification of a 'denial of coevalness' in so many humanities and social science writings is clearly something we want to escape from. Rather, for us, history is multi-levelled and subject to different rates of acceleration and deceleration, and the challenges of telling that kind of immoderate process of many times is itself constitutive (not least of a new notion of 'modernity' as this new understanding) (Koselleck, 2002).

Thus, our alternative account is concerned with the degree and type of *interaction* and *mutual construction* among several coeval time-senses, or what Bloch (1935/1991) called an 'asynchronous synchronicity'. However, this does not mean simply setting up a 'complex and differential temporality' leading to strictly 'conjunctural' analyses of the kind that Perry Anderson (1989), for example, has vouchsafed. In the end, as Osborne (1995: 13) has argued, this strategy is likely to become simply a variation on the very temporal paradigm it sets out to oppose 'since it still works itself out under the sign of modernity'.[13]

Leaving a sense of linear progression behind means that we will not give a full account of the history of clock time in the usual narrative form. That would be at odds with one of the chief messages of this book. Instead we will provide a history that is made up of episodes in the history of clock time. These episodes are temporally and spatially discontinuous, do not always connect to one another or connect only partially, and are not meant to sum up to a narrative whole. We have already argued why this strategy is the appropriate one, but there are two other reasons which also need to be mentioned; one operational, and one methodological.

The first reason concerns the quantity and quality of the historical evidence that we can draw on. Much of the evidence we might call upon to produce a narrative history of clock time is problematic. To begin with, historians have not, on the whole, taken the history of timekeeping to be a key topic. Until recently it has been little studied, so that the general archive that can be drawn upon is still quite small and limited. Though this situation is now changing such a narrative, even if it were desirable, would involve what would often be heroic acts of interpretation and extrapolation. Even the addition of other cognate literatures in fields like horology and the history of science cannot make good this lack. Further, the kind of approach to the history of timekeeping adopted in this book, with its emphasis on networks of embodied practices, brings into play parts of history (such as the history of the body) which are still themselves in a formative stage.

In turn, much of what we want to know about the history of timekeeping must be oblique to the sources. Take the case of embodied practices. Sources on embodied practices are, as Thomas (1991: 5) makes clear, 'surprisingly rich'. But, they are rarely directly on the subject of, say, body tempo: such practices often have to be inferred from sources written for other purposes, which often means triangulation from a whole range of different material. In other words, in pursuit of such information, there is to begin with a need to 'modify traditional historical method in pursuit of data which were not amenable to its treatment, data which did not testify to a primary fact locatable in time and space without reference to other facts' (Wilcox, 1987: 266). Then:

historians of tacit knowledge can [only] study tacit knowledge after it has been reified either in a text or a device. This is not simply to say that the observers knowledge is always *a posteriori*. If that knowledge has not been blackboxed, posterity cannot gain access to it. Also, . . . being a participant does not allow one to spell out his or her tacit knowledge either. One can perform it, but full verbalisation is bound to remain another matter. (Biagoli, 1993: 79)

So, in the final analysis, it has to be realized that certain practices cannot be recovered, and are only open to intelligent surmise. For example, there are whole timing 'vocabularies' (just like, for example, the language of clothes) which have now lost their meaning for us and cannot be reconstructed.

The second reason for using an episodic form concerns the methodological status of linear narrative, especially as it applies to the linearity of clock time. Such a narrative would contradict the kind of theoretical framework which we have already begun to outline, which must imply that 'there are aspects of experience which cannot be treated as simple concrete events existing as points on a time line, even though they have unsustainable temporal dimensions' (Wilcox, 1987: 10). In particular, linear narrative is associated with absolute dating systems which are themselves historically and culturally specific artefacts arising in part from the spread of a particular form of clock time.

It is worth underlining this point. Much of the timing system that we now regard as a part of standard historical practice is of relatively recent vintage (Ermarth, 1992; Grafton, 2003; Sato, 1991). For example, the BC/AD system (which is gradually being replaced by the BCE/CE system) dates only from the seventeenth century. A Jesuit scholar, Domenicus Petavius, published the first formal work setting out the BC/AD system in 1627. His system came slowly into use in the seventeenth century, but with many chroniclers still preferring the older system, the so-called Julian Period. The use of centuries dates from much the same period.

Although Moslem historians had long organised events by centuries, the first major work in the western world to mark periods in such an abstract form was the *Magdeburg Centuries*, an ecclesiastical history written by a group of German scholars in 1559. Though it appeared during the Reformation, the concept of centuries did not come into widespread use before Newton's time. Not until the seventeenth century did most literate contemporaries identify the epoch in which they lived as their 'century', and only in the course of the Enlightenment did the term take on the epochal significance we now attach to it. (Wilcox, 1987: 9)

In turn, these points can be used to make a more general observation, which is that,

for historians before Newton the time frame did not include a group of events; a group of events contained a time frame. This perspective led them to use a variety of relative dating systems, none of which had an absolute temporal significance apart from the group of events that gave it its meaning. Since we use an absolute time line as a basis for our synthetic understanding of the past, modern scholars have often viewed the relative time lines of early historians with condescension, calling relative time simple and primitive. (Wilcox, 1987: 9)

But, of late, such relative time frames, in which particular networks construct their own time, have started to gain credence again in history, as a result of a number of recent historical currents (like the new historicism and 'postmodernism') perhaps, but more as a result of the rediscovery of writers like Proust, of sociologists like Tarde, and of philosophers like Whitehead. Based on the principle of seeing the *process* as the most concrete historical entity, these authors prefigured late twentieth-century historical practice which has been gradually moving away from the notion of an absolute sense of time towards relative time frames which attempt to recover some of the dynamism of temporal framing and experience which is revealed when these frames and experiences are seen as unstable aggregations and not as universes in themselves (Latour, 2005).

This ambition is touched by another: an attempt to acknowledge the sheer *richness* of temporal framing and experience. As Proust (1981: 924) famously put it: 'an hour is not merely an hour, it is a vase full of scents and sounds and projects and climates'. Whatever the exact case, the outcome is clear: orders are produced which can never be written as a narrative whole complete unto itself without doing unacceptable violence to both method and the record.

None of this should be thought of as implying that that these different time frames and experiences cannot be connected to one another. Because they are unstable aggregations which have only gradually developed distinctive properties and degrees of avidity, they will usually contain all kinds of plundered threads from other networks and will, in any case, be only partially connected with each other (Strathern, 1992). In other words, temporal frames and experiences interdigitate in all kinds of ways, in the process often adding in even more differences.

The most clearly worked-out contemporary senses of an episodic form based on the apprehension of partially connected processes can be derived from two sources. One is the work of Michel Foucault. Foucault rejected normalized concepts of time and space for concepts of time and space which would allow him to relate chronologically and spatially scattered discursive practices (see, in particular, Foucault, 1973).

Such a task implies the calling into question of everything that pertains to time, everything that has formed within it, everything that resides within its mobile element, in such a way as to make visible that rent, devoid of chronology and history, from which time issued. (Foucault, 1973: 270)

Using spatial metaphors like the surface, instead of metaphors which indicate some underlying and essential structuring principle, Foucault tried to produce an episodic sense of time as 'a massive mobilisation of small examples' (Probyn, 1996: 35) which would emphasize relation and connection (and disparity and dissension). In this way he could avoid the need to search for deeper meanings which would merge all the segregated spaces that cannot (and perhaps should not) be allowed to coalesce.

The other source is the later work of Michel Serres, now operationalized in the writings of actor-network theorists like Callon, Latour, and Law. Serres and Latour (1995) was concerned to criticize the implicitly allochronical condition of so much western thinking on time which conjures up a historical time made up of tombs and statues continually being laid down as affirmations of the line of progress (Assad, 1999) and insists on making the singularity of the present (understood as 'a given assemblage of particularities' Serres and Latour, 1995: 4) into something special, the now. Burnett sums up Serres' critique thus;

the quirky world maps of the medieval world view... strike the modern viewer as wholly fantastical, since they gathered up the known world and arranged it around a powerful centering point: Jerusalem. We laugh, Serres pointed out, at this and every other ancient cosmography that tried to place humanity in the heart, middle, and origin of everything. And yet, he went on..., are we not the victims of a comparably narcissistic delusion? If Mercator and Copernicus dramatized that human institutions are not at the center of space, the deep cognitive structures have offered us a consolation of considerable power: now, at this moment, we are continuously reassured that we stand at the summit of time.

The idea of progress makes us this guarantee. As Serres put it, 'we conceive of time as an irreversible line, whether interrupted or continuous, of acquisitions and inventions'. And, therefore, continuously abreast of the past, 'it follows that we are always right, for the simple, banal and naïve reason that we are living in the present moment'. From our vantage point at the center of this temporal *mappamundi* we can survey history, secure in the knowledge that we are not only right, but 'righter than was ever possible before'. Moreover, we are guaranteed always to occupy this enviable seat, since each moment simply lifts us higher over all that has come before. (Burnett, 2003: 7–8)

For writers like Serres and Latour, this peculiar cartography of time must be replaced by a 'teeming multiplicity' (Serres, 1982: 53) of diverse and entangled space-times produced by the weavings of circulating actor-networks which constantly but temporarily converge, connect, intersect, interfere, conjoin, dissipate. The trick is to capture the fluctuation and the conjunction and the dissipation of these inventive aggregations in a new more plural sense of time.

What is an organism? A sheaf of times. What is a living system? A bouquet of times. It is indeed surprising that this solution has not been reached more quickly. Perhaps it seemed difficult to intuit a multitemporality. (Serres, 1982: 75)

The ambition we see in the work of writers like Foucault and Serres, then, is the founding of a 'differential history' (Bhabha, 1994), taking place as a 'brecciated', kaleidoscopic set of times (Burgin, 1996). In the rest of this book, we will try to live up to that ambition by reconstructing a number of episodes in the differential history of clock time, episodes which are the result of the interweaving of particular networks of practice, which are laid out on the ground, so to speak, as linked and often overlapping sets of spaces (Thrift, 2003b). These networks, driven by particular kinds of embodied practice enacted, more often than not, through families of devices which are themselves formative, and therefore laden with their own specific skills and competences (which always involve an inseparable semiotic dimension), go to prove Michelet's (1855: 161) famous phrase that history 'is first of all geography'.

2.8 CONCLUSION

It is another common truism of historical work that history must always be history of the present. The past continually turns up in the present: as body stances whose origins have become opaque, as emotions whose grip is unexpectedly strong, as laws and conventions whose motivations are forgotten but still have force, as the frames of measurement and the margins of representation whose arbitrariness is no longer recognized—and as devices whose workings are a palimpsest of different procedures and solutions.

Clock time, more than most, cleaves to this truism because it is a mix of all these different traces of times past still present in the present which come down to us now as something so mundane as to be hardly worthy of notice. But in this book we hope to show, through a series of careful defamiliarizations (Ginzburg, 2001), that it is possible to provide a history of previous everydaynesses when the reproduction of clock time was much less certain and the correspondingly greater effort put in to telling time first produced the outlines of what we fondly imagine to be a modern world.

NOTES

1. 'Sand' is a misnomer here, since lighter powders were used in order not to rapidly erode the containing glass, especially at the neck of the dial.
2. Whilst usually discussed in terms of the Islamic society to which medieval Europe was indebted for the transmission of classical texts, similar practices obtained for several Christian and Jewish feast days.
3. In practice, things are not quite so simple, since the interval between two successive noons is not exactly 24 hours, but fluctuates through the year. The accumulated divergence between mean-time noon and apparent solar noon varies by up to just over a quarter-hour. This was not realized until the 1650s, see Chapter 9.
4. Importantly, mercantile imperatives forced greater and greater discipline on the plantation system which was a feature of much of the period, and they were a

key determinant of more and more rigid time schedules, all of which involved timekeeping devices, sometimes in profusion.

> The labour force, from the overseer down to the field labourers, was organized around the clock. Yet, this organization did not depend on all the members of the workforce watching the clock. Instead the overseer kept time.... Plantations signalled when to start work, take breaks and end work by using horns or bells. This control over timekeeping devices gave overseers 'the power to set the actions of others... against artificial mechanical time'. (Smith, 1997: 14)

Control of timekeeping devices was so absolute that there are reports from Jamaica of hourglasses used by apprentices being destroyed (Birth, 1999). The absolutism of plantation time discipline has led some authors to argue that plantation time discipline may have acted as one model for factory time discipline.

5. Heidegger's downgrading of the ordinary conception of time to an inauthentic form has been frequently contested.

6. Other narratives would have been possible to note here, though they are probably less pervasive. One is the current obsession with iterability and repetition in the philosophical literature (as in the work of writers like Deleuze and Butler) which can be seen as, in part, a coded cultural discussion of clock time. A second narrative is related and concerns reciprocity, specifically in the anthropological literature on the gift. This depends entirely upon a temporal structure which the time of the clock disrupts (Derrida, 1996). Then, finally, there is the narrative of waiting, queuing and the like. Much of this literature is concerned with the incarnation of clock time in timetables, diagrams and the like and with the means of overcoming or evading its supposed hegemony (Adam, 1990). In particular, new notions of the self are seen as emanating from this restructuring of time.

7. These narratives have been influenced by the work of E.P. Thompson (1967) which we have critiqued in detail elsewhere (Glennie and Thrift, 1996).

8. This line of thinking has become so engrained in western cultures that it has even produced a horological counter-culture in the shape of projects like 'the clock of the long now', a 'mechanism and a myth' devoted to drawing time out again (Brand, 1999).

9. It may be significant here that Sudbury was the area's central town for clockmaking and repair expertise from the late fifteenth century at least. Inholde is recorded in several other parishes, including Boxford, West Suffolk Record Office, FB77/E2/2.

10. See Chapter 5 for the substantial working lifetimes of fifteenth- and sixteenth-century parish church clocks.

11. Sanderson, 1994, vol. I.

12. It seems clear that Thompson himself thought the paper all but unassailable. For example, in his *Customs in Common* he remarked that 'while interesting new work has been done on the question of time [since 1967] none of it seemed to call for any major revision of my article' (Thompson, 1991: vii).

    Only in the last decade or so has this situation begun to change, but critical treatments by historians have not really attained the critical mass needed to initiate a thoroughgoing review of Thompson's argument. For example, works by

Whipp (1981), Harrison (1986), Thrift (1988) and Glennie and Thrift (1996, 2002) have been widely treated as free-standing islands of critique, rather than as components of a sustained reformulation. In this regard, it is striking that much of the historical work which has begun to lead towards a recasting of the Thompsonian framework has been concerned with areas other than Britain: the United States (O'Malley, 1990, 1992; Wilnerding, 1999); Australia (Davison, 1992); Japan (Smith, 1986; Shimada, 1994); Germany (Wendorff, 1980); and the Netherlands (Wildenbeest, 1988). Several of these authors initially tried to elaborate Thompson's framework for their own particular societies, reformulating their approaches when his pattern, as it were, 'didn't work' outside the English 'core'.

It is also worth pointing to the historical specificity of Thompson's paper. From the perspective of the mid-1990s, the period from the 1950s through to the 1970s appears as a high point in the industrialized synchronization of societies. In 1967, when Thompson's paper was published, the impending desynchronization of at least certain parts of society was no doubt less obvious than it was to become (Lash and Urry, 1993; Urry, 1991, 1994).

13. Thus, pulling history back into an account of modernity as a 'qualitatively new self-transcending temporality' which has the effect of distancing the present from even that most recent past with which it is thus identified. History is therefore 'as Koselleck puts it, . . . "temporalised". It becomes possible for an event to change its identity according to its shifting states in the advance of history as a whole' (Osborne, 1995: 14).

3
'Not Everyone Occupies the Same Now': Reconceptualizing Clock Time

3.1 INTRODUCTION

Clock time is undeniably important to Euro-American societies. However, it is a more difficult concept to grasp than we might first suppose. In Euro-American societies, clock time tends to be thought of both as a mundane ordering frame—something uniform, ubiquitous, unyielding—and simultaneously as an elemental force, pressing in on our lives and setting us in motion. We want to turn away from this vision of a clockwork world. We will argue instead that clock time cannot be unhinged from the practices of everyday life. Consequently the history of clock time we want to tell differs sharply in both form and content from those that have gone before. Accordingly this chapter is in three main parts which gradually build to an account of that history. Section 3.2 examines new conceptions of times, which have become current in the literature in the last thirty years and whose assumptions are a radical challenge to both standard history and the standard accounts of time. Section 3.3 returns to clock time and sets out how new accounts might be generated against this new background, drawing on work in anthropology, human geography, and the sociology of science, to build an alternative account of clock time as a series of practices that have become general but not necessarily ascendant. Resting on the arguments set out beforehand in this and the previous chapter, section 3.4 then offers an alternative account of the nature of clock time. Finally, some brief conclusions round the chapter off.

3.2 TIME'S TRIUMPH RECONSIDERED

To reiterate: we want to dispute a characterization of clock time as either responsible for all the ills of western civilization, or as at least a willing accomplice to them, which recur in these narratives. We do so based on a burgeoning and now extensive literature on time, developing over the last thirty years or so. This literature rests on actualizing the four main principles introduced in

Chapter 2—that no singular history of time exists, that time stems from diverse practices, that devices need to be taken seriously, and that clock time is ordinary. We will consider each of these principles in turn, and in some detail.

3.2.1 Multiple Times

The first principle is that time can no longer seen as 'something out there' which frames us, as in the Newtonian *sensoria*. But neither is it a perception internal to human beings, 'a purely subjective condition of our (human) intuition' (Kant, 1965/1781, 1787: 82) used 'to frame . . . the multiplicity of beings and entities' (Latour, 1997: 174). Time is, rather, an irreducible stream of unrepeatable events, Samuel Beckett's (1979: 255) 'chaotic conflux of oozings and torrents', which are represented through devices like clocks and calendars as repeatable sequences that come to stand for, and in a sense *are*, time. This vision relies on biological metaphors, in explicit opposition to the mechanical and informational metaphors of industrial notions of clock time. Unsurprisingly, therefore, much attention centres on vitalist writers like Henri Bergson who argue that 'nothing singular can recur. Repetition is possible only in the abstract. What is repeated is some aspect that our sense and intellect have singled out from reality because action can only move among repetitions' (Lloyd, 1993: 105).

In other words, what we call time is an ungainly mixture of time*s*—unfolding at different speeds in different spaces—which intersect and interact in all manner of ways. Many writers have struggled to describe such a world of multiple temporal coexistence. Perhaps the most substantial attempts have been made in France by Deleuze and Serres's development of Bergson's biophilosophical motifs (cf. Lorraine, 2003; Grosz, 2004). For these writers, 'time' cannot be a progressive 'geometric' history of successive events, able to be gathered into one rationalized unity. Rather, the world is conceived as a swirl of times-in-motion produced by many different collectives. In this world each time 'develops more like the flight of [a] wasp than along a line, continuously or regularly broken by dialectical war' (Serres and Latour, 1995: 65) as it constantly interacts with the times of other collectives. The analogy that is used is one of a 'topological' history of folds which wrap around or rhizomatically fold into one another, producing a world of cross-talk, cross-roads, cross-traffic, cross-infection and criss-crossings, a world of broad temporal sweeps, but also of scattered pockets, isolated tatters and possible emergences.

To describe this world of plural times, perturbations, and turbulences, requires diverse textual models—'arborescent growth, narrative motors, chemical transformations, a circulational game involving chance, and so on'—which correspond to the multiplicities of both texts and images (Dixon, 1997: 15). Deleuze likens this heterogenous model of the temporal to 'a geometry of sufficient reason, a Riemannian-type differential geometry which tends to give rise to discontinuity

on the basis of continuity or to ground solutions in the conditions of problems'
(Deleuze, 1994: 210).[1]

Themes in philosophy have been closely paralleled by work in the social sciences and humanities.[2] Such work has several disciplinary origins. First there is anthropology. Over several decades, anthropologists as notable as Evans-Pritchard, Lévi-Strauss, and Leach all wrestled with the notions of time found in other cultures and there is now a substantial archive of ethnographic work which both informs and questions their insights (Gell, 1992). Second, in social psychology authors like James, Mead, and Minkowski all argued for several social realities, each with their own times. Third, sociology has been similarly involved ever since Durkheim's *The Elementary Forms of Religious Life* (1915). Sociologists as eminent as Gurevich, Sorokin, Merton, and Talcott Parsons all wrote on time, each of them cleaving to the general proposition that 'man is not born with "a sense of time"; his temporal and spatial concepts are invariably conferred on him by the culture to which he belongs' (Gurevich, 1964: 27). Fourth, it is hardly surprising that the nature of historical time has been debated in history from Ranke to the Annalistes (Revel and Hunt, 1995; Leduc, 1999). Finally, work on time has been a vital part of human geography since the rise of the 'human ecology' school in the 1920s (Thrift, 1977, 1997b; Parkes and Thrift, 1980, Glennie and Thrift, 1996, 2002, 2005).

In all five of these disciplines, notions of multiple times have become increasingly important since the 1970s, as the sheer diversity of temporal maps and images has been realized and taken into prevailing accounts. Thus, in anthropology Johannes Fabian (1983, 1991) has pointed to a multiplicity of cultural 'chronotypes' (Bender and Wellbery, 1991). In social psychology, writers like Flaherty (1999) emphasize the wide variety of experienced time. In sociology, writers like Adam (1990, 1994, 1995, 2004), Urry (1994) and Bluedorn (2002) have insisted on the multiplicity of time. A similar reorientation in history has involved the growth of virtual history (Ferguson, 1997) and a growing interest in social and cultural histories of time (Poole, 1998; Bartky, 2000; Gay, 2003). Finally, a similar refocusing has taken place in human geography, following the growth of 'chronogeography' and time-geography in the 1970s (Parkes and Thrift, 1980; Hägerstrand, 1975, 1978; May and Thrift, 2001). All these developments bear witness to Bender and Wellbery's contention that:

The thematization of time in contemporary research draws to some degree on the insights of historicism and phenomenology, but is distinguished from those theoretical antecedents by the emphasis it places on plurality and complexity. Time ... is not a single medium of consciousness or a unified movement in history. It is manifold. Numerous chronotypes intertwine to make up the fabric of time. The social and cultural processes of temporal construction rely on and also re-elaborate antecedent rhythms and articulations. These multiple times can become objects of contention because individuals experience them differently and because they bear ideological implications. Time asserts itself in contemporary inquiry less as a given than as a range of problems, the solutions to which are constantly open to renegotiation. (1991: 15)

3.2.2 Times as Practices

A second principle for contemporary studies of time is that time is embedded in *practices*. This realization has many forebears: Heidegger's *Being and Time*, Merleau-Ponty's phenomenology, and Wittgenstein's later works are all important way-stations. So are North American traditions of symbolic interactionism and the Russian tradition of dialogism (Thrift, 1996, 2000b). Regarding the study of time, these forebears come together in Pierre Bourdieu, especially his work on the Kabyle in the 1960s (Bourdieu, 1989, 1977). Bourdieu produced an account of the construction of time as always embedded in practised bodies and bodily practices, famously insisting that practices must unfold in time: there are no goings-back, no second thoughts. So the 'time of strategy' is crucial. Ironically, much academically inclined work tends to forget this simple fact, bending instead towards the elaboration of formal structures and rules in contemplative time that are never tested in the time of practice, and which forget that the interval is its own rule. Hence Bourdieu's two most famous temporal aphorisms: 'to abolish the interval is to abolish strategy'; and 'to substitute strategy for the rule is to reintroduce time, within its rhythm, its orientation, its irreversibility. Science has a time which is not that of practice' (1977: 6, 9).

For Bourdieu, life is bound up with practices which must take place in time and therefore always involve an art of 'necessary improvisation' (1977: 8), what Ginzburg (1980: 28) has beautifully described as 'elastic rigour'. So,

when the unfolding of the action is heavily ritualised, as in the dialectic of offence and vengeance, there is still room for strategies which consist of playing on the time, or rather the tempo, of the action by delaying revenge so as to prolong the threat of revenge. And this is true, *a fortiori*, of all the less strictly regulated occasions which offer unlimited scope for strategies exploring the possibilities offered by manipulation of the tempo of the action—holding back or putting off, maintaining suspense or expectation, or on the other hand, hurrying, hustling, surprising, and stealing a march, not to mention the art of ostentatiously giving time ('devoting one's time to someone') or withholding it ('no time to spare')...

To restore to practice its practical truth, we must therefore reintroduce time into the theoretical representation of a practice which, being temporally structured, is intrinsically defined by its tempo. (Bourdieu, 1977: 7–8)

Furthermore, practice is always *embodied*, where embodiment is understood not as a surface on which the signs of culture are inscribed but as a set of spaces in which the body moves in different ways according to what objects are present and which body parts are engaged. Thus embodiment is

the... way the body 'turns onto' things, onto objective space, onto living things. Here there is a type of communication that is always present, but only makes itself really visible in pathological or magical experiences. Nevertheless the ordinary experience of relations to things also implies this mode of communication. Being in space means to establish diverse relationships with the things that surround our bodies. Each set of relations is determined by the action of the body that accompanies an investment of desire in a particular being or particular object. Between the body (and the organs in use) and the thing is established a connection that immediately affects the form and space of the body . . .   (Gil, 1998: 127)

The body, in other words, is a 'geometer' of practices. It 'believes in what it plays out: it weeps if it mimes grief. It does not represent what it performs, it does not memorise the past, it *enacts* the past, bringing it back to life' (Bourdieu, 1979: 73). Butler (1997: 154) glosses this statement thus: 'here the apparent materiality of the body is recast as a kind of practical activity, undeliberate and yet to some degree improvisational'.[3] As she goes on to point out:

one can hear strong echoes of Merleau-Ponty (1962: 183) on the sedimented or habituated 'knowingness' of the body, indeed, on the indissociability of thought and body: 'Thought and expression . . . are simultaneously constructed when our cultural store is put at the service of this unknown law, as our body suddenly lends itself to some new gesture in the formation of habit'. But one hears as well Althusser's invocation of Pascal in the explaining of ideology: 'one kneels in prayer, and only later acquires belief'.   (Butler, 1997: 154–5)

The link to Butler here is important, for the other main source of work on practices as time is feminist theory and especially its working out of what it means to touch and be touched. Since Kristeva's famous paper 'Women's time' (1981) argued that culture has traditionally associated women with a kind of seriality and worked towards new female temporal sensibilities that were not associated with repetition or eternity, feminist theorists have worked on and with times as points of struggle and identity formation (Langbauer, 1999). Many have argued that women have, in several ways, a fundamentally different relation to time, stemming from different everyday routines (Davies, 1989, 1994), taking in different embodiments (Grosz, 1994, 1995, 2004), and putting more value on experiences like birth and beginnings than on death and endings (Battersby, 1998). Female relations to time have been masked by prevalent masculine temporal practices (including, in some accounts, clock time), which leave little or no space for coherent self-description, often through the simple device of consigning women to the repetitive times of everyday life and (their equivalent) the domesticated spaces of the home, whilst men roam the temporal highways and byways (cf. Felski, 2000).

But there are other models. New syncopations are possible (Clement, 1995). New rhythms can be adopted. New identities can be formed. For example,

Le Doueff's general point is the interdependence of scientific thought and prevalent social and cultural conceptions: she investigates how Galileo could have (correctly) formulated the law of falling bodies using time rather than space as a part of reference; when space seems more self-evident as a fixed reference point within the scientific context of his culture. Her point is that Galileo could not have formulated his law without relying on pre-and non-scientific notions of temporality. And, in turn, the principles governing his law were able to be comprehended and understood by other scientists and educated non-specialists because they accorded with everyday notions of time.

If, however, post-Euclidean and post-Newtonian conceptions are made possible during an historical questioning of the postulates and values of the Age of Reason and the era of the self-knowing subject, these have still, in spite of their conceptual distance from Euclid and Newton, confirmed the fundamental masculinity of the knower and left little or no room for female self-representations, and the creation of maps and models of space and time based on projections of women's experiences. It is not clear that men and women conceive of space and time in the same way, whether their experiences are neutrally presented within dominant mathematical and physics models, and what the space-time framework appropriate to women, or to the *two* sexes may be. One thing remains clear: in order to reconceive bodies, and to understand the kinds of active interrelations possible between (lived) representations of the body and (theoretical) representations of space and time, the bodies of each sex need to be accorded the possibility of a different space-time framework. (Grosz, 1995: 99–100)

Whatever the theoretical niceties, that early modern women and men often had very different temporal experiences is undeniable. Laurence's account of women's daily round between 1500 and 1760 is worth quoting at length:

The word 'work' has for us connotations of waged labour, but most women's work was unwaged... what women did in the household towards the general maintenance of the family. Household manuals laid down how the good housewife ought to spend her day. Thomas Tusser's *Five Hundred Points of Good Husbandry* of 1580 gives us morning activities: cleaning the floors; spinning and carding wool; preparing ingredients for cooking and brewing; preparing breakfast; feeding the cattle; brewing; baking; dairy work; laundry; malting; and preparing dinner at noon. In the afternoon the housewife (the mistress of servants) should chivvy her servants back to work; use up the left-overs from dinner; sew; same features for pillows; and make candles. In the evening, she should feed the hens and pigs and milk the cows; lock up the hens; bring in the washing; and lock up the house. Then she should serve supper; amuse her husband; tell the servants what to do the following day; and go to bed having washed the dishes, laid 'leavens' (prepared yeast) and saved the fire. In summer she was advised to go to bed at 10 o'clock and rise at 4, in winter to bed at 9 o'clock and rise at 5. Many of the jobs were occasional; and in urban households, and some rural households, goods like candles, bread and beer were bought ready-made.

A study of accidental deaths recorded in the coroners inquests for the later Middle Ages gives a fascinating insight into the ways in which real women spent their days. Many accidents were commonly concerned with fetching water and preparing meals, working with livestock and brewing. The number of accidents rose as noon approached and women became tired and hungry. Afternoon accidents were predominantly concerned with laundry or seasonal agricultural work. Drowning and burning, and accidents connected with fetching water, and with laundry and cooking, were much the commonest causes of accidental death for women. They took place both in their own houses and in the homes of neighbours.

The evidence of inquests held in the sixteenth century suggests several changes in women's daily routines. The greater use of chimneys had improved domestic safety. Relatively few women died in fires, and of those who fell into the fire, several were described as ill at the time of their deaths. Accidents fetching water were common; indeed, drowning was the commonest cause of death recorded by coroners. Young girls, often servants, fetching water early in the morning when it was still dark, fell into ponds and down wells. So early were some of these accidents that it is clear that Tusser's times of working were not necessarily just a counsel of perfection. Agnes Ellyot, a Sussex women, drowned in 1554, going to fetch water at a water pit at 4 a.m.; the stake on to which she was holding gave way.

But it was not accidents in the home which were most likely to kill women; it was accidents outdoors. Falling off a branch while picking pears, being struck by falling pieces of timber, being caught by pieces of machinery like the cog of a malt mill or the arm of a horsemill, all caused deaths in the sixteenth century. Travelling, too, was attended by all sorts of risks, especially in the vicinity of a large river . . .

For most of the period 1500–1760 poor women spent their time at home, looking after the children, fetching water, cooking, sewing and doing laundry, spinning and, at certain seasons, agricultural work. They might keep poultry, do petty trading, brewing or gardening, and take part in their husbands' trades. There were variations in this daily round. Extreme poverty cut down both cooking and cleaning, since there was nothing in which to cook or prepare food, and no possessions to need cleaning. Living in the town, especially for the poor, probably meant buying prepared food from cookshops. Sufficient wealth to employ a servant added directing the servant's activities to a woman's work. Even in the eighteenth century, when country houses were often divorced from the agriculture which supported them financially, women took a serious interest in the management of the household, though slightly shamefacedly. Mrs Lybbe Powys recorded in 1756 that Lady Leicester of Holkham Hall was often to be found in the kitchen and was once spotted there at 6 o'clock in the morning, 'thinking all her guests safe in bed, I suppose'.   Laurence (1994: 109–11)

3.2.3 Time as Networks of Interaction

A third principle of contemporary time studies is that timekeeping devices must be treated anew, abandoning the technological determinism that saturates the literature on timekeeping devices, at least since Lewis Mumford (1926). A standard account (typically based on a very few sources such as Thompson, 1967; Le Goff, 1980; Landes, 1983) has emerged in which timekeeping technology has an essence, either that timekeeping devices have inherent qualities which they somehow transmit to people and society, or that they embody a mode of relation to the world based on 'calculative' thinking (Crosby, 1998). Porter provides a near-perfect thumbnail sketch:

The demand for a more rigid and predictable calendar was created by administrative needs of church and state, for whom there was a time to pay taxes, a time to report for military service, and a time to observe Lent or celebrate Easter. Clock time, too, acquired religious significance, and the punctual observance of matins in monasteries was amongst the first incentives for living by the clock. Industrialised work relations had a more pervasive influence, and ever since the beginnings of industrialisation the clock has been amongst the principal agencies of discipline in factories, schools and offices. Its growing sovereignty necessarily came at the expense of natural, diurnal rhythms of light and darkness, hot and cold. It was, in short, part of an artificial regime, the technological, economic and social conquest of time. By the late nineteenth century, with the spread of rail networks, it even began to seem desirable to impose uniform hours on wide swathes of land running from north to south. A bit later, against strong opposition from farmers and others still residually committed to natural cycles, governments first declared that time should be moved forward every spring and set back every fall. (Porter, 1995: 23)

Such simple accounts are nullified by a realization of the multiplicity of times and the richness of the time of embodied practices, bringing us to our third principle: that timekeeping devices—almanacs, calendars, clepsydra, clocks, hourglasses, sextants, sundials, and the rest—must be treated as key elements in *networks of interaction*. Devices do more than act as registers. They act as an ecology (Netz, 2004).

Inspiration for an account of timekeeping devices that avoids technological determinism and essentialism can be supplied by developments in the sociology of science. Early attempts to forge an anti-essentialist view of technology were based on notions that society 'shapes' technology, 'accounts which suggest that the capacity of the technology is equivalent to the political circumstances of its production' (Grint and Woolgar, 1997: 19). In turn, these accounts of technology as socially constructed have spawned a series of accounts that go beyond social shapings of technology by attempting to transcend distinctions between 'social' and 'technical'. That distinction is outdated because 'human capacities are . . . inevitably and inescapably technologised' (Barry, Osborne, and Rose, 1996: 13). Human capacities are a part of *fields of inter-action* which produce action—and the effect of 'agency'—rather than a set of actors each of which possess their own 'agency'.

Three of these anti-essentialist accounts seem particularly germane. One comes from the work of writers like Grint and Woolgar (1997), who suggest that the competing interpretations of devices constructed by different fields of interaction produce different versions of what a device is and will do, and not just contrary interpretations of the 'same device'. Another is provided by writers like Rose who, heavily influenced by Foucault, configure fields of interaction as forms of expertise (Barry, Osborne, and Rose, 1996). For these writers:

human relations to technology are not merely those of a passive 'reduction'; rather technology is an aspect of what it is to be human (Canguilhem, 1994). And if technology is potential, it is because technology carries with it a certain 'telos' of operations, a certain directive capacity. In other words, technology—both in terms of the human side of technology and of the technology of what it is to be human—is integral to those relations of authority and subjectivity that insert ourselves into the space of the present, giving us the status of living beings capable of having 'experience' of the present. (Barry, Osborne, and Rose, 1996: 15)

Third, there is actor-network theory, in which devices are part of the work of constructing *networks* that involve both humans and non-humans. Here, the heterogeneous entities that construct networks are analysed through principles of 'symmetry', which implies impartiality towards the status of actors (or, rather, 'actants' to underline the semiotic point that both humans and non-humans are included), and 'free association', which implies the rejection of *a priori* distinctions between the social and the technical. In Callon's words 'The rule which we must respect is *not* to change registers when we move from the technical to the social aspects of the problem studied' (1986: 200, our emphasis).

Thus, actor-network theory takes a different approach from technological determinist approaches. It points that the world is made up of diverse actor-networks which are more or less successful attempts to associate, mobilize, and then make durable human-nonhuman alliances. The work of association is never done since actor-networks are 'hesitant orders of partial connection'. These networks must be continually active, endlessly reconstituted by intermediaries which keep the networks going. Because the existence of actor-networks depends so heavily upon the work of circulation which defines them, their continuation relies on a whole series of 'immutable mobiles'—devices, types of people, animals, money, and so on—which can be transported from one location to another without changing form and which allow networks to become more durable : these immutable mobiles harden and anneal the networks, making it possible for them to last. It is no surprise, then, that actor-network theory sets great store by the role of objects in the world: objects that are no longer passive. As crucial elements of actants, they have their own partial agency, adding substantially to the capacity of actor-networks to survive and prosper.

These three different but related anti-essentialist stances allow us to position clocks and other timekeeping devices more exactly as components of numerous material-semiotic networks, rather than as general instruments generally applied. Within these networks, particular understandings of time circulate according to the uses to which devices are put in particular practices with their own particular forms of timekeeping expertise. To reprise: in this book *clocks are not a general instrument*. Instead, *clocks operate in many networks of practices at once* but they do so in different ways, resulting from conflicts and compromises made during centuries-long histories of contextualization.

And it is worth remembering that 'contextualisation is fabricated and negotiated like everything else' (Latour, 1997: 143) as part of a prolonged process in which people and things 'exchange properties and replace one another' (Latour, 1996: 61). As Latour puts it:

Although charged by humans with the sin of being 'simply' efficient, 'purely' functional, 'strictly' material, 'totally' devoid of goals, mechanisms nevertheless absorb our compromises, our desires, our spirit, and our morality—and silence them. They are the scapegoats of a new religion of silence, as complex and pious as our religion of speech. What exegesis will have to be invented to provide commentary on the silence of machines? What secular history will ever be able to narrate the transcription of words into the silence of automatons?

Beyond our infinite respect for the deciphering of scripts, we need to have an infinite respect for the deciphering of *inscriptions*. To propose the description of a technological mechanism is to extract from it precisely the script that the engineers had transcribed in the mechanism and the automatisms of humans and nonhumans. It is to retrace the path of incarnation in the other direction. It is to rewrite in words and arguments what has become, what might have become, thanks to the intermediary of mechanisms, a mute function. (Latour, 1996: 207)

Actor-network theory and other non-essentialist accounts allow us to make three arguments about clock time. First, clocks and other timekeeping devices do not just project an idea or experience of time onto society (that is, technological determinism). Rather, they are assemblages of signs and things which are, simultaneously, elements of material-semiotic networks within which particular inscriptions of time vary according how they are used in particular practices, according to particular forms of expertise.

Second, and similarly, clock time is not simply a notion projected by human individuals, but a social process, socializing people, devices and institutions.

In growing up, the individual learns to understand the time signals customary in his society and to regulate his behaviour by them. Much of the idea of time that an individual possesses, therefore depends on . . . institutions representing and communicating time, and on the individual's experience of them from an early age. . . . clocks are . . . so constructed that in one way or another, . . . they are incorporated in the symbolic world of human beings. (Elias, 1992: 13–14)

Third, having abandoned the idea that clocks are simply inert mechanical objects or utensils with no powers of their own, they (like many other mechanical objects) need to be repositioned as 'inorganic organised beings' (Stiegler, 1998). These have their own dynamics resulting from their material-semiotic positioning in particular networks or practices, and clock time is therefore the *result of vital behaviour; not just a metrological frame*.

Likewise we can discard ideas of clock time as an overarching and unproblematic time-space frame (or the idea that western history can be understood as the institution of this frame):

where does the obsession for a time-space frame 'in which' entities would reside, or a frame that the mind would 'impose on' things in order to apprehend them, come from? No amount of labour will ever produce that sort of space and time. It is useless to oppose, as is so often done, the 'lived world' of human subjectivity apprehending space and time and all the rich colours of intention and affectivity with, on the other hand, the scientific and technical objective world ceaselessly cutting a meaningless space-time into isotopic and isochronic units.... The subjectivity of space and time is not what is left when the objective space-time has been thoroughly described....

But... our imaginary frame for all events has to come from somewhere. It again seems to be the particular nature of the *objects* used... [whose] circulation... will locally generate a specific type of space-time, as the circulation of any other body with different properties will generate additional spaces-times-activities. This does not mean that we are *in* an isotopic space and an isochronic time, but that, locally *inside* metrological claims, there are effects of isochrony and isotopy produced by the carefully monitored and heavily institutionalised circulation of objects that remain relatively untransformed through transportation...

... The building of metrological networks for space and time is a crucial feature of western history. It has to be documented, to be sure, studied and respected, but it does not have to be confused with an account of how our mind evolved, or with the understanding that other circulations may have of time, or with the ontology of world making.   (Latour, 1997: 184–6)

Further, we must move away from the idea that an overarching space-time frame is even the goal, because clock time is usually subsidiary to other practices. Only rarely does keeping clock time become *an end in itself*: in making and maintaining clocks, in certain arenas of religion, science and navigation, in certain everyday situations where exact coordination is necessary. Therefore we must acknowledge that clock time is the product of collectives for whom its maintenance may be compulsory or hesitant or incidental or simply irrelevant. The degree of relevance partly depends upon the timing practices of other social collectives with which collectives interact. Clock time becomes a more important metric as more collectives adopt clock time into their practices. And, lastly, it means accepting that whilst clock time is clearly used to promote regularity and control we cannot assume that, in a world of the unanticipated and the wayward, regularity necessarily happens and control is always achieved.

Some of the ways in which clocks are hesitantly inscribed in material-semiotic networks are well illustrated by the early mechanical clocks of major religious institutions, especially cathedrals, which go to illustrate Putter's (2001: 136) proposition that 'the impression of "medieval indifference to time" is only the reflex of our own indifference to medieval time'. Religious institutions, particularly monasteries, had long experience of using water clocks, and even mercury clocks, so it is unsurprising that the earliest recorded mechanical clock escapement in Europe comes from Dunstable Priory in 1283 (Beeson, 1977, 1988; North, 1975; Thrift, 1996). Evidence of similar clocks, and considerable

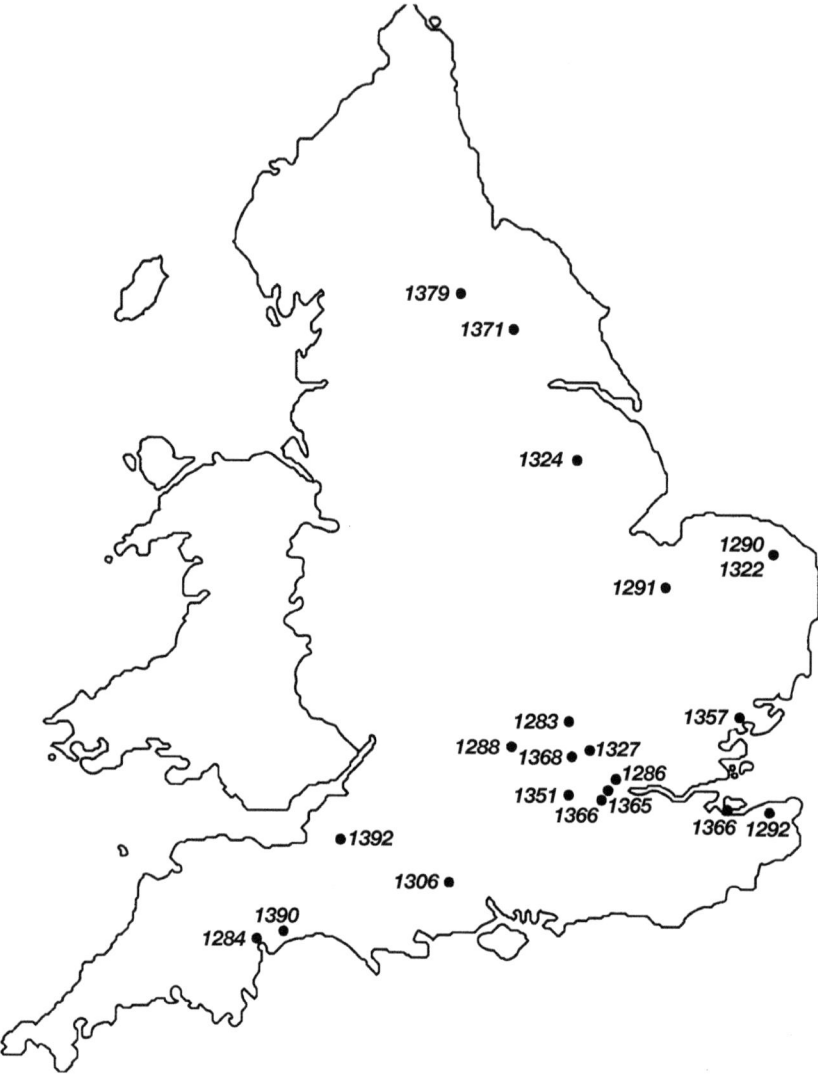

Fig. 3.1. Locations of the earliest documented English clocks.
Source: based on Beeson (1977); updated Thrift (1996).

time and resources invested in them, survives for several other thirteenth-century institutions (Figure 3.1). These soon became seen as normal attributes of important Christian institutions, as in 1324 when the Treasurer of Lincoln Cathedral offered a donation of a new horologium because:

the Cathedral was destitute of what other cathedrals, churches, and convents almost everywhere in the world are generally known to possess. (Beeson, 1977: 10–12)

Many medieval church clocks were highly sophisticated, using special (and probably newly invented) gear wheels to produce all manner of information, most commonly including hours, days of the month, and lunar phases. Planetary positions, perpetual calendars, predicted times of eclipses, and high-tide times might also be shown (North, 1975). The striking of the hours was often enlivened by various automata, which could be fearsomely sophisticated in themselves. Such clocks were large—commonly with an iron frame about a metre across—and the clock at St Albans Abbey (*c*.1329–50) was possibly twice this size (North, 2005). Their construction was a major engineering project, costing maybe £1 million in today's prices, and depending upon an interlocking set of knowledge communities and their expertise, from skilled metal workers and other craftsmen, to the specialized, professional *horologiani* who can be traced working on several clocks. Thus, Roger and Lawrence Stock worked both at St Albans and Norwich Cathedral (1321–5), and the cathedral clocks at Salisbury (1386) and Wells (1392) were apparently made by the same craftsmen (Duley, 1977; Wells Cathedral, 1994). Finally, there were clock theorists and engineers like Richard of Wallingford, a blacksmith's son who rose to be Abbot of St Albans, leaving several extraordinary manuscripts on timekeeping (North, 2005). These communities drew upon diverse craft, scholarly, practical, and theoretical knowledges that were often of considerable historical vintage, including: classical and Arabic sources; practicalities of constructing sun dials, planispheres and astrolabes; the entire spectrum of the Liberal Arts quadrivium (arithmetic, geometry, astronomy, music); mechanical expertise; and so on. Even with these knowledges, the extreme difficulty of constructing monumental mechanical clocks is clear:

The Sacrist's Rolls of Norwich Cathedral from 1321 to 1325 contain the first extensive financial records concerning the construction and installation of a large mechanical clock. The man in charge of the work was one Roger Stoke, who later worked at St Albans, and who was in both places assisted by Lawrence Stoke. The clock had a very large astronomical dial—it was of iron plate and weighed 57lbs—with models of the Sun and Moon, automata, including 59 sculpted images (done by one Adam, a wood carver), and a choir or procession of monks. There was much colouring and gilding. Smiths, carpenters, masons, plasterers and bell-founders were engaged over a period of three years. The competence of most of the craftsmen concerned seems to have been equal to the occasion, but the making of the main astronomical dial went less smoothly than the rest. In 1323 the fabrication of the large plate was entrusted to Robert of the Tower (Robert de Turri) in London, but in his hands the whole work was ruined. The man was himself ruined *(depauperatus)* and only 10 of the 18 shillings advanced to him could be recovered. Other artisans proved to be equally ineffective, ruining the metal in their attempts. Men were sent from Norwich to London for news of progress, but at length it was necessary for Roger Stoke himself to ride to London to supervise the engraving of the plate. The total cost of the clock was in excess of £52. (North, 1975: 385)

These monumental mechanical clocks were not important, however, only or even primarily as timekeepers. This is now hard to see since 'their importance has been obscured, . . . by . . . modern utilitarianism in which the clock is seen merely as a prototype machine of modern industrialisation or as a timekeeper whose principal function is to produce units of time' (Haber, 1975: 400). Consequently, clockmaking from the first mechanical escapements in the 1270s to the mid-seventeenth century adaptation of the pendulum to mechanical timekeeping (enabling much greater accuracy), becomes side-lined. Yet this was 'the very period when monumental astronomical cathedral clocks were flourishing. The reduction of the history of pre-Galilean clocks to the single function of timekeeping makes these clocks seem redundant and a miscarriage of effort. Their trains of automatons have been viewed as "mechanical puppet shows", and their ornamentation has been regarded as irrelevant to clockmaking, . . . [and] left to the interest of antiquarians or the amusement of tourists' (Haber, 1975: 400–1).

But these clocks were a rich amalgam of numerous material-semiotic practices. First, they were grand-scale representations of religious cosmology, with their iconographies illustrating the meaning of time passing in a Christian world. Secondly, they were a means of education about architecture, mechanics, and astronomy, as well as religious principles. Third, they were works of art, integrating central Christian conceptions of 'God as maker and the world as a work of art' (Haber, 1975: 407). Finally, they were central elements of religious ritual, from services to passion plays. In this, there was continuity from the earlier use of water clocks and bells, which gives us some notion of how far clocks had already become a key part of religious life, as in mid-thirteenth-century references to water clocks (*orologia allorium*, 'clocks of earthenware vessels') in St Albans Abbey's chronicle:

'The prior shall enter the church from the chapter house', the rule of John of Hertford tells us, 'to be presented to God and the Holy Martyr Alban at the high altar, with . . . the shawms sounding with the horologe, the tapers lighted round the altar, and the throne uncovered'. Very few of those treatises . . . [giving] . . . precise and complete descriptions of actual clocks, can convey quite as much about humanity as can this passing reference to one way in which a great abbey clock of the thirteenth century was absorbed into the high ritual of the church. (North, 1975: 393)

3.2.4 New Times

The fourth principle of contemporary accounts of time is that 'new' senses of time are continually being born out of the matrix of everyday life. Indeed, we may well be living through a period in which their natality is particularly rapid which, in turn, may account for renewed intellectual attention to time (Thrift, 1996). Three of these new time-senses seem particularly pertinent. The first is produced by *simultaneity*. In Kern's classic *The Culture of Time and Space*

1880–1918, he argues that in that period the West's sense of the present notably thickened:

In an age of intensive electronic communication 'now' became an extended interval of time that could, indeed must, include events around the world. Telephone switchboards, telephonic broadcasts, daily newspapers, World Standard Time, and the cinema mediated simultaneity through technology. (Kern, 1983: 314)

For some commentators there has been a quantum leap in the scope and extent of simultaneity even since Kern's book, as clock time has become general around the world through the spread of timekeeping devices and other forms of keeping clock time like television. This leap in intensity has now become a qualitative shift in time experience. 'Probably never before have so many people simultaneously experimented with time as they are doing today' (Nowotny, 1994: 132). Like Virilio, Urry (1994) analyses the growth of what he calls 'instantaneous' time which has several important aspects: the immediate experience produced by the newspapers, television, and other media (Kovach and Rosenstiel, 1999); the development of a 'three minute culture' resulting from the greater opportunity for instantaneous response associated with new electronic technologies; and the commodification of the past, so that representations of the past are constantly encountered as 'bite-size' consumable chunks. Each of these developments, argues Urry, has produced a greater concentration on the present.

In similar vein, Appadurai (1996: 82) argues that the rise of consumption 'which is something beyond a consumer revolution, something we may call a revolution of consumption in which consumption has become the principal work of late industrial society' has produced a new aesthetic of ephemerality which

expresses itself at a variety of social and cultural levels: the short shelf life of products and lifestyles; the speed of fashion change; the velocity of expenditure; the polyrhythms of credit, acquisition, and gift; the transience of television-product images; the aura of periodisation that hangs over both products and lifestyles in the imagery of mass media. (Appadurai, 1996: 83–4)

This new aesthetic depends on continuous imaginative work by consumers, 'labouring daily to practice the descriptions of purchase in a landscape whose temporal structures have become radically poly-rhythmic' (Appadurai, 1996: 83).

This observation of polyrhythmicity links to another widespread observation, made by writers like Urry (1994), and Nowotny (1994): that a second new time sense is being forged via the general *desynchronization* of individuals' time-space paths:

there is a greatly increased variation in different people's times: they are less collectively organised and structured as mass consumption patterns are replaced by more varied and segmented patterns. There are a number of indicators of such time-space desynchronisation in many areas of leisure activity; the increased significance of 'grazing', that is, not eating at fixed meal times in the same place in the company of one's family or workmates; the growth of free and independent travellers who specifically resist mass travel in a group where everyone has to engage in common activities at fixed times, and the growth of the VCR, which means that TV programmes can be stored, repeated, and broken up, so that no sense remains of the authentic, shared watching of a particular programme. (Urry, 1994: 134)

In other words, as temporal schedules become more flexible, new experimentations with these schedules become possible. But, in turn, this general desynchronization may also be being caused, in several countries, by lengthening work hours amongst middle-class segments of the population, which forces individuals to use what time they have when they have it, and to offload many domestic activities onto others, as they prioritize their own time use (Schor, 1994; Roberts, 1992; Gershuny, 2001).

A third observation is that the gradual cultural birth, in the nineteenth century, of very long-term time senses arising from the discovery of geological time and evolutionary time has its twentieth-century correlate, which Urry calls *'glacial'* time (Urry, 1994). This third new time sense is based on longer time horizons arising in particular from the environmental movement, with its concern for 'the natural world', ethic of planetary care, and reappropriation of certain places, perhaps precisely because of their increasing commodification, which involves their reinstatement as votive containers of a 'historical' time. This reinstatement partly stems from greater emphasis on the importance of reminiscence, made possible by 'instantaneous' technologies through which reminiscences can be recorded and stored.

The glint of yet more new time senses has been spotted by others. For example, many have identified the death of historicism, the rise of new forms of remembrance (Frow, 1997), and new technological innovations like video (Cubitt, 1991) as forcing new forms of narrative structure on to everyday life. Others have espied postcolonial senses of time in which non-western cultures are no longer placed in times other than that of the West, and become fellow time-travellers (Fabian, 1991; Spivak, 1991) and in which national imaginaries can be refigured (McClintock, 1995; Bhabha, 1994). The problem remains, however, with these as with the other new time senses remarked upon above, that they are too often (though not always) based in technological determinisms of various kinds, and that they are rarely the subject of any sustained empirical research which might place them in particular communities of practice, whether these be futures traders or lovers.

3.3 TOWARDS NEW ACCOUNTS OF CLOCK TIME

Here we attempt to use the four principles outlined in Chapter 2, funnelled through the contemporary literatures on time reviewed in the preceding section of this chapter, to produce an account of clock time which embeds clocks and other timekeeping devices in a series of more or less historically durable networks of the social. Drawing mainly on anthropology, human geography, and the sociology of science, with contributions from social history, sociology, and constructionist social psychology, it is possible to argue that any account of clock time must, simultaneously, contain three main elements.

3.3.1 Embodied Practices

First, clock time consists of a series of *embodied practices* which are often specific to particular networks of the social and which can be summarized, using the phrase of Alliez (1996) and Deleuze (1996), as 'conducts of time', by which they mean the characteristic gaits, appropriate to particular spaces. In other words, there is no one clock time, but a series of clock times, multiple cultural performances, brought into existence by communities of practice which soak into their ways of being, so that they become measures of going on that are taken for granted. This is not usually or simply a process of 'inscription', however. Rather, it is a process of 'tuning in' or 'configuration' to particular conditional styles of action which count as the passing of events. The body is a critical element of these styles which works with, but also beyond, spoken words (Radley, 1991).

In turn, the notion that time is a set of embodied practices questions the notion that temporal experience has a fixed, even closed quality. Rather, time takes on an open, mobile texture. This sense of time is perhaps best captured in the work of Mikhail Bakhtin whose concern was precisely to combat the idea of time as closed in which:

> There are no real alternatives, for everything has already been given in the rules or chain of causes. People act out patterns or do what the laws have prescribed, their actions instantiate, but never exceed, rules or pregiven laws. What people do not do is genuinely choose, even though they may imagine otherwise. History and individual lives merely unfold in time, and do not make anything that is genuinely surprising. (Morson, 1994: 21)

Instead, embodied practices produce an 'essential surplus' bound up in the concrete act itself:

> specifically, with those aspects of it that are not the mere product of earlier solutions or timeless patterns. The act exceeds the circumstances that occasioned it... One reason the surplus exceeds timeless rules or laws is that it is essentially related to the irreducible praticalities of the unrepeatable moment in which the act occurred. In this sense, the concrete act, as Bakhtin understood it, is essentially historical. In human action, time is of the essence.

> According to Bakhtin, the concrete act cannot be transcribed in theoretical terms in such a way that it will not lose the very sense of this eventness, that precise thing which it knows responsibly and toward which the act is oriented. Eventness—a key concept for Baktin—is indispensable for real creativity and choice. Without it, the event becomes a

mere shadow of itself, and the present moment loses all the qualities that give it special weight. (Morson, 1994: 22)

Morson (1994) and Bernstein (1995) have tried to systematize this sense of time as embodied practice in the notion of 'sideshadowing'.

In an open universe, the illusion is inevitably itself. Alternatives always abound, and, more often than not, what exists need not have existed. *Something else* was possible, and sideshadowing is used to create a sense of that 'something else'. Instead of casting a foreshadow from the future, it casts a shadow 'from the side', that is, from the other possibilities. Along with an event, we see its alternatives, with each present, another possible present. Sideshadows conjure the ghostly presence of might-have-beens or might-bes.

In sideshadowing, two or more alternative presents, the actual and the possible, are made simultaneously viable. This is a simultaneity, not *in* the time but *of* times: we do not see contradictory actualities, but one possibility that we have actualised and, at the same moment, another that could have been but was not. . . . A haze of possibilities surrounds each activity. (Morson, 1994: 118)

But describing clock time as embodied practices makes it important to make a number of points straightaway. To begin with, as has already been pointed to in the previous chapter, time can be rigorous without the aid of clocks and other timekeeping devices and much of the use of clocks was, to begin with, a means of formalizing tacit temporal knowledge. Take the example of waiting, a basic activity:

We know, proportionate to the amount in the kettle, how long it will take for the water to come to the boil. We know the time-scale of the brewing cycle, and that of the cooling. We know that the successful execution of brewing a cup of (real) coffee depends on the appropriate sequence and timing of all the necessary actions. Knowing it, we can even use that period of waiting for another activity. Far from being incessant, undifferentiated flux depending on humans to get together and structure them, natural processes have an inherent structure. If they did not have time embedded in them, a time proper to them which is demarcated by beginnings and ends, much of human and animal waiting would not exist. (Gasparini, 1995: 36)

Then we need to point out that clock time operates in and on a whole set of sensory registers that belong to the body. Most obviously, through history, timekeeping devices have often had a strong *visual* appeal. Clepsydra, hour glasses, graduated candles, and sundials, as well as clocks; all represent time as a visual record. But the fact that we are now taught to 'read' clocks may well make us blind to the other sensory registers in which clock time operates. Of these, sound is probably the most important.

As we shall show, until and beyond the advent of clockfaces in the fourteenth century, the passage of clock time probably had an almost entirely aural character for most of the population. In particular, it meant the sounds of *bells*. 'Reading' these bells—how soon we slip back to the visual—is a skill we may now only have in part (Corbin, 1998). For example, the towns and cities of medieval England resounded to a cacophony of bells whose different tones and purposes people could 'read' in ways we no longer know:

> bells rang everywhere, and every ringing signalled a time of one kind or another. Hearers could differentiate not only between the bells of occasion and those of measure, but also between the signals for different kinds of occasion, for which the bells performed markedly different functions. Some bells exhorted the hearer to perform an action: to pray, to attend services, to retire for the night, to start work or to stop it. Some ringings were narrative rather than hortatory, in that they registered events or announced news, local or national: weddings and funerals, royal births, imminent dangers, military victories. Bells, that is, could realise 'occasion' either as noun or transitive verb: they might simply announce the occasion, or they might occasion an action by their ringing. In practice, the three categories of function—measure, hortatory occasion and narrative occasion—often overlapped. The narrative and hortatory converged often and easily; the alarm bell, while announcing an emergency, also mustered militia; the victory peal, while declaring triumph, also dictated celebration. Likewise the hortatory and the horological gradually became enmeshed. The bells that rang the hour could also, for large contingents amongst its audiences, dictate action. Of the three categories of bellringing, only two are reciprocally exclusive: the narrative and the horological. The peals that announced extraordinary events—danger, death, victory—had to distinguish themselves by sound and timing from the ones that routinely rang the hour. (Sherman, 1996: 37–8)

This aural vocabulary, often with its own local variations (Thrift, 1996), presupposes a different stance to the world:

> bells were *listened to*, and evaluated according to a system of affects that is now lost to us. They bear witness to a different relation to the world and to the sacred as well as to a different way of being inscribed in time and space, and of experiencing time and space. The reading of the auditory environment would then constitute one of the procedures involved in the construction of identities, both of individuals and of communities. Bell ringing constituted a language and founded a system of communication that has gradually broken down. It gave rhythm to forgotten modes of relating between individuals and between the living and the dead. It made possible forms of expression, now lost to us, of rejoicing and conviviality. (Corbin, 1998: xix)

But clock time could be marked by sound in other ways too. For example, music provides a range of temporal cues which are a part of its attraction (see Epstein, 1998) which both use and abuse clock time, congruent with Johnson's (2002: 65) point that 'music's medium is not just sound; it is also time. Music does not just take place in time, in the literal sense of its duration; its patterns of repetition, transformation, recall, and experience shape our experience of time'. Thus, as de Nora puts it;

> like other cultural materials, music can provide resources for the circulation of temporal processes. At a basic level, music can be used to measure time (Toscanini's reading of Mozart's overture to *Le Nozze de Figaro* is often described as the perfect egg timer—if you want a very soft boiled egg). Music can also be used to map or mark the phenomenological

experience of time's passage (for example, a 'long time' can be turned into 'no time at all' through the introduction of music). Musical translation of time occurs when music is perceived as providing a ground against which time's passing can be observed, (re) evaluated and (re) experienced. Through the ways time is musically 'chunked', linear or serial time can be converted into cyclical time, time can be heard to be repeating itself, recalling, retracting, or retreating. It can also be understood as compressed or decomposed (eg. 'cut time', or 'double time'). Music can make time 'fly', it can make us 'forget time' and it can also 'drag time out'. Music can also animate time: for example, Adorno (1974) considered Stravinsky's use of primitive rhythms (ie. dance rhythms) degraded the bourgeois subject through its abandonment of overarching rhythmic patterns in favour of sheer pulse. In Adorno's view, this was one aspect of the 'sacrifice' of the subject which he so disdained in Stravinsky. In the celebration of the musical unit of the pulsing instant, historical structures and historical consciousness were forsaken. (1997: 54)

Other considerations also intervened, which implies that musical practice had a need for exact time measurement from a comparatively early date (in much the same way as is found in musics from many parts of the world; cf. Clayton, 2000). Thus, during the period in which we are interested, the ability to mark time in music became more and more pronounced because of innovations in the technology of music itself. For example, from the point at which musical notation was invented by Guido of Arrezzo in the eleventh century, what were often vast feats of memory or, at best, primitive means of writing elements of music like neumes, were gradually replaced. Notation allowed new temporal architectures to come into existence, new means of synchronizing attention to regularly recurring events in the environment: temporal architectures like counterpoint, new rhythms and meters, the establishment of the primacy of temporal form, and so on (Goodall, 2000). These inventions were new means of hearing that used different patterns of attunement to events. Equally, the technology of musical instruments was changing as new and more exact means of construction became available and as standardized tuning began to diffuse. Such technical innovations allowed more exact timing to be achieved. (It is striking that many leading early modern mathematicians and physicists—Galileo, Boyle, Kepler, Huygens—were fascinated by music, in part because of the challenges of timing it laid down.) Given this history, it is perhaps a surprise that an explicitly temporal device like the metronome was invented so late (1812).[4] But, by this date, music was already a source of many and complex timings (including clock times) which meant that the metronome was simply a useful prosthetic addition to what was an already rich temporal tapestry.

In recent years more historians have begun to attempt to reconstruct 'auditory regimes', and their own 'soundscapes' of distances and separations (Cockayne, 2000; Corbin, 1998; Smith, 1999; Rath, 2003). Conner argues that one index of modernity is that people are surrounded by noise, whereas those who are not modern are not. But if we return to the example of urban bells, we can show that that making this case unambiguously by comparison with previous historical periods is not as simple as he assumes, as the evidence of foreigners living in early modern London shows only too well.

When Phillip Julius, Duke of Stettin-Pomerania, rode into the city on September 12, 1602, he and his entourage were astounded by what they heard.

> On arriving in London we heard a great ringing of bells in almost all churches going on very late in the evening, also on the following days until 7 or 8'o clock in the evening. We were informed that the young people do that for exercise and amusement, and sometimes lay considerable sums of money on a wager, who will pull the bell longest or ring it in the most approved fashion. Parishes spend much money in harmoniously-sounding bells, that one being preferred which has the best bells. The old Queen is said to have been pleased very much by this exercise, considering it as a sign of the health of the people.

Paul Hetzner, a German jurist who had made the same trip in 1598 and wrote it up in his pan-European *Itinerary*, observes that, in general, English people are 'vastly fond of great noises that fill the ear, such as the firing of cannon, drums, and the ringing of bells, so that it is common for a number of them, that have got a glass in their hands, to go up into some belfry, and ring the bells for hours together for the sake of exercise'. No less impressed with the custom was Oraxio Buzino, chaplain to the Venetian ambassador in London in 1617 and 1618, who notes that the boys made bets 'who can make the parish bells be heard at the greatest distance'. Such wagers sound like deliberate attempts to breach the parish's acoustic horizon, to transcend the boundaries marked out in rogation processions.

Church bells functioned as the most obvious 'soundmarks' in the acoustically dense soundscape of early modern London. In addition to providing recreation for the youths of the parish, they were actually rung to summon people to services—but with ostentatiously Protestant restraint. According to Frederic Gerschow, the Duke of Stettin-Pomerania's secretary, bells were never tolled for the dead, even if they were sometimes rung to spread news of a grave illness in the parish. Loudest of all, apparently, was the bell of St Mary-le-Bow. John Stow, who in the *Survey of London* pronounces of the church 'more famous than any other Parish Church of the whole Cittie, or suburbs', notes how the bell's ringing signalled rhythms of the workday. Any lateness would prompt apprentices to complain, 'Clarke of the Bow bell with the yellow lockes,/For thy late ringing thy head shall haue knockes'. A proverbial association between the bow bell's sound and your true Londoner was already current by 1617, when Fynes Moryson could tell a Europe-wide readership that 'Londoners, and all within the sound of Bow-Bell are in reproach called Cocknies and eaters of buttered tostes'.   (Smith, 1999: 53)

Further, clocks did not always function only in the visual or auditory registers. Some had time-markers directed towards other bodily senses, like the late fourteenth-century clock faces with studs for *touching* in the dark (Borst, 1993), or the incense clocks used extensively in China and Japan (Bedini, 1991). The *smell* of scent shows the possibilities of other forms of apprehension of clock time than were common in Europe.

In practice, clock times will often register across several senses at once, as a series of different but summative cues: a dial, a bell, a neighbour going out, a rooster crowing, and so on. Such cues are registered so often as to be all but unconscious elements of daily time structures. Indeed, one of the most powerful of our time senses may be born out of what Lefebvre (1995, 2004) tried to indicate with his 'rhythmanalysis': the embodied swash and swirl of cities that marks out different times of day: the noise of people going to and from work, the hum of conversations in the town square, the late night carousings, all the sounds of the streets (Schlör, 1998).

A final point about embodied practices is that people *learn* to tell clock time. As they do, this sense of time sinks into their bodily comportment to become part of what they are. People's ways of moving (their gait, their manner, their air) tend to incorporate their stance to clock time. 'Time', to employ Borges (1970: 269) famous dictum 'is the substance I am made of'. Nowadays, most people in the West learn the exact temporal organization of the body at school. Though schools are very exactly timed environments, through which people pick up much of their sense of appropriate and inappropriate temporal behaviour, remarkably little focused research has addressed how clock time is taught (Adam, 1995). But though 'historians cannot observe the bodies of the past in motion,... the sources from which inferences can be made are surprisingly rich' (Thomas, 1991: 5), and this archive of restored behaviour is expanding (e.g. Biagoli, 1993).

Thus, in considering how clock time is, quite literally, incorporated we need to consider the history of body practices which have a strong temporal element and which are an essential part of the history of many forms of clock time. These practices produce what McClary (2000) and others have called a serial culture, a 'culture of the interval' that both prefigures and strengthens the hold of clock times (Langbauer, 1999). Through the period we are considering, these arts of the body increase in number and complexity, often providing a taste of clock time before clocks became general, in the same way that practices of vision were 'set up' for film before film was ever invented (Bruno, 1993). To recap on previous paragraphs, we can see the culture of the interval making its way into music, for example, as rhythm is emphasized, as musical scoring starts to use bars, and as other details of timing become prevalent. We can see the culture of the interval making its way in other activities too. For example, dance becomes more formalized, is written down through new forms of dance notation, and, along the way, its metric qualities are emphasized. The 'geometric' dance of the sixteenth and seventeenth centuries (Franko, 1993), and the formal dance of the eighteenth century with its emphasis on 'Regularity' (Roach, 1996) are both examples of the achievement of figure and depth through rhythm, through metric quality, and also through so-called 'track drawings' (cf. Guest, 1989). Then, and relatedly, we can see the development of etiquette (Elias, 1984/1992; Davidoff, 1982) which, with its emphasis on exact timing and turn-taking equally runs on a kind of metrical quality; etiquette is a kind of social choreography which, by Jane Austen's time, is a ballet of 'trajectories of marriageable bodies, passing through the most strictly regulated social spaces, seeking their erotic fates' (Roach, 1995: 156). Clock time, in other words, is gradually written in to the very gait of the body, aided by changes in the urban environment like pavements and lighting which allow these gaits free range (Ingold, 2004; Joyce, 2003). This process can be considered in more detail by looking at the example of the military community.

As in civilian life, so the everyday life of the military was also affected by the need for exact position and juxtaposition, in particular through the evolution of drill and similar rigid positionings of the body, which came to take up increasing amounts of time in most armies (Holmes, 2001): some of the drills developed by Maurice of Orange from ancients like Aelianus and Vegetius, which became general in much of Europe as a result of example and a series of books, can lay claim to being the first time-and-motion studies in their exact and exacting attention to time.

From Aelianus (the) key borrowing was the simple notion of training soldiers to move simultaneously in response to stylised 'words of command'. Aelianus had listed 22 different 'words of command' used by the Macedonians; but when Maurice's cousin and aide, Johann of Nassau, had analysed the motions required to handle a matchlock, he counted 42 distinct postures, and assigned a fixed word of command to each of them. A simpler drill, far closer to Macedonian precedents, was also derived for pikemen, who were needed to protect the arquebusiers from cavalry attack during the rather lengthy process of reloading.

The practical importance of such pedantry was very great. In principle, and to a surprising degree also in practice, it became possible to get soldiers to move in unison while performing each of the actions needed to load, aim, and fire their guns. The resulting volleys came faster—and misfires were fewer when everyone acted in unison and kept time to shouted commands. Practice and more practice, repeated endlessly whenever spare time allowed, made the necessary motions almost automatic and less likely to be disrupted by the stress of battle. More lead projected at the enemy in less time was the result: a definite and obvious advantage when meeting troops not similarly trained. This was what Maurice and his drill masters had aimed for; and once their success became clear, the technique spread to other European armies with quite extraordinary rapidity. (McNeill, 1995: 128–9)

Thus, by the time that William of Orange arrived in England in 1688, he found 'a small standing army which had considerable and varied experience of active service, which was well-enough armed and equipped, and which was trained to a system of drill and tactics as up to date as those practised elsewhere in Europe' (Houlding, 1981: 172; Childs, 1976). Helped by the circulation of a large military literature which officers could and did read, including drillbooks like Dundas's *Principles of Military Movements* with their accompanying crib cards, and, in the late seventeenth century, by the general drill regulations composed to provide the army with a core of common practice, by the eighteenth century drill had become a carefully defined practice of bodily sequence right across

Europe—and an essential element of battle (Holmes, 2001).[5] For example, so far as marching was concerned:

> The recruit was first taught to be 'master of his person', throwing off the carelessness of civilian carriage and adopting the stiff self-possession of military bearing. Next, having learned the simplest postures, he proceeded to the simplest of the evolutions—dressing to his front and flanks, and making the various turns on the spot. Instruction at marching now began; and the greatest stress was put upon marching... The techniques of movement became steadily more sophisticated as the century wore on, demanding precision, and basic instruction at marching reflected these developments. Until the early 1750s the recruit was taught only to maintain his proper posture and bearing, to take paces either 'long' or 'short', and to step out either 'quickly' when marching in column, or 'softly' when on parade, when manoeuvring, or when advancing in line and these times and distances were measured only against the scale of what was customarily practised within each management. The mid-century was in most respects a watershed in the development of marching technique... Marching style—that is the manner in which the legs were lifted and put down—only assumed a regular fashion in the army at the mid-century, after 1748, with the adaption of the 'Prussian step': taken from the stiff-heeled marching style introduced to Russia during the reforms of Frederick William I's region, this was to be a notable innovation and was to remain the style after which British infantry performed linear drill until late in Victoria's reign. The Prussian step made for great precision at speed; and indeed a rate of 120 paces per minute was considered, for the experienced soldier, 'nothing more than an easy walk'. (Houlding, 1981: 259–60)

The general increase in marching pace was accompanied by experiments with the use of music to keep time.[6] Thus, in the later 1740s:

> drums, and more particularly the new fifes, began to be used in some regiments to set a marching cadence [and by] the end of the Seven Year's War... marching in step to a musical cadence... [became] standard practice in the Army. Once the great advantages of speed, precision, flexibility and simplicity which the cadence and the resulting closed ranks made possible had become apparent, however, it was found that music, itself was sometimes a distraction, sometimes (in action) inaudible, and the cadence difficult for the drummers and fifers to maintain. By the mid-1770s, therefore, the musical cadence was not used in the field, but was retained for training purposes: by the music the recruits accustomed themselves to the standard pace and time, now all important as a foundation of close-order manoeuvres. (Houlding, 1981: 261)

During the same period, the military also put increasing emphasis on using soldiers' time profitably in field fortification: digging trenches, raising embankments, building redoubts, constructing bridgeheads, and the like. The approach was practical, since it was realized that 'the elaborate mathematics and geometry of engineers were subjects too dry for everyone to relish; and indeed there was no need of handling the scale and compass... [nor] of problems, nor tiresome calculations, in order to learn the art of putting all kinds of posts into a proper state of defence' (Houlding, 1981: 224). But the upshot was clear: directed bodies keenly aware of time.

Idleness, in effect, was banished from military life. This was a great departure from earlier custom, since waiting for something to happen occupies almost all of a soldiers time, and when left to their own devices, troops had traditionally escaped boredom by indulging in drink and other sorts of disruption. Debauchery was not banished entirely under the regime Prince Maurice and his imitators established, but it was usually confined to off-duty leave time.   (McNeill, 1995: 129–30)

3.3.2 Skills

Second, clock time consists of different kinds of clock time *skills*, which vary according to the networks of the social in which they are embedded. Clocks take on quite different material-semiotic roles as they settle in alongside other timekeeping devices, each with their own long and intertwined histories, as all manner of communities of practice participate in the constant engineering of diverse materials and signs to achieve their purposes. Chemists, particle physicists, and computer scientists may need the skills to measure and interpret clock time in nanoseconds, but very few communities of practice do, and this nearly infinite regard for clock time may be attenuated or even reversed.[7] Norms of punctuality are a good example, being strongly related to the kind of activity in which participants are engaged. Thus the knowledge required may be simply knowledge of the socially appropriate time to turn up (perhaps 'fashionably late' for parties); or of how to show a studied casualness towards clock time (such as the louche attitudes of certain aristocratic and artistic communities, or specific forms of hanging out, such as 'liming' in Trinidad; Birth, 1999); or of how to resist an imposition of punctuality (as in repeated tardiness at work, various fiddles to do with clocking in and out from work, or altering the time on timekeeping devices, which might sometimes be effective challenges to a boss's power; Birth, 1999, Greenhouse, 1996); or how to enact a complete disregard of clock time (as in raves and other 'oceanic' experiences; Pini, 1998).

Of course, there are cultures where treatment of clock time is purposely lax. For example, for the Kelantanese peasantry on the Malay Peninsula, their easygoing treatment of clock time is a skilled statement of cultural identity.

Villagers value their pace of life and . . . this pace is leisurely. The orientation of villages is toward present time, yet a present time in which different events and activities do not compete with one another for primacy. What is not done one day will be done the next, or the day after that. There is a strong ethic that haste is unseemly, the mark of a person too concerned with material advances who may not be paying sufficient attention to social obligations. People who hurry their activities are often the subject of gossip. They are regarded as less refined . . .   (Raybeck, 1992: 330)

The use made of timekeeping devices like watches and clocks is shaped by this particular social format: though the Kelantanese peasantry frequently own watches they still use a simple 'coconut clock' (a pail of water in which is floated half a coconut shell bored with a small central hole; the measured interval is simply the time it takes for the shell to fill with water and sink) to time their kite-flying contests, because its highly visibility protects the judges from charges of favouritism.

Quite often, temporal skills become evident only when there are breakdowns of coordination. Ways of dealing with the stresses and strains brought on by such events (such as 'keeping cool') may become formative. In many societies coordination may become quite flexible in order to deal with these stresses and strains without necessarily lessening the impact of clock time. Exact timings may still be necessary but they are displaced.

A number of caveats need to be entered at this point. Most particularly, communities of practice are not homogenous entities. Thus knowledge of clocks and clock time use is differentiated and distributed. More often than not, no complete knowledge of clocks and clock time is use held by any one person or institution; expertise is shared (Hutchins, 1995). Indeed, knowledge of the proper uses of clocks and clock time in a given situation may well be restricted to *specialists*. Again, and relatedly, knowledge of clocks and clock time will vary in its character. Whilst there will be some theoreticians who hold something close to propositional knowledge (Mokyr, 2002), others will have the kind of practical knowledge that comes from being involved in installation, maintenance or repair, or in amateur tinkerings that have reached the standard of some professionals. Yet others may have only the vaguest knowledge, enough to ascertain clock time, or alter a clock, but no more. Yet others may find clocks and clock time a challenge or an irrelevance:

> In Trinidad, not...everyone can read the clock. Those who cannot use clocks still understand what two o'clock is, but they rely upon people who can read clocks to know the time, or they associate a particular hour with a time of day—for example, morning, afternoon, evening. In addition, many who can tell time do not in any way structure their lives around the passing of hours and minutes. They see clock time as a reference for parts of the day, rather than something to determine when they do a particular activity. Finally some both understand and apply clock time. When they make an appointment, they will be there at an appointed time, and they expect others to be there too. (Birth, 1999: 111)

So stances towards the usage of clocks and clock time may be contested. Whereas clock time may be important to managers and management consultants, for workers it may be something to dissent from, either directly or subtly: 'doing the absolute bare minimum required when others know that one has the time and resources to do more is a clear statement' (Hutchins, 1995: 225).

3.3.3 Cultural Connections

Third, clock time is a *mentalité*, a set of cultural conventions which have become generalized across many different communities of practice. There is no simple relation between the appearance of clocks and associated metrics and their acceptance as important. Rather, that relation has to be socially constructed via specialized institutions, particular cultural products, new forms of vocabulary, and the corresponding effort that goes into stabilizing their use.

For example, it is quite clear that the importance of clock time can be interpreted as part of a more general historical tendency towards 'interval training' (MacAloon, 1995) based on the expansion of measurement of all kinds. Measuring is valued as a practice for its own sake as the world is increasingly treated in terms of uniform operation: 'lines, squares, circles and other symmetrical forms: music staffs, platoons, ledger columns, planetary orbits' (Crosby, 1997: 10).

The arbitrariness of this process of cultural acceptance is something we now rarely remark upon because it has become a part of our cultural background, but it may be laid bare in other times and places. For example, in fourteenth- and fifteenth-century Europe cultural conventions about clocks were still quite fluid; there were clock faces that worked anticlockwise (Arthur, 1994), and clock faces with 3, 4, 6, 16 or 24-hour dials (Borst, 1993; Beattie, 1952; Loomes, 1997). It is important, then, to appreciate that the new clocks appearing in Europe were not necessarily seen as challenging older temporal orders:

> it was progress enough if the [mechanised] astrolabe... needed readjusting only once in the morning and once in the evening; it would then be as accurate as before, indicating the uneven temporal hours of the following day and the next night that still dominated life and could be read from the curves on the inserted plate of the astrolabe. Specialists no longer had to carry out laborious measurements by day and by night for each time-check, and laymen no longer need to use their hands to find the hour, but simply their eyes, and at night even just their ears. The fact that the new machine was given a striking mechanism, thus assuming the additional function of a bell, did not fundamentally alter people's sense of time. Although shorter measures of time than the seven canonical hours and the twelve temporal hours were now signalled from the tower, it was initially still the bellringer who sounded them by hand as soon as the striking mechanism woke him. (Borst, 1993: 93–4)

The process of acceptance depended, then, upon means of cultural standardization, which were neither obviously apparent, nor easy to achieve. There was no teleological impulse. Rather there were, initially at least, many possible solutions: 'hours of the day were counted differently from place to place, in small hours, large hours, or whole hours, and it was rare to start the numbering at midnight as we do today' (Borst, 1993: 94–5).

Of course, this is not to say that the mechanical clock did not disrupt people's consciousness of time in the late Middle Ages. It did. But it did not create a 'modern' time, or a 'universal' time, as some writers would have it. The degree of possible cultural latitude was too great for the mechanical clock to simply settle in, mechanical hand into cultural glove. The accepted meanings of clock time had to be worked on, but many meanings were possible and a whole semiology of clocks sprang up. For example, following Mayr (1986), Borst (1993) notes at least four ways in which the clock was symbolized: two non-scholarly, two scholarly.[8] The mystic's sense of time, as represented by the German Dominican Heinrich Seuse, in his *Horologium Sapientiae* of 1334, saw the divine mercy of the Saviour in the form of 'an elaborate clock whose melodious bells chimed every twenty four hours. The mechanical clock and the carillon came to be seen as mirroring the soul' (Borst, 1993: 95). A second meaning of time had its roots in skilled trade and centred on the hourglass. This device, when it first appeared in the fourteenth century, was apparently figured as

a symbol of moderation, regularity and satisfaction combined within the moment. It made working people aware of passing moments, silently and without a numerical rhythm. Each individual divided and occupied such movements differently, the scholar in the study, the preacher in the pulpit, the advocate in the courtroom, the sailor on watch, the housewife at the oven. Yet in the hands of Death the hourglass reminded them of their final hour and urged them to make use of the moment for as long as there was still time. (Borst, 1993: 95–6)

Two other meanings of time were of a more scholarly nature.

A third, atomised concept of time adopted fractions of hours that previously could have been calculated but not represented. The tower clock now struck the half and quarter hours too, and people thought in terms of minutes and seconds that up to then had only been used by astronomers. Could the influence of the planets on human destiny be reliably ascertained at last, as Firmicus Maternus had demanded? About 1330, the Oxford mathematician Richard of Wallingford, promoted by then to Abbot of St Albans, not only constructed a planetary mechanical clock but also cast the horoscopes of the small children of the royal family, predicting their entire future beginning at the cradle. He found many imitators. (Borst, 1993: 97)

Then there was one other meaning of time, as a mechanized concept. Typified by the work of Nicholas Oresme in his *Book of the Heavens and the World*, published in 1377, Oresme described the universe as:

an *horologe*, a regular clockwork that was neither fast or slow, never stopped, and worked in summer and winter, by night as well as by day. He drew a direct comparison between the movements of the planets and a mechanical clock that balanced out all forces by means of its escapement... In particular, the planetary clock came to represent the cosmos, the improved astrolabe rather than the accurate time-measuring device; its designers could compare themselves to the creator of the universal machine

Oresme's objection to the astrologers was... that planetary movements were incommensurable with one another and thus never reconverged to form identical constellations. The clock's dial and hand movement, however, visibly confirmed the Aristotelian definition of time as the numerical value of a motion from a prior point to a later one. When Oresme claimed he saw a clock in the heavens, he had in mind the large mechanical clock King Charles V had installed in his palace in 1362. From 1370 onwards, all Parish church clocks had to keep in step with its somewhat capricious chime; it allotted townspeople their working day. It was the king, the designer par excellence, who stipulated how the course of social time was to run. (Borst, 1993: 97)

Thus what became the dominant metaphor of the clock—of the world being like clockwork—was actually a semantic juvenile, a comparative newcomer to clock time's range of meanings.

Thomas Aquinas observed... that all things moved by (man's) reason act as though they had reason, even if they lack reason, and illustrated his point by reference to a moving arrow and a clock. Arrows and clocks maintained their privileged position as images of space and time, respectively, with the clock slowly rising into the position of a peculiar metaphysical device. Nicholas Oresme likened the universe to a mechanical clock that was created and set moving by God, who also saw to the proper proportions of the celestial wheels. The clockwork metaphor was taken seriously by Kepler when he attempted to understand the world as a unit interconnected by actions at a distance. In 1605 he writes

> 'I am much occupied with the investigation of the physical causes. My aim is to show that the celestial machine is to be likened not to a diverse organism... but rather to a clockwork (Whoever thinks that the clock is an organism attributes to it the glory due to its maker). Insofar as nearly all the manifold movements are carried out by means of single, quite simple magnetic force, as in the case of a clockwork all motions (are caused) by a single might.'

Giovanni Alfanso Bovelli (1608–1679), an Italian physiologist, sought to explain the motion of animals on mechanical principles in his *De Motu Animalium*. He also attempted to account for the fertilisation of the ovum by the male sperm by an odd comparison of the process with the workings of a Timepiece made of cogwheels and moved by a weight.   (Fraser, 1975: 65)

3.4 AN ALTERNATIVE ACCOUNT OF THE NATURE OF CLOCK TIME

We have continually emphasized our resistance to the idea of simply replacing one of the grand accounts of the history of time, based on either a series of revolutions in horology or in social discipline, with another. However, at the same time, we do not want to see our work presented as simply a series of partial empirical excursions which have no larger significance. Therefore, in this section, we begin to work towards an account which synthesises our research into a more general model of changes in timekeeping practice between 1300 and 1800, a model which we think traces out a series of important revolutions in timekeeping practice.

Of course, the various changes in the practices of timekeeping that are described in this book may seem so glacial that some readers will think that they should not be dignified by the term 'revolution'. Needless to say, we disagree. These changes may not necessarily have the dramatic qualities and the relative speed of onset that are often associated with the word 'revolution', but insofar as they track changes in the root *anticipations* that people have about how the world turns up, they must surely qualify.[9] Our interest, in other words, has been in longer, slower but no less effective revolutions in everyday timekeeping practice that naturalized new kinds of temporal *attunement*, which promoted new kinds of awareness of temporal objects as temporal, as timekeeping objects were bound into everyday practice. As recent work in 'cognitive phenomenology' (Petitot *et al.*, 1999) emphasizes, this is no simple process since it requires the generation of new bodily dispositions as well as the kinds of cognitive understandings to be found in books and manuals. If we had to look for an example of what we have been trying to get at, it might again come from the world of music: people have to learn to play musical instruments and this is not just about musical notation. It also involves the correct accommodation of lungs, lips, hands, device(s), and so on. In addition, playing musical instruments requires a certain spontaneity. In other words,

The world is comprehensible, immediately endowed with meaning, because the body... has the capacity to be present to what is outside itself, in the world, and to be impressed and durably modified by it, has been protractedly (from the beginning) exposed to its regularities. Having acquired from this exposure a system of dispositions attached to those regularities, it is inclined and able to anticipate them practically in behaviours which engage a corporeal knowledge that provides a practical comprehension of the world quite different from the intentional act of conscious decoding that is normally designated by the idea of comprehension.   (Bourdieu, 2000: 135)

It is the building up of these new anticipations of, or attunements to, time—these new 'as-if naturalizations' if you like—that this book aims at showing as present in the historical record. In many ways we would argue that it is these new kinds of common senses that are the real revolutions in history, revolutions which historians are now beginning to study in considerable detail (as, for example, in histories of shopping practices, Glennie and Thrift, 1996).

Given the research that we have carried out for this book, how would we now frame this process of attunement? We would argue that such learning of new anticipations (and learning what constitutes that learning) takes place at three different levels and involves three different intelligibilities.

The first, which we have already foreshadowed, is corporeal learning. Many timekeeping practices are so deeply grooved into the body that they are intuitive. They emerge without conscious understanding through a kind of osmosis. Claxton (1997, 1999) calls this kind of learning 'know-how without knowledge'. This is obviously difficult to describe since it lacks clarity and articulation and is often not verbalized. But it is nonetheless crucial. Practical mastery of environments (so-called intelligence without reason) often emerges through immersion and experimentation in ways which are not open to conscious understanding or able to be easily turned into expert knowledge (e.g. Ingold, 2001).

However, it is obviously absurd to suggest that all learning is of this kind. So we come to the second kind of learning about time, which we might call 'cognitive'.

However, our sense of cognition is not of a rational choice process. Rather, it consists of sets of ad hoc calculations attached to the moment, 'toolboxes' of simple propositional heuristics that are used to assess situations, rather than general purpose decision-making algorithms (Gigerenzer, 2000; Gigerenzer and Todd, 1999; Gigerenzer and Selten, 2001).[10] These heuristics are 'fast and frugal', often involving the gathering of very little in the way of rich information. They are computationally cheap rather than consistent, coherent, general—and expensive of time. They constitute a kind of 'quick fix' that will work most of the time to the degree necessary to tackle a particular situation; 'tools' selected to be able to read and influence the environment, but with minimal effort.

Such heuristics are bound to be adaptive. They are adjusted to circumstance and can be swapped around. And they are 'constitutively leaky' (Clark, 2001) in that they depend upon constant interaction. In particular, they evolve through what Gigerenzer (2000) calls 'foraging behaviour'. Effort is adjusted to circumstance according to what information is available in the environment. So, for example, in the case of timekeeping, in certain circumstances, considerable effort may have to go into exact time-telling, if the situation calls for it. But most of the time circumstances call only for a close approximation which will involve the expenditure of much less effort in order to extract limited information from the environment.

But, in turn, this brings us to the third kind of learning which we will call ecological rationality. The environment itself speaks—it is a purveyor of large amounts of information and practices resulting from the interaction between corporeal logics, adaptive heuristics, and the information that the environment provides. Generally speaking, if the environment is information-rich, then heuristics will be quick and simple. For example; a quick glance at a clock face may suffice. If the environment is information-poor—or so complex that it is difficult to negotiate—then more complex heuristics may be needed: for example, asking a friend who knows, or knows how to access, the requisite information (Nardi and O'Day, 1999). In acknowledging the importance of ecological rationality, we move closer to Hutchins's (1995) and others notion of distributed intelligence. 'Thinking' is distributed across environments via a range of different divisions of labour and tools. In classical 'post-humanist' fashion, thinking exists as a set of spatially and temporally distributed practices which do not start or end with the individual human being and it is extended through the use of various tools (like clocks) which allow the corporealities of practices to be extended and allow new kinds of attunement to come into existence.

To summarize, we draw on Schatzki:

[The] prioritisation of practices over mind brings with it a transformed conception of knowledge... Knowledge (and truth) are no longer automatically self-transparent possessions of minds. Rather, knowledge and truth, including the scientific versions, are mediated both by interactions between people and by arrangements in the world.

Often, consequently, knowledge is no longer even the property of individuals, but instead a feature of groups, together with their material set-ups. Scientific and other knowledges also no longer amount to stockpiled representations. Not only do practical undertakings, ways of proceeding, and even set ups of the material environment represent forms of knowledge—propositional knowledge presupposes and depends upon them. (2001: 12)

It follows that when we look for revolutions in timekeeping we need to look to all three kinds of learning—corporeal, heuristic, and ecological—if we are to understand the history of telling the time, and how clocks (and related timekeeping instruments) were used by groups to construct the new forms of synchronizations and eventfulness (what Flaherty (1999) calls 'routine complexity') that we now take to be normal.

3.5 CONCLUSIONS

We can now return to the concerns of this and the previous chapter with some degree of confidence that we can see clock time in a different way. Thus, to begin with, clock time is no longer a total, homogenizing, and always oppressive force. It is not one thing. Rather it is, 'open, processual, and eventful' (Morson, 1994: 43). It is, in other words, a set of different networks of temporal practices—often with their own notions of what constitutes clock time. And there is no good reason to think that clock time lacks existential depth. After all, it dwells in and through our bodies as an accumulation of different stances and gaits built up now over many centuries.

Further, clock time is no longer unnatural. This kind of thought can only thrive where nature is figured as the slow and inexorable change of the seasons—as a sylvan, bosky, Euro-American nature—which clock time has taken over, speeded up, and then left behind. But natural processes are often very *fast* and show high levels of complex and self-generative organization. In certain senses, clock time is simply the means by which we try to keep up with nature, *not* vice versa (Thrift, 2000b).

And clock time is not omnipotent or omnicompetent. As we have already seen, clock time has come into existence as a result of the needs of *particular* networks of practice to be able to measure their own and others practices simultaneously. But the values of that process of measurement can be very wide. Furthermore, these networks are spatially distributed, which means both that it takes effort to maintain them (from winding clocks to placing satellites in space) and that all kinds of strange and unpredictable encounters will be staged between the networks which in turn generate new practices. Then, as a final point, clock time cannot overcome the simple fact that the future is always 'absolutely other and new. And it is thus that we can understand the very reality of time, the absolute impossibility of finding in the present the equivalent of the future, the lack of any hold upon the future' (Levinas, 1989: 46). Clock time generates only an *illusion* of omnipotence which life will continue to knock down.

Cracking up, breaking down, hitting the wall, life derailed, the opening of an abyss, being unable to put the next foot forward—these are extreme experiences, but they are all about a rupture in which the future disappears. They bring us face to face with contingency. (Game, 1997: 117)

None of this is to argue that 'abstract' clock time is not a powerful set of systems-in-use. But, perhaps it is to argue that the power of clock time comes from the fact that it can be so many things to so many people. Its power comes from the *difference* implicit in the multiple possibilities it generates, even though it is usually figured as the one and only, or at least its shadow.[11] In other words, not everyone occupies the same now, because both the one and the now are plural events.

The thoughts in Chapters 2 and 3 can be put to one side as we now move to our examination of a series of particular episodes in the history of clock times. But not too far to one side. For they have informed how we have approached our material, what we have looked for in the material, and how we have drawn out various interpretations of that material. They do not therefore act as a background to what follows but rather as a kind of middle ground, weaving in and out of the narratives and case studies that follow.

NOTES

1. Some of the same conceptions can of course, be found in modern relativistic and quantum physics. See the collection edited by Savitt (1995).
2. We have taken little note here of work in literature, but, as the previous comments on writing strategy make clear, literature is both a source and a model for work on time.
3. This is Butler's (1997) attempt to mix Bourdieu with Derrida, introducing Derrida as a source of repetition.
4. The first attempt to produce a metronome was made in 1696, effectively an adjustable pendulum with calibrations but without an escapement to keep it in motion. But it was not until 1812 that Winkel perfected a double-weighted pendulum which would beat low tempos, an idea brought in to the commercial domain by Maelzel in 1816.
5. Different places were used for different military situations. 'The 1786 Regulations were the first attempts to establish a standard pace. Recruits were taught to accustom themselves to a standard pace by practising their marching on long stretches of ground measured off accordingly, and marked with tapes or lines' (Houlding, 1981: 260).
6. There is some controversy over the exact date of onset of marching in step but it seems likely that until the 1750s no exact step was employed, because it seemed difficult, because it looked too much like dancing, and so on. The fife seems first to have been introduced into the Army in the 1740s (Houlding, 1981).

7. Skills will vary across communities, too. And these skills will include the skills associated with manipulating instruments. For example,

> in a sense, a scientist's manipulation of an instrument to have it yield meaningful results may not be unlike the bodily skills expected of a courtier to achieve the appropriate presentation of the self. Similarly a scientist's reading of a given experimental report as confirming or refuting a given theory (one in which her or his professional identity and career may be at stake) may be commensurable to the processes by which a courtier reads the signs of favour, interest or coldness as they are conveyed by her or his fellow courtiers. (Biagoli, 1995: 78). See also Lawrence and Shapin (1997).

8. Mayr (1986) has also fashioned a history of the growth of the clock metaphor in continental Europe after 1300 by using a range of literary and iconographic sources. Mayr identifies a gradual increase in the use of the clock metaphor. In late medieval Europe, the metaphor was used infrequently and for specific purposes. In literary sources, the clock is used as a representation of the complex working of the world in writings as diverse as poetry and philosophical works. Iconographically, the clock is associated with two figures. The first is Sapientia or Divine Wisdom.

> Berthold's *Horologium devotionis* and even more, Suso's *Horolgium Sapentiae* came to be widely distributed. The demand for the first was still large enough after the invention of printing to give rise to twelve incunabula editions. Suso's book, soon translated into the major European languages, is said to have been, by the end of the fifteenth century, second in popularity only to Thomas a Kempis's *Imitation of Christ*... (Mayr, 1986: 33)

The second figure was Temperantia, temperance. As a woman with a clock, Temperantia became a familiar figure in the prints and paintings of the fifteenth and sixteenth centuries, symbolizing key virtues such as punctuality, discretion, circumspection, caution, and frugality. In other words, what were chivalric values were transformed into the values of the bourgeois townsman. 'Temperance had become the basis of the ethics of the urban middle class' (Mayr, 1986: 38).

Generally speaking, Mayr finds that in the fourteenth, fifteenth and sixteenth centuries the clock was not used as a metaphor with any frequency. However, in the seventeenth and eighteenth centuries, in contrast, use of the clock metaphor became 'strikingly frequent, more frequent, probably, than any other' (Mayr, 1986: 29). Clocks were used in two more ways. First, they were symbols of order and regularity. 'The sentiment surfaced in short phrases like, "as true as a clock", "title pages and tables as define as the clock strikes", "a well-armed order, clock and guide-line"...' (Mayr, 1986: 41). It is significant that here we see the clock transforming everyday speech as well as performing in the writings of Donne, Shakespeare, Webster, Middleton, Johnston and the like. Second, they were used as icons. For example, Mayr shows how the clock (along with the sundial and the sandglass) was used as an iconographic accompaniment to Father Time and more generally, 'to illustrate various general aspects of the concept, time; its fleeting nature, the importance of husbanding it, the benefits of a disciplined daily regimen, the virtue of productivity, and so on' (Mayr, 1986: 41).

Significantly perhaps, Mayr detects the most sceptical view of the clock in Europe emanating from England.

> Clocks were not only called 'true' and 'practical' but also 'cold', 'gloomy', and 'long-faced' as well as discordant and dishonest. Various English authors reiterated this point: 'The preaches of England began to strike and the clocks of England, that never meet iampe on a part together';

'He will lie cheaper than any beggar and louder than most clocks'; 'They agree like the clocks of London'; and 'Alas, we scarce live long enough to try/whether a true made clock run right, or lie'. (Mayr, 1986: 50)

In part, this view seems to have been associated with the fact that many clocks were of foreign (and especially German) origin. But in, part, it also seems to have been a more general feeling:

English authors who reflected more deeply understood well enough that their dislike of the clock was in part the familiar resentment of bearers of ill news: what made clocks exasperating was their evident tendency to run fastest when time was most precious and their tactless insistence on striking the hours when that service was not welcome. This observation was expressed in many styles and moods, ranging from light-hearted witticism to high tragedy. According to Robert Greeve, for instance, 'The Usurers Clocke is the swiftest clock in all the Towne; tis, sir, like a woman's tongue, it goes ever halfe an hour before the time; for when we were gone from him, other clocks in the town strooke foure'. Other authors echo similar sentiments, 'Labouring men count the clock oftenest... are glad when their task's ended' and 'The clock upbraids me with the waste of time'. (Mayr, 1986: 51)

However, the sources that Mayr draws on are ambiguous and though straightforward histories of clock metaphors like Mayr's are alluring, they are also problematic since they are open to the charge of riding roughshod over the sources, and necessarily being oriented to the viewpoints of a few elite communities of practice.

9. It is vital to establish the variable chronology of the processes of production and dissemination that precipitated new forms of temporal doing in people's lives, slow and hesitant as these may sometimes have been, not least because they have been under-examined in the historical record precisely because have remained so remarkably under-theorized in history. However, models like those emanating from the history of science, from cultural history, from the history of material culture, and from various other forms of history suggest that this state of affairs is now changing and that the emphasis on discovery, iconic moments and remorseless, almost machinic, processes of transmission typical of much historical endeavour in the field until recently is now being foresworn for something more productive of genuine historical understanding.

10. Though Gigerenzer's approach has been criticized on a number of grounds—and, especially, trying to apply fast and frugal heuristics to complex social situations (see Sterelny, 2003)—it works well in situations where the 'database' is relatively small and is at least partly technical.

11. As Game (1997: 116) puts it, 'we never have one experience at a time: the experience of lived, fleeting, fortuitous time is reliant upon the experience of abstract time (or, at least, the shadow of such a time)'.

4

Clock Times in Medieval and Early Modern Bristol

4.1 INTRODUCTION: STRATEGY AND CONTEXT

4.1.1 Strategy and Evidence

This chapter explores clock times over several centuries in one city, providing a case-study within which to situate, and from which to draw out, several questions on clock times that are then developed further in the subsequent chapters of the book. We adopt a retrogressive approach, starting with the significance of clock time in the late eighteenth century, before tracing emergent infrastructures of timekeeping, and the structures, practices, and uses of clock times in both formal and everyday activities, in early modern and then medieval Bristol. We tell the story 'backwards' so to speak, in order to highlight how much narrower, and more overtly 'disciplinary', is the documentation for the earlier periods, compared with 1750–1800. Throughout, we will have the same range of questions in mind, but the evidence available with which to answer them is, as we have already pointed out, hugely variable.

It can be difficult to avoid the pitfalls of an overly 'genealogical' approach, especially the tendency to rationalize, *post hoc*, the significance of early features that appear 'ancestral' to later features. We need to emphasize how the narrower early documentation makes it increasingly difficult to discuss temporal practices without 'reading them off' from the ordering frameworks of markets, religious services, or whatever. The adage that lack of evidence about past practices not being evidence that the practices did not occur is apposite here. The changing character of available evidence may, but should not in itself, show that everyday temporal practices were necessarily very different from later times, when more direct and more diverse evidence on practices is available.

Acknowledging that brevity can make a definition appear glib, our practice-centred view of time-consciousness encompasses people's capacities to act; to coordinate; to respond to; and to plan temporally structured activities; people competent to, as it were, insert themselves and others in networks of sequences and simultaneity; people able to interpret and communicate temporal information with others; people able to imagine and anticipate temporal cues.

However, we do not see these capacities as necessarily intellectual, but as primarily *embodied, unreflective, improvised*, and as frequently *taken for granted*. These sorts of attributes can make time consciousness an elusive target in the archives. One category of things very rarely specifically recorded is precisely that which is too obvious to be (literally) noteworthy. Where archival traces of time-consciousness are glimpsed at all, they are frequently fleeting. Those fleeting traces, though, are of the utmost importance.

In attempting to characterize the urban temporal environment of Bristol, we are interested in identifying those activities and institutions important in providing everyday time-structurings; in patterning everyday times through work, sociability, public order, and leisure. Such time-structuring did not depend on clocks and clock times, of course, so the extent to which clocks were used to sustain everyday temporal structures itself comprises a second set of important questions. We therefore attempt to identify exposure to, and knowledges of, clock time across the city. How widely was clock time information available, and to whom? These practices and infrastructures had geographies too, and we also examine the mutual impacts of changes in times and timings, on the one hand, and changes in Bristol's spatial structure, on the other.

Despite Bristol's relatively rich archival and secondary sources, though, we do not claim to have been able to answer all these questions even for the better-documented periods. On some dimensions Bristolians have left a good deal of evidence, both through direct discussion, and indirectly through traces of practices and devices that were not articulated explicitly. But even here the *social range* of evidence and findings is varied, and on many topics very uneven by status, by wealth, and by gender, amongst other cleavages.

Sometimes we can find almost nothing that addresses our questions. This variation in evidence is not something to be glossed over, but a recurrent problem in trying to explore topics that were so taken for granted by those involved. The available evidence can never be pursued as far as we would like, and we are challenged to use what evidence there is in ingenious ways, to reveal something of the senses and skills of that 'too-obvious-to-explain' world.

To begin by highlighting such complications may seem pessimistic, but serves to highlight two very important points. The first is that this account inevitably contains many gaps. If research is about weaving strands of evidence together into a fabric, then it is tempting to characterize the result less as a tapestry than as a fishing net. And to reiterate, the picture for the eighteenth century is much less 'gappy' than those for earlier times. Secondly, it is to leave space for those practices of time and timing which have been very largely taken for granted by those involved. It is a cliché to say that nothing is as neglected as the things or practices that are most obvious and familiar, but at a range of two centuries—never mind the eight hundred years back to high medieval Bristol—we need to keep spaces in which to be aware of the matters contemporaries took for granted.

Patchy and inconsistent evidence notwithstanding, the chapter presents a considerable range of new information, especially about what we term 'temporal infrastructure'. By this we mean 'the availability of temporal information' in the city, and its impact on local environments through public clocks, domestic clocks and watches, bells and other time signals, as well as various indirect time cues. Different spaces within (and sometimes beyond) Bristol were both settings for, and products of, several overlapping communities of practice whose senses of times depended on the interaction of long-run and short-run practices and social relations in local settings. Bristol accommodated the plural time-senses of its constituent communities of practice, which were closely shaped by a host of changes whose detailed unfoldings and interplay were place specific: changing technologies of timekeeping; changing technologies of production; organizational changes in distribution; changing technologies of lighting; changes in the geographical structure of the city and the locations of specific activities.

4.1.2 Why Bristol?

To use Bristol as a case study in urban clock time is both opportunistic and strategic. Bristol is comparatively unusual among English provincial cities in having held its medieval prominence right through the early modern period into the nineteenth and twentieth centuries. Compared with many other towns therefore, which either lost their medieval importance (e.g. York, Chester), or grew from virtually nothing into major centres (e.g. Manchester), Bristol affords a long-run picture of urban timekeeping. Through a combination of its county status from 1373, a long tradition of civic record keeping, and that luck required for documentation to survive over several centuries, the city possesses relatively abundant corporation and parochial documentation. Moreover, Bristol is very well provided with various unofficial, but highly relevant, documentary sources, especially from the mid-eighteenth century. We draw on familiar sources like the early directories and the various Bristol newspapers; and on exceptional work by other scholars on eighteenth-century Bristolian culture, leisure, and education, and on the early modern economy; on horological databases of Bristol clocks and clockmakers; and on new and systematic analysis of various Bristol sources, including its uniquely full series of eighteenth-century militia ballot lists, parish and city rates assessments, City chamberlains' accounts, and all extant churchwardens' accounts for the city's parishes.

Beyond its enduring urban prominence, and the excellence of its archives, however, Bristol is a particularly relevant case-study because it has provided the setting for some important studies of time. Bristol has figured large in the historiography of time-consciousness through the work of—to name just a few—Douglas Reid (1976) and Mark Harrison (1986) on working hours and time in the working week, stimulated by E.P. Thompson; Jonathan Barry (1985) on leisure and civic identity; Clive Burgess (1987, 2002) on medieval religion and the Reformation; Simon Penn (1989) on medieval marketing; and Chris Humphrey (2001) on timing in medieval public life, besides work on early modern calendars, and eighteenth-century craft and trade specialization. While some of this work considers time more explicitly than others, and reflects on calendars or senses of historical time, rather than clock time, it remains true that many facets of past times have been as extensively studied for Bristol as for any British provincial city.

4.1.3 Times

While this book focuses on clock times, timing practices, presumptions, and competences there were many important scales of 'time-consciousness' circulating in Bristol, from religious or cosmological times encompassing the whole history of the world and of humanity, to precision in seconds, or fractions of a second. Between the extremes of cosmic time and the stop-watch were many other timescales, all important in certain contexts, for greater or lesser numbers of people. The narratives of foundation, continuity, and destiny that sustained national or other macro-social identities could involve timescales beyond a thousand years, as with the London foundation myth involving Brutus's arrival from Troy, manifest most sharply in strong accounts of England as a (or The) Protestant Nation.

Narratives of Bristol as a civic community also involved foundational times. Several Bristol antiquarians invoked times back to the city's origins before the Norman Conquest, and especially the charter of 1373, and county status, but taking forms dependent on wider national or religious narratives entwined with Bristol's civic identity (Barry, 1995). Other statements invoked a genealogical time of parochial and familial continuities, and accounts of immediate forebears articulate a biographical timescale over several decades.[1] Economic and political commentators dwelt on yet shorter periods, and many agricultural and (again) religious time-senses were strongly seasonal. In Bristol, as elsewhere, numerous calendrical issues are prominent in newspapers and pamphlets.[2]

The clock times that divided the day, and helped to structure daily life thus come towards the short end of a wide spectrum of timescales. It is indeed rash to presume that these co-varied: so far as Bristol is concerned, there are few signs of a general, consistent, and thoroughgoing temporal competence. The overall picture is much more mixed, with Bristolians at any given time showing relatively sophisticated treatments of, and systematic attitudes towards, time at certain scales, but not others. They are *explicitly* aware of some scales and not others, and the waxing and waning of interest in particular sorts of time were commonly out-of-phase with one another.³

4.2 MID- TO LATE EIGHTEENTH-CENTURY BRISTOL

4.2.1 Contested Interpretations

The diverse historical accounts of long-run shifts in timing and time-consciousness generally privilege 'work', broadly defined, whether factory work, prayer, or market trading hours, as the critical source for an instilling of time-discipline amongst the general population, around whose insistent rhythms the rest of everyday social life must fit. What can be said, then, about *work times* in mid- to late eighteenth-century Bristol?

Though among eighteenth-century England's larger cities in England, Bristol had few early factories, and most work retained an artisanal or workshop character. While there was extensive coal mining in Bedminster and Kingswood, respectively south and east of the City, much of the mining was done by relatively small-scale operators, even among the larger employers in brass-founding, sugar-boiling, and other processing activities. The same applies to eighteenth-century harbour activities in general, though the rhythm of rising and falling tides, with their pulses of shipping into and out of the harbour, marked passing time (Aughton, 2000).

Debates on work's impacts on time-senses during the later eighteenth century revolve around E.P. Thompson's celebrated 'Time, work discipline and industrial capitalism' (Chapter 2, above). Evidence from Bristol both supports and disputes Thompson's argument. Lacking direct evidence of work patterns (though see Voth, 2000), historians analysed work as the temporal inverse of better-documented events ranging from public disorder to wedding ceremonies. Thompson received early support from Reid (1976) who identified, from the days of marriage ceremonies, that work in Bristol and several other towns showed a Thompsonian irregularity, with weekly and seasonal spasms of intense activity interspersed with periods of slack, including a strong 'Saint Monday' tradition. Reid said little about clocks per se, but the implicit assumption was that regularity in clock time within the day paralleled the structure of work-days within the week. That is, regularity was a general attribute of work-patterns on different time-scales.

However, very different conclusions were reached when work hours were analysed from the timing of public disorder in Bristol (Harrison 1986, 2000). Harrison's main concern was another Thompsonian theme: crowd behaviour and public order. He reconstructed working patterns as the inverse of the temporal pattern of crowd events. Finding public disorder strongly concentrated at a few specific times attested to a well-established working week and working day, which

Harrison found well defined before the late eighteenth century. This was not caused by factories, though, since little industrialization occurred in Bristol until the 1830s. What Harrison identified was a very strong rhythm to the working week for most craft and industrial occupations in which Saturday was much like a normal working day, and Monday was an integral part of the weekend. This was not the idlers' 'Saint Monday' of Thompson's account, for not working on Monday

> was a fixed arrangement and not merely a by-product of weekend inebriation. These workers worked on Saturdays; their 'weekend' was consequently Sunday and Monday rather than the present-day Saturday and Sunday. In some cases, 'St Monday' was traded off in return for a half-day holiday on Saturday; this represented a move towards the 'modern' weekend, but not a novel regularization. (Harrison, 1986: 140)

This dispute reveals intractable questions about time, work and life in Bristol, even in the late eighteenth century. The use of such indirect evidence underlines the scarcity of direct evidence on working hours in Bristol. Where this evidence survives, it is sometimes clearly atypical—the very regular, standardized hours of bureaucrats and clerical workers at the late eighteenth-century Custom House were unlikely to resemble those of most Bristolians, whose work was less involved with bureaucracy than the fluctuating rhythms of trades and tides.[4]

Ultimately, though, the issue of clock time is more tangential to this debate than it once appeared, or seems desirable now. Because the work–time debate has focused on the day rather than the hour, and on the regularity or irregularity of work, rather than its volume, intensity, or use of clock times, much of the argument has engaged less, and less directly, with clock time than did Thompson's original article.

What is clear is that most people in England were in long and stable working hours:

> between 1750 and 1850 almost all employed people, particularly in the towns, were to be found at work between the hours of 6 a.m. and 6 p.m., Tuesday to Saturday. It is more useful to see individual circumstances as deviating from, but almost invariably adding to, these hours than to emphasize the diversity of working hours between different occupational groups. The crucial point is that there were basic hours shared by a vast majority of the labour force; consequently there was a recognizable working day and working week. (Harrison, 1986: 140)

Work in many sectors of the Bristol economy started at 6 a.m., perhaps give or take 30 or 60 minutes at different times of year (Barry, 1985: 131–2). The midday break in most trades lasted an hour, but with considerable variations in start-time between 11 a.m. and 1 p.m. End of work times, like start-times, varied seasonally, with work tending to go on later in the summer. But there is little clear evidence on which to echo or reject for Bristol the sorts of mid-eighteenth-century changes in London working hours that Voth identifies from Old Bailey depositions (Voth, 1998).

Table 4.1. Regular timetabled coach departures from Bristol, 1775

London	out	4 p.m.	daily bar Saturday
	in	about noon	daily bar Monday
Bath	out	7 a.m.	daily
	in	9 to 10 p.m.	daily
Exeter/West	out	9 to 10 a.m.	daily
	in	5 to 7 p.m.	daily
Birmingham/North	out	7 p.m.	daily
	in	7 to 9 a.m.	daily
Portsmouth/South	out	7 a.m.	daily
	in	9 to 11 p.m.	daily
Oxford etc.	out	8 a.m.	daily
	in	6 to 7 p.m.	daily
South Wales	out	about noon	daily
	in	about noon	daily

Source: J. Sketchley, *Bristol Directory, 1775*.

If Bristol's records are relatively reticent about mid-eighteenth-century working hours, documentation of times in *transport and communications* is much fuller. A particularly clear example is provided by the large network of formally timetabled coach and wagon services that linked Bristol to its hinterland and to other urban centres (Matthews, 1794: 93–4). Post coaches departed daily to London, Bath, Birmingham, Exeter, Portsmouth, Weymouth, and South Wales, at specific advertised times (Table 4.1).

By late century, at least six operators ran daily services to some or all of these towns, mainly from inns in fairly central locations. Departure times were advertised and seem to have been closely observed for both post- and passenger-coaches, unsurprisingly given that speed, punctuality, and comfort were key competitive criteria. Freight wagons provided less-specific cues, since their timing was comparatively imprecise—while wagon services formed at least as complicated a network as coach services; they were timetabled according to the day, rather than the hour, of their departure (and arrival). While there may have been customary times of departure, known to those observing their passage, times were not prominent in their publicity.

Turning from movement of people to information, the postal system provided another source of intra-urban time-structure. Bristol's postal service was organized from the principal post office near the Exchange through seven out-offices, open from 7 a.m. to 9 p.m. as nodes in a system making deliveries to all parts of the city and suburbs thrice daily. Postmen set out from offices at 8.30 a.m., 12 noon, and 5.30 p.m., both to make local deliveries, and to take mail for other offices to the central post office (Matthews 1794: 89–90). The post was also temporally structured by the coach timetable, since letters for carriage outside Bristol had to reach the post office at least half an hour before the relevant mail coach departed. Thus customers required practical knowledge of both the time, and the system's temporal structure, to use it effectively.

Another striking instance of time awareness comes from the hiring system for hackney cabs. These light, two-wheeled, horse-drawn carriages and their drivers could be hired, at least by wealthier passengers, with hire rates calculated according to *time*, not distance. Late eighteenth-century passengers paid one shilling for the first 45 minutes, sixpence for the next 15 minutes, and sixpence for each subsequent 20 minutes (Matthews, 1794: 91–2). It would be interesting to know if similar time-based hiring occurred elsewhere in the city, but nothing in the contemporary press suggests that these hire-terms puzzled potential customers.

Besides work, and communications, clock times were also used to structure everyday *leisure and recreations*, which often involved considerable timing and coordination for elite and plebeian activities alike. Entrepreneurs promoting commercial leisure were quick to use clock times to organize and advertise events; which are conspicuous in early newspapers.[5] Here, too, there were knock-on effects as the scheduling of leisure events affected people's organization of everyday activities.[6] Available evidence is dominated by material generated by organized leisure, so we must emphasize how this skews what can be gleaned. Extant materials chronically understate the scale of any leisure that was not specifically organized, especially that everyday sociality that involved people dovetailing their presence in undocumented and informal—notably ad hoc—ways. Nonetheless, it is clear that the bulk of leisure, even for the wealthy, was informally timed, if timed at all. This included much visiting, social drinking, parading, hanging around meeting places such as College Green, and the like.

As Jonathan Barry shows, timings of mid-eighteenth-century recreations were influenced by four main factors: the nature of the recreation, working times, the availability of public lighting, and the lay-out of the city.[7] Recreations varied in their requirements (space, light, props, formal organization); in their spontaneity; in their duration (for example, long plays tended to start earlier than shorter performances) and in their prospective clientele (for example, lectures aimed specifically at non-working women frequently took place in the morning, those aimed at men in the evening).

Weekday leisure activities were concentrated around lunchtimes and in the evenings. Lunchtime events generally lasted an hour, but varied among 11.30–12.30; 12.00–13.00; 13.00–14.00, or other times between 11 a.m. and 2 p.m. Sunday events were widely distributed through the day, but avoided clashing with morning services or evensong. Few events were arranged for Saturday afternoons, the prime time for Bristol's markets. Popular holiday events like sword- or cock-fights often started at 10 or 11 a.m., but on half-day holidays they infrequently started before 2 p.m. Activities occurring outside these times were generally a statement that those involved didn't need to work. Corporate processions and society meetings usually began mid-morning, followed by a grand dinner, often at 2 p.m.

In the evenings, most leisure activities—whether for wealthy or poor—finished relatively early, going on longest during summer's long, light evenings. Many evening leisure activities were markedly seasonal and so was their timing. Most evening activities began between 4 p.m. and 7 p.m., and auctions and shows generally closed between 8 p.m. and 9 p.m. Clubs and societies usually ended at 9 p.m. in winter or 10 p.m. in summer, which remained the official closing hours for public houses. Theatres generally opened at 6:30 or 7 p.m., slightly later than in the 1740s. Even the fashionable commercial gardens like Vauxhall Gardens near the Hotwells closed at 10 p.m. during the summer and 9 p.m. in September.

From the mid-eighteenth century, evening leisure extended with improvements in artificial lighting. More efficient oil lamps were replacing lanterns as street lights, and the Corporation extended the hours of street-lighting to run from sunset to sunrise in winter, in 1749, and all year round from 1766. This move expanded the possibilities for work, movement, and for elite leisure—where the impact was most evident. High costs made lighting and extended late night leisure prerogatives of the leisured and/or extravagant—the social activities than ran latest were fashionable balls, concerts, and waxwork shows.

The commercial leisure sector grew in size, importance, and complexity. New recreations were introduced and older recreations became more commercialized, especially at public houses whose entrepreneurial licensees used recreational events to attract custom and sales. Promoters of commercial leisure from theatrical entrepreneurs to publicans were major users of clock times in scheduling, timing, and advertising their events, taking advantage of both traditional holidays and new leisure facilities and practices. Clock times that were increasingly used as an organizational tool more frequently entered everyday sociability and self-organization in their own right.

Time was a pervasive dimension of commercializing recreations, and the restructuring of social and gender differentiations. In accessing leisure activities those with relatively fixed work-hours lost out relative to those with lighter workloads or more flexible hours. For example, such shifts and differences are evident in autobiographical memoirs by an accountant and a cabinet-maker, who both note how a shift of leisure venues from the city centre, alehouses apart, constrained women more than men.

4.3 KNOWING THE TIME IN BRISTOL, 1750–1800

4.3.1 Ownership of Domestic Clocks and Watches

The possession of clocks and watches is central to horological history, and Thompson famously argued that clock- and watch-ownership became much more common 'at the exact moment when the industrial revolution demanded a greater synchronization of labour' (Thompson, 1967: 69). Thompson argued from evidence for clock and watch production, but we look at the consumption and ownership patterns of private timepieces from documentary sources, and from surviving clocks as artefacts, before turning to supply-side evidence.

The closest approach to a systematic survey of household possessions is found in probate inventories. As their name suggests, these are lists of a person's movable goods (that is, excluding real estate and perishables) at their death, drawn up under oath by executors to facilitate the disposal of the estate. Inventories survive in large numbers for early modern England between the late sixteenth century and about 1730, although for a longer date range in a few areas. Unfortunately, they drastically under-represent women, except widows (since married women's property would normally pass to a surviving husband without going through probate), and the poor (whose estates fell below the value threshold to be liable for probate). Although numerous inventories survive for women, and for poor men, they cannot be assumed to broadly represent either group. With inventories skewed towards relatively wealthy households and property owners, ownership rates calculated from them will exaggerate ownership among the population at large. Inventories may, though, understate the ownership of possessions that people gifted to others before their deaths. There is anecdotal evidence that clocks and watches had particular status as heirlooms, a sentiment manifest as a personal gift made face-to-face from the deathbed, in which case they would not appear in the probate inventory.

Despite these faults, however, probate inventories survive in sufficient numbers for Bristol, across a sufficiently broad social spectrum, to provide a picture of timepiece ownership through city levels of society. Though relatively few in number, the poorest documented households were unambiguously poor, living in a single room, with basic apparel, household possessions, furniture and bedding, if any. Inventories survive for significant numbers of people below the modest wealth threshold above which inventories were legally required.

Clocks were owned quite widely through the mid-eighteenth-century Bristol population. Taking gross inventoried wealth of £40 or below to define 'very modest worth', it is striking that over 40 per cent of such testators had clocks (and/or watches) by the 1740s. This is obviously not an instantaneous measurement of ownership, since items may have been purchased over several years, and possessions may have been disposed of before inventories were compiled. These results therefore point to a wide range of Bristolian households having acquired clocks well before then.[8]

The social depth of clock ownership was greater than for many other components of new consumption lifestyles. Between 1740 and 1780, clocks were found in more Bristol probate inventories than were mirrors; more often than pictures or prints; more than half as frequently again as silver items; and more than three times as often as china plates, dishes, or cups. Presence/absence measures are simple, but the results are unmistakeable: clocks were comparatively common in Bristolians' probate inventories.[9]

Inventories describe the goods that people possessed: they do not directly show the processes through which they had been acquired, or the circumstances—by purchase, inheritance, or gift. However, sources that can provide direct sales data—clockmakers' shop or dealing books—do not survive, so we cannot analyse the social shape of the market from them. Even where shopbooks survive, though, they exaggerate the social standing of clock-purchasers, because they concern new clocks, whereas the second-hand market was substantial, through specialist dealers, pawnbrokers, or informal trading.[10]

Various sources attest to the humble status of many owners and purchasers of clocks, for example in witness depositions and other court materials. Thus, cases involving stolen watches often involved owners of far from exalted status. Almost stereotypical plebeian owners included labourers, seamen, or servants, who received arrears of wages in a large lump sum, following a season, a voyage, or a contracted period of work.[11] Plebeian autobiographers and diarists were very scarce, but contain some references to clocks and watches, as that by the clerk and account keeper William Dyer in the 1760s.[12]

The descriptions and valuations of timepieces in inventories caution against exaggerating the social selectiveness of ownership. Clocks and watches certainly *could* be prestigious luxury goods, whose fashionable veneered, gold or silver cases were worth substantially more than their working mechanism. Such lavish items are prominent in today's museum collections. At the time, though, much plainer items proliferated. Relatively cheap items accounted for the bulk of the market in both the new and second-hand timepieces. In late eighteenth-century Bristol, most inventory valuations for 'a clock and case' lay between £1 and £3. Higher valuations usually involved more luxurious materials, as in the valuations of £4 10s. and £4 14s. 6d. for clocks with 'Japanned' cases, owned by Stephen Rogers, mariner, in 1764 and Jacob Milsome, victualler, in 1762.[13]

There was no 'standard location' for clocks within inventoried houses. Diverse locations suggest that clocks were used for several purposes. Just over one-third of clocks were recorded in kitchens then, in decreasing frequency, in parlours, dining rooms, and chambers (bedrooms). Some 20 per cent of clocks were in yet other rooms: workshops, retail shops, staircases, or halls. It was unusual but not unknown for households to possess two or more clocks, kept in different rooms. Such cases usually involved victuallers, innkeepers, or brokers, with one clock seemingly oriented to private domestic time (in a parlour or kitchen), with the other(s) in more 'public' areas.

4.3.2 Clockmakers and Clockmaking

Bristol was not of national importance in clock or watch production. London craftsmen dominated English production, especially of watches, by 1750 drawing on a considerable geographical division of labour, in which specialized bulk production of particular parts took place in areas, including Lancashire, for metropolitan-based assembly and finishing of timepieces. Some Bristol makers supplied components such as gear-wheels to makers in and around the city, while others used parts or part-assembled components sent from London (Moon, 1999: 13–17, 43–6).

James Sketchley's first *Bristol Directory* in 1775 recorded twenty-six clock or watchmakers. Six of these men also appear in one or more of four surviving sets of militia ballot lists from the 1770s, which list able-bodied men aged between 18 and 45, plus twenty more men described as clockmaker or watchmaker not listed by Sketchley.[14] Taken together, the directory and militia lists name forty-six clock- and/or watchmakers active during the 1770s, though not all may have been active in the trade simultaneously. These clockmakers were widely, but unevenly, dispersed through the city (Figure 4.1). Between 1750 and 1775, a steady stream of new clock- and watchmakers were trained in Bristol (Table 4.2). In most years three or more appear in

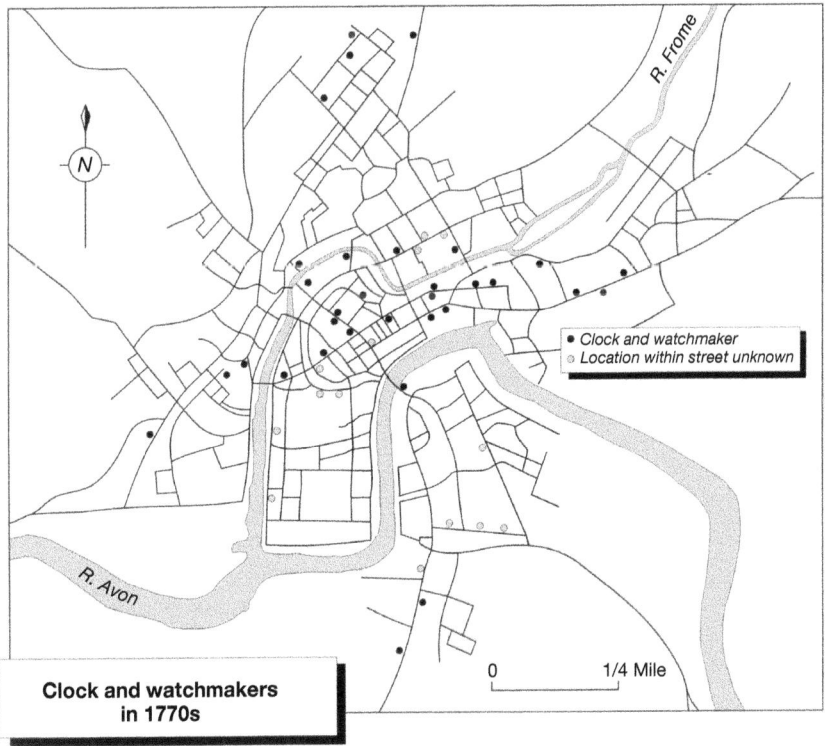

Fig. 4.1. Locations of clockmakers and watchmakers in Bristol in the 1770s.

Table 4.2. Clock- and watchmaking apprenticeships begun in Bristol, 1640–1819

Year	Apprentices	Year	Apprentices
1640–4	2	1730–4	4
1645–9	3	1735–9	4
1650–4	2	1740–4	9
1655–9	1	1745–9	2
1660–4	1	1750–4	9
1665–9	—	1755–9	11
1670–4	3	1760–4	9
1675–9	2	1765–9	18
1680–4	1	1770–4	12
1685–9	—	1775–9	14
1690–4	1	1780–4	18
1695–9	2	1785–9	12
1700–4	4	1790–4	11
1705–9	2	1795–9	4
1710–4	1	1800–4	7
1715–9	2	1805–9	9
1720–4	1	1810–14	11
1725–9	1	1815–19	6

Note: Total number of new apprenticeships registered in each five-year period.

freedom and apprenticeship records. Many later appear in directories and militia lists, though some moved back home or elsewhere, rather than working in Bristol.

The broad consistency of the figures points to 'clockmaker' being a recognized occupational designation, likely to be meaningful for men ascribed it in militia and other lists, and not therefore an arbitrary and unreliable ascription.

Important as apprentices' training and freemen's privileges were, an over-reliance on freedom and apprenticeship papers would paint too formal a picture of clockmaking as a trade. Learning the trade could be a much less structured experience, as it was for Thomas Page, around 1750. Page's biography was recorded on 24 November 1755, when the parish authorities in Axbridge, Somerset, examined his right to reside there.[15] Page deposed that he had been born in Chew Stoke (7–9 miles south of Bristol). When seventeen, he became a journeyman to Thomas Neighbours, a clockmaker, in Tower Lane Bristol, where 'he wrought by the week almost two years at several times, but never above three quarters of a year at one time'. Then, he 'lived in Axbridge for fifteen months, married to Anne Horler with no child'. Page's case was that, not having resided in Bristol continuously for a year, he lacked the right to settle there, but was entitled to a settlement in Axbridge. Given the complex circumstances, however, on 15 June 1758 Page deposed again, describing his life and places of residence in greater detail:

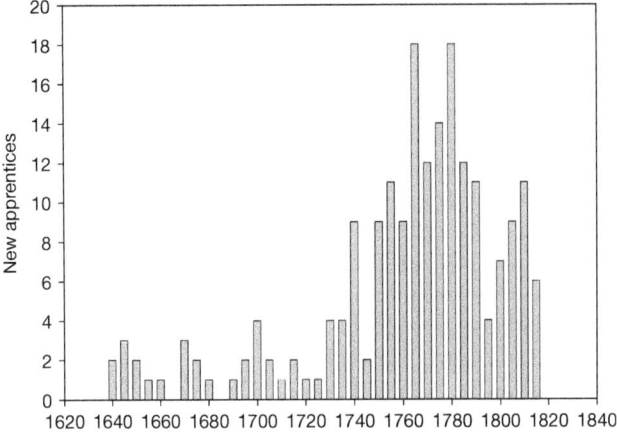

Fig. 4.2. Clock- and watchmaking apprenticeships begun in Bristol, 1640–1819.

he . . . lived with his parents to the age of sixteen years when he agreed with Joseph Nabors of Tower Lane in Bristol, clockmaker, to serve him as a journeyman in their trade but for no certain limited time, & was to have half of the wages which he earned by his work with him, and was to find himself with diet and lodging at this own expense during such his continuance in the said service.

[H]e lived with John Nabors for about a quarter of a year being paid his wages weekly, then returned to Chew Stoke . . . for half a year, then to Bristol and another agreement with John Nabors, now to receive full wages . . . and continued . . . three-quarters of a year, who duly paid him his wages sometimes weekly and otherwise. And afterwards removed to Chew Stoke and lived there with his father for a short time, and then came to Axbridge and rented a shop there at a yearly rent of 20s., living there about a quarter of a year, afterwards rented another shop of Mrs Martha Taylor in Axbridge . . . rent 20s., and was to pay her sixpence a week for his lodging and continued there about three-quarters of a year, and sometime afterward married Anne Horler, the daughter of Elizabeth Horler of Banwell . . . and lived in the house of Thos. Corner of Axbridge for 1s. a week where he and his wife lived for about three-quarters of a year, afterwards hired a room of one Richard Lamapell there at ten pence by the week and stayed there about a quarter of a year and then went to sea and remained there about nine months and then returned to Bristol and served . . . John Nabors in like quality as before for about 9 weeks.'[16]

Page narrates a life-course much less tidy than the standard seven-year apprenticeship. Moving to Bristol aged sixteen or seventeen, Page was clearly a trainee working for Nabors, without the security of formal apprenticeship, though this flexibility may have suited Page: he bargained for his wage, and lived outside rather than within his master's household.

Such full documentation is rare, but if Page's experience was unexceptional, Bristol's pool of clockmaking expertise was larger than will appear from lists of freemen and apprentices, and makers' names on surviving clocks. Overall, though, it is relatively formal records that dominate surviving documents, and from which the history of clockmaking in Bristol has largely to be written.

As artefacts, the timepieces themselves provide considerable information about makers. The numbers of known makers can be significantly augmented by inscriptions on clocks themselves. Artefactual evidence is less haphazard than might be thought because legislation of 1698 required that 'no person shall make or cause to be made any clock or watch without engraving their own name or place on every clock or be fined £20'.[17] This legislation formalized existing practices, which intended to advertise makers, to inhibit forgeries, and to facilitate the identification and recovery of stolen items. Sometimes, newspaper advertisements describing stolen property provide the main evidence for makers (Table 4.3).

The named maker might, especially at the top end of the market, be the jeweller or silversmith who made the case and dial using a bought-in mechanism, rather than the actual maker, but such instances account for a small proportion of surviving Bristol clocks. Almost invariably, though, it can be shown, using rate-books and militia ballot lists, that the men named were resident in the city.

Horologists' compilations of surviving timepieces, and extant descriptions of timepieces now lost, built up over many decades, had long identified numerous Bristol clockmakers active between 1750 and 1799. Systematically extending the search to draw on various documentary sources, A.J. Moore (1999) raised the number to some 372, with some subsequent additions.[18] Clocks or watches by 171 Bristol makers have known work surviving. Some of the 171 are represented by a single surviving clock, others by several dozens; 168 of these men also appear in documentary sources, but three are known only through signed clocks or watches.

Table 4.3. An advertisement to recover a stolen watch

A ROBBERY

Last Saturday between 11 and 12 at night, Samuel Selick was stopped in the Rope Walk and robbed of a SILVER WATCH by a middle siz'd man wearing a dark cut brown Wig on short hair, is supposed to be a seafaring man as he had on a dark coulour'd Waistcoat or Jacket.

The Watch is of a middle Size, has an Enamell'd Dial Plate a silver Cock a blue Figure Piece and opens with a common Nick, makers name George Baunem Baristol, engrav'd on whole side and pierc'd in the form of a fly. Roll no. 1762 on the Edge in very small figures, not easily discovered at first sight, with a steel chain and a Silver Seal, the impression 3 Wheatsheafs and a Flying Dove with an Olive Branch in his mouth. Whoever will give any intelligence of the Watch at the Printing Office in Small Street, so that it may be had again, shall receive ONE GUINEA reward, and if the Offender is thereby brought to Justice, a further reward of TWO GUINEAS shall be paid on his conviction of the said Robbery (over and above the 40s. allow'd by Act of Parliament) by SAMUEL SELLICK.

If offer'd for Pawn or Sale it is requested to be stop't and NOTICE given as above—the Villain is sppos'd to have an Accomplice as a Man was heard to Halloa at a considerable Distance about a Minute or less before the Robbery.

Source: Moore (1999: 59), quoting *Felix Farley's Bristol Journal*, 12 August 1773.

Table 4.4. Known Bristol clockmakers, 1750–1800

	No.	%
Extant clock(s) and maker documented	168	45
Extant clock but maker undocumented	3	1
Documented maker without extant clock	201	54
	(372)	

Sources: Extant clocks from Moore (1999), supplemented; documented makers from author's database (PG).

The limitations of an artefactual basis for identifying clockmakers are made evident by the many other makers who are documented, often repeatedly, but for whom no work survives. Those with extant work account for fewer than half of known Bristol makers (Table 4.4); 201 other makers are documented, in some cases over long periods, without any of their products having survived. Obviously, continuing horological and historical work means that these figures may increase in the future. All in all, over 99 per cent of known Bristol clock-watchmakers are identified in documentary sources, compared with just under half known through surviving work.

The number of makers suggests total annual output of several hundred pieces. We cannot determine how many timepieces were retained by purchasers in Bristol itself, for it is clear that Bristol-made clocks were purchased by many households in the environs of Bristol, and across an extensive hinterland reaching into south Wales. What is impossible to say is how many clocks ending up in these areas were produced in Bristol, rather than in London, on the one hand, or in local workshops, on the other. It is clear, however, that during the eighteenth century there were several recorded clockmakers in local market towns, and in some of the larger villages, within Bristol's regional hinterland. In none of these places, however, were there more than one or two craftsmen operating at any one time.

4.3.3 'Bristol-Style' Clocks: Distinctiveness and Regional Distinctiveness

Bristol makers shared in national trends such as the rapid uptake of the pendulum and anchor escapement, and the appearance of minute and seconds hands. So did makers in the surrounding area, so seconds-indication was no big-city clockmakers' monopoly, even if the gear-work had been obtained as ready-made, 'standard' components, from Bristol or London. Within this sharing in national design trajectories, horologists have identified the mid-century emergence of several features distinct to the Bristol area. Distinguishing characteristics of 'Bristol-style' long-case clocks included decorative shaping and carving, especially on the case hood and the main (trunk) door. Again these also characterized a surrounding area extending more than twenty miles, including both villages such as Chew Stoke and Astwick, and local towns like Frome (Robinson, 1981; Moore, 1999).

A second distinctive feature was the prominent inclusion of a dial indicating 'High Water at Bristol Key', usually in the arch above the main face. Tide-indicators were produced around several of England's major ports at the time, so it is unsurprising to see them made in Bristol. They were also being made, and bought, in several places around the region, besides Bristol itself.

4.3.4 Public Clocks and Time-Cues

The domination of the horological literature by a concern with domestic clocks and watches severely marginalizes public 'turret' clocks, and thereby chronically underestimates the temporal competencies of eighteenth-century people and environments. Bristol's public clocks were important sources of clock time for Bristolians, whether or not they owned a timepiece themselves. It was public clocks, located in churches and other major buildings, whose bells, chimes, and dials provided time indications central to the 'feel' of town life. A majority of Bristol's parish churches housed clocks by the mid-eighteenth century, as we discuss in a moment, and several secular public buildings also featured clocks, often as prominent elements of architectural design. Those in late eighteenth century Bristol included the Exchange, the Assembly Rooms, various market halls, the Council House, and Colston's Hospital.

An eighteenth-century Bristolian did not, in other words, need to possess a clock or watch to know the time, since much knowledge of the time came from public clocks, or other public timepieces. The latter were of various types, besides the mechanical clocks on which we concentrate here. There were public sundials on several city churches, on Broad Quay, and on several public and private buildings around the city. Their accuracy should not be exaggerated, especially for comparatively crude 'scratch dials', but the number of clocks and watches in use, and their increasing accuracy, created a demand for more precisely constructed and marked sundials, which could be used to set or check the time shown on clocks and watches.

By the mid-eighteenth century, at least a dozen of Bristol's parish churches contained clocks. The most important was that on Christ Church in the city centre. This had a complex mechanism of chimes ringing at one o'clock, six o'clock and eleven o'clock, and exhibited two 'jacks' (mechanical figures) who struck each quarter hour with hammers on a bell visible from the street. These and other bells were central to time judgement for eighteenth-century Bristolians. Church bells were rung at set hours, particular churches being clearly associated with the bells rung at certain hours. Thus the 9 p.m. winter curfew was rung on a bell at St Nicholas. There was also a 9 o'clock bell at St James, 8 o'clock bells at St Werburgh's and St Thomas', a 7 o'clock bell at St John's, and 5 o'clock bells at St James, St Thomas and St Philip's. The striking of bells was not confined to signalling full hours, but was both more frequent (with several church clocks chiming the quarters) and more intricate (with chimes playing different tunes at specific hours).

These signals spilled out to reach more than a parochial audience, because most of the central parishes were so small that a particular bell was audible over large parts of the city. In this way, direct knowledge of the time of day was communicated audibly, and did not necessarily require that people had direct sight of public clocks (or, if weather conditions permitted, sundials). Similarly, midnight and other times were rung and/or called out by the night bellmen that accompanied the city Watch on their nightly rounds of the city streets.

The extent to which people registered and relied upon these signals is nicely conveyed by a correspondent in 1772 who noted, in the course of deploring the inconsistent chimes of St Philip's and St Mary Redcliffe, compared with St James and St Stephen's, that quarterly bells were important as providing 'a constant monitor of time' to all the inhabitants of the four quarters of the city. (Note the implication here that these four bells provided broad coverage of the city.) Even if some chimes and clocks were defective, then, the city's clocks were providing many more audible time cues, and Bristolians increasingly relied on knowing the time for the everyday routine.

Taking this argument a stage further, to discuss public timepieces solely in terms of bells and clocks with outdoor dials is to overlook the many clocks *inside* public buildings. Those attracting specific note from contemporaries were in churches, the Exchange and the Assembly Rooms. An interior clock in St Mary Redcliffe church is shown in an engraving of *c*.1730 (Smith, 1994), and there were others inside Bristol churches but pictorial evidence is sparse. Nor should we overlook clocks located in inns, taverns, and alehouses. Not all licensed premises had clocks, but licensing hours made time of relevance to proprietors, customers, and the City authorities. Innkeepers' probate inventories suggest that many licensed premises had a clock, usually plain, in their taprooms. Though not specific to Bristol, drawings of alehouse interiors also suggest that clocks were common, as do references to clocks in plays and other literature.

This matters because of the enormous numbers of licensed premises in Bristol. Like other ports, Bristol contained what today seems a massive number of inns, taverns and alehouses relative to its population. Early eighteenth-century Bristol probably had between six and seven hundred licensed premises, estimated by Jonathon Barry to equate to one alehouse, tavern, or inn for every fifty-six inhabitants. By the 1770s directories and militia ballot lists, numbers of licensed premises had fallen somewhat, partly because of tighter regulation, and were maybe more specialized. But even though the latter sources probably omit unlicensed and very small alehouses, they were still very widely distributed.

Important as public timepieces were, they were not the only way in which the time of day could be told from the environment. Such information did not only come from clocks, sundials, or bells. The routines of urban life themselves provided many time cues from which clock time could be inferred and circulated into other urban timed spaces. There were many events whose usual times were widely known, and from which clock time could be inferred with reasonable accuracy, though we should not exaggerate the precision of many routine events. Most obviously, times could be inferred from events or known starting and/or finishing times such as church services, civic processions, market activity, working days, or leisure events.

One important facet of the temporal information in post and coach timetables was its geographical structure (Figure 4.3a). Transport services were not merely functions at points, they were activity along routes, as when the nine o'clock London-bound coaches moved through the streets as they departed. The noise of (more-or-less) regular stagecoaches passing along their routes provided an approximate temporal cue in those localities. This was geographically uneven: termini and routes were both affected by considerations of street congestion and space for storage yards and stabling, and so concentrated in certain areas.

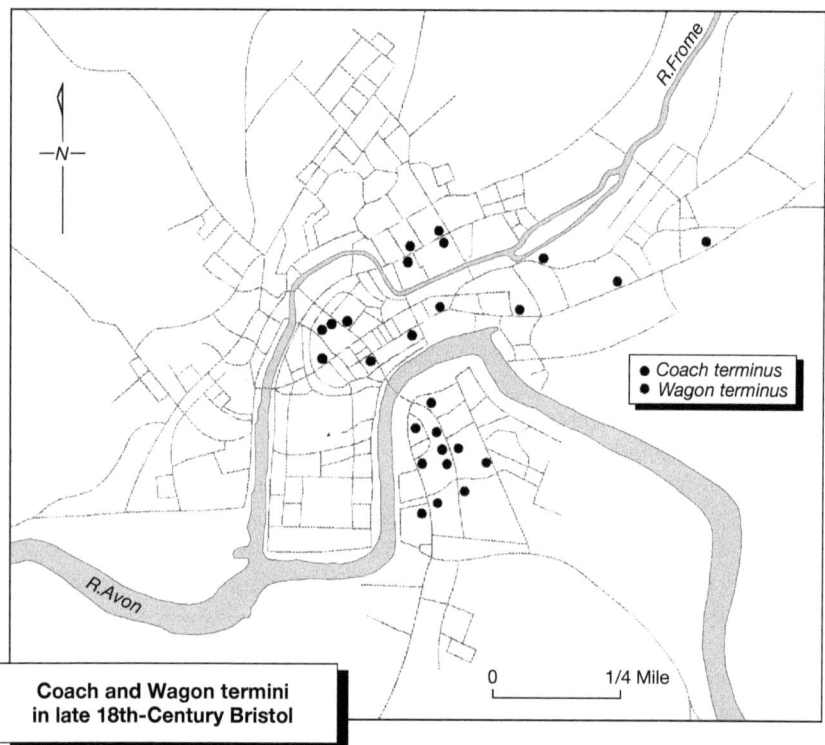

Fig. 4.3. Communications as time-cues in late eighteenth-century Bristol. Fig, 4.3(a) Coach/wagon termini. Fig 4.3(b) post offices.

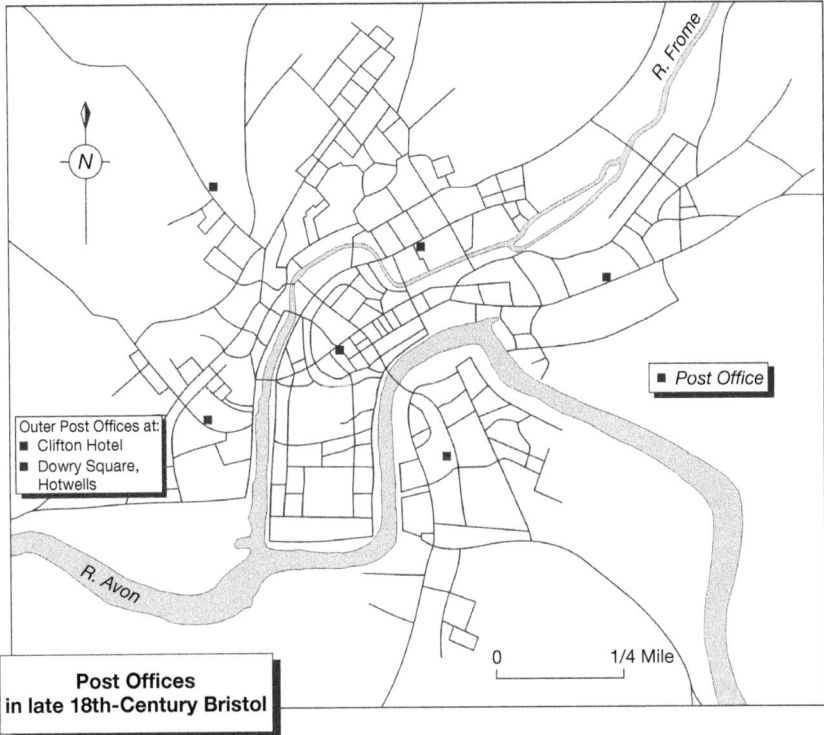

Fig. 4.3. continued

Consequently there was a considerable geography to such distributed time cues, as well as a denser sequence of cues at certain times of day.

This example of an 'indirect time cue' involves temporal cues flowing from particular timed spaces into the urban environment more generally. Once disseminated, whether intentionally or no, they were then available for other uses, whether general and everyday, or highly particular and specialized, for which they were not principally intended. The movements of postmen and post-coaches also provided indirect time-cues, especially through office-departure and delivery times.

Moreover, the posts illustrate how fixed times for the departure and delivery of post provided reference points around which people timed other activities involving the despatch or receipt of goods or information. The timing of posts assumed particular importance for businesses needing to coordinate activities with partners, clients, or others. Once again, both effects were geographically uneven, as both economic activities and social status varied across Bristol. The strikingly uneven distribution of post offices carried with it an uneven distribution of both time-cues and wider temporal integration with the posts (Figure 4.3b).

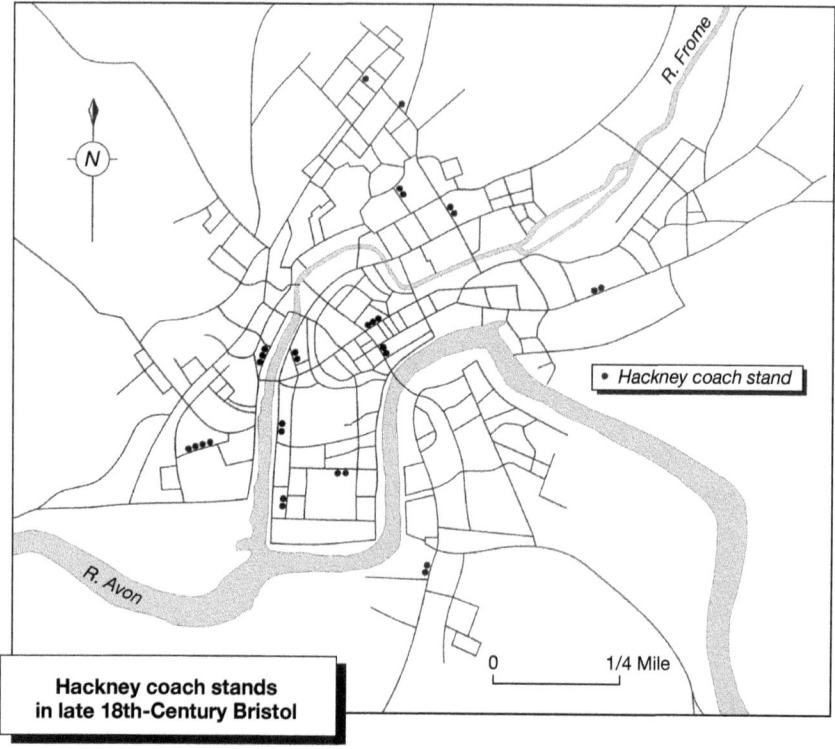

Fig. 4.4. Hackney coach stands in late eighteenth-century Bristol.

The availability of hire-coaches was similarly related to social space. Hackney coaches, given the level of their time-dependent fares, clearly catered for the relatively well-to-do, and this shows in the concentration of the thirty hackney coach stands, which broadly correspond to districts of high property rents (Figure 4.4). The social implications of changes in lighting are more complicated. While the provision of lighting was best in wealthy districts, all classes benefited from better lighting in the main streets.

In the course of the sixteenth and seventeenth centuries the sustained demographic and topographic growth of the city had profound impacts on leisure, quite apart from those that may have been connected with changing patterns of work. Between 1640 and 1750 Bristol's population roughly trebled from about 15,000 to over 40,000, having at least doubled in the century before 1640. The earlier phase of growth was mainly accommodated by growth within the medieval built-up area, but the latter period involved not only much inner city development (not least on the former Castle) but also a steady extension of house construction along roads to nearby villages, and on hills surrounding the town (such as Kingsdown). Unlike the common nineteenth-century pattern, as richer citizens gradually left the inner city parishes, the city centre was *not* converted into cheaper housing, but redeveloped with wider streets and classical architecture for business and public purposes. As a result, both entrepreneurial and professional households and the labouring population became more suburban in residence.

All in all, these changes had significant impacts on open spaces hitherto used for recreation (including some churchyards, quaysides, and College Green) and on former marshland (such as what became Queen's Square). These spatial changes had several consequences for relationships between time and leisure, even if the actual conduct of work and leisure remained unchanged.

First, the loss of central open spaces to popular leisure made it more difficult for working people to dovetail work and certain kinds of leisure (particularly walking and games) in a temporally informal way. Both serious and satirical commentators stressed the loss of amenities to Bristolians as public land was used for private houses and gardens. Popular sports needing extensive facilities for participants or spectators were squeezed out of their seventeenth-century locations. Bowling greens, tennis courts, and cockpits were built over; spectator sports such as boxing, wrestling, and backsword moved out of town. Space-consuming sports appealing to the prosperous, such as bowling and shooting, were no more able to keep their central locations than less-respectable pastimes. In all these cases, informal recreation became more formalized because of its distance from homes and workplaces. Participation or spectatorship now required much greater attention to timing and to temporal coordination.

Second, remodellings of urban space created new arenas for socially exclusive leisure, such as greens and walks for congenial parading, especially in the richer suburbs, from St Augustine's to St James's and the Frome, and also by Redcliffe church. Although we should not mistake aspirations for realities, and ordinary Bristolians—workmen and apprentices—continued to use Queen's Square or College Green for meeting and walking, and the latter was the venue for a prize fight as late as 1748, the trend was absolutely clear. Such events were the exception rather than the rule. Only on official civic festivities did these spaces maintain their character in accommodating huge crowds to watch bonfires and fireworks, to hear speeches, or to particpate or watch processions. In everyday use, leisure activities were simultaneously becoming more closely time-structured, more formal and more targeted on certain sectors of the population. Different people and groups experienced pressure for an explicit temporal organization and planning of leisure activities to differing degrees.

Third, the centrifugal movement of recreation venues out of the city stimulated, and was in turn reinforced by, the increasingly central role of the public house as a venue for organized leisure. As venues for both formal and informal recreations became more peripheral to the city, the market this offered was exploited by a ring of major public houses around the city and in nearby beauty spots. Entrepreneur-publicans took an active role in re-establishing sports leaving the city centre as part of the facilities of public houses. They sought, and attracted, great crowds for sporting spectacles and the various sideshows that broadened their appeal. Similarly, public houses close to riverside, meadow, and downland walks offered refreshments and additional entertainments. Even the most informal recreations moved towards greater temporal structure, as in the case of swimming and bathing. The number of free places to bathe diminished, and commercial developments formalized swimming into a much more regulated time-pattern: the practice became less improvised. And just as the rural public houses came to dominate and formalize outdoor sports, so public houses within the city did the same for indoor sports and activities.

Overall, then, the dispersal of arenas for recreation affected peoples' ability to participate, and increased the social and gender differentiations in recreational activities. Time was one pervasive dimension of that restructuring and differentiation. Those with relatively fixed work hours lost out relative to those whose workload was lighter, or whose hours were more flexible. Gender differences in access to leisure were exacerbated. In at least two autobiographical memoirs, one by an accountant and one by a cabinet maker's apprentice, authors note that the flight of leisure venues from the city centre, apart from alehouses, constrained women more than men.

In conclusion, leisure calendars and timetables at any given time resulted from the intersections of civic, religious, political, and seasonal recreational calendars. The expansion of commercialized leisure, though often discussed as corrosive of old, common calendars of leisure, sought rather to exploit their character as holidays, independent of specific religious or civic meanings. Traditional holidays did not become outmoded alternatives to new, continuous leisure facilities, rather they were themselves opportunities for commercial development, and in that development clock time was increasingly used as an organizational tool. More than that, clock time had in its own right become an important element of customary practice.

Overall, in early Georgian Bristol, then, numerous dimensions of everyday life were riddled with clock times, with mutually reinforcing impulses from work, from transport, from leisure, and from everyday sociality. Dense networks of various kinds of timepieces, and other devices, made 'knowing the time' a routine matter for people in most parts of the city. The making of distinctions in hours and minutes was pervasive, if very largely taken for granted. There are few indications that telling the time was an unfamiliar practice, or that people might be disadvantaged by an inability to do so—that too was taken for granted. It remains to be seen whether these features come about relatively recently: did Bristol see a transformation of temporal practices in the preceding fifty or hundred years?

4.4 EARLY MODERN BRISTOL

4.4.1 Clock Times in Practice

A much narrower range of documentary evidence is available for the sixteenth and seventeenth centuries compared with the eighteenth. In particular, the sorts of sources that shed light on leisure pursuits, such as newspapers, club accounts and minutes, diaries, and correspondence, are either non-existent or very scarce for the earlier period. To some extent this does indicate that the activities themselves lacked the same degree of formal organization as was later the case, but differences in the motives and means for the preservation of these kinds of material are also important factors. One outcome is that work on earlier periods has to rely on a narrower range of evidence, and evidence in which 'official' documents of one form or another are preponderant.

The comparative poverty of harbour and ship-related source materials is disappointing, since we would expect the expansion of Bristol as a port for long-distant and oceanic voyages to have had some significant impacts on time-awareness at least among mariners and those who knew them. But compared to either the Middle Ages or the eighteenth century, early modern documentation provides little information on the topics in which we are interested here.

For landsmen, the start-times of work implied by city regulations are similar to those evident in the eighteenth century, but the overall volume of evidence available is limited. Records of wages paid normally refer to durations of work in days and half-days. Looking at references to fractions of days does give us some idea about the sensitivity to distinctions in times worked in that they made a difference to what the work cost. The main source of the amount of work being carried out by individuals is in the accounts of the City Chamberlain and the churchwardens of around a dozen of Bristol's parish churches.

The Chamberlain's accounts, which are fairly detailed down to about 1640, but thereafter so aggregated as to be useless for this purpose, do not mention, or calculate, using hours of work. But they do distinguish time-worked in quarters of days, from the earliest years of the surviving accounts, $c.1540$. One week after Michaelmas 1542, an unnamed labourer was paid for $4¾$ days work. About two months later, another unnamed labourer was one of 4 men working on a job organized by John Bacock, two men worked for 4 days and one for 3 days, with the labourer for paid for $3¾$ days. And just before Michaelmas 1543, an unnamed man worked for $2¾$ days paving a kitchen.[19]

Such examples can be multiplied during the rest of the sixteenth century, for skilled and unskilled men alike: $1¾$ days to a carpenter just before Christmas 1545; a labourer $¾$ day in early March 1551; John Morys, mason 14d. for $1¼$ days a fortnight later. Payments for times including $¼$ or $¾$ day are never more than a small percentage of all entries recording wages, but they appear in accounts compiled by many different people, and cover men including tilers, plumbers, carpenters, masons, tilers and sawyers, besides their assistants, and labourers, both working with skilled men and working alone. Sixteenth-century payments for fractional days' work are correctly proportional to the full day's wage for each activity in the relevant year's accounts. And similar cases can be found in parish churchwardens' accounts, not just among men employed by the corporation. We do have here a general sense of work-time and day wages as both being seen as bases for calculation of what shorter times were worth.

Administration and regulation in the city routinely used clock times for arranging meetings, opening the city gates, ringing the morning bells and curfew, regulating markets, licensing hours, and the like. We see the use of clock time to organize, outside the context of work or markets, in sixteenth-century regulations for Bristol Grammar School. The school's rules required boys to attend from 6 a.m. to 5 p.m. in summer, from 8 a.m. until 4.30 p.m. in the winter, and from 7 a.m. to 5 p.m. in the autumn and the spring. The formal school hours exemplify the use of clock time to organize everyday activities, and school life and the content of lessons also contributed to the socialization of children into a clock-time world. The Cathedral School and Bristol Grammar School are the best documented of Bristol's several grammar schools, but the city also contained numerous elementary schools, several supported and housed by the individual parishes. If similar use of clock time characterized the numerous parish and private schools in Bristol, the emphasis often placed by historians on the need for employers to teach clock time to employees has been fundamentally misguided. Having learned to operate within clock time for school attendance, it was not a concept that had to be relearned when it came to work later in life.

4.4.2 Ownership of Domestic Clocks and Watches

Beginning once again with domestic clocks and watches, it is clear that the levels of production and ownership of clocks and watches were much lower than in the mid- to late eighteenth century. Even making allowances for the much less comprehensive documentation, there cannot have been anywhere near the same number of makers in the decades around 1700 as there were in the 1770s. Whereas more than three makers a year appears in Bristol's freedom and apprenticeship records between 1750 and 1775, an average of just one a year appears between 1700 and 1750. Between 1640 and 1700, the average is one every two years, and no clock or watchmakers at all are recorded in Bristol's freedom and apprenticeship records before 1640, although some London-trained makers were present alongside makers whose skills had been acquired in the making of locks, guns, or other precision metalwork items.

Less than one-third as many clockmakers are known for the first half of the century as the second. Fifty-four clockmakers have extant work (eight otherwise undocumented), and sixty known makers do not. Once again, to judge from

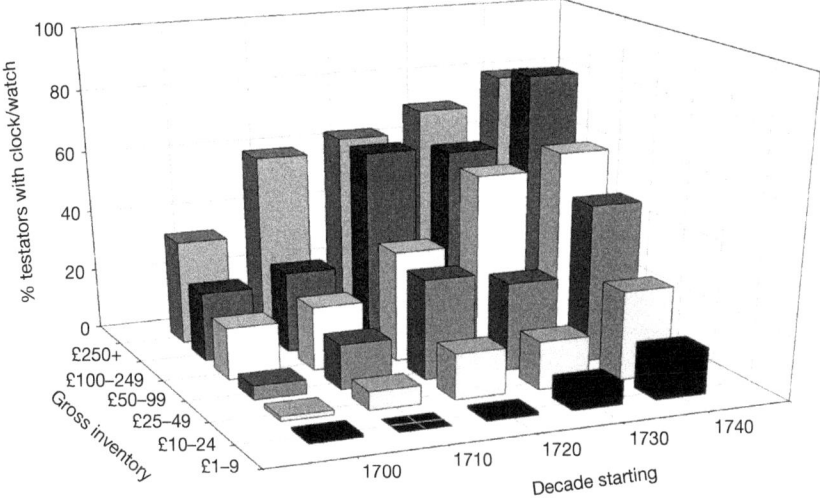

Fig. 4.5. Increases in clock/watch ownership by gross wealth of inventories.

addresses, up-market clockmakers are over-represented among those with extant work. A substantial majority of makers with surviving work are known only from one or a very few pieces, although a handful have many surviving clocks and watches, and clearly produced many hundreds of pieces over their careers.[20] The stable proportions of men known from documents and from clocks reflects two offsetting changes: there are fewer surviving clocks, but also fewer men documented in the absence an earlier equivalent of occupational lists like directories or militia ballot lists. In the period before 1700 (a longer period, obviously, than the other two) there are fewer known makers, and far fewer—less than a quarter—of them have any surviving work.

Measures of clock/watch ownership from probate inventories reinforce this picture. Bristol clearly saw a sustained increase in domestic timepieces during the late seventeenth and early eighteenth centuries. An expansion in the ownership of timepieces occurred at every level of wealth (Figure 4.5). After about 1720, even the poorest group of inventories include some men with clocks or watches.

There would appear also to have been significant demand for other time-keeping devices providing clock time, notably sundials and hourglasses (we use 'hourglasses' here as a generic term covering all the time intervals mentioned for similar instruments: these range from one minute to eight hours). Probate inventories are a less useful source here, since fixed sundials (that is, attached to buildings or plinths) were real estate, and hence outside an inventory's scope. Hourglasses and portable sundials were movable goods, and hence liable to be inventoried, but most cost only a few pence and, as with most very cheap items, appraisers only occasionally valued them separately. It is not possible to say how

Table 4.5. Types of evidence available on Bristol clockmakers before 1800

Period	Documented makers with no extant clock	Documented makers with extant clock(s)	Extant clock but maker undocumented	
pre-1700	76	21	1	(98)
1700–49	60	46	8	(114)
1750–99	201	168	3	(372)
% known from...		pre-1700	1700–49	1750–99
surviving clocks/watches		22.4	47.4	46.0
documentary evidence		99.0	93.0	99.2

Sources: see Table 4.4.

frequently such items are concealed by valuations for 'other small things', or a similar phrase. Portable sundials are, however, occasionally mentioned from the supply side. Thus in 1661, the inventory of a chapman from St Mary Redcliffe included '2 dozen pocket sundials'.

4.4.3 Public Clocks and Time-Cues

If the evidence is conclusive that domestic clocks and watches in Bristol were unusual until the late seventeenth century, we certainly cannot say the same for church clocks. Clocks did become more common on the city's churches during the seventeenth and eighteenth centuries, but these filled in the gaps within quite dense existing provision. In four cases, the evidence points towards churches having had clocks by the early eighteenth century, but probably not for a long period before that time. Otherwise, most central churches had maintained working clocks long before the eighteenth century (Table 4.5).

The cathedral and at least seven Bristol parish churches had clocks by 1640, and these offered wide coverage across the city. Several signalled times with bells, and a growing number had dials with either one or two hands. The functional orientation of visual signalling is indicated by the orientation of visible exterior dials. Typically these faced towards the centre of the city and/or towards the docks, as with the dials on the north side of St Mary Redcliff's tower, or the west side of St Michael's-on-the-Hill. Many church sundials were probably unrecorded since documentation for them is scarce, mainly because their maintenance costs, once installed, were virtually negligible, but both the installation and maintenance of church clocks could be expensive and hence tended to be documented in churchwardens' accounts.[21]

Despite the much narrower range of evidence available, early modern Bristol still looks like an environment in which clock times were a familiar general idea, that was used both to order the city and the public, and was a resource for people to order at least some of their own activities. There were progressively fewer

Table 4.6. Evidence for church clocks in early modern Bristol

Cathedral	Clock in St Augustine's Abbey (compotus accounts) dissolved in 1540 to become the cathedral. The cathedral clock bell marked time for Bristol Cathedral School, as for its pre-Reformation precursor.
All Saints	No clock in accounts (extant 1446–1661)
St Ewen	No clock in accounts (extant 1454–1639)
St Nicholas	Account destroyed in 1940 bombing. Extracts 1520–1727 made c.1906, show clock with quarter-striking throughout. City by-laws show St Nicholas's clock as indicator of trading hours for adjacent market, and on the wharves.
Christ Church	Clock from the earliest accounts (1531) and into the eighteenth century. A series of elaborate chiming clocks with bells struck by armoured human figures.
St John Baptist	Clock from the earliest accounts (1532) past 1700. Chimes added in 1570s. Ringing of daybell/curfew from 1620s through 1650s.
St Mary Redcliff	Clock from the earliest accounts (1532) past 1700. Clock dial from 1590s; chimes from 1690s. Ringing of daybell & curfew.
St Thomas	Accounts from 1544. Clock with exterior dial from 1596.
St Werburgh	No mention of clock in accounts 1548–1700, but payments for ringing of set hours from 1580s through 1670s, implying that clock time was available to the men paid for the specified ringing.
Ss Phillip & Jacob	Accounts from 1564, clock installed in 1670s, dial in the 1690s.
Temple	Accounts from 1582, clock installed in 1650s.

Sources: Churchwardens' accounts, all at Bristol City Record Office unless stated. Cathedral: Beachcroft and Sabin (1938); All Saints: P/AS/ChW/3 (See also Burgess (2000)); St Ewens: P/StE/ChW/1–2; St Nicholas: P/StN/*miscMSS*; Christchurch: P/Xch/1a–1c, P/Xch/HM/4(a); St John the Baptist: P/StJB/ChW/–2; St Mary Redcliffe: P/SMR/ChW/1–6; St Thomas: P/StT/ChW/1–102; St Werburgh: P/StW/ChW/3(a)–3(b); St Philip and Jacob: P/StPandJ/ChW/3(a); Temple: P/Temp/ChW/1–24.

domestic clocks and watches, so the environment was less dense with private time. Even so, the ringing of the main public clocks had been established since the Middle Ages, and provided a generally available sequence of time-cues, which people could use to relate to un-timed moments or periods. Sundials and the broad rhythms of the port and of city life also provided more continuous, if coarser, measures of the day.

4.5 CLOCK TIMES IN MEDIEVAL BRISTOL

Not only did certain elements of clock times in early modern Bristol have medieval antecedents, but certain stimuli to the structuring of everyday life through clock times were stronger in medieval Bristol than they were several centuries later. However, the problems that arise from a narrow evidential base are even more acute than for early modern Bristol. For medieval Bristol we are almost totally dependent on formal documentation produced to regulate specific activities. We can discuss most topics only anecdotally, if at all, unless they were associated with formal city regulation, which is comparatively well documented. We want to state as clearly as possible that we regard the narrow range of discussable topics as a property of the available evidence, *not* as intrinsic to clock time in the fourteenth and fifteenth centuries. Given the evidential limitations, we consider times in medieval Bristol under two overlapping themes, respectively concerned with the formal uses of clock time, and with indications of people's general capacity to operate within a framework of clock time.

The framework of monastic and ecclesiastical institutions was critical for timekeeping in medieval Bristol. There were several religious houses in and immediately outside the city, with foundations particularly numerous in the mid-twelfth century, and continuing through the thirteenth and early fourteenth centuries. Besides these abbeys, friaries, and nunneries, there were several hospitals, almshouses, and other charities in the city, many administered through religious establishments.

There were also the parish churches, although too little documentation survives for a systematic survey of fourteenth- or fifteenth-century church clocks. The only surviving fifteenth-century accounts are from the two tiny central parishes of All Saints and St Ewens, both in close proximity to the two major time-signalling parishes of Christchurch and St Nicholas. There was clearly a striking clock at St Nicholas, since this was the clock routinely cited for the indication of hours that regulated various facets of marketing, as discussed below. And Christchurch's first surviving churchwardens' accounts, from 1531, show a well-established clock and chime there too. Several other parishes have extant accounts from the 1530s and 1540s, and these reveal several other church clocks and associated striking bells (Table 4.5, above), but we cannot say when they were installed, and with what particular objectives.

That the known mechanical clocks were all in religious buildings did not mean that clock times were an exclusively religious preserve, for they permeated everyday life through several channels. Perhaps their farthest-reaching everyday impact was in the control and legitimation of the operation of public markets, especially for food (Penn, 1989; Humphrey, 1997). Ordinary market customers were protected from forestallers, wholesalers, and outsiders largely through the use of clock time to define reserved trading periods, which were sounded by bells at specified hours. Thus Bristol's butchers and fishmongers were not allowed to buy produce from boats or market stalls before 9 o'clock in the morning; cooks (proprietors of small cook-shops selling pork, chicken, and other hot food) were forbidden to buy poultry before the hour of ten 'smyten at St Nicholas' (Penn, 1989: 118), or to buy fresh fish before ordinary consumers; and bakers and brewers were barred from buying grain or malt in the market until after the appropriate market bell had been rung (Penn, 1989: 114, 125).[22]

It was not just the open markets that were temporally regulated, however. Time-limits were part of the wider regulation of exchange. They involved dealers outside the open markets and away from the quaysides, and they regulated some kinds of retailing. Thus brewers were to sell beer only in daylight, but tapsters (that is to say, the retail alemongers) were only to buy beer from the brewers in the latter part of the day, with earlier buying confined to householders (Penn, 1989: 125). Tapsters themselves, along with tavern keepers, were not to have customers in their houses after the ringing of the curfew bell at 9 o'clock in winter and 10 o'clock in summer. Other time-based regulations dealt with different scales and means of trading. For example, fishmongers' sales of salt fish were restricted to the hours between 8 a.m. and 11 a.m., and between 2 p.m. and 4 p.m. in winter (but until 7 p.m. in summer), whereas an ordinance of 1344 stipulated that salt fish could be hawked round the streets 'from the ringing in of the day, prime then beginning, to sunset' less an hour for dinner (Penn, 1989: 121).

Several aspects of urban production were also restricted through prohibitions on activity outside specified hours. Some guild regulations were, for a variety of reasons, designed to limit the times at which work could be done, although most of the guilds' restrictive regulations intervened more directly in who could or could not exercise a particular trade. Thus prohibitions on night-work had long been in place, partly to maintain standards of work (in cases like shoemaking and gloving), and partly to restrict non-guild production, but it is telling that clock time was readily drawn to provide a practical measure through which to convert abstract regulatory priorities into practical everyday measures.

Similarly, clock timing was called on in general civic regulation and maintaining public order (as, essentially, were the market rules). The evening curfew bell, at 9 p.m. in winter, and 10 p.m. in summer, not only marked the hours to which tapsters' activity was restricted (Penn, 1989: 130), but functioned more generally to clear the streets for those, street cleaners and the like, whose work there was facilitated by relative peace and quiet.

Indications from other than normative sources, of a general capacity to operate within a framework of clock time, are scarce, but some are available. In the opening decade of the fifteenth century Thomas Knappe, mayor and shipowner, endowed a substantial chapel on the Back where mass was to be said at 'five a clock in the morning' every day for merchants, mariners, craftsmen, and servants. A public library had been endowed in 1464, situated in an attic of the guild (of Calendars) house above the north aisle of All Saints church. 'The library was to stand open every weekday, for two hours in the morning and two in the afternoon, for anyone who wished to enter for the purpose of study' (Orme, 1989: 214).

Although we lack detailed regulations for schools in medieval Bristol, it is clear from some early fifteenth-century translation exercises that clock time was important in schools' daily routine. Almost unique as an instance of such material from before 1500 is a book believed to be the work of a man called Thomas Schortt, begun in Bristol in the 1420s (Orme, 1989). The sentences for translation in Schortt's book are mostly about everyday life outside school, but three (numbers 70, 97, and 110) mention daily school life and organization. School begins early: 'I come to school in the morntide sorry, but I go to dinner merry and glad' (spelling modernized); there is a pause for breakfast at or about the canonical hour of prime: '[I am] in the school every day fasting until nearing prime of the horlage[23] clercke, to whom it falls to be busy in ringing hours of the day'; and another at midday for dinner: 'I smote flint with the smiting iron and there fell a spark into tinder whereof I tend the light in the lamp, while the son has roasted bacon to his noonmeat' (Orme, 1989: 107–11, spelling modernized). Both formal school hours and the content of lessons thus reinforced the prior socialization of children into clock times.

Several new almshouses were endowed in the late fourteenth century. They lack surviving regulations, or details of daily regimes, but surviving regulations from similar foundations elsewhere at similar or slightly later dates show that clock time was used to organize the daily routine of almspeople in ways resembling those used in schools.

Finally, several more general currents in late medieval Christianity fostered greater temporal awareness, especially through doctrines concerning Purgatory, and the calculative attitude towards the pardon associated with devotional acts, and being prayed for after one's death. Such attitudes were crucial to the considerable attention paid to marking anniversaries, so clearly shown for Bristol by the work of Burgess. Bristol's parish calendars 'became densely packed with temporally-specific occasions, events and practices, which could reach a considerable number of people'. Anniversary celebrations involved not only prayers, but also the distribution of doles, principally comprising bread or money. The distribution of doles at set hours on particular dates meant that calendar- and clock-awareness were not beyond the experience of plebeian households.

It may also be significant that a number of Bristol churches added or rebuilt belfries in the fourteenth or fifteenth centuries. This activity was probably geared both to greater ringing on the anniversaries of benefactors, and to a greater ringing of hour signals (Burgess, 1987). It seems too early for the belfries to have been built for the (distinctively English) practice of change-ringing, which emerged from the early sixteenth century (Sanderson, 1994).

The point of these various examples is to suggest that time constraints were familiar everyday experiences for medieval Bristolians. As traders, as consumers, as employers, and as employees, various activities in medieval Bristol were regulated through clock time. One need not claim that these regulations were universally adhered to, nor that actual practices were as temporally regular as ordinances and by-laws imply, to see important facets of everyday life as structured through clock time. But there is virtually no evidence of plebeian actions or speech that can either support or refute notions that low-status medieval Bristolians could use clock times to narrate or plan events, as there is later on.

For the most part, the presence of clocks is attested to by items of expenditure in fabric or other accounts, but occasionally, the availability of clock time in religious establishments is evident in other ways. On 24 August 1535 Richard Layton, one of the Commissioners appointed by Thomas Cromwell to enquire into the state of religious houses, wrote to his master in London, from St Augustine's Abbey in Bristol (the present-day Bristol Cathedral, a status it gained in 1542). In one letter, Layton listed relics collected from monasteries in Wiltshire and Somerset, and finished:

From Sainte Austin's withouts Bristowe, this Saint Batholomew's day, at iiii of the clocke in the morning, by the speedy hand of your moste assured poor preiste, Richard Layton. (Bettey, 1986: 35–6)

The source of Layton's knowing the time is likely to have been the abbey's own clock, spending on which is recorded in so-called *compotus* rolls (Beachcroft and Sabin, 1938). What needs stressing is that the absence of similar material from earlier dates means that we cannot say when, or among whom, these sorts of usage began. 'Exceptional' groups of monks, merchants, and administrators dominate the evidence, but we lack the evidence to say whether or not they monopolized the practices. As for later on, once clock-time practices and signals were established in a specific setting, they moved into wider public circulation, and could be used in many different contexts.

4.6 CONCLUDING COMMENTS

We want to discuss our conclusions from this chapter under two main points. The first concerns the largely empirical topic of the place of clock time within Bristolian life. The wide range of evidence presented here attests to its prominent presence over a long period of time. From the Middle Ages through the eighteenth century, Bristol was increasingly saturated with information about clock time, based on numerous clocks and watches; on sundials—whose improving precision was largely stimulated by the proliferation of mechanical timepieces; and on a host of direct and indirect time-cues. By the mid-eighteenth century, a large proportion of households owned at least one timepiece, but knowledge of the time was a routine part of daily life for a population far beyond those who themselves owned timepieces. E.P. Thompson, in the passage quoted early in this chapter, presumed that 'the availability of precise clock time' depended on 'who owned... clocks and watches', but this view now looks badly misguided. One did not need a clock or watch of one's own to both work and play by clock time. In other words, those possessing timepieces constitute a core from which information about the time of day spread: they do not define the outer boundaries of a community of knowledge, as has often been implied.

The 'explosion' of private clock and watch ownership from the late seventeenth century was a significant episode in many ways, but it did not constitute a sea change in the everyday availability of information about clock time. From very much earlier the use of clock time was both assumed and promoted by activities in several facets of life: the regulation of production and markets; the conduct of church and civic events; the scheduling of communications; and the organization of recreational and leisure events (often entangled as they were with civic or guilds identifications of specific communities). It was facilitated by the widespread availability of time-signalling associated with public clocks and bells. Work, communications, and leisure all shaped time-use in everyday lives; all three clearly used clock times to do so; and all three effectively acted as potential time-cues for people who understood or were familiar with their temporal patterning.

All in all, the evidence from early modern Bristol strongly suggests that E.P. Thompson was much too cautious in his conclusions so far as temporal infrastructure is concerned. There was a much wider distribution of both private and public clocks than he suggests, and at earlier dates than the late eighteenth century. There was greater access to knowledge about time; and much wider ability to treat with time. Not owning a timepiece meant neither 'lacking information' nor 'lacking ability' to reckon with time. None of these things awaited industrialization or factory work discipline (in any case most rapid in Bristol during the second and third quarters of the nineteenth century rather than the last quarter of the eighteenth).

The second major dimension of our conclusions is methodological. One striking feature of the retrogressive approach adopted here has been a sharp sense of how the volume and nature of documentary source materials has narrowed as the focus of our attention has moved from the eighteenth century to the Middle Ages. At first glance it would be easy to construct the three periods discussed as a sequence in which successively wider dimensions of everyday Bristolian life were affected by, or explicitly structured through, clock time. It may indeed be that this was so, but we have not shown this to be the case. That impression is produced by the very different, and increasingly broad, arrays of sources available for each of the periods. Put crudely, we simply do not have the sources that would allow us to gauge the use of clock time in relation to (say) alehouse-based recreations in the fifteenth century. Such evidence, though scarce, survives for the Bristol of the 1750s, but none survives for the 1450s or the 1550s. That earlier gap might be significant in itself. However, our view is that we cannot safely interpret this absence of evidence as negative evidence.

We would be equally guilty of overinterpretation if we took the picture we have painted for Bristol as presenting experiences typical of England as a whole. Bristol, obviously, was not England as a whole, nor was it 'typical' or 'representative' of the country, however we consider this. Throughout the medieval and early modern centuries, Bristol was one of England's largest provincial towns, a major port in domestic and international trade, a major centre of economic, political, and religious debates, and had pretensions as a regional cultural capital: in a well-known phrase it was 'the Metropolis of the West'. The wider English context therefore requires serious consideration, and we embark on describing this consideration in the chapters that follow.

NOTES

1. Ranging from published 'biographies of the city', such as Latimer (1893–1900), to unpublished autobiographies, such as that of William Dyer, Bristol Central Library, reference section, manuscripts collection, 20095–6.
2. In various manifestiations *Felix Farley's Bristol Journal* was the most prominent of the Bristol newspapers.
3. P. Glennie, 'Times in Georgian Bristol', paper at Georgian Geographies conference, Institute of Contemporary Art, London, September 2000.
4. Sketchley, *Bristol Directory* (1775).
5. See also Table 4.3.
6. For example, Dyer arranged attendance at meetings or matches to dovetail with work and travel is clear from his diary, Bristol Central Library, Manuscript Collection 20095–6.
7. The following paragraphs are based on Barry (1984).
8. The same can be said of households in the surrounding area, contrary to Estabrook's argument that urban/rural consumption patterns were very different in Bristol, and its immediate surroundings. Of all the consumer goods Estabrook discusses, clocks show the smallest town-country differentials in ownership: Estabrook, 1999: 138)
9. We note that some timepieces may have been heirlooms whose owners could not tell the time, but do not envisage this as important on a large scale.
10. According to Earle (1998) this was especially a characteristic of ports.
11. As was the case for many goods on both sides of the early Modern Atlantic, Lemire (1988); Roche (1989) especially chapter 12; Smith (1991); Styles (1994).
12. Bristol Central Library, reference section, Manuscript collection, 20095–6.
13. Bristol Record Office, inventories 1764/CP (Rogers), 1762/CP (Milsome).
14. Surviving militia ballot lists were compiled in 1771, 1779 (?June and November), and 1780. For discussion of the ballot system and nature of the source, see Glennie (1990).
15. Somerset Record Office, D/P/ax 13/3/4.
16. Somerset Record Office, D/P/ax 13/3/16a.
17. 19 William III *c*.28.
18. Moore's (1999) list contains some 372 Bristol clockmakers, and this is supplemented from others recorded for Gloucestershire (Dowler, 1984), The Chew Valley (Moore, Rice and Hucker, 1995), Somerset (Moore, 1998), and Bath (White, 2002).
19. Bristol City Records Office, F/Au/1/3 unpaginated.

20. Among makers who numbered their clocks, the Yorkshireman William Snow (junior) and George Goodall reached at least 1,299 and 639 respectively. At Winster (Cumbria) James Barber (junior) produced at least 1,435 clocks in the second half of the eighteenth century (Loomes 1998: 424–31; 1997: 83–5, 127–9, 242–4).

21. It may well be that an increasing display of time inside churches was a late seventeenth century feature, since clocks with large and expensive interior dials, often on screens or above entrance doors, had been an important component in several of the London churches built under the direction of Sir Christopher Wren between 1666 and the turn of the seventeenth century.

22. Little Red Book, volume 1, p. 39; volume 2, pp. 221–7; Great Red Book I, pp. 134–40. In the last of these, though, neither guild nor market regulations are explicit as to the exact time at which this might occur.

23. *Horlage*: more often rendered horloge, meaning mechanical clock (Dohrn-van Rossum, 1992: 46), the etymological root of horology. Prime was the first of the canonical hours of the medieval monastic/church day, probably around 9 a.m. at this time.

5
The Provision of Clock Time in Pre-Modern England

5.1 INTRODUCTION

This chapter develops one of the major themes from the previous chapter, namely the nature and availability of clock time prior to the late eighteenth century, the point at which E.P. Thompson argued that the everyday use of clock time was becoming widespread. We will return to other themes raised in Chapter 4 in subsequent chapters. For the moment, we take as read the plurality of sources of temporal structure, that is the many different purposes for which social activities were organized using clock time (discussed in Chapter 6), and the wide range of uses to which clock time was put (among the subjects covered in Chapter 7).

We concentrate here, then, on the empirical evidence relating to the availability of clock time information derived from clocks, public and private, and other timepieces. Already, in Chapter 3, we have explained the publicness of many early clocks and of time-signalling through bells and chimes rather than dials.

That 'public' is important, for much of this chapter concerns clocks that were public in their main purpose, and their location. An emphasis on the publicness of clock time is particularly necessary, because of deep-rooted—but inappropriate—assumptions made by many writers since the nineteenth century about the indication and recognition of clock time. Much of the historiography of timekeeping centres on private time, in other words on domestic clocks and personal watches. This is indeed an important topic, and we discuss domestic clocks and watches in the third (and last) section of this chapter. However, for several centuries after the first development of mechanical timekeepers, watches and household clocks were *not* the forms in which mechanical timekeeping was usually encountered. Before the later seventeenth century, and for parts of the population a good deal later, the 'typical' clock was a much larger machine, located in a (quasi-)public building such as a parish church, a market hall, or a town hall. While some of these clocks had large dials on which the time was indicated by either one or two hands, bells or chimes provided the main means of signalling the time. As will become clear, clock time in previous centuries was a decidedly different thing from twentieth-century clock time, and we consider historical perceptions of what was necessary to, and required for, such temporal signalling as was deemed usefully precise.

The chapter therefore provides a well-documented overview of the historical availability of both public and private time, leading on to a systematic picture of the density and distribution of public and private timepieces in late medieval and early modern England. We present much new evidence on 'the infrastructure of public time', by which we mean the distribution of public clocks and time-signalling, from the fifteenth century until *c*.1700. As discussed in Chapter 2, this account involves more than just the existence of mechanical clocks as objects, since we conceive of clocks as networks involving not only timekeeping devices themselves (the mechanical clocks and other objects used in keeping track of times of day), but also both signalling devices (clock-bells, chimes, dials, and so forth), and the personnel associated with timekeeping and signalling. Consideration of personnel, in turn, involves not merely 'who did what', but also the definition of tasks, and the creation of social roles through which the devices were brought into routine practices of everyday life.

Only after discussing late medieval and early modern public time do we turn to the early availability of private time, through the household or individual ownership of domestic clocks and watches. This discussion also draws on new research, and examines the possession of timepieces, the character of 'ordinary' early clocks and watches, and where they were used in the home. Thus, the chapter departs from the focus of most horological literature on the physical character of timepieces. Important as that topic is, for us it is subsidiary to how contemporaries used and perceived clock time information, and to the spatial pattern of available temporal information.

5.2 ATTITUDES TOWARDS PROVISION OF PUBLIC TIMEKEEPING AND SIGNALLING

Long-run changes in the public indication of time involved major sensory shifts, as well as technological change. To present-day perceptions, telling the time is a visual activity. We most frequently establish the time of day by looking at a clock or watch with either a dial, where the time is indicated either by the position of hour, minute, and (perhaps) seconds, hands, or a digital display, which may display units down to tenths of a second. Reading the time of day with our eyes we assume this activity is intrinsic to time-telling, but the historical specificity of the practice needs to be emphasized. The dominance of visually signalling and telling the time is a considerably more recent development than are public clocks and time-signalling as such.

Of course, instruments such as sundials were a longstanding visual means of apprehending 'clock' time (at least during daylight hours and provided the sun was shining), so we are not suggesting a complete switch between the senses in time-signalling and reading. But sundials did not readily signal the equal-hour divisions of the day that are fundamental to mechanical timekeepers. The variable lengths of time through the year for which the sun was above the horizon, and the varying angle and length of shadows cast, meant that—even assuming uninterrupted sunlight (in England!)—equal-hours time indication required sundials of highly sophisticated design and construction. Most early medieval time-reckoning, especially that of the church, continued to work with 'unequal hours' defined as the twelfth part of the day (or night). Other than at the equinoxes, daytime and night-time hours were of unequal lengths. The twelve hours between sunrise and sunset were longer than the twelve night hours in the summer, and shorter in the winter. Only when the day and night were each exactly twelve hours long would the length of the daytime hours and the night-time hours be equal, and equivalent to the 'equal hour'.[1] It can certainly be argued that the adoption of equal-hours reckoning (hitherto confined to specialist groups such as astronomers), rather than reckoning in unequal solar hours, was a more significant change in medieval European timekeeping than any specific advances in timekeeping technology (Dohrn-van Rossum, 1996).

Daytime 'hours' that were longer than equal hours in summer, and shorter in winter, were obviously problematic for mechanical timekeeping. This is more appropriately phrased the other way round: mechanical timekeeping was comparatively unhelpful for reckoning in unequal hours whereas, notwithstanding the numerous imperfections of early clocks, mechanical clocks provided immediately applicable information within an equal hours framework. At all events, it is worth stressing just how widespread was the use of equal hours in late medieval England, as in much of western Europe. It is worth stressing, too, that what later became the standard way(s) of signalling the time of day emerged over a long period, rather than as a finished whole.

Various configurations of clocks and time signal mechanisms were to be found in medieval and early modern England. It would be inappropriate to present them as a series of stages, each configuration necessarily following from its predecessor, but it is nonetheless convenient to consider them as a spectrum ranging from simple to complex, and from the exclusively aural towards the visual.

At their most spartan, clocks might not have any associated signal mechanism, though most had at least a rudimentary indirect signalling capacity from a very early date. Many early monastic and church clocks seem to have acted mainly as prompting devices, producing a signal that prompted a watching official to ring a bell. Such a device involved three main elements, namely a regulator, a going train, and a striking train. A regulator was a mechanism able to produce a regular movement. Before pendulums, developed from the mid-seventeenth century, clocks usually used verge and foliot regulators. These devices were powered by the fall of a suspended weight, restrained by gear mechanisms affecting the back-and-forth rotation of a vertical rod carrying a weighted bar. The going train was the system of gears and cogs that converted the oscillation of the regulator into the cumulative movement of a mechanism, keeping a count of oscillations. The striking train apparatus caused given numbers of movements to produce a signal, in early examples through the sounding of an alarm device set off by the release of a restraining device.

Since any such alarm device required a striking train as well as a going train, it was a comparatively simple matter to add a larger scale mechanical linkage to a bell reserved for the striking of the clock. Direct, mechanical linkage of the clock mechanism to a clock bell or to a chime barrel was the most common fifteenth-century configuration of clock and indicator. The use of a dial and pointer was a later development, and the use of multiple pointers, with the introduction of a minute hand, occurred still later.

Many uses of time signalling, before the invention of mechanical clocks, were to provide unambiguous signals for particular groups of people. These signals were often for relatively specific purposes—to open a market; to convene a meeting; to sound a curfew; and so on. The importance attached to specific signals outweighed that attached to continuous timekeeping; the latter was not an end in itself, so much as an unintended consequence. However, in large cities the very many time signals sounding at different points of the day, and distinguished by their different numbers or patterns of strikes on particular bells could become environments of 'acoustic chaos', as Dohrn-Van Rossum characterized fourteenth-century Milan (1996: 130). The crowding of complex bell-striking patterns became unhelpful for people needing to be attentive to one or a few particular signals, and this was the context for the rationalization and simplification of time-signalling practices into a sequence of hour-strikings whose time-indications were used for many different purposes, a process Dohrn-van Rossum (1996) sees as general among large Italian and German towns.

Similar trends are apparent elsewhere in Europe. England was very much part of these trends, far more prominently than was recognized by Dohrn-van Rossum. Rather than his handful of early English clocks, horologists have long recognised more than twenty in late thirteenth- and fourteenth-century England, and other examples have emerged quite recently (Table 5.1).[2]

These figures relate to European cities, monasteries, and royal sites, but it is abundantly clear, at least for England, that clocks and clock times were not restricted to these exalted social centres. As we shall show, clocks were found not only in St Paul's Cathedral and Westminster Abbey, but in London's ordinary parish churches, and in English towns more generally.

Early mechanical clocks became widespread before there was any standard way of counting hours, or signalling them: it took time for 'modern' hour-striking practices to come into being. Initially the now-familiar twice-daily striking, from one to twelve anchored to noon and midnight, was not self-evidently 'the way to count time', and various fourteenth- and fifteenth-century examples can be

Table 5.1. Documented public clocks in medieval European cities

Selected modern states	by 1360	1370	1380	1390	1400
Italy	10	14	22	26	33
France	4	7	34	50	74
Germany	1	7	18	26	32
Low Countries		3	12	21	30
Other mainland Europe	1		22		68
England (updated)	14	20	21	23	25

Sources: Mainland Europe from Dohrn-van Rossum (1996): 125–72; 'other mainland Europe' includes Belgium, Netherlands, Spain, Austria, Switzerland, Czech Republic, Slovakia, Poland, Ukraine, and Croatia. For details of the revised figures for England, and any future additions see the 'late medieval timekeeping' sections of the website <www.ggy.bris.ac/clocks>.

found of different striking cycles. In some, hours were counted from sunrise rather than noon/midnight. Sunrise always occurred at the same time of day, but noon varied—from about four o'clock local time to about eight o'clock. Such systems were eventually discarded in favour of a noon anchored at midday. Hour-numbering provides another instance in which various systems were deployed, before more uniform practices emerged. Besides reckoning one to twelve twice in every twenty-four hours, late fourteenth- and fifteenth-century examples include counting from one to twenty-four, counting from one to six four times, and counting from one to eight three times. Hence such seemingly strange records as a chronicler recording the death of a civic notable 'at the twentieth hour' (Dohrn-van Rossum, 1996: 109). In this case, hours were numbered one to twenty-four from sunset, so the twentieth hour is about 3 p.m.

It is striking, given the very sparse documentary evidence, that some discussion of the relative merits of different systems survives, discussion which recognizes for example, the damage to bells, hammers, and gearing done by the 300 daily strikes in a one-to-twenty-four system, as against the 156 of a one-to-twelve system, or the possible ambiguity of one-to-six and other shorter systems that entailed fewer strikes in total (Dohrn-van Rossum, 1996: 114–15). One later example of a similar calculation in England comes from the churchwardens' account book from New Windsor (Berkshire), where daily totals of strikes are counted as an arithmetic exercise.[3]

While evidence is restricted since most extant sources are elite in character, the relative coarseness of hourly striking attracted some comment. It is interesting to reflect on how early these comments first appear, as do various means of more precise signalling from the fourteenth to early sixteenth centuries, such as striking quarter-hours. The precision of such signalling was limited by clocks' mechanical shortcomings (modern reconstructions of verge-and-foliot clocks find it difficult to achieve accuracy better than several minutes per day), but early attempts to indicate quarter-hours attest to the latent demand for more frequent signalling. More detailed signals were not only more frequent, but more differentiated, especially where particular times could be accentuated by more sustained ringing (especially the tolling of daybell and curfew bells), or the playing of distinctive tunes on chimes (commonly every three hours), as we saw for Bristol in the previous chapter, or through more elaborate chiming of the half and quarter hours.

More refined signalling was not necessarily aural though, but could also, or instead, be visual through the use of dials. Dials were not, to reiterate, regarded as integral parts of early mechanical clocks: they took several centuries to become part of the definition of 'what a clock is'. Even in the second-half of the nineteenth century some parish churches still commissioned clocks with chimes, but without a dial (for example, in Bedfordshire, Pickford, 1991). Some church clocks retained single-hand dials into the late twentieth century, as at Tytherington, Gloucestershire (Beeson, 1978: 42–4). Even where clock dials incorporated only one hand, subdivision of the dial's hour ring enabled the distinction of at least quarters of hours, and often 'half-quarters'. Note here that one-handed clocks tended to have three marks and hence four subdivisions between numerals, whereas minute-hand clocks replace this with four marks and five subdivisions. The timing and frequency of different signalling devices, and their distribution, are discussed later in the chapter. For the present, we want to stress that more refined signalling systems were not simply dependent on changes in clock mechanisms, such as the more accurate timekeeping enabled after the invention of the isochronic pendulum.

Over time, then, both urban and rural church clocks were more frequently equipped with chimes, in addition to their clock-bell mechanisms, and with dials, even in the countryside, which often lagged one to two hundred years behind most towns. By the early eighteenth century, a dial (even if it possessed only one hand) had become integral to how people thought of 'a clock'. Time-measurement, in other words, was being generally understood as something visual as well as acoustic. And following quite closely behind the proliferation of dials was a sustained increase in the proportion of parishes—again, not all of them 'urban'—that installed quarter-striking clocks, either to replace or to augment their existing hour-striking clocks.

A further impact on visual timekeeping came from the impact of clock time on sundials. Of course, sundials were of considerable importance for clock time, since the observation of local noon provided a (potentially) daily cue for clock-setting. This apparently unproblematic task became very much more complicated from the late 1650s, with the realization that the interval between noon on successive days varied substantially through the year, so that the intervals between two successive noons fluctuated through the year by a quarter of an hour over or under twenty-four hours. Thus an accurate clock by definition required frequent altering to maintain correspondence to the inconsistent apparent movements of the sun. Clockmakers compiled and circulated 'equation of time' tables, showing how far ahead of, or behind, local noon a clock should be, at different times of year. Not only that, but precise sundial construction became much more difficult, as engravers sought to produce dials that could indicate 'mean' or true clock time, in addition to, or instead of, solar time. Only a sundial of considerable accuracy allowed the 'correction' of a clock at other times of day, so the increasing numbers of mechanical clocks, and the multiplying uses of clock time, were powerful stimuli both to the construction and maintenance of more sundials, and to the provision of better sundials.

Attempts to provide signals more frequently than hourly, and the more sophisticated use of sundials, both illustrate ways in which individuals or groups of people expended considerable effort in maintaining clock timekeeping. This effort might involve complicated systems of meeting the financial and logistical demands of keeping clocks working and corrected to apparent solar time. In both cases this could involve them or their parishes in significant expenditure, and in turn this sometimes stimulated exceptional arrangements for raising the necessary money.

The key point here is that time-signalling by the clocks of medieval and early modern England was time more public than private; and more aural than visual. Clocks were more about making a noise than showing a spectacle. When dials and the visual indication of time became more commonplace, aural time-indication through bells and chimes still retained considerable importance, and telling the time involved hybrids of aural and visual skills. In other words, it is misleading to suppose a once-and-for-all transformation between 'aural' and 'visual' time.

Differentiating aural and visual indication of the hour has important implications when clock time is considered as a complex of practices, and we therefore consider the everyday skills that people required to 'tell the time'. Telling the time was not always the comparatively complex task we nowadays think of, interpreting the position of two (or three) hands on a clock face. Fewer and less abstract skills were required by a bell-based system as opposed to a dial-based system. Signalling hours (and possibly their subdivision into quarters) arguably required only that listeners be able to count the strikes, to determine their position in a sequence of hours, so long as they comprehended counting one to twelve from noon or midnight. However, interpreting a clock face, especially one with both hour and minute hands, was intrinsically more complicated. Its abstract element was greater, the minute hand indicating numbers other than those to which it points. Of course, visual time indication possessed certain advantages, particularly in its clearer depiction of the continuous passage of time. A dial, whether with one hand or two, offered an ongoing observable presence in a way that periodically striking bells did not. But this advantage came at the cost of making time-telling a more difficult task.

Thus 'telling the time' comprised differing repertoires of skills and activities in different places, at different times, and with respect to different timekeeping devices. A moral we draw here is that telling the time in the late twentieth century has become a more complicated procedure than it was several centuries ago, and that the numerical and other skills required then were fewer than we might assume. Accordingly, presumptions that few English people could tell the time in, say, 1600 are likely to be suspect if they are based on modern ideas of what telling the time entails.

Concern over reliable temporal provision was occasionally discussed directly, though explicit statements are relatively rare. Our discussion of necessity draws extensively on two indirect indications of concern with times and time-signalling. One is to show the lengths to which people and communities went in order to maintain time-signalling, and the other is the sheer variety of ways in which clock times were used, and on which the routine conduct of everyday life relied. We consider these two themes after briefly examining early explicit assertions as to the importance of public clocks.

Although explicit discussions of clock time are, as we say, rare, they can occasionally be found. The town ordinances for Godalming, Surrey, asserted in 1620 that:

the use of a clock in the said town is very necessary for the inhabitants thereof for the keeping of fit hours for their apprentices, servants and workmen.[4]

In the same year, 1620, John Rashleighe, the mayor of Fowey in Cornwall, sent his son in London a note accompanying the components of the malfunctioning Fowey town clock:

wth the bark of otes I send then our old Towne Clock wth all his nessaryes belongyng. I wish you to sell yt & to bye a new to be sent hether in the bark because *we can not want on long*. . . . also lett yt have and Index [ie dial-hand] fitted . . .[5]   [our italics]

Practical concerns with the local maintenance of clock time and provision of time signals are clearly and strongly implied by the elaborate and sometimes costly actions taken to maintain clocks and signalling apparatus. The efforts directed to the provision of temporal information testify to prevailing, though often unarticulated, attitudes prevailing to clock time. From documentary sources these efforts can be traced both in the 'normal' working of local arrangements (almost always unaccompanied by associated discussion) and when such arrangements were disrupted or threatened, for example by the serious breakdown of a clock, or the death of a clock-keeper. The latter can be particularly revealing, since as so often when priorities were too widely known to require explanation, 'normal' attitudes were most often made explicit when established timekeeping or time-indicating methods broke down. It was just such circumstances that led to the writing of John Rashleighe's letter, just quoted.

Examination of the arrangements implemented for the maintenance or replacement of broken clocks and signalling arrangements clearly brings out the urgency that surrounded such repairs or purchases, and revealingly testifies to the significance that parishioners attached to reliable time signals and information. That many parishioners did attach considerable importance to the public availability of information about clock times is indicated by the many lines of evidence concerning parishes' efforts to keep clocks going. Running through scores of thousands of accounts from hundreds of parishes is a general seriousness about the provision of clocks and clear temporal signals.

First, parishes were willing to undertake periodic large-scale spending for the construction, replacement, or major overhaul of a clock. Although the performance of individual clocks clearly varied greatly, parishes could generally reckon on getting more than a century of service from a well-maintained clock. A central element of maintenance was pre-emptive replacement of wearing parts. Again it is difficult to generalize, but the interval between such refurbishments was frequently around thirty years.

The second, and complementary, dimension of maintenance spending is of routine low-level routine monitoring and maintenance, often taking the form of long-run agreements with clockmakers, scores of which documents survive today. These set out, in some detail, the obligations of clockmakers in return for either an annual fee or a lump sum payment; the period over which the obligation was valid; and the bond that the clockmaker would be liable to forfeit if the specified tasks were not performed.

Third, alternative arrangements were rapidly implemented when clocks did break down and could not immediately be replaced, whether borrowing a clock to use in the interim, or mounting a rota to watch and turn hourglasses (examples in use varied up to eight hours). In each case, and often separately from these measures, men were paid to ring the hours of the day normally struck by the clock (especially when the mechanical fault lay in the striking train rather than the going train).

Fourth, parishes made arrangements to clearly signal particular hours, especially through the extended ringing of day and curfew bells (Figure 5.1). The signalling of the beginning and end points of the working day, of curfew, and of other times related to particular local activities, were a constant focus of attention. They were widespread outside town and market centres, as well as within them.

It was precisely because parishes were prepared to pay for all this that expenditure was recorded and preserved in churchwardens' accounts. Parishioners clearly grasped both what they sought in terms of signalling, and the circumstances through which this was achieved. Among hundreds of illustrations, a handful will suffice to indicate how a need for temporal information was appreciated, and acted upon in a variety of ways. At St Mary's in Reading in 1599, the vestry meeting agreed that

And whereas William Marshall, Clerke, at the Accounts holden the One and Twentithe daie of March 1599, was then allowid iijs. iiijd. towardes the increasinge of his wages, and for the same hee is to Ringe the Eighte of Clocke Bell everie Eveninge both holie daie and workinge daie thorowe the whole yere. And because it is a great paine and troble to the said William to continewe the same, At this Accounte it is Decreed and Orderid, That he shall have iijs. iiijd. more augmentid to his wages to be paied unto him for his wages and Paines takinge in the Ringinge of the Bell, as is aforesaid, And so hee is to have xls. per Annum. (Garry and Garry, 1893: 105)

A more specific measure was taken at Woodbridge, Suffolk, in 1660, when the townsmen were moved to spend 11s. 10d. on

5 yds of Cloath to make [Thomas] Blocke a Coatt to ring ye [8 of the clock] Bell in in Cold Nights.[6]

This was two years after they had bought an eight-hour glass to assist him in ringing, at four o'clock in the morning, noon, and eight o'clock in the evening.

Individually, these instances may be small examples, but their very widespread occurrence is telling.

When available, clock time was put to extremely diverse uses. Many are familiar ends relating to the working day, market activity, public meetings, church services and prayers, alehouse regulation, musters, guild, fraternity or vestry meetings, social drinking, and so on. Particularly striking in the terms in which they are mentioned is that the tone is almost invariably matter of fact. There is a massive and taken-for-granted familiarity with times, hour-bells, and what they could be used for. We shall quote from some unusually explicit statements about public clocks, and people's attitudes toward them, but these statements are not the norm. A second striking feature is that early writers commonly identify the particular clock on which the relevant time was signalled (thus fourteenth-century regulations for Bristol market specified 'the smiting of twelve of the clock at St Nicholas's). We interpret this sort of specification as blending the idea of a general metric with the actuality of a particular signal. This eliminated ambiguity in practice, where clocks did not show the same time, by assigning authority to one of them as 'right' for the given purpose. Many town and market regulations elsewhere similarly attach authority to particular clocks: this was clearly seen as an important matter (Riley, 1872:348; 1874:432–3; Barry, 1984; Dohrn-van Rossum, 1996 *passim*; Humphrey, 2001).

We cannot overstress the degree to which time signals were commonly used resources for the conduct of everyday life. The signalling of particular times may have originally served quite specialized trading or religious functions, but those signals became a more general resource for everyday organization going far beyond their original purpose. The particular hours for which early clocks were installed—church services, market opening, curfew, morning daybell, and so on—became component parts of more thorough ringing of hours. Signals provided for one institution or purpose were used by others, as where clock times become the framework for schedules in schools, almshouses, and hospitals, and time-regulation of common agricultural rights such as grazing and gleaning corn.[7]

Not least, clock times become extensively used by 'ordinary' men, women, and—though the evidence is very sparse—by children, in making arrangements and in narrating their activities. Very little documentation for these humble lives survives, but ability to handle clock times in this way is found both in official sources, such as witnesses' court depositions, and in the sorts of informal correspondence that has occasionally survived purely by chance. Such examples involve many sorts of people and some surprising places, and they will receive fuller discussion in Chapter 7.

5.3 INFRASTRUCTURE: THE NUMBERS AND DISTRIBUTION OF PUBLIC CLOCKS

Then, as now, the distinction between 'private' and 'public' clocks was not entirely clear cut in early modern England. Our concern here is with clocks either maintained or used beyond the particular households or institutions within which they were located. This includes some clocks largely dependent on financial support from private individuals, that provided time-signalling for significant public audiences. Often the situation was part- or 'quasi-public': only sometimes were 'public' clocks financed by the public as a whole, rather than by certain groups of people. For example clocks in parish churches might be funded from general rate income, by a flat-rate levy on householders, or by particular individuals or bequests. We count all these as 'public clocks'.

Though we focus on parish churches these were not, of course, the only or the earliest locations for timepieces that were to some extent 'public' (Figure 5.1). From the late thirteenth century, there were numerous mechanical clocks at monasteries and other religious communities, and clocks were often found in cathedrals.[8] At their dissolution in the late 1530s, monastic and other religious houses were inventoried by King's Commissioners. The resulting inventories, which are lists of unsold items, rather than snapshots at the moment of dissolution, are chiefly concerned with plate, vestments, and such like. Some contain details of clocks, but these had often been sold relatively promptly and, in any case, were usually regarded as items that happened to be present, not as part of the apparatus of Catholicism. So they provide anecdotal evidence, rather than the potential for a systematic survey. For example, Commissioners reported on 20 October 1538 that the goods sold for £5 10s. from St Thomas priory near Stafford included a clock, and that a clock in the church was among goods sold for £6 from a dissolved monastery at Darley in Derbyshire (Walcott, 1871: 211–21). Since it was not standard practice to itemize goods sold, such cases may reflect no more than a conveniently available list of receipts. Where clocks remained to be inventoried, though, some Commissioners felt a need to explain why, such as the clock explicitly described as 'oulde' at Barnwell Priory in Cambridgeshire, or the former Cistercian house at Sawtre, whose remaining possessions included an old clock valued at just 3s. 4d, or the clock at Dale Priory, Derbyshire, said to be worth 6s. (Walcott, 1871: 221–40).

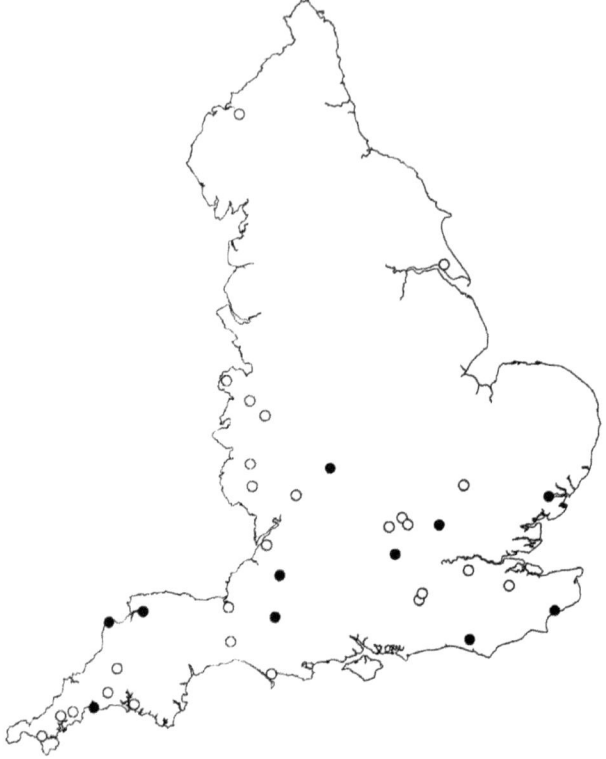

Fig. 5.1. Clocks on civic buildings in early modern England.
Indicative rather than definitive, the map shows the locations of clocks on town halls and market halls identified by Tittler (1991), marked with open circles, and other town clocks, marked with closed circles, from ongoing research (updated regularly at <www.ggy.bris.ac.uk/clocks>).

Not that clocks were confined to religious sites. During the fourteenth century, clocks became familiar features of English royal palaces, especially under Edward III (reigned 1327–77), who provided clocks for various of the palaces and castles among which the royal household moved. Besides causing large striking clocks to be built at Westminster, Windsor, Queenborough, Kings Langley, and Sheen, he met and communicated with the accomplished scientist and clock-designer Richard of Wallingford (Beeson, 1989; Brown, 1959; Mortimer, 2006; North, 2005).

In the fifteenth and sixteenth centuries, increasing numbers of clocks were added to town and market halls, and the like. Robert Tittler's *Architecture and Power* (1991) lists over twenty small towns with town hall clocks before 1640, a list we can more than double (Figure 5.1).[9]

Figure 5.1 is far from comprehensive, but shows very clearly that several of the places for which chance mentions of town or market clocks have survived are at the outer margins of anything we might call the English 'urban system'. Towns such as Chard in Somerset, Fowey in Cornwall, or Hedon in the East Riding of Yorkshire, were all locally important, but had few wider roles. Their possession of clocks testifies to a general concern with marking the time of day. Whether this concern was specifically urban, of course, cannot be investigated from the corporate or anecdotal urban documentary sources in which these examples were found, since only towns—and only some towns at that—produced these sorts of records in the first place. And there were very likely other clocks in quasi-public spaces, as in Bristol.

Overall then, public clocks were found in many places besides the parish churches on which we concentrate here, so this discussion will inevitably understate the overall numbers and distribution of public clocks. But what makes church clocks an apposite focus for us is the relative abundance of documentation for parish spending, and its comparatively systematic character. We may have some confidence in negative evidence. In other words, documentation of parish spending enables relatively secure identification of parish churches *without* clocks, and hence discussion of trends and patterns in the proportion of churches with clocks and various methods of time-signalling.

5.3.1 A Systematic Analysis of Clocks in Parish Churches

That there has been little exploration of parish church clocks, and timekeeping hitherto has not inhibited some bold statements in print, since the J.C. Cox's exuberant proclamation in his pioneering *English Churchwardens* that 'there was hardly a clockless church to be found in either town or country in the fifteenth century' (1913). Cox's books rested on detailed documentation, but only for some thirty parishes, mainly urban. It remains widely cited among horologists notwithstanding its vigorous contradiction by C.F.C. Beeson's *English Church Clocks, 1280–1850*.[10] Beeson emphasized the sparsity of Cox's evidence, surveying surviving church clocks nationally to conclude that Cox's 'point of view is not endorsed statistically' (1971: 25). It may have been Cox's statement that led E.P. Thompson to suggest (1967: 63) that perhaps 5,000 (about half) of English parishes had a clock by 1500, though Thompson cites no source. Thompson did not develop the implications of his suggestion at all—perhaps just as well given his picture of most pre-eighteenth century English people as living in a clock-free environment!

With so many clocks and documents failing to survive to the present day, horological histories of public timekeeping mainly comprise lists of clocks, and their recorded dates. In other words, they compile positive evidence, ranging from the survival of clocks themselves to mentions of clocks in documents generated by urban, market, or other regulatory authorities (Le Goff, 1980, 1988; Dohrn-van Rossum, 1996).

For horologists, the key questions are about positive evidence: of clocks themselves, of records of clocks and their makers. Issues of the treatment of negative evidence were much less important. However, for systematic surveys of the density and geographical patterns of timekeeping negative evidence is crucial. We need to be able to identify places without clocks, and distinguish them from places lacking evidence pointing one way or the other. We would also like evidence from England's hundreds of small, unincorporated towns, and from rural parishes, which between them contained most of the population.

Two key questions need to be separated: the first about the reliability of churchwardens' accounts as evidence for the presence and absence of clocks, and the second about the range of places covered by extant accounts. We shall consider these in turn.

First, then, the reliability of accounts in recording clocks, if present. It was certainly possible for a parish to have a church clock without it being wholly responsible for financing it. Sometimes clocks were provided by benefactors. Occasionally these benefactions included land or money to generate an income from which routine maintenance costs or wages could be met. But, even in these cases, churchwardens' accounts are likely to include some indication there was a clock. The very frequency of winding and maintenance means that it was likely to enter accounts, albeit occasionally, and property benefactions were likely to figure on the income side of wardens' accounts.[11]

For the later Middle Ages, problems are more likely to arise because of the array of fraternities and guilds in many parishes, each with their own responsibilities and accounts. Where all facets of a clock were the specific responsibilities of a particular fraternity, churchwardens' accounts are likely to contain little more than a brief summary of aggregate fraternity spending, and a clock might well be invisible.[12] In the main, though, complete devolution of the clock to a fraternity seems to have been unusual.

We cannot identify a single case in which an early modern clock survives alongside a full run of accounts, but where the former is nowhere evident in the latter. More problematic (and more usual) are cases where only a few accounts survive, especially where these are heavily summarized or abridged. Towards 1700, there is a general tendency for accounts to become much more summary and terse, sometimes to the point where it is increasingly difficult to be confident that a gift of a clock, and its subsequent upkeep, could not pass unnoticed. Even here the implications are clear: a systematic analysis of churchwardens' accounts may, because of a combination of local arrangements and patchy documentation, understate the presence of clocks.

Second, the range of places with accounts. It is straightaway apparent that the range of parishes whose accounts survive changes over time. The available parishes are always unrepresentatively urban in character, but this lessens over time. The

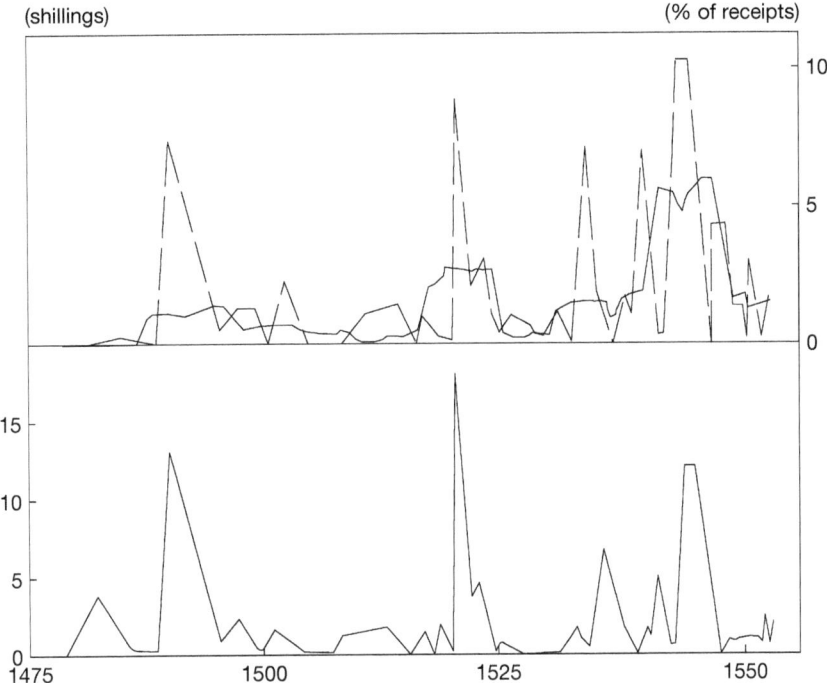

Fig. 5.2. Annual spending on the church clock at (Bishops) Stortford, Hertfordshire, 1470–1560.

range of documented rural parishes likewise increases, as accounts begin to survive from more of the smaller parishes, whose internal organization had little of the intricacy required to manage large parishes containing several settlements. For the sixteenth century there are some very small rural parishes represented, but they become really numerous only in the late seventeenth century.

The analysis and maps that follow come from a systematic, and so far as possible comprehensive, survey of the churchwardens' accounts that survive for English parishes between c.1400 and 1700 (Figure 5.3), a task much facilitated by Ronald Hutton's listing of accounts he located by the early 1990s (Hutton, 1994: 263–93). Including our further discoveries, churchwardens' accounts, detailing churchwardens' administration of parochial income and expenditure, survive from just over 1,000 parishes before 1700 (Figure 5.4).[13]

Their interpretation can be complicated, especially in cases in which parishes have only a few surviving accounts from odd years, but they are, in the main, relatively reliable indicators of whether churches contained clocks or not, and of what operating the clock entailed in terms of resources and effort.[14] With some exceptions (where accounts are incomplete or lack detail, or where special

Fig. 5.3. English parishes with surviving churchwardens' accounts, 1400–1700. Churchwardens' accounts survive for more than 1,000 English parishes, 1500–1700. Even so, over 85 per cent of English parishes are without extant accounts from before 1700, and the pattern is regionally uneven.

arrangements provided for clocks from endowments rather than from parish funds), most of the time 'no clock in the accounts' does mean 'no clock in the church'. In practical terms, what this means is that analysis of churchwardens' accounts is likely to underestimate the numbers of churches with clocks, but only where special (and its seems relatively unusual) arrangements were in place.

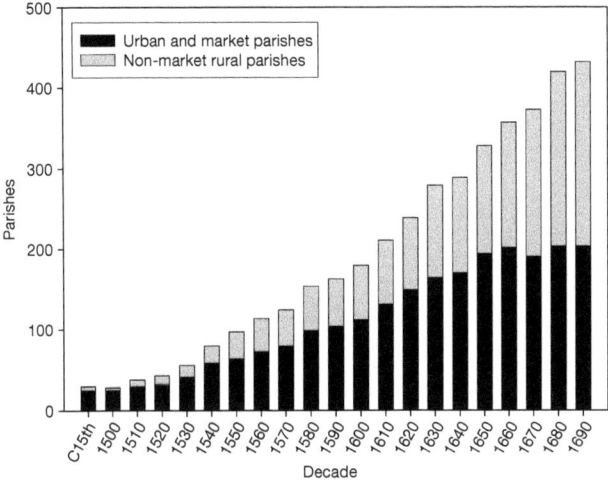

Fig. 5.4. Numbers of parishes with surviving churchwardens' accounts.
Numbers increase steadily, but from a very low base in late medieval England. Initially, documented parishes are overwhelmingly in towns. Only towards 1700 are rural and non-market parishes a majority, even though they accounted for nearly 90 per cent of all parishes.

Moreover, the significant numbers of parishes which have detailed surviving accounts covering more than two centuries means that it is possible to trace their operation in very considerable detail.

A more serious issue is the representativeness of parishes with surviving churchwardens' accounts. First, their numbers vary considerably over time from c.1400 to c.1700 (Figure 5.4). The rarity of early accounts is obvious, as is sustained growth in the accounts available over time. So too is the fact that many parishes' accounts are available only for limited periods. Outside the City of London, the majority of parishes with pre-1540 accounts have no accounts at all after 1650. So the composition of the 'sample' of documented parishes varies greatly over time. This makes tracing long-run changes in clock possession among parishes a complex task. Second, documented parishes are more 'urban' than in England at large (Figure 5.4).[15] It is well into the fifteenth century before any documented non-town parishes appear, and not until the 1630s do documented 'rural' parishes outnumber 'town' parishes, although they were far more numerous in the landscape.

Third, the geographical coverage of churchwardens' accounts is also uneven (Figure 5.5). The City of London is altogether exceptional, since pre-1700 accounts survive for three-quarters of its hundred-plus parishes; most from an early date, and most spanning long periods. Elsewhere, just five counties—Somerset, Devon, Gloucestershire, Durham, and Cheshire—have extant accounts

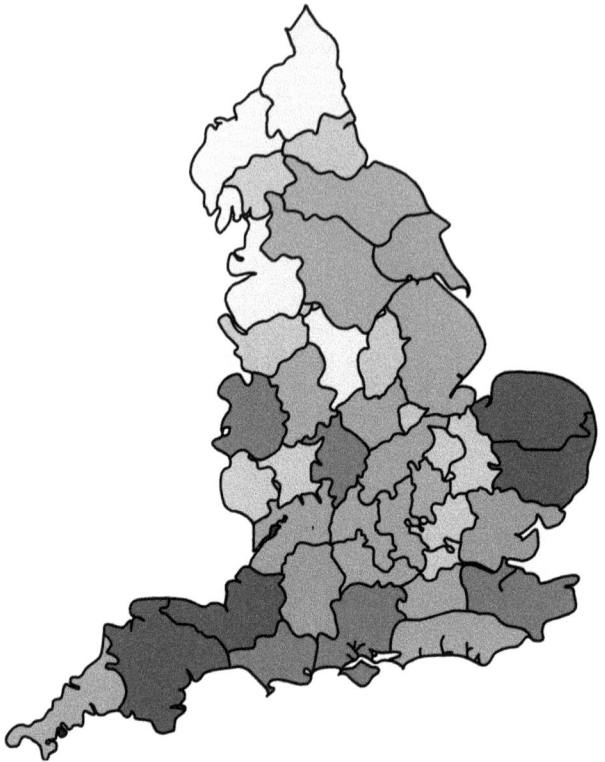

Fig. 5.5. The uneven geographical survival of churchwardens' accounts for rural and non-market parishes.
Although parts of south-west England and East Anglia are conspicuously well documented, extant accounts are very sparse in several northern counties. Fewer than one in twelve; between one in twelve and one in eight; between one in eight and one in six; more than one in six.

for between one in six and one in eight parishes. Many counties in southern England and the West Midlands have accounts for between one in eight and one in twelve parishes. While only Derbyshire, Cumberland, Northumberland, and Worcestershire are practically undocumented, accounts are few and far between for eastern and northern counties, which have no documented parishes at all in some decades. In short, we can discuss church clocks more authoritatively for some parts of England than others.

One factor in regional variations, given the greater survival of urban parish accounts, is the density of cities and towns. But going beyond this, Figure 5.5 shows that the geography of 'rural' coverage is also highly uneven. The best-documented counties, where over 10 per cent of 'rural' parishes have surviving accounts, are Somerset, Suffolk, Surrey, Bedfordshire, and Durham, and the

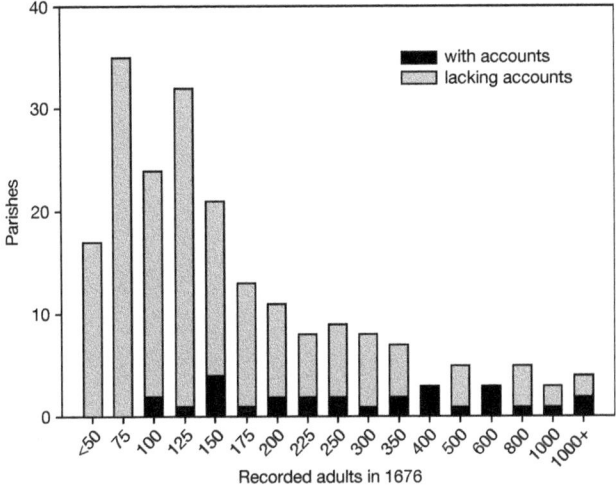

Fig. 5.6. Parishes with surviving accounts, by recorded adults in 1676. Although churchwardens' accounts survive more frequently for larger parishes, extant accounts have survived from parishes spanning a wide range of sizes and populations.

last two of these do not involve many parishes. Contrariwise, ten counties have accounts for under 3 per cent of 'rural' parishes, and for much of northern England the account base for 'rural' church clocks is slender indeed (albeit that in Yorkshire and Lincolnshire 2 per cent of parishes may consist of between ten and fifteen places (more than in the much smaller Durham with its high percentage of documented parishes).

More encouragingly, the surviving churchwardens' accounts cover a wide range of settlements. While towns are better documented than rural parishes, sixteenth- and seventeenth-century coverage includes many rather small, low population, rural parishes, in most parts of England.[16] For example, several dozen parishes with fewer than 100 adults recorded in the 1676 Compton Census, do have surviving churchwardens' accounts (Figure 5.6). So while the changing characteristics of documented places need careful consideration when numbers and patterns of church clocks are analysed, such analysis is certainly possible.

5.3.2 Numbers and Distribution of Parish Church Clocks

Baldly stated, the main finding is that a majority of documented parishes possessed working clocks at some stage between 1400 and 1700. Many parishes, rural as well as urban, possessed clocks from an early date, though exactly how early remains obscure since many clocks were installed before that parish's earliest surviving churchwardens' accounts. Over time, the proportion of documented

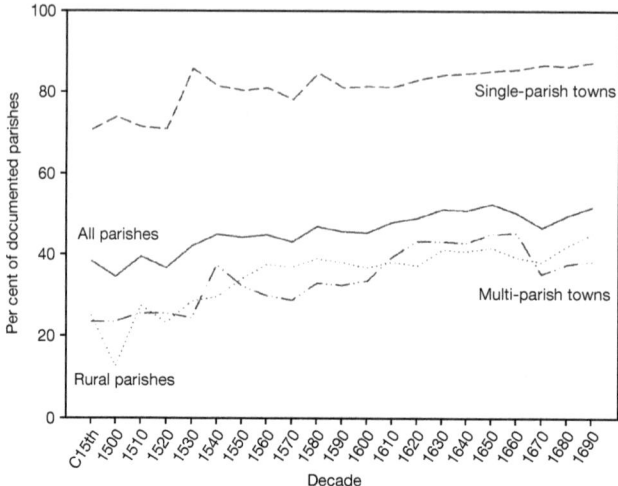

Fig. 5.7. Proportion of documented parishes with church clocks, overall and for different settlement types.

parishes with clocks rose considerably. Although at first sight this increase appears modest (Figure 5.7), this observation can readily be explained by the changing character of the documented places.

Chief amongst these is the fall over time in the proportion of town parishes among those documented. Since these possessed clocks more often than small rural parishes, proportions of church clocks are considerably overestimated in the early decades, but this source of overestimation gradually declines over time. So, other things being equal, Figure 5.7 would show a downward trend as more documentation from rural parishes becomes available. That it instead shows an increasing trend means that a real and substantial increase in the provision of church clocks was occurring.

This is shown by Figure 5.7, which shows the possession of clocks for parishes in different types of settlements, which exhibits several interesting and telling features. It is striking that the urban parishes most likely to have clocks were those in single-parish towns, whereas in multi-parish towns individual parishes were less likely to have clocks (although there certainly are numerous instances of rampantly competitive behaviour among neighbours). While there were town parishes without church clocks, however, a town with no church clocks was almost a contradiction in terms. The London pattern is also interesting, particularly changes during the sixteenth century, and the long-lasting impact of the Great Fire.[17]

Figure 5.8 also makes the increase in 'non-town' clocks look smaller than was actually the case. At first glance, after sustained sixteenth-century growth the

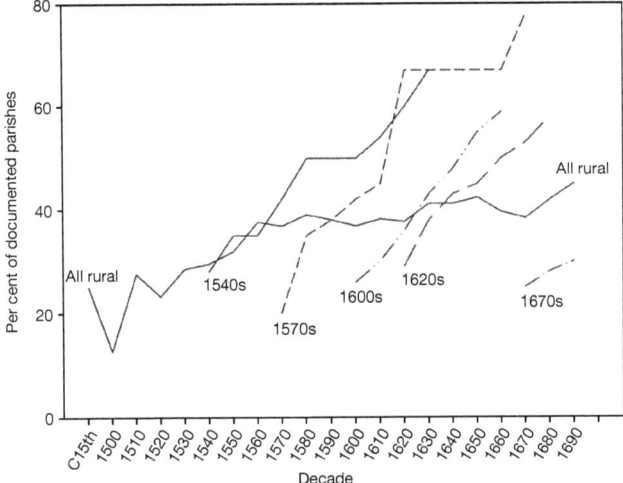

Fig. 5.8. Trends in church clocks for a selection of parish cohorts according to their earliest surviving accounts.

proportion of non-market parishes with clocks took a century to rise from the high 30s to low 40s per cent. This is another artefact of changing documentation, as is clear from examining parishes grouped according to when they enter extant documentation (Figure 5.8).

Each decade a new cohort of parishes enters those whose churchwardens' accounts have already started. In every decade, clocks are initially less common amongst newly documented parishes than among those already documented, but thereafter numbers of clocks grow steadily. With each decade though, comes a new cohort of parishes characterized by a lower proportion of parishes with clocks than the parishes already in view. Simultaneously, some parishes disappear from view each decade, because their records no longer survive. These parishes are broadly typical of parishes whose documentation continues, that is to say, a higher proportion of the parishes leaving documentation each decade possess clocks than do the parishes entering documentation. Thus, of parishes with extant accounts from the sixteenth century or earlier, 96 disappear from observation before 1620 and have no further accounts until after 1700: 43 per cent have clocks at the time of their disappearance, compared with between 20 per cent and 30 per cent of parishes newly documented. For the rest of the seventeenth century, the figures for each decade's hitherto undocumented parishes change very little, but from 1620 to 1700 over 56 per cent of parishes leaving documentation already have clocks.[18]

All in all, then, levels of parochial clock ownership needed 'to run to stand still'—to keep increasing—even to maintain an apparently stable overall level as

Fig. 5.9. Percentage of parishes whose churchwardens' accounts refer to church clocks.

the number of documented parishes increases. Significant numbers of already-documented parishes had to acquire clocks to offset the disparity between the loss of parishes many of which already had clocks, and the newly documented places in which clocks were much less common. And indeed, there are many scores of parishes that lack clocks in their earliest accounts, but install them during the sixteenth and seventeenth centuries. Given the changing size and composition of documentation, the striking finding is not that the increase was slow, but that any overall increase appears at all while so many features of the documentation act the other way.

There were significant regional variations in the density of church clocks. In part, these reflected regional differences in the density of small towns and markets, in which church clocks were pretty well universal features (although, as noted, not found in every parish in multi-parish towns). They also reflected regional differences in the prevalence of clocks in 'rural' parishes (Figure 5.9). Among the better-documented counties in the late sixteenth century, clocks were found in more than half of the 'rural' parishes of Berkshire, Devon, Hertfordshire, Suffolk, and Somerset. Between one-third and one-half of such parishes in Cambridgeshire, Durham, Lincolnshire, Sussex, Lincolnshire, and Cornwall had clocks. Church clocks were least widespread in the central-southern counties of Hampshire and Wiltshire, at about one in five 'rural' parishes.

A century later, towards 1700, a majority of documented 'rural' parishes in Durham, Hertfordshire, Cambridgeshire, Devon, and, most strikingly with thirteen out of fourteen, Staffordshire, contained clocks. More than one-third of such parishes in Gloucestershire, Somerset, Berkshire, Cornwall, Lincolnshire, and Bedfordshire now maintained clocks. They were still comparatively scarce in counties like Wiltshire, and there had been a significant drop in the proportion of rural parishes with clocks in Norfolk and Suffolk. This last feature partly reflected the increase, right across the country, in the numbers of small parishes for which seventeenth-century documents survive. In other words, over time the documented parishes become less biased towards larger and more populous parishes. However, there is more than this to the declines in Suffolk and Norfolk. Many East Anglian parishes experienced a long-term drop in population from the later Middle Ages onwards, and there are several examples of dwindling parishes being unable to maintain clocks for either financial or logistical reasons, usually coming to a head when facing substantial mechanical overhaul of the clock.

Elsewhere, at times of disruption such as the English Civil War in the 1640s, several clocks 'disappeared' temporarily, before being reinstated during the 1650s and 1660s. And, at any given time, a small minority of church buildings were experiencing severe structural problems, through collapsing towers, shifting foundations, and the like. With clocks usually situated in towers, and heavy ringing contributing to structural weakening, it is unsurprising in such cases to see clocks being removed while repairs or rebuilding were effected, and this could take ten or even twenty years. For a variety of reasons, therefore, over both the long and the short term, the number of clocks in an area could go down as well as up.

Nonetheless, it is clear that even some very small places maintained church clocks over long periods.[19] An analysis of the possession of clocks compared with parish populations in 1676 shows this to be the case for both 'town' and 'rural' parishes (Figure 5.10). We need not imagine these examples as representative of all small parishes nationwide—specific local factors such as gentry patronage were likely at play—but small size was no bar to parish timekeeping.

In just one instance can evidence from churchwardens' accounts be assessed against comprehensive local documentation of church clocks. This early and unusually systematic enquiry formed part of an archdeaconry visitation of Bedfordshire in 1708 (Figure 5.11) (Pickford, 1991). The 116 parishes detailed in the visitation were compared with the bare dozen churchwardens' accounts surviving at the same period. The outcome is reassuring, insofar as similar

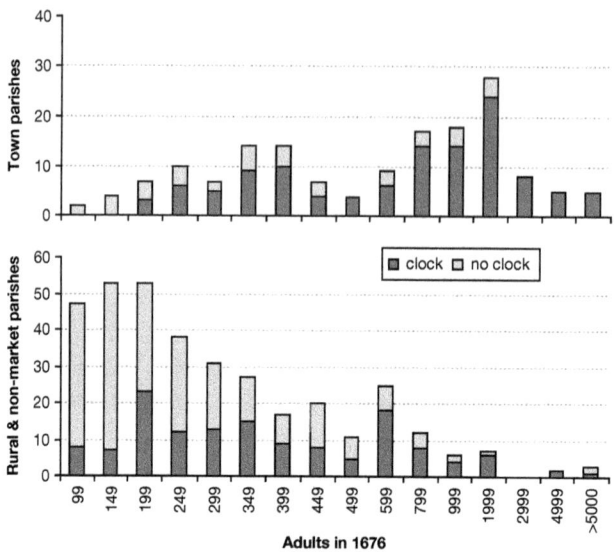

Fig. 5.10. The frequency of church clocks by recorded adults is a parish in 1676.

proportions of parishes have clocks (9 of the 10 towns, and 36 of 106 non-market parishes) to the churchwardens' accounts.[20] If anything, the visitation figures appear underestimates, since at least three parishes were maintaining clocks, according to their accounts, but omitted them in the visitation return. It is a great pity that this Bedfordshire visitation seems exceptional, but its corroboration of findings from surviving accounts is encouraging.

5.3.3 Forms of Time-Signalling

Turning to time-signalling, we see increases in various forms of communicating the time, which combines increases in both provision of signalling facilities, and numbers of documented parishes. Despite initial ambiguities in the terminology for describing signalling mechanisms, churchwardens' accounts enable an exploration of the nature of time-signalling, and of its time-space pattern (Figure 5.12). Hour-striking was common from the earliest recorded church clocks. More sophisticated forms of time-signalling were discretionary: only later, for example, did dials come to be regarded as inherent features of clocks.

The steady growth in the number of parishes with chimes compares with a rather steeper seventeenth-century growth in clocks with dials. It is necessary to remember that the great majority of clock dials had a single hand only, variously referred to as a hand, a finger, an index, or a point. Compared with the episodic signals from bells, single-handed dials offered a continuous indication of the time,

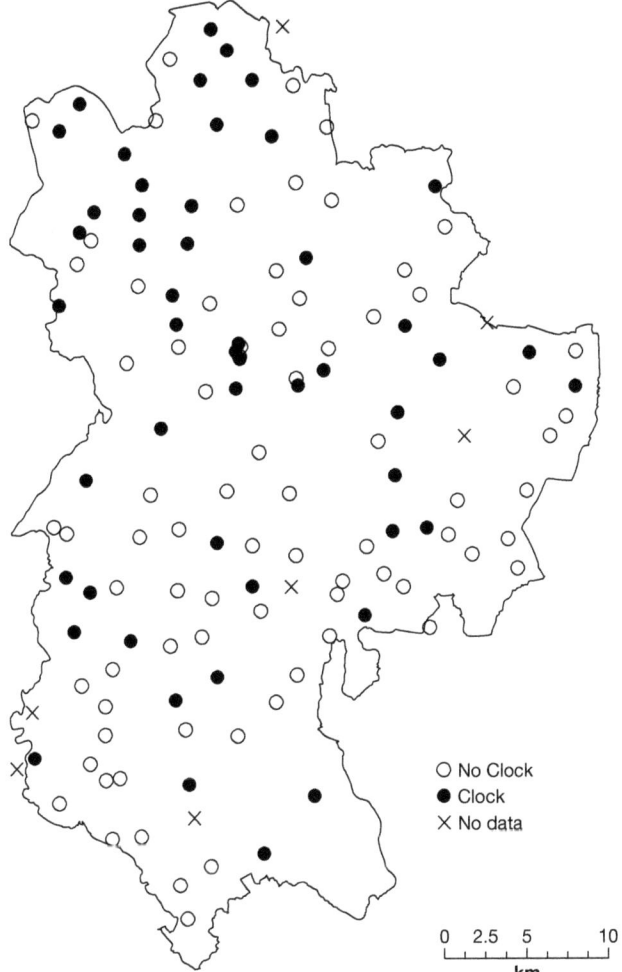

Fig. 5.11. Clocks recorded in the Visitation of the Archdeaconry of Bedford, 1708.

though their having only one hand limited its detail. The major significance of clock dials at that time lay less in their practical use than in their emphasis of the visual as a medium of timekeeping. For most users of public clocks, time continued to be something you heard, not something you read. Providers of new clocks and bells were quite explicit about this. Clock bells 'provide a constant monitor of the time to all inhabitants', said a Bristol petitioner in the 1670s, while a new clock bell at Boston in 1709 was authorized 'to tell the hour to the people loudly and clearly, and place[d] . . . in the lantern on the highest part of

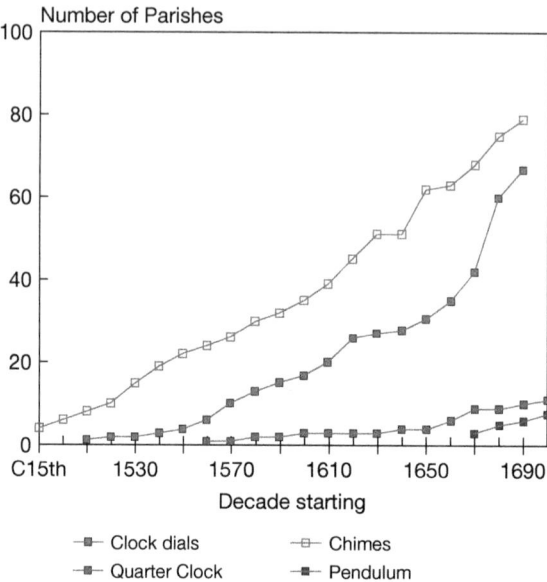

Fig. 5.12. Trends of various means of signalling clock times.

the tower... for the better and more audible hearing of the sound thereof.'[21] The shift to mechanical clock time as something that was primarily *seen* was a long, slow process, although devices such as sundials were visual.

The rising number of parishes in each category reflects both the new addition of signalling to an existing or newly constructed clock, and the entry into documentation of parish clocks already so equipped. Further analysis shows that the former substantially outnumber the latter. The increase in dials and chimes are thus genuine extensions to parish provision of time signals.

Quarter-striking clocks enabled parishes to refine their aural signalling, or if linked to dials, both their aural and visual signalling. The history of particular parishes could be quite eventful: at St Lawrence, Reading, a fifteenth-century clock had struck the hours with hammers held by at least two 'jacks', which can be taken to be automata modelled as human figures. This clock appears to have been of considerable age, since it was replaced in 1521 by a new clock with a 'clock dyall', one of the earliest recorded anywhere. The dial was repaired by local joiners and smiths in 1560, and again renewed in 1587 following damage caused by bells falling on the transmission mechanism from the clock to the dial. By 1627 that clock had been modified to strike the quarter hours, and this modification lasted until the installation of a completely new purpose-built quarter clock in 1673. At that same time, a new set of eight chimes bells were installed, to replace simple striking of the hours, and set up so as to offer a choice of two different tunes on each day of the week. It is possible that the works in 1673 also involved the fitting of a pendulum, but the account's wording is vague. By the end of the century, the addition of a minute hand created a 'modern looking' clock, but through a very long process of acquiring the various elements of hour-striking, single-handed dial, quarter striking, musical chimes, and minute hand.[22] Most parish clocks had less eventful histories and lacked one of these elements, often more, well into the eighteenth century, but their histories also comprised long series of piecemeal changes.

Though the pendulum was still a recent development, it was appearing in parish church clocks within twenty years of its first use in Coster's 1658 clock for Huygens.[23] Tracing early pendulum clocks, whether completely new or as an adaptation of an existing clock, is again problematic since it depends on the word 'pendulum' (or some recognizable phonetic variant!) actually appearing in an account. Often phrases such as 'new making the clock' or 'altering the going of the clock' *may* cover the addition of a pendulum and modification of the escapement—as we noted at Reading in the preceding paragraph—but such cases have been excluded from Figure 5.12.[24]

Pendulums were prominent in new clocks being installed in London in the late 1680s and early 1690s, as Wren's rebuilding of City churches after the 1666 Great Fire neared completion.[25] The Fire's extensive destruction, however, means that London is relatively marginal to the earliest diffusion of church clock pendulums in the 1670s, since most of its churches still lay ruined or cleared. Clearly, though, the potential of greater precision and reliability was seized on elsewhere, and not just in towns. The village of Yarnton in Oxfordshire paid for a pendulum clock as early as 1680, less than a decade after the first uses in public clocks in Cambridge, London, and Oxford.[26] Sometimes, perception of the precision and status deriving from a pendulum's greater precision was explicit in wardens' descriptions of new clocks (and their greater expense), but the language of many churchwardens was vague enough for it to be uncertain just how many late seventeenth-century clocks were of the new type.[27]

The early examples at Yarnton and South Newington in Oxfordshire seem to reflect contact between surrounding villages and the county town, where several colleges of the university, and the churches of St Mary the Virgin and St Martin, had early pendulum clocks. Likewise, pendulums at Bishops Stortford (Hertfordshire) and South Weald (Essex) appear linked to London and Cambridge. The pendulum was added to Bishops Stortford church clock in 1673 by a man named as 'Mr Clement', very likely the William Clement who was building clocks with a pendulum and an anchor escapement in Cambridge and London in 1671 and 1672.[28] Further afield, though, pendulum-regulated clocks attest to a broader sense and grasp of the timekeeping possibilities. Unambiguous examples come from parishes in the West Midlands (Southam in Warwickshire, and Hanbury in Staffordshire); north-west England (Macclesfield in Cheshire); and the West Country (Newland in Gloucestershire). The process of converting older verge-and-foliot clocks to pendulums and anchor or dead-beat escapements accelerated through the eighteenth century, in all parts of England. Signalling was still mainly aural, but the visual element was growing; more places were paying attention to timekeeping below the level of the hour. Relative precision was, for some parishes at least, seen as highly desirable and worth paying for; and the appearance of dials, chimes, and the like are important indexes of these concerns.

5.4 PRIVATE TIME: DOMESTIC CLOCKS AND WATCHES

Domestic clocks and watches, rather than large public, 'turret' clocks, have formed the backbone of horological study. This interest in clocks and clockmaking, like many other areas of material connoisseurship, is by no means a recent phenomenon, having emerged by the late eighteenth century, and certain elements of horological interest, especially in 'clocks as visual machines' and 'clocks as beautiful objects', are observable considerably earlier (Mayr, 1986). The diary of Samuel Pepys, that mainstay of so many historians of everyday metropolitan life, provides some useful examples, including his description of seeing the Queen's night clock in her apartments in 1664; his fascination in seeing Lord Brouncker (president of the Royal Society) dismantling and reassembling a watch; and his arranging to see clockmakers and others work on clocks and other mechanical gadgets. Pepys was not exceptional in this mechanical interest, which can be seen in other periods and places, and among women as well as men (Chapter 6, below).

A great deal of horological history has been written from the supply side, that is from the perspective of innovations in mechanical design or aesthetic dimensions (such as cabinet making and decorative techniques), and the production of their components (such as the bifurcation of bespoke and mass-produced components). Accordingly, the late seventeenth-century diffusions of the pendulum and the balance spring as regulators in clocks and watches respectively, have lain at the centre of historical accounts of private time, through their effects on the reliability, costs, and uses of timepieces.

The recent upsurge in equivalently vigorous enquiries into 'demand side' questions has begun to explore the numbers of clocks and watches in use throughout the country; the sorts of households in which they were to be found; and the social status of individuals liable to take their watches from their pockets for inspection. It must straightaway be said that this work has substantially changed earlier views on the ownership of timepieces.

In Chapter 4, we used probate inventories to examine the social profile of clock ownership in Bristol. Since the 1980s, inventories have been extensively and systematically analysed to trace changes in social structure, farming, craft industries, and consumption, and this work provides some indications of how far the story told for Bristol applies elsewhere in England, too (Weatherill, 1989; Evans, 2001; Overton *et al.*, 2004). At its simplest: how far does the pattern of timepiece ownership in Bristol presented in Chapter 4 serve for England as a whole? Or was Bristol exceptional—perhaps because of the city's large size and status compared with circumstances experienced by the rural majority of English population? Perhaps Bristol's near-coastal location raised the possession of clocks and watches compared with inland areas, where communications were overland and hence slower. Or maybe southern England in general was more involved with clock time than relatively sparsely populated northern counties, where much craft-industrial work was done within households?

Detailed geographical comparisons of the apparent levels of ownership of clocks and watches in probate inventories needs to be approached with some care, since complications arise from several sources, notably (i) reasons for which objects could be omitted from inventories; (ii) regional variations in the intricate administrative frameworks for probate; (iii) regional differences in the compiling of inventories, and (iv) their subsequent survival. These issues have been much discussed, but are not disabling for the present argument concerned with broad comparisons.[29]

The most accessible starting point for resolving such questions is Lorna Weatherill's (1988) systematic analysis of eight areas of England, analysing some 3,000 probate inventories spanning the 1670s to the 1720s.[30] Weatherill was interested in clocks as just one among many new consumer goods in early modern England, both durables (including clocks, cutlery, mirrors, lighting utensils, imported fabrics, and ceramics) and groceries (tea, coffee, chocolate, tobacco, imported fruits), whose ownership was becoming more widespread. Weatherill aimed to trace the social and geographical spreads of some of these new items (Table 5.2).

Weatherill's findings highlight how, compared with many goods, domestic clocks spread relatively rapidly between the 1660s and 1730s. By definition, the rise in clock ownership precedes their appearance in inventories, which were compiled after deaths of owners. Goods would have been acquired somewhat earlier, though this may be less serious a complication than it would be today, because adult deaths were not then so dominated by the elderly.[31] The higher death rates among young and middle-aged adults in early modern England, though, mean that early modern inventories refer to adults with a wide range of ages at death. We should imagine the growth in possession of domestic clocks as occurring perhaps one or two decades in advance of its appearance in inventories.

One horological surprise is that all over England clocks were found in households of widely varying wealth. They were by no means as specific to elite households and to towns, as most historians have supposed. Comparing clocks with the other goods analysed by Weatherill produces two particularly striking

Table 5.2. Consumer goods in English probate inventories, 1670s–1720s

(a) Per cent of probate inventories with selected consumer goods, eight areas combined

Item(s)	1670s	1680s	1690s	1700s	1710s	1720s
Clocks	9	9	14	20	33	34
Pictures	7	8	9	14	24	21
Window curtains	7	10	11	12	19	21
Cutlery	1	1	3	4	6	10
Looking glasses	22	28	31	36	44	37
Earthenware	27	27	34	36	47	57
Silver or gold	23	21	24	23	29	21

(b) Possession of clocks by area

Area	Period	% inventories analysed with		
		Clock	Pictures	Looking Glass
London area	1675–1725	29	37	74
East Kent	1675–1725	36	16	47
North-west England	1675–1725	33	9	31
North-east England	1675–1715	15	25	44
Cambridgeshire	1675–1725	14	9	27
Hampshire	1675–1705	7	3	19
North-west Midlands	1675–1725	7	4	14
Cumbria	1675–1725	7	3	6

Source: Weatherill (1988), tables 2.1, 3.1.

results. First, the spread of clocks and watches was relatively rapid compared with many consumer goods, especially given their relatively high cost. Clocks and watches spread at least as fast as did other goods in Table 5.2 that could be acquired for considerably less money, such as pictures, cutlery, window curtains, or china.

Second, clocks were found in households across the full range of settlements that Weatherill distinguishes. She classified settlements into four broad categories, namely London, major towns, small towns, and villages or hamlets. The likelihood of finding most new goods in probate inventories was strongly dependent on the status of the settlement in which the inventory originated, but for clocks the geographical disparities were muted compared with most new items. Things like window curtains, cutlery, tea and coffee equipment, were all comparatively scarce outside London and the larger provincial towns. Looking glasses were relatively common in small towns as well as large towns, but they too were much less common in the countryside. The contrast is brought out by standardizing ownership percentages for each settlement type to the average ownership level for England as a whole, expressing goods' occurrence proportionately to that (Figure 5.13). The shallow gradient of the 'clock ownership' line, denoting small differences among settlements, contrasts

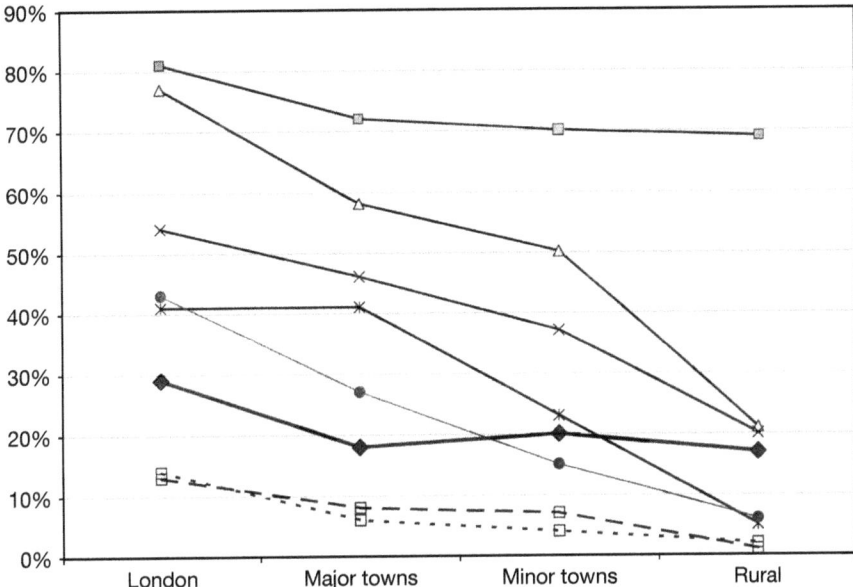

Fig. 5.13. Spatial variations in inventoried consumer goods, 1660–1730. The goods are those in Table 5.2. Here, reading from top to bottom in London, they are: earthenware; looking glasses; pictures; window curtains; clocks; cutlery.

strongly with those for cutlery, mirrors, window curtains, tea and coffee. Instead it resembles the much more muted settlement-type differentials that characterized cooking pots, the most mundane and everyday of the items Weatherill analysed.

Weatherill's results, though exploratory, are interestingly suggestive of a substantial penetration of clock time, and of private timepieces, into domestic life. Her survey can be extended and supplemented in various ways, to examine these issues in more detail. Here, we elaborate four themes, two extending Weatherill's exploration of patterns of clock ownership, first chronologically, and then socially. The remaining two themes go beyond presence/absence measures to consider what probate inventories reveal about early domestic clocks as objects, and their locations and uses within households.

First, then, we are interested in the longer-run history of private clock ownership, especially before the substantially growing clock ownership uncovered by Weatherill. Earlier probate inventory material certainly emphasizes how dramatic was the explosion from the late seventeenth century onwards in timepiece ownership. However, private clock ownership already had a long history by the 1660s, although never accounting for more than a small percentage of inventoried persons, with a scattering of clocks from the early sixteenth century or before. Earlier domestic clocks were found within a more limited social range of owners, mainly among gentry and aristocratic households, on the one hand, and wealthy urban traders and professionals, on the other.

Second, one of the most vexing (and persistently uncertain) questions for historians of early consumer society is how deep the market, and the consumption of durable goods and groceries, penetrated English society. More specifically, at what social levels was the bottom of the market for clocks and watches, and how did this change over time? Here the social selectivity of probate inventories is an obvious handicap: they do not provide comprehensive social testimony because they were not required below a threshold estate value.[32] The estates of significant numbers of people with only modest possessions were not drawn into the formal probate procedure, and so inventories were not compiled for them, although in practice executors might compile them for other reasons. So probate inventories, even from the archdeaconry courts that served testators of modest wealth, are biased towards wealthier households. For example, of the 3,000-plus households in Weatherill's regional samples, summarized in Table 5.2, only twenty-eight relate to men styled 'labourer'. Yet this puny figure is not because labourers were uncommon. Estimates of the numbers of Englishmen so described in *c*.1700 usually include around 25 to 30 per cent of the entire English population (Wrigley, 1987: 169–94).

Nevertheless, Weatherill argued that the poor households, marginal to the probate process, had remained almost entirely outside the new consumer culture. She detected the significant social break within slightly richer and higher-status groups of craftsmen and husbandmen who are more regularly represented within the inventoried population: 'As with husbandmen, it was very unusual for [labourers] to own any of the new and decorative goods, and their houses were similarly, and sparsely, furnished' (Weatherill, 1988: 176). Other commentators, however, reached differing verdicts on the social depth of consumer goods markets, although they have not been specifically concerned with clocks, and have often focused on cheaper items, especially textiles: Joan Thirsk (1978) and Margaret Spufford (1984) each concluded that consumption of new items extended far into the mass of labouring households. Most members of such households worked discontinuously or irregularly, relying on the improvised assembly of what historians term 'a jigsaw of makeshifts' (King, 1997) to sustain themselves and their dependents, but it did not follow from this that they were either ignorant of, or inexperienced with, new commodities (Thirsk, Spufford, King *et al.*, 2004).

One might object, though, that while such households could occasionally have afforded such inexpensive items as ribbons, tobacco, or food, the situation was different for costlier items such as clocks and watches. At the very least, the onus is on those taking a relatively positive view of the consumption experience of the English labouring poor to produce further evidence, yet the types of evidence central to consumption debates, notably probate inventories and household accounts, are rarely available for the relevant households.

What we want to suggest here is that poorer households were not left outside the clock time population. First, poorer households shared in the common acoustic and visual environments of time-signalling: sound waves made no discrimination according to income. Second, labourers' (and craftsmen's) working times were frequently regulated by clock time, especially—but not only—in towns, so clock time could impinge directly on everyday living. Third, they were involved with churches and markets, musters and fairs, inns and entertainments, all of which were organized using clock times.

More specifically, we also want to dispute the exclusion of poor households from the possession of clocks of their own. To begin with, even among Weatherill's figures very poor inventoried men and women did sometimes possess clocks: just over 2.5 per cent of those with gross inventory values under £10, 6 per cent of those between £11 and £25, and 11 per cent of those with total gross valuations between £25 and £50 (1988: 107). These numbers are modest, but they are not the zero that would be expected on the most pessimistic interpretation. Despite their rarity in Weatherill's samples, surviving labourers' inventories are less scarce than she implies for several parts of England, and here clocks and occasionally watches appear among the lists of goods. While they could conceivably be produced by inventories for 'retired' people, who had passed on almost all their property to their heirs prior to their deaths, other evidence does show genuinely very poor households with clocks (King, 1997: 157).

Weatherill's sampling strategy was largely dictated by the essentially comparative character of her main questions. When the investigation of inventories from an area is more intensive and comprehensive than in Weatherill's study, the relationships between wealth and clocks can be pursued in much more detail. For example, Figure 5.14 illustrates how the considerable increase in the proportion of East Sussex probate inventories with clocks, between those for people dying in the 1710s and in the 1730s, was produced by increasing recorded clock ownership right across the spectrum of household wealth. For every single one of the twenty-eight wealth categories shown there, the proportion of households with clocks increased. The increase was smallest among the richest households, where levels of clock ownership had already been high in the 1710s. It was very marked for the poorest groups of testators, especially those with movable goods whose gross inventory value (that is, ignoring debts that they themselves owed) was less than £40.

Analysis of probate inventories may be supplemented by household inventories compiled for the very poor on occasions other than their death. For example, Overseers of the Poor recorded the assets of pauper householders whose welfare was about to become a parish responsibility, because of their incapacity through age, injury and/or illness. Arrangements were made to maintain them in their own homes rather than in parish workhouses or almshouses, which would very likely prove more costly to the parish. Goods were therefore appraised to assist the parish in drawing on both parish and household resources. Where they survive at all, pauper inventories typically survive as isolated documents in parish archives, not gathered together in ecclesiastical court archives like probate inventories, and have only recently received historians' attention—especially for Essex (Sokoll, 2001; King, 1997). Because of their timing within household life cycles, they potentially provide a better insight into the 'normal' life of very poor households, whereas probate inventories may represent an estate already part-way through transmission to the next generation.

It is therefore striking that, in nine eighteenth-century rural parishes in north Essex, about one in five pauper inventories contained a clock or a watch (King, 1997: 162–79). The overall sample of inventories is small, but there is no suggestion that they were unrepresentative—in wealth or otherwise—of pauper households. Rather, they illustrate how even very poor households could, at least at certain life-cycle stages, acquire relatively expensive consumer goods. It must be stressed that King is not seeking to paint a rosy picture of their material culture:

while the pauper inventories... challenge, to some extent, our assumptions about the austere lives of the poor and about their lack of material reserves in times of need, the picture that emerges is by no means an optimistic one. Indeed, given the rapid increases in material wealth experienced by many other farming and other middling households during this period, it seems likely that in relative terms the labouring families of Essex were getting poorer rather than richer. (King, 1997: 183)

For the present discussion, our main point is that clocks were found in a significant proportion of pauper households, independent of trends in the relative consumption levels of different social groups. Indeed, that point is the more striking for coinciding with increasing material disparities between pauper and richer households. We contend simply that significant numbers of poor households, households whose poverty was sufficiently deep and enduring as to necessitate support from parish authorities, were nevertheless clock owners. If this point holds for communities across England at large similarly to those studied hitherto, the poor cannot be seen as existing outside 'private time' any more than they existed outside 'public time'.

The third and fourth themes can be explored because the appraisers of inventories sometimes provided additional description of the clocks listed. While many items are no more detailed than 'a clock', others describe the style of the clock or the materials from which it is made: for example, 'a clock with brass wheels and lead weights', or 'a clock in a japanned case'. Many inventories contain a separate valuation for a clock or watch, the appraisers' estimate of the monetary value of the clock to the heir, that is, its second-hand sale value. Post-mortem sales of house contents were common and executors' probate accounts, which sometimes accompany inventories in ecclesiastical court records, frequently list sale prices similar to inventory valuations, so that valuations can be taken as broadly accurate. Finally, a clock's location within the house is often specified, since appraisers often listed and valued items as they proceeded from room

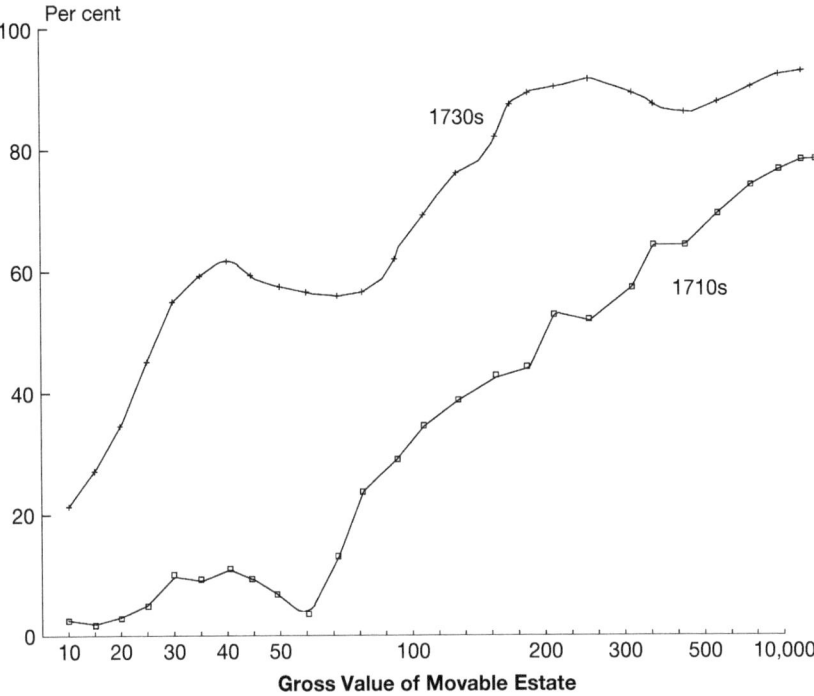

Fig. 5.14. Testators' ownership of clocks by probated wealth, East Sussex 1710–39.

to room through a house, and room names appear as headings within the inventory.

Consideration of early domestic clocks as objects is handicapped by the very few examples that survive, especially compared with the proliferation of eighteenth-century and later clocks. Although intense horological attention has been devoted to the few surviving clocks from before *c*.1700, almost all clearly originate from the upper levels of the domestic market. Indeed most have survived precisely by being treasured as beautiful pieces of furniture, as aesthetic items with important supra-utilitarian meanings. Items of this type and quality are not necessarily any guide to the plainer clocks owned by less exalted households, a claim borne out by the ways in which the clocks in humbler households are described and valued.

Appraisers' slightly fuller descriptions such as 'a clock and case', 'a clock with weights', 'a clock and jack', 'an old silver watch', and so on, were not based on any formal system, and ought not to be over-interpreted, but the broad attributes of the clocks being described and valued are reasonably clear. First and foremost, very few were the elaborate and expensive items that represent early modern clocks in today's museum collections. The latter are best

Table 5.3. Valuations of clocks made by probate appraisers: East Sussex, 1730–7

Valuation	Total	Number of valuations of...		
		'clock'	'clock & case'	'watch'
10s.	2	2	–	–
10s. 6d.	2	1	–	1
16s.	1	1	–	–
17s. 6d.	1	1	–	–
18s.	1	–	1	–
20s.	3	3	–	–
21s.	5	3	1	1
25s.	6	4	1	1
30s.	6	3	3	–
35s.	3	2	1	–
40s.	6	5	1	–
42s.	1	–	1	–
50s.	5	2	3	–
55s.	1	–	1	–
60s.	2	1	–	1
65s.	1	–	1	–
70s.	2	–	2	–
80s.	1	–	1	–
84s.	1	1	–	–
90s.	1	–	1	–
100s. +	1	1	–	–

Note: Valuations are given in shillings (s.) and pence (d.), 12d. = 1s., 20s. = £1.

Source: East Sussex Record Office, Archdeaconry of Lewes probate inventories. W/INV 1–4900.

envisaged as the elite tip of an iceberg of mostly quite plain clocks, utilitarian in purpose and of relatively prosaic design and construction. Despite their modest appearances, though, the mechanical sounding of hours indicated by descriptions of jacks, bells, and/or hammers, was common. Descriptions of more elaborately constructed or decorated items, such as 'a clock and case' indicating a long-case ('grandfather') clock, or clocks in 'japanned' or 'inlaid' cases, were relatively unusual. Clear patterns in appraisers' valuations of clocks show that this does not result from randomly inconsistent detail in appraisers' descriptions, which also show many relatively cheap items. More elaborately described clocks are consistently associated with higher valuations.

The overall pattern of appraisers' valuations bears out the accessibility of clocks to purchasers of humble status. The archdeaconry of Lewes in East Sussex, where the survival of early eighteenth-century inventories is particularly good, provides a good example (Table 5.3). The mean valuation of the fifty-two timepieces separately valued between 1730 and 1737 was 36s. 4d. (£1.81½p), the median was lower at 30s. (£1.50p). Just over two-thirds were valued at 40s. or less. Those described as 'clock and case' seem to have been more elaborate or newer (very likely both), with a mean valuation of 47s. (£2.35p) and median of 46s. (£2.30p). By the 1730s, valuations appear slightly lower than those being made at the start of the century. In the same part of Sussex, the most common valuation for a clock between 1700 and 1710 had been 40s., again with more clocks valued at lesser than higher amounts. Among a wealthier group of testators in Buckinghamshire between c.1660 and c.1710, 40s. had also been the most frequent valuation made for clocks (Reed, 1988).

The impression conveyed by valuations in inventories is consistent with the extremely rare evidence from the sorts of clockmakers who supplied these parts of the market. Seeking such material forced us to look beyond seventeenth-century England for appropriate material. In fact, the best-documented 'humble' clockmaker is a Welshman named Samuel Roberts, working in the small town of Llanfair Caereinion, near Welshpool (Pryce and Davies, 1985). The sales recorded in Roberts' extant working notebooks provide a wealth of information about the materials and costs of his clocks, the social and geographical range of his customers, and the arrangements through which his clocks were bought.

Loomes suggests that, in general,

the numbers of clocks 'known' or 'recorded' by any particular clockmaker or family of clockmakers... can be misleading, as it by no means reflects the total number of clocks that the maker produced. A full-time clockmaker could make one clock every fortnight, roughly twenty-five clocks per year, so that a potential from a forty-year working life could be well over 1,000 clocks—many more if his workforce included several apprentices and/or journeymen.   (Loomes, 1997: 8)

Roberts's output is slightly lower; around 1770 he produced over twenty clocks a year, and seems to have produced about 300 clocks in just under twenty years (Pryce and Davies 1985: 108–15).

One hundred and seventy seven clocks, or 61.6 per cent of Roberts's output over this period, sold for 45s. or less, and only nine clocks (3.1 per cent) cost 60s. (£3) or more. His most frequent selling prices were 45s. (96), 44s. (51), 50s. (32), 48s. (19), 42s. (17), and 52s./6d. (13), accounting for over three-quarters of his documented output between 1755 and 1774. Thus the 40s. clock valuation often found in probate inventories represented the bottom end of Roberts's price range.

Roberts's prices were much lower than those paid for the highly decorated products of leading London clockmakers for clocks combining high accuracy and aesthetic beauty, with precise joinery and elaborate veneering. Roberts's products and prices certainly drew in a broad clientele, and the maker was clearly aware of the difference made to the affordability of his output by his parsimonious use of brass in his clock mechanisms: many of his innovations or idiosyncrasies in production and design were driven by a desire to economize

Table 5.4. The clock sales of Samuel Roberts, 1755–74

Price	1755–9	1760–4	1765–9	1770–74	TOTAL
40s. or less	–	–	1	1	2
41s.–44s.	13	23	30	13	79
45s.	21	20	35	20	96
46s.–49s.	4	1	14	21	40
50s.	2	15	5	10	32
51s.–59s.	2	2	8	17	29
60s.–80s.	3	–	–	1	4
above 80s.	1	2	–	2	5
Total	(46)	(63)	(93)	(85)	(287)

Source: Calculated from information in Pryce and Davies (1985: 325–401).

on brass (Pryce and Davies, 1985: 67–79, 93–115). Nonetheless, 40s. (£2) represented a significant fraction of a smallholder's or labourer's annual income. Farm labourers earning some ten to fifteen pounds a year must have found 40s. a significant sum, even if paying by instalments over a long period. That they were usually paid in arrears, in relatively large sums, though, could create consumption cultures more amenable to comparatively large purchases than where wages comprised small but frequent amounts.

Roberts's records also clarify what customers received for their money. The basic clock included an hourly strike, a dial with one hand, and ran for about thirty hours per winding. More expensive models (although the additional expense was relatively modest) could also have a minute hand, appropriately different dial engraving, a set of chimes to sound tunes at certain hours, or could run for eight days per winding.

It is also interesting that, notwithstanding that Roberts lived and worked in an out-of-the-way place, he had a number of local competitors (Pryce and Davies, 1985: 53, fig.15). These clockmakers came not only from the towns of Welshpool, Newtown, and Llanidloes but also from several villages within a radius of 10–20 miles. They may, like Roberts, have supplemented their incomes from clockmaking by various other activities, but their numbers indicate the considerable availability of clockmaking expertise even in this notably lightly urbanized part of Britain.

The lack of comparable sources prevents a comparably detailed analysis of any seventeenth- or early eighteenth-century English clockmaker, but we envisage their production and clientele as similar to those of Samuel Roberts. They may perhaps have been better linked to networks supplying parts, such as gear wheels or dial plates, produced faster and more cheaply by makers with tools like gear-cutting engines, as was seen in Bristol.

Neither should we assume that early domestic clocks were necessarily used in similar ways to those seen later. Again, probate inventories are useful, because

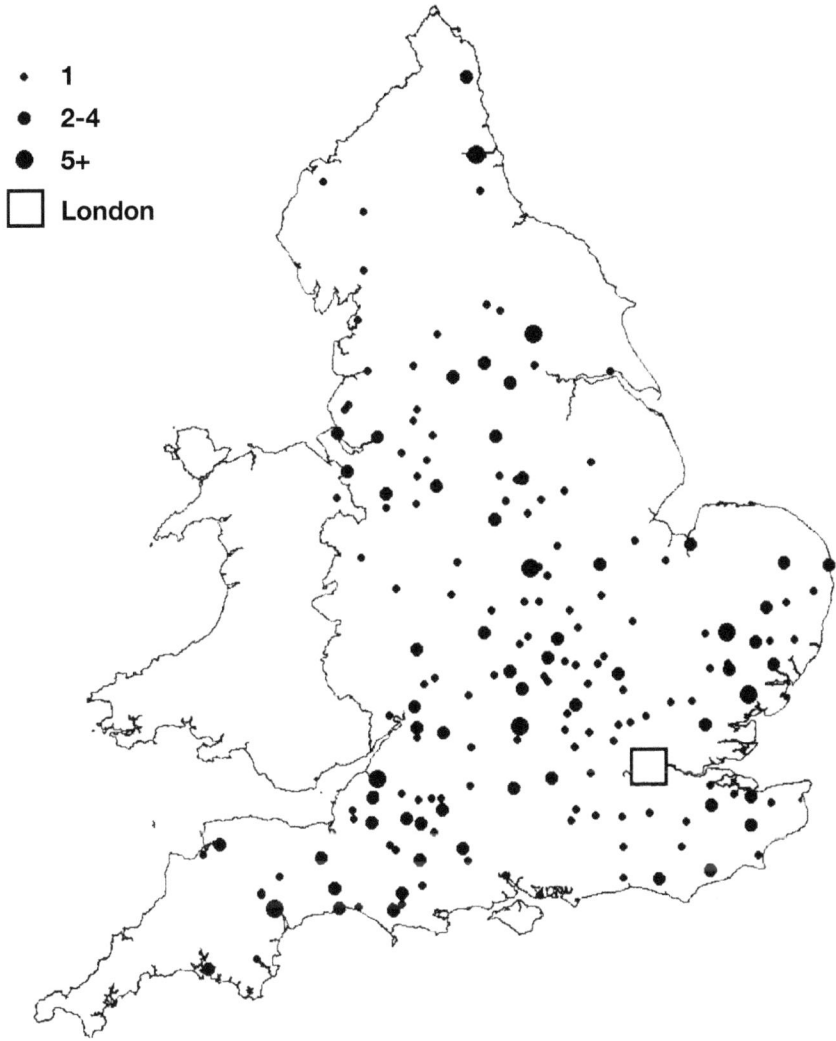

Fig. 5.15. Locations of clockmakers active before 1730 for whom one or more clocks survive.

appraisers often compiled them room by room, and were specific about goods' locations within houses. Clocks' locations can hint at their uses within household life. Commonly over 80 per cent of inventories situate the clock in a specific room,[33] though watches' intrinsic portability makes their locations much less informative. The very clear results for clocks are not what might be predicted from the horological literature (Table 5.5).

Table 5.5. Rooms in which clocks were located in selected inventory collections

Area (date)	[% of houses with clocks having a clock in this room:]						
	Hall	Parlour	Chambers	Kitchen	Stair	Other	(n)
Buckinghamshire (1660–1715)	39	17	15	5	3	17	(59)
East Suffolk (1700–25)	34	28	15	21	2	0	(56)
East Sussex (1700–10)	6	1	9	80	1	3	(80)
East Sussex (1730–37)	4	3	7	84	4	2	(245)
Bristol (1680–1770)	13	22	8	33	4	22	(47)
around Bristol (1680–1770)	4	13	0	84	0	0	(23)
Shropshire	28	26	12	25	6	3	(78)

Notes: (1) (n) = number of inventories specifying location of clock within house.
(2) Totals may sum to over 100% where some houses contained more than one clock.
(3) 'Other' rooms are mainly dining rooms (Bristol), galleries or passages (Bucks).
Sources: Buckinghamshire–Reed (1988); East Suffolk–Evans (2001); East Sussex–East Sussex Record Office, Archdeaconry of Lewes probate inventories W/INV 1–4900. Bristol–Evans (2001); Shropshire–Trinder and J. Cox (1980), Cox and Trinder, (2000); Kent–Overton *et al*. (2004).

The commonest early modern clock location was the kitchen. Parlours, halls, and staircases did contain clocks—in particular, staircases did sometimes contain long-case clocks *à la Tristram Shandy* (Sterne, 1759–63), but not nearly so often as kitchens. This applied across the social spectrum, and in several parts of England. By comparison, the much lower numbers of clocks found before about 1640, mainly in high status households, appear mainly in 'front-space' rooms such as parlours, drawing rooms, and halls. The presence of clocks in particular rooms was obviously limited by whether households possessed a room given that name. So dining rooms and galleries were not 'options' that were available to middling and poor households. Higher up the social scale the kitchen became a more specialized room, and less likely to contain the household's clock, at least until one reached households that possessed two or more clocks.[34]

The overwhelming majority of surviving early modern clocks are long-case ('grandfather') and table clocks, which conjure up images of parlours, dining rooms, and stair-heads as the main household locations for clocks. That impression may be appropriate for these surviving clocks, but in their ornamentation and cost they are atypical of clocks at that time, whose plainer character, and cheaper prices, were related to their more utilitarian locations within domestic interiors.

That in 1700 many clocks were located in kitchens has implications about the uses made of clock times, and the people involved. Clocks in kitchens have rather different implications for the gendering and social structuring of clock times from those in parlours or dining rooms. Clocks in studies can be interpreted as part of an intellectual, male space, and the Sussex barber with a clock 'in the trimming room' carries less intellectual but equally male associations, whereas the large number of kitchen clocks connotes clock time as much less exclusively masculine.[35] In short, late seventeenth- and early eighteenth-century clock ownership had a substantial utilitarian component, though the worth of even kitchen clocks was more than solely utilitarian. Very many household servants were routinely exposed to clock times, and it is implausible to suppose that this was forgotten the moment they left household service.

5.5 CONCLUSIONS

Our central finding has been the sheer density of temporal infrastructure in early modern England: the numbers and density of both public clocks and private clocks and watches. Overall, William Harrison in 1577 provided a broadly accurate description of English clock time:

the common and natural day [being] observed continually by clocks, dials and astronomical instruments of all kinds. (Edelen, 1968: 379)

We have traced the social and geographical range of the availability of timepieces through several lines of evidence. Public clocks existed in considerable numbers, especially those in (and, increasingly, on) parish churches, civic buildings, cathedrals (and earlier monasteries), and 'quasi-public' buildings from inns and schools to post offices and almshouses. We have added to recent work establishing the early origins and rapid late seventeenth-century growth of privately owned clocks and watches, and the presence of clock time in many domestic interiors. And we have attempted to indicate the variety of ways in which a generally recognized need for clear time-signalling was realized, either aurally, visually, or through a combination of the two.

These findings help us to place our Bristol case-study (Chapter 4) in the broader English context. In the facets of temporal infrastructure examined here, the temporal environment in Bristol was not radically different from that found in many other parts of England.[36] There were differences in the density of timekeeping and signalling, but these was no dramatic contrast between clock times present in Bristol but absent elsewhere. The differences among different parts of England, and between town and country, were differences of degree not differences in kind.

Numbers of church clocks increased substantially during the sixteenth and seventeenth centuries; they increasingly incorporated more frequent and elaborate ringing or chiming, and continuous time-indication using large dials on church towers. These trends were marked in both urban and rural parishes, and over much of England. Most public clocks were sufficiently maintained to keep time fairly well, and were at least adequate to a variety of practical uses. The reputation, in historical literature, of church clocks as constantly breaking down has largely arisen through the erroneous interpretation of money spent on routine tasks as indicating the need for significant repairs. The point here is that daily winding, the disconnection of striking mechanisms between the curfew and day bells, oiling, and periodic cleaning were integral to the clock running normally, and not a sign that it was malfunctioning. Later commentators have here been imposing their own (taken-for-granted) definitions of 'a working clock'. The same is true of concepts of accuracy. Clocks did not necessarily need to be extremely accurate to be useful. More important in early modern England was that communities and households could handle the running of clocks to produce signals that had become built into everyday conduct, either through conscious design or implicit habit.

It is also clear that many contemporaries attached considerable importance to public information on the time of day, and that this served a wide range of motives. Time signals played roles in numerous activities besides those for which clocks had originally been constructed.

The oft-made distinction between urban and rural timekeeping, or rather between timed urban life and natural rural life, has been drastically overdrawn. There were urban–rural differences in early modern English timekeeping, but they were differences of degree, not differences in kind. Urban 'chronotypes' or 'temporal orderings' were in several ways more complex than their rural equivalents, but they were not utterly foreign to the non-townsman or woman (Bender and Wellbery, 1991; Dohrn van-Rossum, 1996; Glennie and Thrift, 1996). In any case, very large numbers of country people regularly visited towns for fairs, markets, work, recreation, to see kin, courts, visitations, and on other occasions (Glennie, 2001: 132–55). Significant numbers of rural dwellers spent some years, especially as servants or apprentices, living in towns, before returning to live in villages and hamlets, to which they may have been native or not.

The long history and dense infrastructure of public timekeeping, and the wide range of people on whose lives clock time impinged, significantly affected private timekeeping during and after the horological revolution. Although the actual workings of clocks remained unfamiliar, even mysterious or magical, to many people, the notion of linking one's conduct to certain hours was something so common as to have attracted little explicit comment. This has often been interpreted as indicating that most people could not comprehend clock time, but to us the moral is different. It was precisely *because* clocks were a familiar part of many everyday landscapes that they did not need to be explained.

We therefore conclude that the focus of so much work on domestic clocks and watches, and on factory work-discipline, monastic timetables, and 'merchants' time', has led many writers to severely underestimate the temporal sophistication of pre-industrial English society. Church clocks were both an integral part of temporal structures (that is, the uses of time to organize or discipline social activities), and a familiar part of the temporal infrastructure of everyday life. By the last phrase, we mean that church and other public clocks had become an effective part of practical knowledges about time-reckoning in a wide range of everyday activities.

That clock times were already involved in many familiar practices for much of the population helps to explain the otherwise puzzlingly rapid diffusion of clocks and watches between 1660 and 1730. They were newly cheap(er) items, newly reliable items, available to a population able to deal, and in many places long-used to dealing, with clock time. Not the least important factor, we suggest, in Weatherill's 1670–1730 boom in clock and watch ownership, was that large parts of the population already had relatively sophisticated understandings of everyday clock time and of what this could entail. Far from the private clock (or much later the factory) *initiating* a grasp of clock time, as for Thompson, that practical grasp had been present for generations.

Horologists' preoccupation with surviving domestic clocks has also had much wider consequences for prevailing understandings of histories of clocks and times. For example, it pervades Dava Sobel's best-selling *Longitude* (1996). That Sobel thought it so astonishing for a joiner's son from a small, north Lincolnshire market town in 1700 to be deeply interested in precision clockmaking is revealing of her assumptions about the scarcity of clock times in provincial England. This book paints a somewhat different picture.

NOTES

1. In practice, things are not quite so simple, since the interval between two successive noons is not exactly 24 hours, but fluctuates considerably through the year between less than 23.72 hours and more than 24.25 hours. This was not realized until the 1650s, see the discussion in Chapter 7.
2. For example, at St Mary's Abbey, York, in 1324 (Humphrey, 2001: 110), at Ramsey Abbey (pers. comm. from Anne and Edwin de Windt). Northallerton (pers. comm. Richard Britnell).
3. Berkshire Record Office, D/P 149/51/1, unpaginated, on blank page between the accounts for 1561 and 1562.
4. Quoted by Tittler (1991): 137.
5. Cornwall Record Office, Truro, DDR(S)1/902 Letter from John Rashleighe of Fowey to Jonathan Rashleighe of London, January 1620.
6. East Suffolk Record Office. FC25/E1/1, unpaginated.
7. See the following chapter.
8. It is important here to distinguish mechanical clocks from the earlier development of other devices, most particularly large water clocks. These were clearly present at several monasteries before the rapid expansion of mechanical monastic clocks from the 1270s (Beeson, 1977; Dohrn van-Rossum, 1996).
9. Updated maps of early modern town clocks are online at: <www.ggy.bris.ac.uk/clocks>.

10. Cox's arguments are reproduced, for example, in Bellchambers (1969): preface; Haggar and Miller (1974). The rebuttal is Beeson (1988). Bizarrely, Bellchambers cites Beeson as supporting for Cox's claim.
11. For example, land reserved to support costs and wages relating to a clock were sometimes became known as 'clock land', as at 'clock and chimes cottages' at Green End in Aylesbury in 1691 (Legg, 1976: 9).
12. As at Stratford-upon-Avon, where the guild of the Holy Cross maintained both the town clock, 'called the Clock House' in 1472, and the chapel clock, that the Guild had provided in 1463–4, from the 1440s into the sixteenth century (Shakespeare Centre Archives, Stratford: class BRT1/3, especially BRT1/3/74).
13. Details and a mapping facility for this information may be found at: <www.ggy.bris.ac.uk/clocks/parishclocks/animationII/index.html>.
14. For fuller discussion, see Kümin (1995, 1999); the dispute between Burgess (2000, 2002a, 2004) and Kümin (2004); Glennie (forthcoming, chapter 3).
15. Here we make a coarse division into five categories: (i) rural and non-market parishes; (ii) market towns containing (or contained within) a single parish; (iii) market towns containing more than one parish; (iv) county and provincial towns with several parishes; and (v) London, with its more than one hundred parishes. Ascriptions of town and market status are based on Eversley (1967), Clark and Hosking (1993), and Langton (2000).
16. We have used the edition and commentary of Whiteman (1986). We sought parish-level information about communicant age adults, for Anglican, Catholic, and Non-conformist faiths, and (in theory) for non-communicants. The returns for a small minority of places have been disregarded, as where commentators have made good cases for the numbers returned having been households, rather than communicants. Hearth taxes from about the same time offer a less suitable source, because of the inconsistent criteria for exemption on grounds of poverty, and the inconsistency of information about the numbers of households exempted: Husbands (1985).
17. London church clocks are the topic of Glennie (forthcoming, chapter 7).
18. The exact figures for parishes leaving observations are:

Period	Clock	No clock	% with
Pre-1620	41	55	43
Post-1620	147	115	56.1

19. Given the absence of systematic population censuses over this period, we have relied on the major national counts of relatively large groups of the population, in particular communicant-age adults in 1676 (Anglicans, Catholics, Non-conformists, and non-communicants. The last group were not always specified): Whiteman (1986).
20. Three of the five Bedford parishes had church clocks in the visitation. It may be possibly be significant that the apparent exception, Biggleswade, was a peculiar court jurisdiction.

21. The Bristol writer is quoted by Barry (1984): 134–5; for Boston, see Thompson (1856): 189.
22. Berkshire Record Office, D/P 97/5/1–4, 9 (copy of D/P/5/2 in MF 97020); Kerry, 1883.
23. Some debate surrounds disputes over the earliest pendulum clock (Yoder, 1988), which we will not explore here.
24. There are nearly one hundred 'new made' clocks in the documented parishes between 1670 and 1700. Many seem too cheap to be of the latest technology (some may have been bought second-hand from churches or gentry that were upgrading their clocks to pendulums), but there are still a couple of dozen where the cost is similar to pendulum clocks, but with only a vague description.
25. In the rebuilding of parish churches after the 1666 Great Fire, several London parishes bought new clocks from men such a William Clement whose turret clock installations elsewhere in the 1660s and 1670s included pendulums. Clement supplied the new clock for St Martin Orgar in 1675, and maintained it into the late 1680s: London Guildhall Library, 959/1.
26. Oxfordshire Archives, DD Par Yarnton b.7, see also Beeson (1989: 74–5).
27. For example, the rebuilding of St Benet Gracechurch finished in the early 1690s, included over £37 on the workings of a new clock, again from an unnamed clockmaker, in 1691, and a considerably larger sum to "Mr Mitchell" for the associated dial-work: London Guildhall Library, 1568.
28. Hertfordshire County Record Office, D/P 21/5/1–2, (1680). For Clement, see Beeson (1989): 66, 88–9.
29. Van de Woude and Schuurmann (1980); Overton (1985); Orlin (2002); Overton *et al.* (2004): 14–18.
30. Weatherill (1988) analysed possessions in random samples of 65 probate inventories per decade for each of eight parts of England.
31. Apart from the possibility that elderly testators may have acquired goods long before their deaths, obvious complications also include the possible reluctance of older people to buy new types of items, especially those with an explicit technological component. The question of whether older people actually were comparatively slow to purchase new items is itself debated.
32. For overviews and discussion see Overton (1984); Orlin (2002); Overton *et al.* (2004): 22–8.
33. For example, 245 of 276 inventories with clocks in East Sussex in the 1730s, East Sussex Record Office, W/INV.
34. Hertfordshire County Record Office, chapter 4. East Sussex Records Office, Probate inventories, Archdeaconry of Lewes, 1–3254.
35. East Sussex Record Office, W/INV. Laurence (1994): 144, suggests that clock ownership among female testators was significantly lower than among males, her source (Weatherill).

36. The clearest contrasts come in areas where Bristol's sheer size and population density created a denser 'feel' to everyday environments, and the overlapping of hour-ringing public clocks was only experienced in larger cities. However, the general familiarity of public clocks and hour-signalling, and the prevalence of domestic clock ownership (and their locations) all stand as national-level generalizations.

6
Clock Times in Everyday Lives

6.1 INTRODUCTION

This chapter and the following one explore everyday clock-time practices in early modern England, focusing respectively on clock-time uses and concern with precision. In Chapter 2 we outlined a series of strategies to tackle the near-invisibility in the archival material of everyday practices that were usually almost entirely taken for granted. Here we shall touch only briefly on either specialized temporal communities, using specific clock times in pursuit of particular technical objectives, or the disciplinary institutions (government, religion, trade) that dominate formal documentation. Rather, we will explore what we envisage as an 'everyday' temporal community—the bulk of the population—whose clock-time practices unfolded in 'non-disciplinary' circumstances as part of their general awareness of time and society.

In referring to 'everyday life' and 'ordinary people', our intention is a broad definition of people, activities, and places that can encompass something like 'non-elite people, carrying out tasks that do not inherently, or out of necessity, possess a high degree of temporal specialization'. Much of 'the everyday' was taken for granted, raising critical questions about where, and how, it can be pursued archivally (above, section 2.7): about clock times as an everyday resource for people's expressing times of day; about how frequently this was done; and for how many purposes.

We think of clock time, in other words, as a practice whose societal and geographical dimensions are to be recovered, as best we can, from an array of documentary sources that, without exception, were underlain by very different concerns from attesting to the use of clock time.

Our central question is how, and when, clock times were important in the 'everyday' practices of 'ordinary' people. What range of clock-time practices and purposes can be excavated from traces in the historical record? Obviously, 'everyday' practices and 'ordinary' people raise questions of definition and methodology, and we elaborated our approach in Chapter 2. We have sought sources beyond the archives of official regulation that dominate so much timekeeping history. Of course, this is not to ignore these archives but they have to be read both with and against the grain, with a sensitivity for the unsaid.

The two chapters draw on various types of narratives, personal diaries, court depositions, autobiographical memoranda, letters, almanacs, and marginal annotations, using chance survivals of incidental material as our windows onto everyday practices. We cannot claim statistical representativeness. Rather, we take these survivals as evidence of specific practices. In several short vignettes we explore rare or unusual survivals of sources that offer particular insight into the occasions and specifics of clock-time usage, both in explicit descriptions of activities or anticipations, or in the implicit assumptions underlying particular usage.

This is not to turn our backs on the various disciplinary purposes of clock times: urban regulation, controlled markets or working hours, church times, and more. Neither the existence nor the immediate effects of such disciplines is in doubt, but we discard the usual assumption that they were necessarily the pre-eminent basis for everyday clock times. That is an important substantive question, and in tackling it we accord at least as much significance to clock-time infrastructures, and to less formal and less disciplinary clock-time practices.

Accordingly, the body of the chapter is divided into seven substantive sections. First, in section 6.2, we identify some indications and uses of clock times in late medieval England, to show the uptake of clock time in a range of milieux. We then jump forward in section 6.3 to the late seventeenth century to show the very many ways in which clock times entered into descriptions of people, events, and environments in John Aubrey's collection of biographical material in his *Brief Lives*. We briefly review the resources available in section 6.4, while section 6.5 explores the broad contours of clock times used by diarists. Section 6.6 comprises ten brief vignettes of uses of clock time in specific contexts, many of them plebeian, and concerning activities that involved a good deal of self-organization, rather than the clear imposition of outside authority. Here, too, we register some important methodological lessons, especially regarding the interpretation of apparently negative evidence of temporal competences. Given our earlier emphasis on the high density of clocks, how contemporaries understood clock times and time-signals is of considerable interest. Section 6.7 therefore explores people's acquisition of clock-time skills, in schooling and in everyday life. Section 6.8 integrates the preceding sections into a long-run account of changing clock-time practices. A brief concluding section returns to a general critique of current literature.

6.2 CLOCK TIMES BECOME ORDINARY

Relatively soon after the first development of mechanical clocks in the late thirteenth century, clock times are documented in the course of regulating, or facilitating the conduct of, various activities. Since regulation was a major motor

Clock Times in Everyday Lives 183

Fig. 6.1. Locations mentioned in this chapter.

of recording, it is no surprise that many of the earliest documented clock times were regulatory, as in Bristol. Similar uses can be found in many other power centres: the church, cities, towns, the court.

We must be wary of uncritically projecting normative regulations as accounts of everyday practical conduct. In several cases, though, there are signs of regulations in action. Regulations are adjusted or modified, which would not have been necessary unless they were having effects; people were prosecuted for transgressions, many regulations were periodically renewed, rather than staying on the books purely through inertia. As we shall see, some of the regulated towns must have used time signals, because those signals were put to secondary uses, including other regulatory uses.

As we have seen, bells were the chief instruments conveying clock time in all sorts of accounts, from normative legislation or by-laws to descriptive narrative. We can discern how clock-bells 'made sense' in everyday life through people's apparent comfort with using hour-struck signals for their own purposes.

	Ecology of DEVICES	Embodied PRACTICES	Timekeeping SKILLS
DISCOURSES of measurement and precision			
Urban SPACES (multiple communities)			
COMMUNITIES of practice			

Fig. 6.2. Components of researching clock time practices.

These involve a more explicit use of time as an ordering metric of the everyday compared with, for example, earlier opportunistic use of particular monastic service-times. The names used can make some references to clock-bells disconcerting—as where churchwardens paid a sexton for 'ringing the eight o'clock bell at four o'clock'—but that such uses appear routine and sustainable shows contemporaries' growing comfort with clock times.

Sources of knowledge about the time are made explicit from time to time in a set of different mileux. First, there are *legal cases*, since particular courses of actions could be established, or called into question, with respect to public information. Thus, a deponent at York in 1395 explained that he knew the time at which he approached York because he heard 'the striking of the bell in the Minster popularly called "clokke"' (Goldberg, 1995: 106). We shall return to the law later.

Much of the monastic impulse to timekeeping revolved around prayers and services, rather than the mechanics of monastic living in itself. But a well-defined order facilitated smooth routines of both worship and institutions containing it, and clock-time based timetables passed quite readily into the *royal court*, and *elite households around the court*, in connection with administration, piety, and legitimacy. It is in this second milieux that some of the earliest and most specific timetables were constructed for royal children, to cultivate regal conduct and skills.

Arrangements survive for Prince Edward (later, as Edward V, to be murdered in the Tower of London) as a three-year-old in 1473, and again in 1483, and for John Mowbray, Duke of Norfolk, in about 1435.[1] The royal ordinances stipulate the powers of Edward's 'governers', what he was to be taught, and a timetable of hours for different activities and prayers, that sometimes seems more concerned with schedules than with specific skills:

that he rise from his bed . . . betwixt 6 and 7 of the clock, and say with his chaplain or some other honest person matins of Our Lady with prime and hours [that he] shall rise every morning at a convenient hour according to his age, and till he be ready no man be suffered to enter into his chamber

that he go to his rest every night at latest by 10 of the clock

[that he] be in his chamber and for all night livery [food and drink] to be set, the traverse [curtain] drawn, anon upon eight of the clocke ['ix' in 1483], and all persons from thence to be avoided. (Orme, 1989: 178, reordered)

Edward's upbringing went on in the already timetabled household of Edward IV, where ordinances laid down a said mass for household officers at 6 a.m., matins at 7 a.m., and a sung mass with the children at 9 a.m. Servants were to dine at 10 a.m. (11 a.m. on fasting days) and have supper at 4. p.m., before the royal household ate an hour later. Porters opened the gates between 6 and 7 a.m. and closed them at 9 p.m. every day, except in summer when they opened an hour earlier and closed an hour later (Myers, 1959; Society of Antiquaries, London, 1790).

A handful of households centred on devout upper-elite women also had well-defined timetables, focused on piety and prayer. The relatively complex routines of Edward IV's mother Cecily (d.1495), and Margaret Beaufort, Countess of Richmond (d.1509), the mother of Henry VII, suggest a remodelling of existing practices. Both followed carefully timed routines of business, meals, and prayers in units as brief as quarter-hours.[2] About 1490, Cecily's household ordinances set out the duchess's daily routine (Table 6.1).[3]

Table 6.1. Daily routine of Cecily, Duchess of York, *c*.1490

She is accustomed to rise at seven o'clock and her chaplain is ready to say with her matins of the day and matins of Our Lady. When she is fully ready, she hears a Low Mass in her chamber and after Mass she takes some breakfast. She goes to the chapel to hear the divine service and two Low Masses, and from there to dinner. During dinner, she hears a reading on a holy subject, . . . After dinner she gives audience for an hour to all who have any business with her. She then sleeps for a quarter of an hour. After sleeping she continues in prayer until the first bell rings for Evensong. Then she enjoys a drink of wine or ale. Without delay her chaplain is ready to say both evensongs with her, and after the last bell has rung, she goes to chapel and hears Evensong sung. From there she goes to supper where she repeats the reading which was heard at dinner to those who are in her presence. After supper she spends time with her gentlewomen in the enjoyment of honest mirth. One hour before going to bed she takes a cup of wine, and then goes to her private closet where she takes her leave of God for the night, bringing to an end her prayers of the day. By eight o'clock she is in bed. . . .

On eating [i.e. meat-eating] days at dinner there is a first dinner by eleven o'clock during the time of High Mass for carvers, cupbearers, sewers and officers

On fasting [i.e. fish-eating] days by twelve o'clock, a later dinner for carvers and waiters

On eating days, supper for carvers and officers at four o'clock, and for my lady and the household at five o'clock.

The schedules of monasteries or devout households were disciplinary in a different sense from the factories described by Thompson (1967), where time-discipline was simultaneously standardized, regular, and coordinated (Glennie and Thrift, 1996b: 285–88). Different people had to be coordinated, certainly, but that was because only in certain very specific respects were roles standardized. Several household officials and many servants were important precisely because their routines and flexibility catered for the timetable to which the focal individuals of the household aspired.

A third mileux was the urban. Most medieval English *cities* were small by European standards, and few may have been large enough to have experienced the 'acoustic chaos' that Dohrn-van Rossum (1996) describes as motivating the implementation of orderly hour-ringing. Yet public clocks spread rapidly in English monasteries and towns, attesting a more general belief that clock time brought some combination of practical, political, and/or self-presentational advantages.[4]

Chapter 4 showed clock times being used to organize and police fourteenth-century Bristol's markets, quays, and inhabitants. Similar uses of clock times pepper other sources, from elite judicial proceedings like the trial of Joan of Arc (1431), which began at 'eight of the clock in the morning', to small-town Maldon (Essex) where market ordinances switched from canonical hours, prohibiting food sales 'until the hour of prime, when the bell is rung' and restrained butchers until 'after matins is rung on the bell of All Saints church', to clock times. Fifteenth-century Maldon's day ran from the Angelus bell (4 a.m. summer, 5 a.m. winter) marking the first mass and standing-down the night watchman, to the curfew bell at 10 o'clock (summer) or 8 o'clock (winter).[5]

Humphrey (2001: 108) has christened such self-regulatory uses 'aspirational time'—a broad desire to systematize an urban temporal framework—as opposed to the 'practical time' of specific activities. Such exercising of power—by the church, market authorities, and employers—is central to the disciplinary histories of Le Goff and Thompson. So far, so familiar, but Humphrey also emphasizes that 'bottom-up' community self-organization was a powerful motive for time-signalling. Thus, similar uses of clock times appear in late medieval agricultural by-laws in *rural* settings, weakening the utilitarian argument that clocks were implemented to simplify the complex acoustic environments of large cities (Dohrn-van Rossum, 1996).

But if particular interest surrounds the instances where time-regulation and enforcement were *not* imposed by authorities, they are intrinsically unlikely to be documented. Formal time-regulations and policing generate records of rules, infringements, investigations, and punishments. Without this document-generating infrastructure, evidence is necessarily more fragmentary. Tracing 'bottom-up' clock-time restrictions is problematic because evidence about them is inevitably haphazard and fragmentary in the absence of overseeing institutions that generate documentation. Commonly, clock-based community self-regulation is first recorded only when threatened or suspended, as when 'gleaning bells' were documented when common gleaning rights were extinguished during nineteenth-century enclosures. Gleaning, the post-harvest scavenging arable fields for grain, was a practice usually restricted to women, children, and sometimes elderly men. When longstanding practices are spelled out, it becomes clear that communities used clock time to define where and when certain households could glean, with time signalled by church clocks.

> gleaning operated in most Essex parishes to strictly controlled and well-observed timetables... governed by the ringing of a bell, usually the church bell, or, as it was called locally, the gleaning bell.   (Hussey, 1997: 63)[6]

Gleaning bells in nineteenth-century Essex generally sounded at eight or nine o'clock in the morning, and at five or six at night, to keep gleaning within the common working day for men.[7]

These comparatively recent expressions of local time-restrictions among agricultural households have close resemblances to those found in manorial by-laws several centuries earlier. Thirteenth-century village by-laws mixed natural, opportunistic, and canonical hours in restricting harvesting, grazing, gleaning, and moving livestock (Ault, 1972). Restrictions at Newton Longville (Bucks.) used canonical hours from the 1290s to the 1330: 'anyone who wants to gather beans, peas or such like shall gather them between sunlight and prime..., and this after the feast of the Blessed Virgin Mary'. Identical stipulations survive from other Buckinghamshire villages, like Halton in 1295 and 1329, varying slightly at Great Horwood, where 1316 by-laws fined tenants 6d. for gathering beans 'except between mid-prime and prime', becoming 'between sunrise and prime' in 1368 (Ault, 1972: 82–94, 105, 171–4).

Such instances demonstrate how systems of equal, rather than canonical hours, and the practice of numbering hours, after the late thirteenth-century appearance of mechanical clocks, took place in the context of pre-existing regulations and time-markers. Clock time was not the source of regulation per se, but a ready means of reformulating existing restrictions. Regulations using clock time were novel only in that mechanical clocks provided the metric, otherwise the regulations paralleled earlier restrictions drawing on 'natural' markers (sunrise, noon, sunset, and finer qualitative distinctions in daylight), and on available time-signals, especially the audible ringing of canonical hours around monastic communities.

The degree to which 'community' restrictions in manorial court proceedings and by-laws should be seen as the exercise of lordly authority, or as the self-government of tenant communities, has been a perennial debate. So, too, has been the recognition that communities were very rarely communities of equals, and that particular restrictions and practices were manifestations of very unequal power-relations. Community restrictions like gleaning regulations operated unevenly in a whole series of ways, depending on people's local entitlements, on household composition, on activity patterns, and on available time. The outcome, potentially, was that people's immediate experiences of, and relations to, clock times as a part of those entitlements and restrictions, were extremely diverse both quantitatively and qualitatively.

In both town and country, the activities regulated using clock time in the late fourteenth century resembled those regulated through other cues a century earlier. Clocks and equal-hour reckoning were mutually reinforcing. As mechanical clocks became more common, times of day increasingly refer to clock times, rather than natural cues or canonical hours, whether in formal restrictions of marketing or production hours; informal descriptions of events; or community self-organization. As the new ways of describing times of day circulate, clocks become more and more taken for granted as a component of civic life. Churchwardens' agreements with clocksmiths emphasize makers' responding quickly when repairs were necessary, and a similar concern with restoring provision was explicit in John Rashleighe's later letter from Fowey.

The distinction between 'formal' regulatory uses of clock times, on the one hand, and substantially under-documented 'informal' clock-time practices on the other, is obviously simplistic. People, groups, or key inhabitants acting for whole communities took it upon themselves to use clock time in framing specific regulations for particular people. Often these were not independent, but seem modelled on practices in larger or more prestigious cities. For example, Robert Reynes, in fifteenth-century Acle, Norfolk, has material in his commonplace book drawn from Norwich, some ten miles to the west. These range from a list of fires to a note of instructions to the night-watch in a town, running from 'ix of Ye clok at evyn' until 'iii of the clok aftyr mydnygth' (Louis, 1980: 185–7). After listing the dates of several major fires in Norwich over the previous half century, Reynes turns to Acle's own, in 1485, a

gret, dredfull ffyer... the vii day of May the Sounday nest aftyr Crowichemesse vpon the Mayday at iiii of the clok at aftyrnoon begynnyng. (Louis, 1980: 358–9)

Were it not for the chance survival of Reynes's book, we might be struck by the early date of a similar record on 25 June 1600,[8] when one of the North Walsham churchwardens wrote an eyewitness account, for future churchwardens, of that morning's 'Great Fire':

it began about six of the clocke in the morning and went on so fiercely that in two hours the whole body of the town being built chiefly round the market place was in one flame and so in two or three hours were burnt down to the ground.[9]

The likes of rural commonplace books, or churchwardens' accounts might not ordinarily be viewed as the cutting edge of temporal description, but they illustrate not just the normality of clock times in narratives by 1600, but also that the practices and capability went back very much further. The comment in the early fifteenth-century dialogue *Dives et Pauper* that 'in towns and cities men rule themselves by the clock' (Barnum, 1976: 120, 2005) should not invite the supposition that they were not used in narrative elsewhere.

6.3 CLOCK TIMES IN JOHN AUBREY'S *BRIEF LIVES*

The range of ways in which clock-time uses in diaries and other writings or statements appear can be conveniently overviewed and illustrated from John Aubrey's *Brief Lives*, his late seventeenth-century biographical notes on contemporaries.[10] The information is mostly quite compressed and terse, but it is abundantly clear that clock times were an effective descriptive resource. Aubrey was obviously among the social and educational elite: we use him here to illustrate eleven ways in which clock times entered his biographies.

First, as for the North Walsham churchwarden, or Robert Reynes, clock times were second nature in narrating events, sometimes in contexts where the particular hour mattered, but more often in routine description. General Monck's Restoration army 'came into London . . . about one o'clock p.m.' on Saturday 10 February 1660, while Aubrey heard from Robert Hooke on 15 November 1673, of Christopher Wren's being knighted at 5 a.m. the previous day; the wife of William Holder, a noted physician, was sent for by the King 'at eleven o'clock at night'. On Saturday 26 August 1682, 'about 4 o'clock p.m. William Penn went to Deal before sailing for Pennsylvania.' Aubrey also records clock time being noted by earlier writers, telling how he found Francis Potter's Greek testament inscribed 'In 1625 at 10 o'clock on December 10, the number of the beast was discovered'[11] (Barbor, 1982: 207, 330–1, 164, 240, 255).

Second, the most prominent occasions for which clock times were recorded were births and deaths. Aubrey was told the birth-time of John Milton, 'half-an-hour after 6 in the morning' on 9 December 1608, and of Christopher Wren, 20 October 1631 at '8 p.m.—the bell rang VIII as his mother fell in labour with him'. The motive was astrological: William Petty's 'horoscope was done, and a judgement upon it' after his birth 'on Monday 26th May 1623, eleven hours 42′ 56″ afternoon'[12] (Barbor, 1982: 201, 239, 330, 341–2). Aubrey also gave several death-times: Francis Villiers killed on 7 July 1648 at 'six or seven o'clock in the afternoon'; Sir Robert Moray suddenly on 4 July 1673 'about 8 p.m'; Jonathan Goddard at 'eleven o'clock at night, he fell down dead of an apoplexy in Cheapside'; Robert Pugh 'at Newgate on January 22, 1679, Wednesday night, 12 o'clock'; John Partridge, a shoemaker, astrologer, and publisher, 'December 12th, 1685, between 4 and 5 p.m.' (Barbor, 1982: 308, 212, 199, 260, 237–8).

Third, Aubrey's clock times could also convey moral standing, with early rising indicative of dedication and selflessness. John Milton rose at 4 a.m., going 'to bed about nine'; John Haskyns (1566–1638) rose early 'that is, at four in the morning'; the Revd Edward Davenant, a secret mathematician, 'rose at 4 or 5 in the morning, so that he followed his studies till 6 or 7, the time that other merchants go about their business; so that, stealing so much and so quiet time in the morning, he studied as much as most men'. However, William Oughtred lay 'abed till eleven or twelve o'clock with his doublet on... [but] studied late... went not to bed until eleven o'clock'. Late nights conveyed dedicated study, not wantonness, for the young Milton 'studying until 12 or 1 o'clock at night', and Thomas Hobbes's schoolmaster at Malmesbury, who instructed 'him, and two or three ingenious youths more... till nine o'clock'. Similarly, routines denoted commitment to moral time-use. When old, Hobbes 'rose about seven', walked and meditated till ten, with dinner 'provided for him exactly by eleven'; William Prynne was brought food 'about every three hours'; the physician-astrologer Richard Napier spent two hours daily in family prayers, besides telling 'his own death to a day and hour' (Barbor, 1982: 201–4, 172, 87, 234, 150, 259, 219).

Fourth, Aubrey contrasted the greater number and reliability of timepieces c.1680 with their rarity in the early seventeenth century, through an anecdote describing the astrologer-mathematician Thomas Allen's experience as a guest at Holme Lacy, Herefordshire. Allen (1542–1632):

happened to leave his watch in the chamber window, watches were then rarities; the maids came in to make the bed, and hearing a thing in a case cry 'Tick, tick, tick' at once concluded that this was his devil, and took the string with the tongs, and threw it out of the window into the moat, to drown the devil. It so happened that the string hung on a sprig of elder that grew out of the moat, and this confirmed to them that it was the devil. So the good old gentleman got his watch again. (Barbor, 1982: 16–17)[13]

Fifth, though watches had been rarities, public hour-striking had long been familiar. Aubrey's informants sometimes specified how time was known, as in Wren's mother going into labour as 'the bell rang VIII'. Especially in towns, clock times indicated by bells were generally apprehended.

Sixth, the presence of watchmakers was another significant urban attribute. The young William Petty (born 1623) growing up in Romsey, 'a little haven port [with]... most kinds of artificers', watched watchmakers and other skilled craftsmen at work (Barbor, 1982: 341–2).[14]

Seventh, the construction of sundials indicating clock time was a significant technical achievement, through which Aubrey emphasizes mathematical and astronomical talent. Some makers were professionals like Francis Hall (1595–1675) Professor at Liege Jesuit College, and Samuel Foster (died 1652), Professor at London's Gresham College, on whose lodgings wall was 'his own hand drawing, the best sundial I do verily believe in the whole world... shows the time... in Jerusalem, Gran Cairo, etc.'. Other noted dial-builders were non-mathematicians, including the Bristol poet William Holder, and Francis Potter whose 'fine sundial' at Trinity College Oxford, was 'did by Samminitatus's book of sundialling'. Dialling particularly evoked precociously talented teenagers.

Around 1640, a young Robert Hooke on the Isle of Wight 'made a sundial on a round trencher; never having had any instruction, his father was not mathematical at all', while Edmund Halley, among Aubrey's youngest subjects, likewise demonstrated his geometric talent: 'He... at 16 could make a dial, and then, he said, thought himself a brave fellow.'[15] Showing both mean and solar time was rare expertise: Nicholas Mercator presented Charles II with a clock showing 'the inequality of the sun's motion from the apparent motion, which the King did understand... and commend'. However, 'this curious clock was neglected, and somebody of the court happened to become master of it, who understood it not; he sold it to Mr Knibb, a watchmaker, who did not understand it neither, who sold it to Mr Fromanteel (that made it) for £5 who asks now [1683] for it £200' (Barbor, 1982: 124, 114, 162, 255, 167, 124–5, 200).

Eighth, intriguing questions of the limits to meaningful precision are raised by William Petty's very exact birth-time: 'eleven hours 42 minutes 56 seconds afternoon'. Is this intended to be taken at face value, or as a joke (Aubrey's own joke, or as Petty's joke on Aubrey)? Even if frivolous, the anecdote recognizes that: events or durations might be timed to a particular second (even if this instance raises practical questions about defining a moment of birth); that, seconds were identifiable and measurable as distinct moments; and that, differences of seconds possibly had astrological significance.

Ninth, watches readily became heirlooms or special gifts. On his death-bed after a stroke in 1657, the physician William Harvey gave his watch to a nephew. Immediately prior to his execution, Charles I gave watches to the two men assigned by Parliament to the King's bedchamber during his final custody. Clocks also had memorial connotations: Seth Ward, Bishop of Exeter, gave the Royal Society a 'noble' pendulum clock in 1662, in memory of Laurence Rooke, Gresham College's professor of astronomy (Barbor, 1982: 132–4, 127, 273, 318).

Tenth, implicit reckoning of clock time and speeds was revealed during the late sixteenth-century 'Littlecote Murder' trial, where the midwife brought blindfolded to treat a later-murdered infant gave evidence in which she testified that during the journey 'she considered with herself the time that she was riding, and how many miles might be ridden at that rate' (Barbor, 1982: 252).

Eleventh, clock times are mentioned in contexts involving horse-racing and gambling. Aubrey describes two exceptional racehorses owned by the fourth Earl of Pembroke, recalling that 'Peacock used to run the four mile course in five minutes and a little more; and Delavill came since but little short of him' (Barbor, 1982: 142).[16] Aubrey's 'four-mile course' is probably that set up at Newmarket for the King's Plate in 1665. Four of the original articles for the race mentioned timing. Horses were to be 'led out between eleven and twelve of the clock in the forenoon, and shall be ready to start by one'. The competition comprised three heats, every horse having 'half an hour's time to rub between each heat'. Then 'the Clerk of the Race is to summons the riders to start again at the end of half an hour by the signal of drum, trumpet, or any other way, setting up an hour glass for that purpose'. A horse that won all three heats had won the Plate, otherwise there was to be a further race: 'those horses that after the running of the three heats shall run the four mile course, shall lead away, and start within an hour and half, or else to win no plate or prize' (Hore, 1886, II: 246–9). Though not mentioned in the 1665 articles, Aubrey's remark shows that races were at least informally timed, probably as a further basis for gambling: timing became more formal during the eighteenth century. Gambling and timing were similarly prominent in foot-racing (Radford and Ward-Smith, 2003).

Aubrey's repertoire of clock-time occasions was wide, and versatile. Predictably, there are few individuals for whom there is sufficient documentation to identify all the clock-time practices detectable in Aubrey. But neither was Aubrey exhaustive, and the following sections explore the range of circumstances in which his, and further, clock-time practices can be identified, and to gauge their incidence over space and time. After introducing the range of material involved, we survey Pepys and other diarists, before broadening the range of source materials through a series of brief vignettes.

6.4 THE RANGE OF AVAILABLE EVIDENCE AND SOURCES USED

However convenient it is to use Aubrey's brief biographical notes to indicate some of the range of ways in which an eclectically curious mid–late seventeenth-century Englishman might happen to incorporate clock times for various reasons, John Aubrey was not everybody any more than Bristol was all England. It is critical to attempt to assess the extent of the practices evident in Aubrey's *Lives*, and this section identifies some relevant archival resources.

We have sought to trace a broadening range of sources with a shifting centre of gravity, from ecclesiastical, governmental, and civic regulators, through intra-community regulation, to explicit self-regulation by the relatively educated or propertied, to a host of routine, narrative uses seemingly taken for granted. Further extension of social coverage raises even more pressing documentary problems.

To judge from court depositions, the most important near verbatim source for non-elite utterances, similar uses of clock time to locate events within narratives were common in late medieval and sixteenth-century plebeian speech. Obviously, there are likely to have been conventions in deposing, not least because temporal specificity provided one criterion for recognizing versions of events as authentic. General use of hour-times, or sub-divisions of hours, may consequently have been more frequent in depositions than in everyday speech, but plebeian usages mark clock times as significant constituents of everyday life, not just an abstract grid retrospectively imposed on information presented to outside authorities.

Despite the pervasive gender inequalities of early modern science and society, and several impulses to temporal precision being specific to male activities, clock time was no intrinsically masculine preserve. Much early use of clock time involved women, and a Thompsonian dichotomy between time-oriented men and task-oriented women who had little grasp of, or use for, clock times is simply unsustainable. Women's work was often intense, attentive to saving time, and coordinated diverse activities to achieve multiple goals. That their work in and around household activities was rarely overseen by official bodies, or formal authority beyond the household, in no way precluded their using clock time.

Clock times can be found in normative accounts of women's activities like Thomas Tusser's *Points of Huswiferie* (1570, Grigson, 1984). Tusser advocated wives' rising in summer at 4 a.m. and retiring at 10 p.m. (5 a.m. and 9 p.m. respectively in winter).[17] Tusser's prescriptive scheme derived from women's work itself, not knock-on effects from male work. He saw no need to explain clock times to female readers.

Various descriptive accounts, from witnesses at inquests or court cases, show women's narrative use of clock times, so much so that historians have used them in analysing long-run shifts in women's daily time-patterns. Like manorial by-laws, inquest descriptions slowly shift to clock time for locating moments in the day, superceding natural cues and canonical hours. In 1276 a Bedfordshire coroner heard how Emma Sagar fell into a well and drowned whilst carrying a load of straw at 'about prime' and York coroners in the 1430s recorded deaths 'after the common hour of dinner' while Agnes Crispin died 'between three and four o'clock in the afternoon' (Goldberg, 1995: 81–2;[18] Laurence, 1995: 220, 240). Similarly in early sixteenth-century Sussex, female suicides were stated as occurring at 'between 5 and 6 a.m.', 'about 9 p.m.' and 'between nine and ten o'clock before midnight' (Goldberg, 1995: 144–5).[19]

Though many temporal practices were rarely scrutinized or remarked on by outsiders for whom they were not taken for granted, when they are sought out one readily finds that clock times formed part of women's descriptive repertoires. This is unsurprising given the density of public clocks and time signals, and the practical time-handling issues for both women and men in everyday life. Clock-timekeeping impinged upon early modern Englishwomen's domestic lives, from gentry women, where time was important for household management and social interaction, to middling sort and labouring women who were enmeshed in temporalities of everyday trading and domestic activities, though the latter have begun receiving fuller attention only relatively recently.

We look first at diaries, autobiographical memoirs, and letters to illustrate the variety of writers and their uses of clock times, and the range of circumstances in which it occurs. We then focus more tightly on ten vignettes of everyday time-use. Some of these hinge on circumstances in which specific temporal knowledge was being claimed or disputed, while the central point of others is the implicit presumption of familiarity with clock time on the part of particular people or groups. Others focus on calculated timing in relation to work and wages, to tracking the movement of information, and to cooking and recipes.

6.5 CLOCK TIME USE AMONGST DIARISTS

In early modern England, diary-keeping became a more widespread and familiar practice of private or quasi-private writing, and the physical survival of diaries becomes less exceptional, though still uncommon. McKay (2005) identifies over 370 English diaries surviving from before 1700. The best-known early diaries, of men like Samuel Pepys and John Evelyn, are atypically full and sustained compared with most contemporary diaries. Diaries' form varies widely, from dedicated log-books written-up from notes, to opportunistic note-keeping using loose paper sheets, or annotating the margins and blank pages of almanacs or shop-books. This section draws on some twenty diaries and related material, including autobiographical memoirs compiled later in life by diarists reviewing several years' diaries (generally now lost), and compilations such as the commonplace book of Robert Reynes, *c.*1480 (Table 6.2).

Diary-keeping was highly uneven by both gender and social status, reflecting the very uneven contours of writing skills; 'the more complex ability to assess and analyse one's own life and environment' (McKay, 2005: 194), and the range of possible motives for recording. Fewer than 10 per cent of extant diaries were kept by women, who were even more likely to be upper-class than male diarists. However:

The diary may have been primarily limited to certain sections of the population, but that does not necessarily mean that what was observed and recorded was in any way limited to a particular social group. . . . The diaries may have been ego-documents, personal notations of individual lives, but these texts also tended to document the current beliefs, opinions and events of the wider society.   (McKay, 2005: 212)

The growth of diary-keeping reflected both the strength of various motives for recording and the technological and material constraints to recording and document preservation. Without the ability to record—the ability to write, access to writing materials—people's utterances or beliefs might still be recorded by administrative institutions, but not through their own recording practice. There were a broader range of underlying motives, felt in different combinations by particular diarists and at specific times. Piety and self-monitoring of piety, or senses of being caught up in historic events could stimulate people to record their personal participation in devotion, faith, or politics (Delaney, 1969; McKay,

Table 6.2. Authors of the early modern diaries and memoranda analysed

Robert Reynes	late C15th	Acle, Suffolk	commonplace book
John Dee	1577–1607	astrologer & mathematician, Mortlake, London, Prague	annotated almanacs
Lady Margaret Hoby	1599–1605	gentry	short prose diary
Simon Forman	late C16th	physician, London	short autobiography, summarizing diary
William Honeywell	1596–1614	yeoman farmer, Ashton, Devon	short prose diary
Richard Napier	early C17th	physician, Great Linford, Bucks	case notes
Lady Anne Clifford	1603–75	aristocrat, London, Cumbria	short prose diary
William Whiteway	1618–1635	merchant, Dorchester	short prose diary
William Brereton	1635	gentry, Cheshire	short prose diary
Nehemiah Wallington	1618–1654	turner, Puritan, London	short prose diary
John Greene	1630s–1650s	lawyer, London & Essex	annotated almanacs
Elias Ashmole	1640s–1690s	antiquary, administrator, Worcester, London	short prose diary
Samuel Jeake	1650s–1690s	merchant, Rye, Sussex	autobiography & diary
Samuel Pepys	1661–1669	Naval bureaucrat, London	very extensive diary
Roger Lowe	late C17th	shopkeeper, Ashton-in-Makerfield	very short prose diary
Claver Morris	early-C18th	doctor, Wells, Somerset	short prose diary
Nicholas Blundell	1702–28	Catholic gentry, Cumbria	very short prose diary
Jeffrey Bratton	1739–41	village schoolmaster, Bratton, Wilts	short prose diary
Thomas Turner	1754–1765	village shopkeeper, Sussex	short prose diary
James Warne	1758	farmer, Wool, Dorset	short prose diary
Anne Lister	early C19th	minor gentry, Shibden, Halifax	extensive prose diary

Note: The dates are those of periods from which diaries (etc.) are extant.

2005). Plebeian diary-keeping or the accumulation of mundane jottings was fostered by the expansion of education,[20] and greater familiarity with print made the idea of writing more ordinary (Capp, 1979). The particular motivation of diarists matters much more for investigating some facets of clock times than others. Emphasis on moral dimensions of time is likely to appear in diaries explicitly concerned to monitor time-use or time-wasting, whereas any diary may illustrate incidental circumstances of narration.

If plebeian diaries are extremely rare, near verbatim utterances, though similarly rare, can appear where statements were recorded other than through the speaker's own choosing: especially in court or other depositions. We have particularly sought out such material as virtually the only source of plebeian engagements with clock times independent of normative regulation (and hence the aspirations of superior power), and so will need to assess how far the context may distort the content. Nonetheless, we shall place particular emphasis on evidence of plebeian practices in arguing that clock times were widely constituent of everyday contexts and sociality, largely independently of formal structures and regulation.

For Sherman (1996), the mid–late seventeenth century sees a sea change in the time-senses of diarists (and many other kinds of writers), epitomized by Samuel Pepys's diary. Over and above its specific political, social, and personal content, Pepys diary is a critical emblem of transformed senses of diary-recording. Between the pendulum clocks of the horological revolution and the triumph of Thompsonian factory time-discipline, but obscured by those narratives' attachment to metaphors of revolutionary change, Sherman argues, English literary culture witnessed a major reorientation of time-senses (Sherman, 1996: 19).

Sherman dichotomizes *kairos*—an episodic time constituted by significant events, that characterized prevailing textual time-senses—and *chronos*—the empty, passing, homogeneous time of clocks and calendars, contrasting the temporal templates of John Donne's *Devotions upon Emergent Occasions*, and Pepys's *Diary* as different in kind. Donne's temporality is structured around occasions like deaths and an 'overwhelming sense of metaphorical present time'. *Devotions*' lack of *chronos* is an omission 'so systematic as to look like policy'. On the other hand, the 'calendrically-minded' Pepys subsumes *kairos* to a *chronos* as his narrative fills the set clock times and dates of 'homogenous empty time', that both 'require[s] filling (by experience in life, by inscription on the page) and also [facilitates] it' (Sherman, 1996: 21–35, 40–1).

Private diaries, like newspapers and novels, were central documents in new textual time-senses, as better timekeeping and more continuous time-indication saw 'punctiform' or 'point and stretch'[21] time-senses being eclipsed by new 'ongoing' time-senses. These:

were new precisely in that they called attention away from endpoints and invested it in middles—of the current hour, of ongoing life—that were sharply defined and indefinitely extended. . . . Recording time in a secret book, like measuring it on a personal chronometer, is a way of owning it.   (Sherman, 1996: 21)

Pepys (and his successors), writing at a time when:

precise clocks and watches, conspicuous but not yet ubiquitous, proffered to their owners a temporality closely calibrated but not yet controlling, when the metaphorical societal 'metronome' was audible but not yet dominant. . . . the gentry and the merchants. . . were socially positioned to see in these instruments a figuration of personal time—and by extension of the private self in time.   (Sherman, 1996: 21)

Thus 'in Sherman's reading, Pepys is a privileged representative not only of a time period, but of temporality itself' (Kümin, 2004: 215, n.27).

Samuel Pepys's diaries survive for most of the 1660s, when he was in his late twenties and early thirties. He was a rising, and by the decade's end

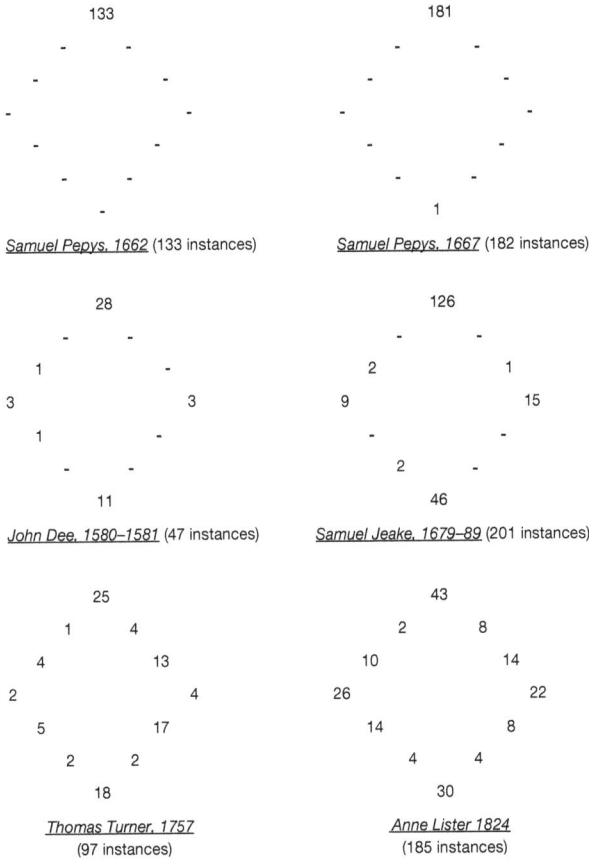

Fig. 6.3. Clock time references by Samuel Pepys and his contemporaries.

modestly wealthy, London bureaucrat, in the early stages of his career as a naval administrator, enjoying an increasing income, and an expanding range of business and social contacts.[22] Pepys is a consistent and energetic diarist by contrast with his predecessors and contemporaries (and indeed successors), quarried for commentary on a myriad facets of seventeenth-century London life, albeit that he is sometimes uncritically treated as a transparent narrator (Kümin, 2004).

Sherman emphasizes both calendar and clock as dimensions of Pepys's temporality in the diary. Though he did not write up his diary every day (Latham and Matthews, 1980: I, XLi–XLviii), Pepys retained a clear day-by-day structure of entries. This discipline in building a record with an entry every day, part-cause and part-consequence of his eclectic approach to recording, was largely unprecedented. For example, April 1617 was an unusually heavily recorded month in Lady Anne Clifford's diary, with entries for twenty-two different days: usually she recorded less often. Other diarists between the 1580s and 1660s quite frequently had spells of no entries for several consecutive days.

However, it is not so clear that Pepys differed dramatically from his contemporaries regarding clock time. Within his narrative for each day, events, activities, plans, and so on appear in broadly chronological sequence. But Pepys's explicit references to clock time are infrequent, in sharp contrast to his explicit recording of dates.[23] Here, he is not obviously different from other diarists (Table 6.3).

Of course, 1667 was the year after the Great Fire, when about three-quarters of the City's churches had been burnt down, and their clocks and bells with them. Pepys was by then accustomed to his minute-watch, which he had had since mid-1665, offering us the opportunity to examine whether Pepys's recording of clock times changed. In practice, though, there are only minor differences in Pepys recording of clock times in 1662, 1665, and 1667.[24]

More than half his entries in 1667 contain no specific clock time, and of those recorded 'noon' is the most common, which might be excluded as not solely a clock time. Pepys only rarely uses clock times in complex sequences within a day, so his overall density of clock-time usage in the diary is relatively low. Where he does refer to clock time it is to refer either to specific times of day (almost always in whole hours) or to durations.[25] His entries often included vaguer words or phrases, such as 'morning', 'afternoon', 'evening', and 'night', sometimes qualified by 'early' or 'late'.

This relatively loose language can hardly be taken to imply an indifference to, or incompetence in, scheduling, or that Pepys did not have a close regard for clock time. That competence is implied by his complex sequences of meetings with different people in different places, the making and keeping of appointments, his (and others) sometime impatience at delays, and so on, but is very rarely the subject of explicit comments. An interest in timing is explicit when he compares the time taken at several points on the walk to Greenwich (13 September 1665, Glennie and Thrift, 2005: 186), but it is an isolated reflection. Pepys's novel temporality, 'new within the traditions of self-recording current when he wrote' (Sherman, 1996: 31), was made possible by Huygens's pendulum technology:

In the 1660s, Huygensian chronology and Pepysian narrative were new, and new in similar ways. They were not wholly new in conception, . . . diarists had certainly thought to write in daily instalments before Pepys. But both were new in execution. (Sherman, 1996: 35)

Huygensian chronometry's most novel and salient feature is its continuity. The continuous movement of hands on a dial and the ticking of domestic clocks and watches creates a new experience of passing time, as it replaces turret clocks' intermittent striking of hours (and, perhaps, quarters). 'Technically Huygens accomplished . . . a sixtyfold improvement in accuracy (precisely the proportion of the old hours to the new minutes, and of the newly measured

Clock Times in Everyday Lives   199

Table 6.3. Uses of Clock times in early modern English diaries and memoranda

	Narration uses clock-time durations	Gives times to hour	Gives times to ½ hour	Gives times to ¼ hour	Plans use clock times	Mentions own clock time-signals	Mentions own clock	Time as moral issue
Robert Reynes	yes	yes	no	no	no	no		
William Honeywell	yes	yes	no	no				
John Dee	rare	yes	yes	yes	yes	yes	yes	yes
Lady Margaret Hoby	–	yes	no	no	no			central
Simon Forman	yes	yes	yes	yes	yes		yes	
Richard Napier	yes	yes	yes	yes	yes			
Lady Anne Clifford (to 1620)	yes	yes	rare	no		no	no	
William Whiteway	yes	yes	yes	yes	no	no	no	
William Brereton	yes	yes	no	no				
Nehemiah Wallington	yes	yes	yes	yes	yes			yes
John Greene	yes	yes	yes	no	no	no	no	weak
Elias Ashmole	yes	yes	yes	yes	no	no	no	no
Samuel Jeake	yes	yes	yes	yes	yes			yes
Samuel Pepys	yes	yes	rare	no	no	yes		weak
Roger Lowe	yes	yes		no	yes	no	yes	yes
Claver Morris	–	yes	yes		yes	yes		
Simon Forman	yes	yes	yes		yes			yes
Nicholas Blundell	rare	yes		no	no			
Jeffrey Bratton	yes	yes	yes	yes	no	yes	no	
Thomas Turner	yes	yes	yes	yes	yes			yes
James Warne	–	yes	no	no	yes	no	no	no
Anne Lister	usual	usual	yes	yes	yes	yes	yes	central

minutes to the newly measured seconds)' (Sherman, 1996: 3–5). In other words, more accurate timekeeping made possible Pepys's 'narrator embedded within the narrative' style, whereas the coarser time-cues available to Donne prevented 'time as ongoing flow' from supplanting 'time-as-occasion'.

There are other claimants though, as influences on 'inside the narrative' temporalities in the mid–late seventeenth century. One was the way in which political publication had been changing over the preceding three decades, through crisis, regicide, Commonwealth, and Restoration, and had stimulated habits of thinking in terms of 'news', and of oneself as part of events of broader historical significance. This is one feature that Pepys shares with several other diarists. Though individual motivations varied, there are wider senses of personal involvement in historical currents.

Early modern English writers from many standpoints were motivated to consider their situation within long-run historical changes. The contested relations among providence, church, Crown, state, and the people, all gave mid-seventeenth-century Englishmen strong incentives to pursue analogies between past and present in their political or historical writing, including poetry (MacLean, 1990: 127–76). As

'the need to write about the present became more urgent, . . . the present intruded upon the former dominance of the past to become itself a valid subject' there emerged 'a general—and by no means fully conscious—sense that current events themselves are as much a part of history as past events'. (MacLean, 1990: 139)

Historians of both politics and the press stress expanding political communications from the 1620s, and rapid developments in public debate through the 1640s as tensions mounted towards the Civil War. Politicians, historians, and poets compared past, present and future in discussing stability, civil war, government, and monarchy. Particular writers and publications varied greatly in their emphases on improvement or decay, stability or recurrence, dynastic inheritance or institutional progression, but diverse political-historical views interpreted current events using analogies between past and present. Legitimating events like the regicide or Restoration depended on setting those events as 'natural' within wider ideologies and public discourses, in which printed materials and commercial publishing played growing roles.[26]

During the 1640s and 1650s, political newsbooks and periodicals fanned Englishmen's sense of themselves as located within events of major historical-cosmological significance (Raymond, 1999; Delaney, 1969). Newssheets also shaped new temporalities because of their periodicity. Their distinctive dynamic of periodicity allied to commercial production's need for regular news, changed public consciousness, with 'decontextualizing and deconstructing effects' on their readers that eroded 'social, cultural, and even political values' and 'Early editors learned very quickly how to make knowledge disposable to ensure a steady market . . . periodicity allowed information to become a business, where it had formerly been an excuse for social interaction' (Sommerville, 1996: 3–4; 6, 161–2).

Sommerville stresses periodicity's impact on the concept of 'facts', and his account of new senses of self brought by periodical news culture parallels Sherman's stress on Pepys's impulse to record and his apparent need to produce daily entries.[27]

Thus public political publications gave mid- and late seventeenth-century diarists both models and motives for their recording, embedded as they were within growing social acceptance that regular news (political and commercial intelligence, science, arts, ethics, theology) comprised 'the stream of society's consciousness' (Sommerville, 1996: 5; McKay, 2005). Many writers point to significant changes in early modern publics and circulations of ideas, stressing the growing 'public sphere' and new venues for face-to-face discussion (Habermas, 1992; Calhoun, 1993), and the interpretive 'work' of discussion done through everyday sociality (Thompson, 1990). Embeddedness in religious or political debate, fanned by popular almanac literature (Capp, 1979), intensified diarists' self-scrutiny, alongside the host of reasons for them to think of themselves as living within uncertain but momentous times.

Sherman's argument is an important one bearing on our argument in various places. At this point, though, he exaggerates both the coarseness of time-telling and sensibilities prior to the horological revolution, in suggesting that people were unable to tell time further than the hour, and in hailing Pepys as manifesting a novel and distinctive temporality compared with his predecessors and contemporaries. The Pepys–Donne comparison is an effective juxtaposition of opposites, but a poor comparison of diary-recording. That Donne could not grasp *chronos* is unclear, because *Devotions'* concern is with mortality and loss. It does not establish the incapacity of Donne or his contemporaries to operate in a Pepysian way in contexts where that temporality was more appropriate.

Compared with other diary-keepers, the scale of Pepys's entries is distinctive: his daily entries of several hundred words are far longer than those of all his predecessors and contemporaries. Lady Anne Clifford's diary had entries for twenty-two different days in April 1617, but only one entry exceeded 100 words, and that included the names of fourteen people. Forty-word entries are uninformative about everyday temporalities, but Lady Anne could be more explicit about clock time than was Pepys, as on 1 August 1662 (when aged seventy-two):

whilst I lay in this Brougham Castle and some 4 or 5 howre before my remove from thence to Appleby Castle, about 7 o'clock in the morning died my Deare Grandchild James Lord Compton, . . . being but 3 yeares and three months old and 16 daies over at his Death.   (Clifford, 1990: 158)

Very similar clock-time usages had appeared in her diaries in 1617: for situating actions, 'between 10 & 11 the Child went into the litter to go to London', 'my Lord . . . did not come home until 12 o'clock at night'; for describing durations,

'Presently after Dinner my Coz. Clifford came & sat in the Gallery ½ an hour'; and for rates of movement, '3rd [May 1617] my lord went from Buckhurst to London, and rid it in four hours, he riding very hard' (Clifford, 1990: 32, 74, 74, 55).

The scale and focus of recording varied greatly both from day to day, and among diarists (McKay, 2005). John Dee's diary entries range from a few dozen words a month to several hundred a day, with immediate consequences for the visibility and content of temporalities.[28] Other and earlier diarists than Pepys are explicitly situating themselves in ongoing events, and using clock times more explicitly and in greater detail than did Pepys himself (Table 6.3)[29]

Yet even given the exceptional scale of his recording, and the range of minutiae taken in by his pen, Pepys's was less explicit about clock times than predecessors or contemporaries that Sherman sees as before or outside the new temporality. Eighty years earlier, John Dee's brief notes showed a scattering of times specified to half- and quarter-hours, and in the 1680s so did Samuel Jeake's—even though this was outside his period of detailed recording—presumably indicating audible quarter-striking clocks. Dee's is perhaps the most detailed surviving Elizabethan diary. Born five years before Elizabeth, and outliving her by six years, Dee's life spanned Elizabeth's own. Many of his interests fostered considerable time-awareness: some, like astrology, apparently quite widespread in the population, others, like navigation, shared by small and specialized communities of practice (Chapters 7 and 8, below).

On Dee's own account, contained in the 'Compendious Rehearsal' he presented to Queen Elizabeth in 1592, and transcribed by the antiquarian Elias Ashmole in the 1670s, time was central to his diaries from their inception in 1547, containing 'some thousands [of observations] . . . very many, to the hour and minute'. He may be exaggerating, of course. Around 1680 Ashmole reported seeing other diaries, now lost, including Dee's weather observations from May 1547 to 1551, and made a single-sheet extract of sixteen very terse entries through 1554, none exceeding two dozen words. Most entries in Dee's surviving diaries from 1577 to 1607 are similarly short, mixing English and Latin in recording events on pages of almanacs—for example, his 'diary' from 1585 to 1601 comprises annotations on Maginus's *Ephemerides* (Fenton, 1998: 16, 304–7). Though at certain periods Dee's diaries are very extensive, much of the surviving diaries consist of only one or two short entries a week, including those for 1577–81. A minority of entries are much longer, mainly relating to Dee's astrological and occult preoccupations.

Clock times appear throughout. The sixteen entries transcribed by Ashmole, despite their brevity, include five specifying the hour of events, and one more detailed.[30] From 1577 to 1581, usually at Mortlake, Dee wrote over 250 entries, most of under twenty words: about one-quarter report clock times, rising from 19 per cent of entries in 1577 to 34 per cent in 1580. The increase follows partly from Dee's marriage to Jane Fromonds on 5 February 1580, because he routinely recorded occasions of sexual intercourse for astrological, and possibly sentimental, reasons.

Dee commonly refers to events as happening at particular hours, as in 1585–86: 'horam circiter 7', 'mane circa horam 9', 'a meridie horam circiter 6', 'circa octavum horam', (Fenton, 1998: 173, 175, 179, 186). He was frequently explicit that the hours recorded were hours 'of the clock', but this specification does not appear to carry any particular significance. Dee did not restrict himself to hours, often using $\frac{1}{2}$, $\frac{1}{4}$, or $\frac{3}{4}$ to qualify hours, just as his earliest entries had referred to 'After noon, hor. $3\frac{1}{2}$ '; '$10\frac{1}{4}$ a mer'; '$9\frac{3}{4}$ a merid'; 'hor. $6\frac{1}{4}$ a mer'; '$1\frac{3}{4}$ after midnight'; and 'hor $8\frac{3}{4}$ nocte'. Use of half and both quarter-hours probably implies an audible quarter-striking clock, but Dee also mentions other fractions of an hour, that probably involved calculation rather than a direct time-cue, in phrases such as 'hora $8\frac{2}{3}$' (1581); 'hor. $10\frac{2}{3}$' (24 August 1588) and 'hora $2\frac{2}{3}$ a meridie' (Fenton, 1998: 13, 236, 11). Quarter-hours also appear in describing durations or intervals. Usages such as 'At a quarter of an hour end'; 'Nothing appearing, or being heard in a quarter of an hour space' and 'pausa $\frac{1}{4}$ horae' convey time treated as more than a sequence of points, but as of measured duration (Fenton, 1998: 59, 146, 221, respectively 29 March 1583, 13 September 1584, 6 May 1587).

Even though Dee's recording is often very brief and the entries for very many days run to only a few words, some entries comprise a procession of clock times:

Arthurus Dee... was born towards hor. 4 min 30 or min. 25, just as the sun was rising, according to my calculations. After 10 of the clock this night my wife's father Mr Fromonds was speechless: and died on Tuesday [the next day] at 4 of the clock in the morning. (13 July 1579)

Edward Maynard born in the morning between 2 and 3 after midnight. Arthur fell into a quotidian ague at 9 of the clock in the morning as he was at service in the hall. Francys christened after noon. (9 January 1592) (Fenton, 1998: 5, 254)

Some times 'of the clock', whether whole, half-, or quarter-hours may have been taken from either clocks or sundials, 'in the afternoon about $4\frac{1}{2}$ of the clock' (10 April 1583), but others from ten o'clock at night, midnight, or 3 o'clock in the morning imply a clock, since the sun had long set, 'circa $8\frac{1}{2}$ hora noctis' (11 January 1588), or 'hor. noct $11\frac{3}{4}$' (12 June 1596; Fenton, 1998: 109, 162, 67, 232, 278).[31]

Predictably, one factor in detailed recording was available time-signalling, most clearly shown by a preliminary comparison of times recorded in five diaries. In particular, we want to try to set Samuel Pepys in a comparative context, focused more on the form of his references to times of day, and specifically his use of clock times. Table 6.3 compares Pepys' diary in the 1667 (after he had acquired the minute-watch emphasized by Sherman), with those kept by John Dee in 1580–1581, Samuel Jeake in 1679–1689 (after the period of Jeake's most detailed recording, Thomas Turner in 1754–1765, and Anne Lister in

1817–1826. The aims are preliminary; to contrast the times recorded, and explore patterns of recorded clock times in relation to likely sources of time-indication, both aural (the striking of clocks) and visual (fineness of indication on clock and watch dials).

The later diaries of Turner and Lister are very much more explicit about clock times than Pepys, quite understandably given the then much greater numbers of domestic clocks and watches. Thomas Turner (1729–1793, diary 1754–1763) lived in the small Sussex village of East Hoathly[32] on the road between East Grinstead and Eastbourne, about twelve miles from each, and seven miles north-east of Lewes and its market. Like Pepys, aged 25–34 whilst diary-keeping, Turner's main livelihood came from shop-keeping, augmented at times by providing mathematical instruction (on arithmetic, decimals, the slide rule) to fellow-villagers.[33] The published edition is heavily abridged, but Turner is a similarly eclectic diarist (Vaizey, 1984). Turner's diary conveys the intersecting rhythms of local life, from farming (lambing, sheep-shearing, sowing grains, harvesting, hop-gathering), and local society (local rates, taxes, elections, office-holding, beating the parish bounds, disbursing local charities) to the less formal but equally embedded seasonal rounds of recreation and gambling: race-meetings at Lewes, local cricket matches, cock-fights, bowling, and footraces.[34]

All these facets of Turner's diary are saturated with the routine use of specific clock times to position and record events, and vaguer phrases appear less often. Clock times are particularly associated with travel, and when narrating events. They occur in a wide variety of settings, rather than confined to particular places. While Turner seems more comfortable than Pepys in specifying times between hours, the pattern is suggestive that single-handed clocks were his chief source. The lack of quarter-hour times strongly suggests that he seldom heard quarter-chiming clocks, and may have been estimating times past the hour from the position of an hour-hand.

A striking feature is that Turner uses clock times in recreational contexts, as on 12 June 1760 when he bet that a horse, steadily ridden, could complete a specific journey, 'between nine and ten miles and very uneven ground... within the hour' (Vaizey, 1984: 206). Such small indications convey clock-time's comfortable incorporation into everyday disposition. Whether his times were accurate is less important a point than that for Turner they were useful to record, and that the voluntary context of their use suggests an active recognition of their use, not merely passive acceptance.

Another seventy years on, Anne Lister's recording of clock times shows someone in an environment dense with clocks: clocks with minute hands, and usually seconds hands; clocks sounding both hours and quarters. Anne Lister (1791–1840, diaries 1817–1826) was of more exalted and leisured status than Turner, a gentry family residing at Shibden Hall near Halifax. She also spent time in York, occasionally visited Manchester, Cumbria, and London, and travelled in France in 1826, and subsequently in eastern Europe. She holds a prominent place within women's history, largely through her diaries (like Pepys's kept partly in code) chronicling sexual and other relationships with several women, valuable to historians of lesbian feeling in the early nineteenth century.[35]

She uses precise clock times with great facility to locate events, to emphasize departures from expectations, and to make advance arrangements. The preponderance of whole and half-hours in times given in her diary partly reflects the many events arranged in advance, by herself or others, or social occasions arranged at fixed times, and advertised accordingly. Nevertheless, her pattern of recorded time differs from that of Thomas Turner. Though both record all five minute times after the hour, there are several contrasts. While Anne Lister's lifestyle involved more formal public events and specific rendezvous than Turner's, the greater prominence of half and quarter hours in her account suggests that she used minute-hand clocks rather than single-hand clocks, and probably that several household clocks struck the quarters as well as the hours. It is also striking that she is precise about times between the half hour and hour about as often as she is precise about times between the hour and half hour, whereas Turner recorded the latter much more often than the former. Again, one part of the explanation may be his lack of, and her access to, quarter-striking timepieces.

From 1817 to 1824, Anne Lister routinely recorded day and date, and was often precise in using clock times, sometimes specifying the relevant clock or watch used, within an environment dense with various time-signalling devices. These were often, but not always, clocks: sometimes private and portable, as when she is able, in one of the earliest surviving entries, to record the hour of disturbances during the night (15 April 1817). At other times, the cues were more public. On 14 December 1812 in York she 'Sat talking till 11 [p.m.] when the chair came to take me into Petergate... Took all with me in the chair & got to Dr Belcombe's as the Minster clock struck ½ past 11', while in London on Monday 2 September 1822, 'the watchman is just crying half-past eleven' as she finished writing.

As she grew older, and diary-keeping became more ingrained in her life, times and durations recorded in her diary became much more precise. In 1824 twenty-five of thirty-seven durations are exact multiples of five minutes, and three more specific still ('six minutes'). But context was always a factor: she *could* refer to happenings at '7.37', '8.07' and 'a minute or 2 before 9', but expressions like 'a little after seven', 'between 5 and 6', or 'first thing this morning' often sufficed. Over time (comparing 1817 and 1824, for example), precision becomes more routine.

Documenting God's blessings, and demonstrating that one had made good use of God's time, motivated many diarists over the centuries, certainly not all Puritans, though for some Puritans—like Wallington—it was spectacularly prominent, and a matter of explicit reflection for many non-conformists, including Samuel Jeake. Lady Margaret Hoby put it frankly in April 1605, 'they are unworthy of god's benefits and special favours that can find no time to make a thankful record of them'. Gradually, the motives were slowly internalized, so that

Anne Lister could write 'If I was once to give way to idleness I should be wretched' (22 May 1817). She was specific about early rising: 'I am dissatisfied with myself for not having lately got up in a morning so early as I thought. It grieves me that I am ever in bed after 5' (9 June 1817). The morality of diary-keeping, for many diarists, was closely tied to retrospective reflection on their diary, and self-monitoring. Again, Anne Lister was the most explicit, musing over:

2½ hours copying out from 13th July up to today of the index to this volume... The looking over & filling up my journal to my mind always gives me pleasure. I seem to live my life over again. If I have been unhappy, it rejoices me to have escaped it; if happy, it does me good to remember it. (24 August 1822)

Another parallel across the centuries between Margaret Hoby, a late medieval elite woman, and Anne Lister, was their construction of schedules for moral living.

Tewsday 28 [August 1604]. In the morninge, after privat praier, I Reed of the bible, and then wrought tell 8 a clock, and then I eate my breakfast: after which done, I walked in to the feeldes tell 10 a clock, then I praied, and, not long after I went to dinner: and about one a clock I geathered my Apeles tell 4, then I Cam home, and wrought tell almost 6, and then I went to privat praier and examination, in which it pleased the lord to blesse me : and beseech the lord, for chrict his sack, to increase the power of this spirite in me daly Amen Amen: tell supper time I hard Mr Rhodes read of Cartwright, and, sonne after supper I went to prairs, after which I wrett to Mr Hoby, and so to bed. (Moody, 2001: 11–12)

Periodically Anne Lister is similarly specific about a worthwhile daily schedule:

I mean to devote my mornings, before breakfast to Greek, & afterwards, till dinner, to divide the time equally between Euclid & arithmetic till I have waded thro' Walkingham, when I shall recommence my long neglected algebra. I must read a page or 2 of French now & then when I can. The afternoons and evenings are set apart for general reading, for walking, ½ an hour, or ¾, practice on the flute. (13 May 1817)

My average hour of getting up in a morning is half-past-five. Dressing & . . . going to look at my horse, takes me an hour and a half. From 7 to 9, I read a little Greek & a little French. From 9 to 10, looking after the workmen. From 10 to 11, breakfast. From 11 to 2, out of doors looking at 1 thing or other, workmen, etc. From 2 to 3¾, walk out Isabella... From 4½ to 6, at dinner & sitting afterwards. From 6 to 6½, dawdle, trifle, call it what you please, with Isabella. From 6½ to 8, write letters or notes, or *'the book'* [diary], or take a *little* miscellaneous reading, & indeed, it is a little, for some days I never open an English book. At 8, go downstairs to coffee & we all spend a sociable evening together till 10. Isabella retires about ½ hour after me, & my uncle and aunt sit up till 11. Such is the model of my present day; such it must continue till we have done with the workmen; & then I shall have about 4 hours a day more time; only just enough to keep one's mind in proper order. (7 February 1824)

Timetables had a moral character through the sense of 'rightness' accruing to their 'correct' performance. Mealtimes, in particular, possessed this normative quality: 'Went without dinner today, not having felt well since I came home (bilious & heavy) which I solely attribute to dining at 3 p.m. & which certainly never agrees with me' (26 December 1817).

Almanacs were timetables, of course, and Dee was not alone in writing diary entries on their pages: John Greene, an Essex-born lawyer living in London, did likewise from the mid-1630s to the 1650s (Symonds, 1927: 386).[36] Capp lists some twenty others, observing that one reason 'for both the popularity and survival of almanacs was their suitability as diaries' (1971: 61–2). From the 1560s, almanacs included blank pages specifically for 'the almanac's secondary role as a diary' (Capp, 1979: 30). The juxtaposition made the times of predicted celestial positions, a potential model for the material added by the diarist. It would be perverse to see the likes of Dee, and other almanac-based diarists, as lacking a strong sense of time as ongoing. In Sherman's terms, almanacs anticipated Pepysian diary-keeping in that they were devices filling time with the events of planetary and other celestial positions. Most almanac-annotators, like Dee and Greene were terse, totalling just a few lines per month, yet clock times are a recurrent feature, especially in connection with six sorts of events.

When Dee described times using minutes, usually rounding to a multiple of five, the motive was usually astrological, as at the births of his own children:

Arthurus Dee... born towards hor. 4 min 30 or min. 25, just as the sun was rising, according to my calculations. (13 July 1579)
    Hora 11 a meridie. 50 minutis vel 40. Natus Rolandus Dee. (28 January 1583)
    Natus Michael, Pragae, hora. 3. min. 28 a meridie, ascendente {Mars}. Locus {Sun} {Pisces}. 3.32'.39. (22 February 1585)[37]
    Madimia nata hor. 6 min. 20 a meridie. (25 February 1590) (Fenton, 1998: 5, 52, 169, 247)

Dee reverted to less-precise timing for Michael's baptism on 14 March, 'A meridie, hora 2½', and precision mattered less for non-family births, Mr Laward's son Thomas 'born at noon or a little after, ¼ or perhaps ½' (Fenton, 1998: 273; 173). Similarly, John Greene's most explicit concern with precise times involved the births of his children, with the most specific timing for the first 'On the 8th of March [1644], about 10 minutes after sunset, being a little after 6 of the clock', and the fourth in 1648, 'about half an hour past 12 of the clock at noon' (Symonds, 1928: 599, 1929: 109).

The London astrologer-physician Simon Forman recorded his own daughter Dorothy's birth on 'the 10th day of July 1605 at forty minutes after 4 of the clock in the morning at Lambeth' (Cook, 2002: 178). Forman was, around 1600, consulted by over 1,000 'ordinary' Londoners a year, most women (Cook, 2002: 72, 152). His most precise specifications of time related to births: on 4 May 1592 he noted that a patient's child had been born at seven minutes to ten, though births even of his own children were not always so precisely recorded. In 1606 his son, Clement, was 'born the 27th day of October, it being a Monday, in the forenoon' (Cook, 2002: 178). Exact birth times mattered for the calculation of horoscopes, and as a physician Forman routinely sought this information for patients, even if approximate. So, too, did his protégé Richard Napier (MacDonald, 1981), another facet of the record-keeping which astrologer-physicians like Forman and Napier seem to have maintained more diligently than the 'regular physicians' of the time while investigating symptoms and recording observations, and monitoring treatments and their effects (MacDonald, 1996; Cook, 2002).[38]

Memories of birth-times were cultivated by diarists themselves. The physician Simon Forman was born, he recorded in autobiographical notes written about 1600, in 1552 'the 31st December, being a Saturday and New Year's Eve, at 45 minutes after nine of the clock at night' (Cook, 2002: 1). Samuel Jeake likewise recorded in 1694, his own time of birth in 1552; Hunter and Gregory 1988: 85). Not that the motive was always solely astrological: parents remembered birth-times affectionately. The ageing Elias Ashmole related his birth on 23 May 1617, at nearly half-past-three in the morning 'as my dear and good mother has often told me' (Ashmole, 1717: 2). Diarists recording their children's births were mirroring the longstanding practice of recording births in family bibles, as by the London mercer Richard Hill, just after 1500 (Table 6.4).

Some men with interests in astrology, and/or calculation more generally, recorded times of sexual intercourse, including Dee, Forman, and Robert Hooke (Fenton, 1998; Cooke, 2002; Robinson and Adams, 1968).[39] Like Dee using the term 'halek', Forman often recorded clock times of sexual intercourse in his casebook, alongside notes on patients, as in 1593 when 'halek Avisa Adams primus, the 15th December at 5 p.m.' initiated a new stage in his stormy four-year relationship with Avisa Adams, a neighbour's wife. He continued to record sex with other women, specifying having 'haleked ... Joan Wilde at a quarter past

Table 6.4. Part of Richard Hill's records of his children's birth times, 1510s and 1520s

Md yt John Hill my first child was borne
the xvij of Novembr Ao MCCCCxviij
at Hillend afforesayd on the day of Seynt
Hewe ...
Md that Thomas Hyll my second child was born
the xxx day of May aoMCCCCCxx yere at
viij of the clok in the morninge
Md that William Hill my third child was
born in brigge stret in the p(ar)ishe of Seint Margarette
the xix day of Octobr ao MCCCCxxi yere
abowt xi of the cloke affore none

Source: Orme (2003), 47.

eight on the morning of Tuesday the 25 February', and this practice persisted into later affairs (Cook, 2002: 79, 151, 180),. He also noted or estimated patients' times of conception, for example that Joan Horton's child had been 'begotten at night about the fourth hour' (Cook, 2002: 72).

Precise times of conception potentially had astrological value, but Forman also timed other potentially significant moments in relationships, 'at 3 p.m. Avisa Adams and I first osculavimus' [kissed] (29 November 1593). Recording specific times went beyond prognostication to a general drive to record, when Forman, after another dispute with Avisa Allen, recorded receiving her letter 'at 11 p.m.' on 12 June (1595), another the following day 'at 52 minutes past 2 p.m.' saying she would no longer visit him. But Forman went to the Allens' house, she returned with him, and 'at 30 minutes past four' they 'haleked'. Following further conciliatory gestures and the gift of some black stockings 'at 55 minutes past 3 p.m.' on 26 September, they were reconciled. Forman subsequently used his record of their sexual activity to satisfy himself that their son was conceived on 27 September at 5.30 p.m, though the infant was weak and died at 6 a.m. on 9 July (Cook, 2002: 70–83). The episode conveys a strong impression that specifying times helped establish Forman's sense of events or feelings as significant.

If more for emotional than astrological reasons, deaths were also a common occasion for more precisely recorded clock times; Dee noting on 6 April 1588, 'Good Sir Francis Walsingham died at night hora 11', Greene recording the hour his father died in 1653, and the Wiltshire schoolmaster Jeffrey Whitaker giving the time of his aunt's death (Fenton, 1998: 247; Symonds, 1929: 113–14). Anne Lister recorded a maid's account of her uncle's death, which she took down for copying into her diary:

the last words my master ever said & he spoke them very plainly . . . was about 10 minutes after 12. He was ever after quite composed & went off without a groan—one would have thought he was only falling asleep—at exactly 5 minutes past 1 by his own watch that was hanging up in the room. (9 November 1817)

Also notable here is the servant reporting the precise time of last words and death, from reading clocks and a watch.

Diarists' use of clock times in describing eclipses, other astronomical events, or striking natural events, followed from the predicted times given for such events in almanacs. Dee recorded the solar eclipse of 15 February 1599: 'A cloudy day, but great darkness about 9½ mane' (Fenton, 1998: 286), and John Greene that of 1652: '29 day [March], when the eclipse was, it was a misty morning, but about the time of the eclipse it was a very clear sky, so that the Eclipse was very visible'. Eclipse predictions were a staple of almanacs, and used to assess their accuracy, as when Greene continued: 'it was darkest about eleven of the clock, but nothing so dark as was foretold and expected. One might very well see what a clock it was by the sundial when it was darkest, as I did by ours in our garden. So that the Astrologers lost their reputation exceedingly' (Symonds, 1929: 112).

Clock times were also particularly likely to be specified for extreme natural events, for a range of motives from discerning divine intent, to casting horoscopes, to detailing memory of striking events. John Dee noted the time in of the earthquake of 1580, and also of dramatic weather, like the thunderstorm 'after 3 of the clock after noon' on 29 May 1597. When Whitaker, on 20 April 1739, described a hailstorm, 'tho' it lasted no more than 20 minutes, yet the water run down Tinkers lane almost enough to drive a mill' (Reeves and Morrison, 1989), he was following an entrenched practice of specific timing. So too with fires. Just like Robert Reynes and the North Walsham churchwarden, John Greene reached for clock time in describing a fire by the Bell tavern in Newgate market 'about 3 o'clock in the morning' (Symonds, 1928: 599).

Likewise, Anne Lister's use of clock times relating to travel and communications, is present among many earlier diarists. John Dee routinely gave clock times for departures and arrivals both in England, and abroad, including his arrival in several cities and ports in the modern-day Netherlands, Germany, Poland and Czech Republic in 1583–84, until 'Wednesday, on Christmas Day morning, we came to Stettin by 10 of the clock' (Fenton, 1998: 106–9).[40] Whitaker and Turner were among those doing likewise. In London in 1644, John Greene's 'father hired a coach for all this terme from 8 in the morning to 12' (Symonds, 1928: 598). Greene was also aware of posting times for mail, noting in September 1643 that the deadline for letters to his brother and sister in Rotterdam was 4 p.m. on Fridays (Symonds, 1927: 393).

Clock times were also particularly noted in various legal contexts. William Honeywell, recorded the hour of his paying debts (23 June 1608), and of his own arrest:

I was at Riddon; about 4 of the clock I was arrested with a warrant of the peace (from Mr Reynell) by two bayliffs, at Kingwell's suite; I went with them that night to Exeter. (29 July 1602)[41]

Similarly, Elias Ashmole was specific that on 14 December 1652, 'I was served with a Subpoena at Sir Humfrey Forsyter's Suit, Three Hor. Forty Minutes post Merid' (Ashmole, 1717: 28). Ashmole recorded relatively precise times for key career events: his taking the oath as an Excise Commissioner in Worcester in 1645 at 11:15 in the morning, and a few months later he was chosen Register to the Commission at 'One Hour 32 Minutes post merid' (Ashmole, 1717: 13–14). Greene, Honeywell, and Turner were among those recording the times of civic meetings and ceremonies (Symonds, 1929: 113–14). Emotional and legal impulses to specific timing sometimes coincided regarding marriage. Ashmole recorded that Lady Mainwaring accepted his proposal of marriage on 6 November 1649 'Eleven Hor, Seven Minutes ante Merid.', and sealed the marriage contract four days later at 2.25 p.m. (Ashmole, 1717: 20).

The equivalent of specified meeting times in personal-scale interaction was the appointment, and several diarists made advance arrangements in those terms. Later diarists recorded a variety of humdrum appointments, with peers, tradesmen, and social superiors, but the fullest account is from Dee, in Prague (Fenton, 1998: 154–5, 163):

about 7 of the clock came Dr Curtz his servant from his master to tell me, that his master would come unto me at 9 if the clock... At 9 of the clock came on horseback Dr Curtz to me.   (Dee, 27 September 1584)

At afternoon came to me one of Dr Curtz his servants from his master to tell me that his master would come to me tomorrow in the morning about 7.8 or 9 of the clock, as I would, etc.   (Dee, 5 October 1584)

Before 7 of the clock I thought good rather myself to go to Dr Curtz, than to suffer him come to me so far, and that for divers causes. So I went to him, and came before he was ready.   (Dee, 6 October 1584)

Dee seems relaxed about the etiquette of times, but Anne Lister used the rigidity of everyday domestic meal-times to test her social standing with one Mrs Saltmarshe without enquiring explicitly:

One thing that struck me was that frequently as she used formerly to ask me to stay to dinner or tea, she had not done so for the last year past. I had once or twice staid till just past their dinner hour for the mere purpose of trying whether she would ask me or not. (25 March 1824)

Appointments were, almost by definition, recorded using clock time, whereas informal sociality generated only incidental mentions of clock time, where there was another motive for doing so, as on twelfth night 1643, when John Greene was with 'my Coson Glassecock and his wife... We play at cards till 4 a clock in the morning. I loose £3 14 at 1^d and 2^d stakes' (Symonds, 1927: 391).

Beyond a couple of references to bells, a night-watchman, and his pocket-watch, Pepys very rarely records how he knew the time. Very rarely did he regard this as notable, probably an eloquent comment on the density of clocks and temporal information in London. This taken-for-grantedness is common, even among the earlier diarists. Only once, in Prague, does Dee refer explicitly to finding out the time:

About 2 of the clock after noon came this letter to me, of the Emperor his sending for me.

'The Emperor has just signified to the Spanish ambassador that he will summon your Lordship to him at 2 of the clock, when he desires to hear you'

Hereupon I went straight up to the Castle: and in the Ritter-Stove or guard chamber I stayed a little. *In the mean space I sent Emericus to see what was of the clock*:... (3 September 1584, Fenton, 1998: 142, our italics)

Where diarists like Jeake did think it relevant to mention the source of time-information, it was in relation to the times that mattered most, especially births, where Jeake recorded the typs of dial by which his father timed Samuel's birth. Once again, the information was otherwise recorded only incidentally, as when

John Greene noted in 1652 that, contrary to astrological almanac predictions that an eclipse would produce total darkness, 'One might very well see what a clock it was by the sundial when it was darkest, as I did by ours in our garden' (Symonds, 1929: 112).

Diarists did, though, have a recurrent interest in the construction and workings of machinery. Before Christmas 1665, Pepys went

> to my Lord Brouncker and there spent the evening, by my desire, in seeing his Lordship open to pieces and make up again his Wach, thereby being taught what I never knew before; and it is a thing very well worth my having seen, and am mightily pleased and satisfied with it. (Latham and Matthews, 1980: VI, 337)

Foreshadowed by Aubrey's description of the young William Petty at Romsey, this was in turn echoed by Claver Morris, and by Anne Lister:

> Got a new watch-glass at Pearson's & asked him to let me see a clock taken to pieces. I am to call next Tuesday for this purpose. (9 January 1824)

To summarize, several different indications show how comfortable almost all the diarists were with clock times. That so many mentions were incidental to other objectives highlights how thoroughly taken for granted were clock times, among diarists both before and after the horological revolution, though clock-time practices took different forms in different situations. This is epitomized by John Dee in 1607. Routinely using clock times in his own activities and recording led Dee to take them for granted. A motive for Dee's longer diary entries was his wish to record conversations, through 'skryers' such as Edward Kelly and Bartholomew Hickman, with various angels. In Dee's record of these revelations, he sees nothing unusual in angels using clock times:

> Hora 11½. As I sat at dinner with Barthilmew Hickman, my daughter, Patrick, and Thomas Turner, about the end of the dinner Barthilmew heard a voice, saying: 'Tomorrow half an hour after 9 of the clock, give your attendance to know the Lord's pleasure.' (Fenton, 1998: 299, 14 July 1607)

That Dee supposes it natural or self-evident for angels to use clock time nicely demonstrates how thoroughly he took clock times for granted.

6.6 TEN VIGNETTES: EVERYDAY CLOCK TIMES IN EARLY MODERN ENGLAND

In various ways, Pepys and Aubrey were atypical of the English population, not least in keeping diaries and biographical notes. However, we have virtually no diaries, detailed autobiographical memoranda, or letters for the overwhelming majority of English men and women. Their writings, or, more likely, utterances, could nonetheless enter the documentary record when they were recorded for some immediate purpose, typically to do with regulation, punishment, or the disapproval of those of higher social standing. For example, we have already seen the deposition in late fourteenth-century York of John de Akom, showing his capacity to understand particular hour-ringing, and the audibility over a wide area of 'the bell called "clokke" '. In particular, court depositions can provide evidence not only about people's understanding of time-signals, but about the uses made of temporal information, and about people's ability to improvise. Critically, because these were statements given under examination, it is often possible to gauge how listeners judged the plausibility of what they were hearing, or reading.

(1) A notorious scandal in early sixteenth-century London, a focal point in deteriorating relations among ecclesiastical authorities, anti-clerical Londoners, and the Crown (Henry VIII), surrounded the death of Richard Hunne, a London merchant tailor, in 1514 in the Lollards Tower at St Paul's Cathedral. A seemingly minor squabble, about a priest's entitlement to the christening robe of a dead five-week-old baby, escalated into a fifteen-month struggle between ecclesiastical and secular legal authorities (Cooper, 1996). Hunne was imprisoned and died in custody. Despite considerable political pressure to confirm churchmen's testimony of Hunne's suicide, a coroner's inquest returned a verdict of murder.

The inquest's 'extraordinary length... [was] due to the complex political issues involved rather than the settling of any knotty forensic problems' (Cooper, 1996: 236), but timing evidence undermined some witnesses' versions of events. The sequence of events was established through at least nine witnesses, of whom seven referred to clock times at least once. The alibis and narratives provided by the Bishop of London's summoner, Charles Joseph, who was dead before the inquest concluded, his servant Julianne Little, John Spalding alias Bellringer, and a priest William Horsey, were, said the jurors, 'purgation which we have proved all untrue' (Anon, 1537).

People in the environs of St Paul's were familiar with clock times, since the cathedral had contained a clock since the 1280s (Beeson, 1971: 7), but the diversity of the witnesses using clock time is nonetheless striking, including church officials, a tailor, a barber, a stationer with a stall in the churchyard, and further afield, an innkeeper and his wife at Shoreditch, east of the City. Clock times were mostly approximate ('betwixt 8 and 9 o'clock in the morning', 'at 7 of the clock at night'), but both the tailor and the stationer described the finding of the body 'within a quarter of an hour' after 7 o'clock in the morning. Such specific timing was enabled by St Paul's clock striking the quarter-hours, and witnesses' everyday capabilities to interpret them was taken for granted, unquestioned throughout the proceedings.

(2) One of the most valuable facets of court or inquest material is that they shed light both on what witnesses said, and on how their accounts were received. The plausibility of claims implicit in testimony is a valuable insight into the skills that

people presumed other people to have, or to lack. For example, on 23 September 1615 Thomas Shott, a farm labourer at Cerne, Dorset, was examined by Sir Frances Ashley, the Justice acting as Recorder of Dorset. Examined about the burglary of John Wheadon's house at Cerne several weeks previously, Shott provided a bland account:

the 8th of August his master, William Toolage of Cerne, having commanded him about 8 of the clock at night to go to his bed, yet nevertheless went out of the house about eleven of the clock at night to his father's house for a shirt because he was to go abroad the next day to reap, but his father being a-bed he would not rise whereupon he returned to his Master's house again and there lay till the next morning.

That this was less than the full story emerged from enquiries by Richard Bartlett, the Cerne parish constable, and Bartlett's evidence of what Shott told him:

[Bartlett] says that shortly after John Wheadon's house was robbed, one Thomas Shotto of Calne was had in suspicion thereof, whereupon the said Richard Bartlett being then Constable, had the said Thomas Shotto in examination and asked him about what time of the night he was abroad in the street when the said John Whedon's house was robbed, who answered the folk told him it was near about midnight, who asked him again what folk told him so who was thereupon mute, but in the end answered again that he heard the clock strike eleven, whereupon he asked him what business he had there so late in the night, who answered he carried a peck of wheat to the widow Northover's, whereupon the said Constable sent for the said widow, and asked her whether Thomas Shotto brought her a peck of wheat at the time aforesaid, who denied it and said hee brought her no wheat, but said he was with her in the yearning of the same night before candle-lighting and brought her grist, but denies that he was with her at any other time. (Bettey, 1981: 15–16)

Shott's shifting story, and alterations of how he knew the hour, hardly bolstered his protestations of innocence. But it is striking that neither Bartlett nor Ashley thought either method of knowing the time was, in principle, implausible for people of low status. The specific occurrence was doubted, not the claim's general feasibility. This conforms with our picture of clock-time uses and practices being thoroughly routine and taken for granted.

(3) Shott's routine use of times 'of the clock' was common to many early seventeenth-century deponents: gentlemen, urban tradesmen and farmers, servants and rural labourers, wives and singlewomen (Table 6.5). A similar picture emerges slightly further east, in surviving depositions from Southampton in the 1620s, where deponents used clock times in a huge range of circumstances, even though many were brief statements about single points of evidence (Anderson, 1931). Depositions referred to years, dates, and religious festivals, rather than hours, in disputes about conditions of dowries or apprenticeships over several years, or vessels' seaworthiness several months earlier. But where disputed events involved specific events or commitments, on timescales of under a day, clock

Table 6.5. Dorchester deponents using clock times to locate narrative within day or night

Place	Occupational or status labels
Dorchester	dyer, weaver, groom, butcher's wife, servant, singlewoman
Fordington	gentleman, labourer, labourer's wife, tailor, 'man'
Bere Regis	carpenter's wife, husbandman's wife
Cranborne	groom
Maiden Newton	man, wife
Blandford	husbandman (2)
Mawpowder	shepherd
Cerne	servant, farmer, labourer, husbandman
Cudworth	servant
Buckland	butcher
(beyond area)	baker, mercer

Source: J.H. Bettey, *The Casebook of Sir Frances Ashley JP, Recorder of Dorchester 1614–35*, Dorset Record Society publication no. 7 (Dorchester 1981).

times were very widely used. The dozens of deponents who quoted clock times were being examined in cases as diverse as piracy, sexual assault, theft from houses or ships, sale of stolen goods, a stillbirth in suspicious circumstances, assault, murder, recusancy, illegal trading, bastardy, receiving stolen goods, a death overseas, pick-pocketing, burglary, highway robbery, negligent loss of a ship, and uttering seditious words. The social status and occupational designations of those using clock times to fix events in time were similarly diverse: landmen and mariners, masters and servants, skilled workers and labouring men, women of both respectable and more marginal status, apprentices and other adolescents, local men and visitors from elsewhere.

It is plausible that clock times were more prominent in depositions than in everyday speech, becoming conventional in deposing because precision in locating events in time helped establish the credibility of witnesses, whether or not time was central to the case. This may have made clock times more prominent in court than in everyday life, but it shows that a capacity to narrate events in clock times was widespread by the early seventeenth century. This harked back to long-established links between time and trustworthiness, in which temporal specificity raised the value of evidence.[42]

(4) Among the earliest body of English letters is the late fifteenth-century correspondence of the Stonor family, of Stonor in Oxfordshire, and London. Some 250 letters sent or received during the 1470s and 1480s survive with related documents, ranging across personal, estate, and commercial business, social introductions, and legal-political matters (Carpenter, 1996). The typicality of correspondents remains unclear, because only two other fifteenth-century

family correspondences survive,[43] but letter-writers routinely used hours 'of the clock' in several contexts when referring to times of day.

Hours refined the time of letter-writing, as for the merchant Richard Bryan, 'written at London the ij^{de} daie of Octobre at iiij a clock', three days after his previous letter written 'at viii of the Clok' in autumn 1479, and for Henry Colet in 1480: 'Written at London opon Seint Thomas Day at nyght at vij of Clok'. The practice was *not* confined to merchants: William Stonor's first wife, Elizabeth, wrote similarly in 1476: 'At London the ix day of Octoburer at ix a Cloke at Nythe'. Nor was the practice particularly metropolitan. In early January 1481, Sir William Stonor's steward, Walter Elmes sent his master a letter from Andover, Hampshire, 'scribbled with hand of him that is your servant to the extreme of my little power... the Wednesday after New Year's day[44] at 6 of the clock in the morning' (Carpenter, 1996: 373). This implies Elmes's taking the time from a clock, since it was before sunrise. Besides merchants and servants, hours of writing were mentioned by gentry correspondents, and an official of the Duke of Suffolk.

Clock times were mentioned with reference to particular events for various reasons. One was to reinforce social or business ties by emphasizing the recipient's swift action or response: 'I R. your letter on Tuysdaye betwixte vj and vij of the kloke' or 'my seid lord was wreten unto by a pryve seall which was delivered to him on Munday last passed at vJ of the Clokke withynne night at Ewelme'. Specific times were used to make arrangements ('sende a servante of yours to mete with me at Wallyngeford on Moneday by vij of the clokke'); to prepare expectations (Elizabeth Stonor's telling her husband in 1476 that a bargeman with letters for him had left at 'viij of the Cloke in the morning'); and to complain about broken arrangements ('According to the commaundement... we were at Stebenhith by ix of the Clok').

Alongside practical and administrative uses of clock time, it is telling to find the use of less literal treatments in which mechanical timekeeping is a source of metaphor. When in 1474 the merchant Thomas Betson, courting Sir William Stonor's ward Katherine Rich, whom he married in 1476, wrote to her from Calais, her familiarity with clock time was taken for granted (Table 6.6). Betson

Table 6.6. Ending of letter from Thomas Betson to Katharine Stonor, June 1474

I praye you, gentill Cossen, commaunde me to the Cloke, and pray hym to amend his unthryfttie maners, for he strykes ever in undew tyme, and he will be ever afore, and that is a shrewde condiscion. Tell hym with owte he amend his condiscion that he will cause strangers to avoide and come no more there...

[Written] At greate Cales on this syde of the see, the ffyrst day off June, whanne every man was gone to his Dener, and the Cloke smote noynne, and all oure howsold cryed after me and badde me some down; come down to dener at ones. and what answer I gaveffe them ye know it off old.

Be your ffeiythfful Cossen and loffr Thomas BetsonI sent you this rynge for a token.

Source: Carpenter (1996): 263–4.

also incorporated time into moralities of work and rest when he finished a letter to Stonor in 1478: 'At London, on our Lady day in the nyght, when I deme ye were in your bede, ffor my nyne smerttyd, so God help me'.[45]

(5) If the Stonor's elite status and their part-time location in London limit what their letters can reveal about England at large, certain early modern sources offer glimpses of clock times in some at first sight surprising activities and places in northern England. One particularly informative case involves Henry Best's manuscript *Farming and Memorandum Books*, written in c.1642 to guide his son in farming at Elmswell, East Yorkshire, and numerous uses of clock times in conducting and describing everyday life. In a context far removed from familiar religious or trading activities, it displays no entrenched adherence to 'folk' times of day, or to natural seasonal markers or religious calendar festivals (though other farmers may have matched the stereotype). The document's scarcity and value come precisely from Best's recording of routine knowledges, practices, and improvisations, normally taken for granted.

Clock times are prominent and recurrent in Best's instructions, falling into three main groups of uses. First for describing times and durations such Best's son would readily understand—activities' timing, sequencing, and coordination, and why they varied. Almost every aspect of farming is described using clock times, from harvesting to thatching, from feeding sheep in snow to drying hay, to taking sheep to market, from castrating ('libbing') male lambs to describing the behaviour of bees (Table 6.7).

Another group of uses of clock time involved coordinating activity with institutions that operated by clock time, like markets. Best notes the importance of timing at Malton horse fair, and annual hiring-fairs for farm servants (Woodward, 1984: 120, 142), but he was chiefly concerned with grain markets, spelling out the coordination of trading hours, journeys, and seasonal variations. However, Best's use of clock time went beyond adjusting to institutional times elsewhere. He coordinated skilled workers and labourers in and around Elmswell. Particularly telling is that he dealt with shepherds' seemingly self-evidently task-oriented work, tasks like washing, shearing, and gelding, through instructing shepherds about times (Table 6.7). Best's instructions would be worthless if, to a shepherd, 'bring the sheep to the fold at 5 o'clock' was a meaningless phrase.[46]

The ensuing sentence demonstrates a third theme—the calculative view of work-rates in 'one that is ready at it will easily geld a hundred lambs in 3 hours'. Best both appreciated adept workers, and sought to exploit time-structures of work to his advantage. Clock time, though not a prerequisite for calculative valuing of time and work, was embedded into Best's reckonings, especially at harvest and other peak periods. Best comments specifically about carting grain, mowing grass, and reaping grain, in the last case to his labourers' disadvantage.

Table 6.7. Some uses of clock times in Henry Best's 'Farming Book', 1642

Carters:	If the morninge be faire, the waines are yoaked by 7 of the clocke and not afore, because of the dewes, and then doe they fetch every waine a loade afore they come to breakefast. But when they leade the wolds and fetch it as farre as Doghill flatte then doe they yoake att sixe a clocke or aboute sun-rise, and then will they bee heare againe att breakefast aboute 8 of the clocke and sometimes halfe an houre afore, and then for loosinge. The waine that is teamed within a quarter or halfe an houre after sunsette is to goe againe if the night bee faire and the moone likely to give light.
Thatchers:	Thatchers...have 3 meals a day, viz. their breakfast at 8 of the clock or betwixt 8 and 9, their dinner about 12, and their supper about 7 or after when they leave worke.
Shepherd:	he gave them a bottle [bundle of hay] att Sun-rise or afore sun-rise,...another about 10 of the clock,...another bottle again about 2 of the clock, and the 4th and last bottle of hay he allwayes gave them after sun-set, and usually about the time that our Threshers leave work. We libbed our lambs this 6th of June... We sent word to the Shepherd to bring them down to the fold betwixt 4 and 5 of the clock. About 5 of the clock we went and carried our foreman to hold the lambs while they were libbed.
Bees:	The usual time of bees swarming... is betwixt 9 of the clock and 3, but especially betwixt 9 and eleven... Bees will flourish and make profer of casting [swarming] 4 or 5 dayes before they caste indeed, and that usually about halfe an houre after 10, and halfe an houre after one of the clocke.
Markets:	In winter time when our folkes goe to Beverley, they are neaver stirring above 2 houres before day because they are soone enough if they gette but thither by eleaven of the clocke.... When our folkes goe to Malton, they are usually stirringe 4 hours before day, which is aboute 3 of the clocke, and then will they bee aboute Grimstone by the springe of the day, and att Malton by 9 of the clocke att the furthest; for in winter time, that markett is the quickest aboute 9 of the clocke, or betwixt 9 and 10, because the badgers comme from farre, many of them, whearefore theire desire is to buy soone that they may bee goinge betimes for feare of beinge nighted. all summer longe, when our folkes are to goe to Beverley-markette, they goe out of our owne yard aboute halfe an houre after 4 of the clocke.
Instructing shepherds:	When yow intende to wash sheepe, yow are to give warninge to your Shepheard the night before that if hee like the morninge hee may bringe downe the sheepe *by 8 of the clocke*. Yow are allwayes to make choice of a fair and hotte day, if yow canne, because of the washers, and likewise to have your sheepe ready to throwe into the dyke betwixt 8 and 9 of the clocke, and not afore, because the mornings are airish.... The washers are to have warninge given the night afore, that they come aboute sixe of the clocke in the morninge. When wee intende to clippe our sheepe, wee allwayes sende worde to the sheaphearde the night afore that if hee like the morninge hee shoulde bringe downe his sheepe next morninge by sixe of the clocke or saoone after, that they may bee gotten penn'd up and bee ready for the sheares aboute seaven.

Table 6.7. continued

Work rates:	When you intende to gelde your lambes yow are to sende worde to your Sheapheard to bringe his sheepe down to the folde aboute 5 of the clocke [in the afternoon] or soone after, and from that time till sunne-set yow may easily gelde an hundreth lambes, for one that is ready att it will easily gelde an hundrethe lambes in 3 houres.
	[he] began not to clippe till after our ploughfolkes had gotten their breakfasts, and hee had done the aforesayd 32 [sheepe] soone after 12 of the clocke.... Aboute 11 of the clocke wee sente in for a Canne full of the best beere for him.
	Haymakers will cocke as much in one hour as they will rake togeather in two, wherefore they seldome beginne to cocke afore three of the clocke, and then doe they beginne there first wheare they beganne first to rake.
	Hee gave them [sheep] as much att a time as they could eat in halfe and houres space.

Source: Woodward (1984): 144, 79, 103, 64, 71, 106, 19–20, 23–4, 25, 100–1, 35, 79.

Best's advice on work-rates has a precursor, prior to mechanical clocks, in advice on hiring harvest labour, in an anonymous thirteenth-century French farming treatise:

you ought to know that five men can easily reap and bind two acres of any kind of corn in a day, sometimes more and other times less... You should engage reapers as a team, that is to say five men or women... And twenty-five 'men' can reap and bind ten acres in a day, working full time... Ascertain how many acres there are altogether which ought to be reaped and see whether they agree with the number of working days... And if they charge you with more working days than is correct according to this calculation it should not be allowed them, because it is their fault that the proper amount has not been reaped and that they have not done the work as well as they ought to have done. (Herren, 1999: 71)

Finally, Best illustrates how clock times were so familiar in everyday imagination as to be a tool in describing other things, like the location and orientation of beehives (Table 6.7). All in all, therefore, Best shows how diverse were clock-times' uses, and presumes that both his son (the intended reader) and various farm workers, both full-time and casual, can understand and work with clock times. Even in what might appear as a classic 'task-oriented' pastoral agriculture, people showed a comfort with clock times, and a capability to be calculative with it.

(6) Turning to the more clearly 'time-oriented' work of urban wage labourers, one obvious source through which to explore attitudes towards time are employers' records of wage-payments, long used by economic historians to analyse trends in wages, income, and living standards (Rogers, 1884: 425–54). The detail of expenditure records varies, and analysis is complicated by the various authorities that from time to time sought to enforce maximum wages, affecting forms of

payment and recording (Roberts, 1981), though these complications are less important here than for analysts of wage-rates.

Work durations are most commonly recorded in whole and half days. Not every such form of words necessarily refers to exactly that time: one source of variation in apparent pay-rates will be unmentioned adjustments where work took more or less time, but was still described as 'a day'. Seemingly arbitrary variations in wage-rates could hide more calculating adjustments to reckonings based on standard day-rates, or other complicating factors.[47] Relationships between daily and weekly wage-rates show a normal six-day working week in early modern England, where sufficient work was available. Daily work hours also appear broadly stable, though becoming more complex in eighteenth-century London (Voth, 1998, 2001). Wage-rates and working hours seem stable compared with sharply fluctuating total volumes of work.

Before the late fifteenth century, several London churchwardens (and some elsewhere—including Bristol were already reckoning work in quarter-days rather than half-days. Possibly this was prompted by one or both of higher wage-rates, and more intense auditing.[48] In the sixteenth and seventeenth centuries, quarter-day payments are explicit in several smaller towns, from Yarmouth and Chelmsford in the east, to Stroud and Tewkesbury in western England.[49]

Payments for hours show a common 'working day', even though accounts were rarely explicit about precise hours or wage-rates. The earliest account entry for work in hours survives in London around 1500,[50] and examples elsewhere include a man at Chelmsford in 1634 paid 6d. for helping a carpenter 'for an hower or 2'.[51] The wardens at Bressingham, Norfolk, in 1638 clearly envisaged the hours in a whole day, when they paid George Barteram, carpenter, for four days work by his son and '2 days worke wanting some hower & halfe of his owne', after paying another carpenter 9s. 3d. for work lasting three days 'and almost another whole day'. On 2 February 1639 they paid Thomas Barteram 4d. 'for 2 or 3 hours work'; and 14d. for one day's work, having paid him 4s. 8d. for four days on 31 January, consistent with a working day longer than seven hours but shorter than ten and a half.

Such explicit and detailed recording is extremely rare among English churchwardens and other employers, but these examples are likely to represent the capacity for more calculative scrutiny of wages among employers whose own recording was much less informative.

(7) Clocks are more visible in documentary sources than other timekeeping devices, but there is some evidence that both sundials and hour-glasses were quite widely distributed. Sundials ranged from precise astronomical instruments to large architectural objects, to small portable devices for practical individual use. Customs records for goods imported by English and Hanseatic merchants through London in 1567–8 include five ships declaring a total of 2,376 dials, and seven declaring a total of 624 hourglasses, mainly from Antwerp (Table 6.8).

Table 6.8. Time-items imported through London by native and Hanseatic merchants 1567–8

Date	Vessel & port (weight) Merchant	Cargo item	Master	Shipping port
2 Oct 1567	*Mary George* of London (45 tons)		Robert Osburn	Rouen
	Richard Hodchet (mayor of Newcastle)	12 **hourglasses** also **compasses**		
23 Oct 1567	*Prym Rose* of Milton (80 tons)		Harry Church	Antwerp
	John Boorne (leatherseller)	half gross **hourglasses**		
6 Dec 1567	*Cock* of Antwerp (40 tons)		Andrew Adrianson	Antwerp
	John Boorne	1 gross coarse **hourglasses**		
15 Dec 1567	*Barsaby* of Lee (35 tons)		Wiliam Syms	Antwerp
	Anthony Scoloker (milliner)	90 dozen **dials**		
14 Feb 1568	*Primrose* of Milton (–)		Henry Church	Antwerp
	Thomas Eaton (grocer)	2 gross **wooden dials**		
25 Jun 1568	*Falcon* of Antwerp (30 tons)		Christian Johnson	Antwerp
	William Cooper (haberdasher)	5 dozen **hourglasses** (£ 8)		
25 Jun 1568	*Ellen* of London (60 tons)		William Spender	Antwerp
	Thomas Eaton	1 gross **wooden dials**		
25 Jun 1568	*Barthelemew* of London (60 tons)		Richard Cokes	Antwerp
	Thomas Eaton	1 gross **wooden dials**		
9 Jul 1568	*Greyhound* of Lee (26 tons)		Henry Rawlins	Antwerp
	John Borne	1 gross coarse **hourglasses**		
14 Jul 1568	*Andris* of Antwerp (–)		Jacob Cornilis	(?)Antwerp
	William Hobson (haberdasher)	5 gross **wooden dials**		
2 Sep 1568	*Edward* of Milton (60 tons)		William Harison	Antwerp
	John Borne	1 gross coarse **hourglasses**		
10 Sep 1568	*Swallow* of London (60 tons)		Richard Poulter	Antwerp
	John Eliets (leatherseller)	4 dozen coarse **hourglasses**		

Notes: There were separate accounts for alien merchants, i.e. neither English nor Hanseatic.
Source: Dietz (1972). The originals are at TNA E190/4/2. Items searched for included clocks, hourglasses, astronomical or mathematical instruments, dials, and technical books.

Domestic production of these items was also significant, with many hour-glasses reaching their customers through chapmen, or small shopkeepers like Roger Lowe in Ashton-in-Makerfield, as he recorded in 1664:

Tuesday. Henry Feildinge, an Hower glasse maker whom I had hower glasses of, came, and I was ingagd for 1 dozen and ½ of hower glasses, and this day I payd hime and made meet with hime and upon 4th May, being Wednesday, I tooke 30 glasses more, and . . . 4 half hower glasses. (Sachse, 1938: 60)

(8) Through the expanding Elizabethan postal system, the government produced greater attention to time over a potentially wide area, both through letter-writers needing to 'catch the mail', and by requiring postmasters to endorse items with the town name and time, and to keep a record of items handled, and times of handling. Of many hundreds that must have existed, just one postmasters' book survives to illustrate a postmaster's recording of items handled under the Elizabethan system, with information copied from the packets themselves (Table 6.9). Among those handled on 29 August 1585, was one at 7 a.m. addressed to Sir Francis Walsingham that had been despatched by Sir John Foster at Alnwick at 1 p.m. on 26 August. Rather faster had been a packet from Lord Burghley to Sir Amyce Pellet, received at 4 p.m. on 30 August, having only left London at 1 p.m. the previous day (Brayshay *et al.*, 1998: 280).

(9) By the late seventeenth century, expectations of promptness are explicit, not least in preventing large public occasions from getting out of control, or

Table 6.9. Entries on a page of the Huntingdon postmaster's book, late August 1585

28	more 1 leter to S*i*r Frances Walsingham
28	*the* 28 of August 1 p. at 2 afternoone to S*i*r Wyllm Russell from S*i*r Frances Wallsynghm dd. at Barnellmes *the* 27 Auguste at 2 in the afternoon
28	*the* 28 of August 1 p. at 6 in the afternoon to S*i*r Frances Wallsyngham from S*i*r Amyes Pellet dd. at Tudbury *the* 27 Auguste at 12 of the daye
29	*the* 29 of August 1 pat at 7 in *the* morninge to S*i*r Frances Wallsyngham from S*i*r John Foster dd. at Allnoweke *the* 26 of Auguste at 1 afternoone
29	*the* 29 of August 1 pacquy at 4 afternone to S*i*r Frances Wallsyngham fro*m* Mr Wotton dd. at Dumbarto*n* the 15 of Auguste at 7 in *the* mornynge
30	*the* 30 of August 1 pat at paste 4 afternone to S*i*r amys pellet from my Lord Bourleye dd. at London the 29 of Auguste at 1 in *the* afternoon
31	*the* laste of August at paste 3 in *the* afternon from my Lord Scrope to S*i*r Frances Wallsyngham an to my Lord & on to may Lord Bowrleye dd. at the 28 of Auguste at 1 in the afternoone.
31	more 1 pacquyt to my Lord from S*i*r Henry Woddwyng dd. at Berweke *the* 28 of Auguste at 7 at nyght
31	more 1 leter to Mr Robart Warckrope from my Lord Scrope
31	more 3 leter too to my Lord Bowrleye & on to S*i*r Thomas Sessell
31	more 1 leter to Mr Robart Bawes

Source: Illustration in Brayshay *et al.* (1998: 280). The original is at Hatfield House, Salisbury MS CP 138.204v.

where late arrival would undermine the character or dignity of an event. Thus, the 12 o'clock start of Queen Mary's funeral in 1695 was 'To be punctually Observed by all Persons therein Concerned'.[52] Promptness was stressed in similar terms in handbills for plays, concerts, other entertainments, and public auctions. For Edward Millington's sale of 'A Curious Collection of Prints and Drawings' at York Street in Covent Garden on 12 November 1691, the sale would 'begin at half an hour past two of the clock precisely'.[53] Conventions about promptness formalized responses to an informal quantification of impatience, and these early everyday exactitudes, like Anne Lister's testing of Mrs Saltmarshe, were opportunities to gauge people's commitment to new forms of punctuality-focused civility.

Publishers of playbooks and promoters of entertainments also showed an appreciation of durations, in recognizing that potential audiences had limited time to devote to an event, and that their interest in seeing a performance might be stimulated by its compact length:

a new interlude and a merry of the 4 elements declaring many proper points of natural philosophy and of diverse strange effects & causes which interlude if the whole matter be played will contain the space of an hour and a half, but if you like you may leave out much of the sad matter . . . and then it will not be past three quarters of an hour of length.[54]

(10) The embedding of clock times in descriptive repertoires is exemplified in recipes collected around 1700 by Diana Astry, prior to her marriage into a Bedfordshire gentry family. She amassed 239 food dishes, 52 wines or cordials, 38 preserves, 21 medicinal remedies, and 25 pickles (Stitt, 1947). She gives sources for over 90 per cent of recipes, which clearly assembled the descriptive styles of people of diverse social standing, particularly from paid housekeepers to her, or her relatives, and from older gentry women including one from Robert May's *Accomplisht Cook* (1678).

Clock times appear in numerous recipes. 'Lady Hewet's cordial water', a particularly complex (and expensive) remedy, incorporated some eighty ingredients and would 'preserve life a few minutes longer in any dieing person' (Stitt, 1947: 143–4). Despite the lack of a standard temperature scale, the variable construction and materials of ovens, and variable fuels, cooking times were frequently given in whole and part-hours. Times involved cooks' skills and experience, rather than slavish adherence to recipes, being conditional on, say, an oven being as hot as for baking bread, or for a venison pasty. Even so, 'task-oriented' instructions like throw your lamprey, unskinned, 'in boiling hot water as long as you can say a Avemary' were less common than clock times:

take 10 eggs, leave out half the whites, beat them very well & strain them & put in the sugar & beat them ¼ hr. . . . Then put in the flower & beat it all together, & put in colliander seeds or carryways, . . . & beat it well together ¼ hr. more.

Likewise, for French bread the cook mixed specific quantities of fine flour, milk, eggs, and salt, then 'knead it and let it stand 1 hr., then mould it up & ½ hr. will bake them' (Stitt, 1947: 151, 136, 114–15). Besides baking, beating, and standing durations, clock times were given for boiling, mixing, salting, marinating, stewing, and soaking, which along with the range of people cited as sources of recipes using clock times, points to clock times as a generalized resource in describing actions, not an idiosyncratic practice.

Because Diana Astry collected recipes used by others, her collection avoided one problem with relying on mention of clock times in recipes in published cookery books, which may be prescriptive rather than descriptive of kitchen practice. Books like Hannah Woolley's *The Accomplish'd lady's delight in preserving, physick, beautifying, and cookery*, of 1675, made extensive use of clock times. Many were timings in whole and half-hours, but more delicate syrups and infusions received much closer attention, as when the baking of a cream tart involved 'half a quarter of an hour' in the oven (Woolley, 1675: 87).[55]

Possible parallels between cooking and industrial processes of heating, dyeing, or the like, are intriguing, but publications on 'plebeian' industry are few, and this remains an area for further enquiry. The evidence for scientific experiments is much fuller. When John Dee recorded several alchemical experiments in 1581, his use of clock times was limited to recording when he had done them rather than their duration, or how quickly changes were produced (Fenton, 1998: 308–9). In the upsurge of science associated especially with the new Royal Society established in 1660, timing was much more central to the description and interpretation of experiments, though access to these spaces was largely restricted to elite men and women.[56] So too were publication of accounts and findings, as in Robert Boyle's description in 1682 of experiments with air pumps. Boyle's *New Experiments Physico-Mechanical, Touching the Air* show both his routine use of clock times, and the limited social range of those present:

having diverse times tried the experiment of killing birds in a small receiver, we commonly found, that within half a minute of an hour; or thereabout, the Bird would be surprised by mortal Convulsions, and within about a minute more would be stark dead, beyond the recovery of the Air, though never so hastily let in. Which sort of experiments seem so strange, that we were obliged to make it several times, which gained it the advantage of having persons of differing qualities, professions and sexes, (as not only Ladies and Lords, but doctors and mathematicians) to witness it. And to satisfy your Lordship, that it was not the narrowness of the vessel, but the sudden exsuction of the Air that dispatched these creatures so soon; we will add, that we once inclosed one of these birds in one of these small receivers, where, for a while, he was so little sensible of his imprisonment, that he ate very cheerfully certain seeds that we conveyed in with him, and not only lived ten minutes, but had probably lived much longer, had not a great Person, that was spectator of some of these experiments, rescued him from the prosecution of the trial. Another bird being within about half a minute, cast into violent convulsions, and reduced into a sprawling condition, upon the exsuction of the air, by the pity of some fair Lady's (related to Your Lordship) who made me hastily let in some air at the stop-cock, the

Plate 1 A clockmaker at work, c.1450–75. From part of an astrological treatise showing occupations involved in astronomy and prognostication, the drawing shows a part-disassembled domestic wall clock, with its foliot bar projecting above the clock case, weights and cords suspended beneath it, and a gear wheel, parts, and tools on a bench or shelf. Though turret and domestic clocks differed considerably in scale their mechanisms could be very similar in other respects. Reproduced by permission of The Bodleian Library, University of Oxford, MS. Rawl. D1220, fol. 32.

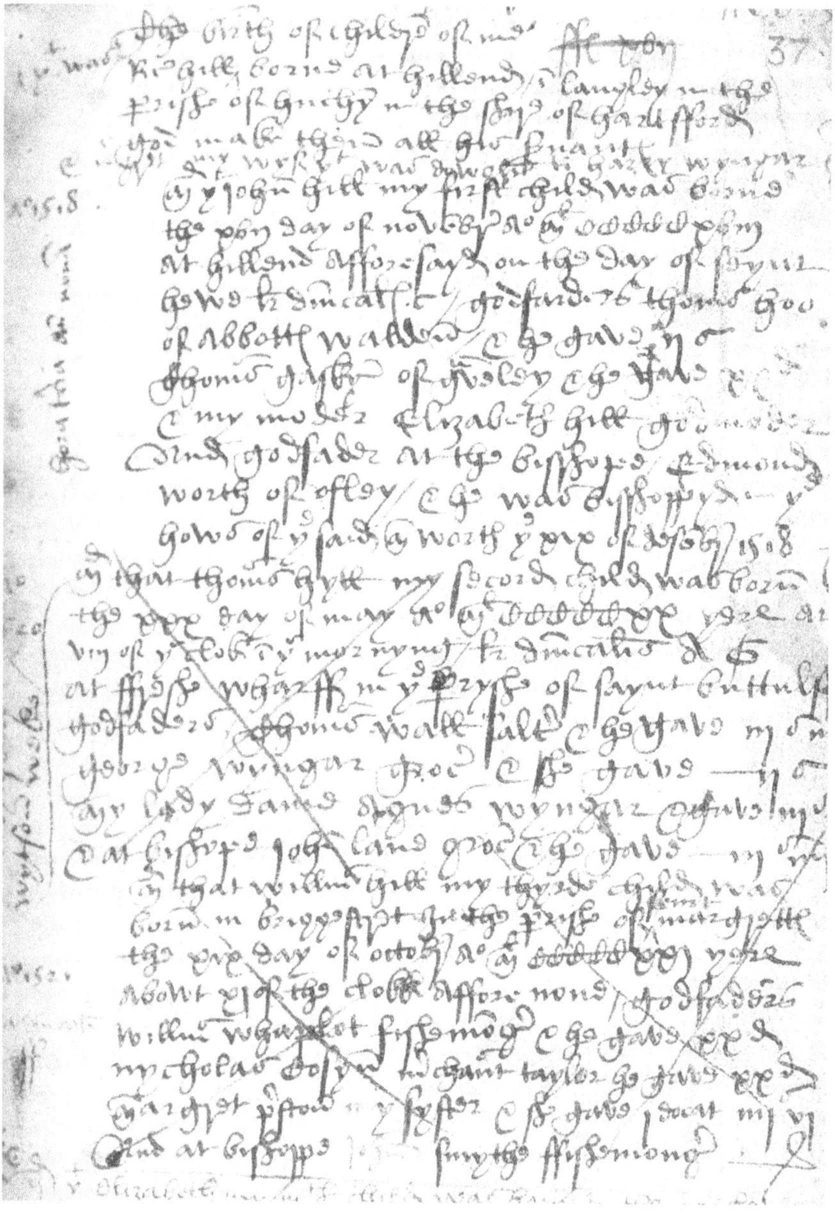

Plate 2 Birth dates and times of Robert Hill's children, 1517–21. Hill, a London mercer, divided his time between the capital, and property in Hertfordshire. Here, on a blank leaf of his family bible he records the date and hour of the births of three of his children, once in the country and twice in different London parishes © Master & Fellows, Balliol College, Oxford.

Plate 3 A sketch by Hans Holbein for his portrait of the family of Sir Thomas More. In this sketch of c.1530, the weight-driven clock is prominent, just as in the painting produced by Rowland Lockey in 1594, now in the National Portrait Gallery, London, which adds later family members to the group. Reproduced by permission of the National Portrait Gallery.

Plate 4 Mechanism of a turret clock by Leonard Tennant of London, c.1600–20. Tennant was one of London's most important builders of church clocks and other large turret clocks in the early seventeenth century. Several of his clocks were destroyed in the Great Fire of 1666, but this example escaped, having been made for Cassiobury Park in Hertfordshire. The gearing towards the right of this view is the going train, with the foliot bar and its movable weights uppermost, and the asymmetric teeth of the crown wheel visible below. The second set of gearing towards the left of the illustration forms part of the striking mechanism. Photograph © The Trustees of the British Museum.

Plate 5 Churchwardens' recorded spending on parish church clocks, 1540–1700.
(a) the 1540s (b) the 1590s (c) the 1640s (d) the 1690s
Parishes with (solid circles) and without (hollow circles) hour-striking clocks.
Rising numbers of documented parish church clocks in early modern England are produced by the combination of improving survival of evidence and growing numbers of clocks, as many new clocks were built, and almost all existing clocks were maintained.

Plate 6 A London clock for export markets, *c*.1700. By the late-seventeenth century clocks were being exported from London to many parts of the world. This example came from an established maker in the City, with a sideline in clocks for the Turkish market. Though with some features directed to that context, on the dial and on top of the clock, this is a basic single-handed design of pre-pendulum type, but benefiting from larger scale production, and more specialized production of components. Photograph courtesy of the private collection of Mark Taylor.

Plate 7 Cottage interior, illustrating plebeian clock ownership. The painting's subject is 'the poor widow's cottage', but among the consumer goods shown with this eighteenth-century widow are not only joined furniture, crockery, and tea-ware, but a pendulum clock. The clock is working, though missing its case, and has a large brass dial with hour and minute hands. © Victoria and Albert Museum.

a

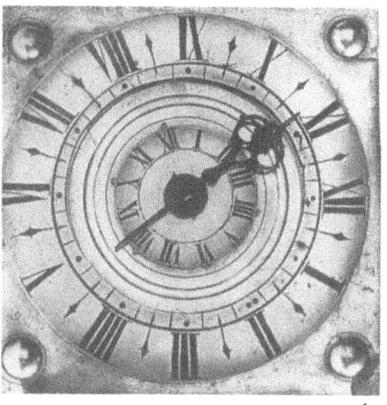

b

Plate 8 Fine time-telling from single-hand clocks. (a) at the top end of the market: Joseph Knibb, Oxford, 1690; (b) in the provinces: Walter Archer, Stow-on-the-Wold, c.1700.

The combination of a finely pointed hand, and a detailed and clearly laid out hour ring, enabled single-handed clocks to be read and interpreted to small fractions of an hour. While early-modern church clock dials were much more coarsely crafted, their large size and the gilding of hand and dial meant they could be read from considerable distances. Many parishes across England were content to retain single-handed clocks well into the twentieth century. Photographs reproduced by permission of Brian Loomes.

Plate 9 Fine time-telling from a sundial. Greater accuracy in clocks and watches increased the demands made of sundials in showing time by the sun. In 1769, a dial-maker in York marked his dial in minutes, using the diagonal lines of a vernier scale to permit even finer reading. The accuracy with which the dial was set up may not have matched the precision of its engraving, but shows a valuing of fine distinctions. Photograph reproduced by permission of Brian Loomes.

Plate 10 John and Joseph Harrison's long-case clock of 1728. Completed two years before John Harrison's move to London, the clock indicated the importance of precision both in its performance, and by displaying an 'equation-of-time' table in the frame on the door of the case. © Worshipful Company of Clockmakers, London, and the Bridgeman Art Library.

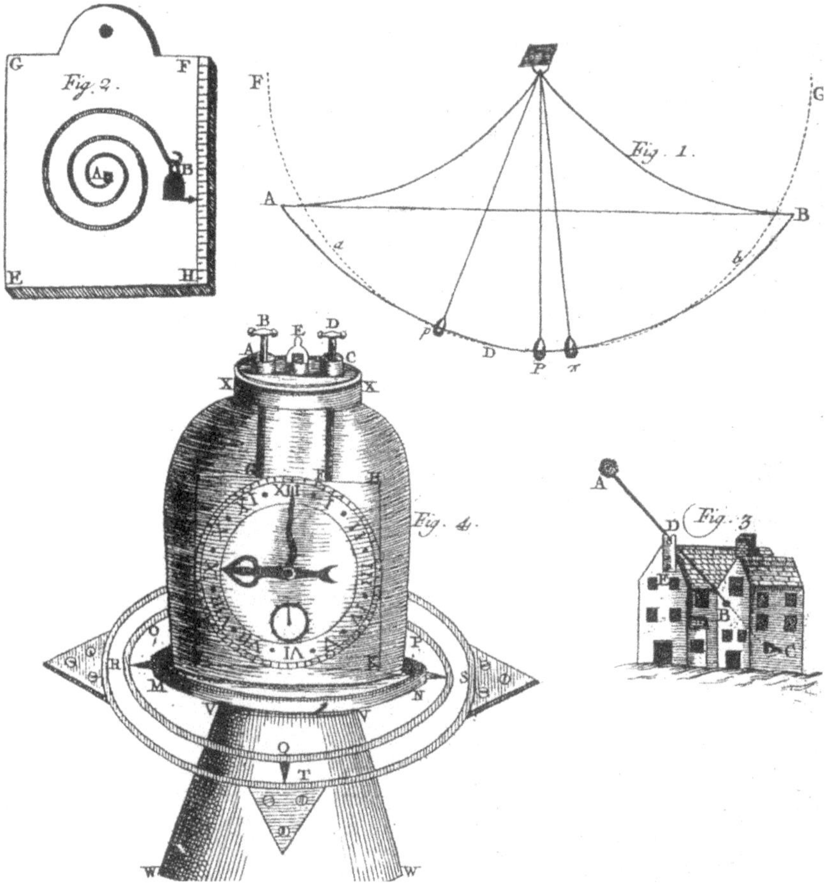

Plate 11 Illustration from Jeremy Thacker's The Longitude Examin'd, 1714. Thacker's design foreshadowed some elements of John Harrison's method, including temperature compensation, and maintaining-power in the chronometer, and an improvised star-sighting method for determining the accuracy of the clock against sidereal time. © British Library Board. All Rights Reserved, 533.f.22.

gasping animal was presently recovered, and in a condition to enjoy the benefit of the Lady's compassion. And another time also, being resolved not to be interrupted in our experiment, we did, at night, shut up a bird in one of our small receivers, and observed that for a good while he so little felt the alteration of the air, that he fell asleep with his head under his wing; and though he afterwards awaked sick, yet he continued upon his legs between forty minutes and three quarters of an hour; after which, seeming ready to expire; we took him out, and soon found him able to make use of the liberty we gave him . . .

If to the foregoing instances of the sudden destruction of animals, . . . perhaps Your Lordship would suspect, with me, that there is some use of the air which we do not yet so well understand, that makes it so continually needful to the life of animals. (Boyle, 1682: 184–5)

Boyle's account demonstrates both how thoroughly embedded description in minutes had become for experimenters and their readers, but his populating the distant end of the social spectrum with 'doctors and mathematicians' shows how incomplete a picture of clock times can be gained from elite sources alone. What we think we have demonstrated through these vignettes, though, is that many glimpses of less exalted social circles also show familiarity and comfort with clock times.

Several broader points arise from these vignettes, of which we emphasize two now. One is to urge proper caution in attempting more than broad inferences about people's skills or attitudes from their use—or avoidance—of times of day in utterances or writing. *Contexts were critical.* The second issue involves the general relationship of timekeeping technology and the societal impulses to timekeeping. Technologies clearly do *not* dictate the sophistication of time-awareness. Earlier historians tended to write the history of time-consciousness from the technology of timekeepers, and, though much criticized, the tendency persists in writing addressing both academic and general audiences (Chapter 2). However, social disciplinary accounts often simply reverse the causality, to make timekeeping technology driven by social structural impulses to timekeeping. The direction of causality is reversed, but the relationship remains simplistic. Our aim has been to transcend this dichotomy.

6.7 SCHOOLS AND SOCIALIZATION

The preceding emphasis on how thoroughly clock times and skills pervaded everyday life in early modern England poses important questions about the acquisition of these capacities. It is telling, if frustrating, that questions of where and when clock-time skills were acquired is a difficult question to answer. After briefly considering the range of skills involved, we focus on schools and temporal skills, on schools as timed and socializing environments, on school hours, on learning to tell the time, before considering socialization within specific work experiences, and the broader socialization of children within adult society. We are conscious of merely scratching the surface of a major research topic, but aim to establish some outlines of a research agenda.

Clock-time skills involve much more than 'telling the time', in the sense of interpreting the appearance of a clock, whether we are considering dial and hands or a digital display (the present discussion will disregard digital displays, since these were rare before the 1970s). Successful cue-interpretation also depends on understanding of an abstract division of time into hours and minutes, and on durational skills—knowing what 'half an hour' is. Digital instruments aside, contemporary telling the time means reading the positions of hands on a clock-dial: 'little hand' for hours, 'big hand' for minutes, perhaps another hand for seconds.

This is, however, anachronistic for late medieval or early modern England. Early clocks were physically very different: much larger, heavier, more expensive, needing much routine attention, and in different settings. Most were public turret clocks, signalling the hours with bells rather than dials. Though a minority of public clocks had dials, even in 1500, clock-dials became common only with the proliferation of domestic clocks from the mid-seventeenth century. Almost all sixteenth-century dials had a single hand. Consequently, for centuries telling the time from public clocks entailed skills different from, and simpler than, those required today. Telling the time involved hearing rather than seeing; recognizing the hour-striking bell, counting strikes up to twelve, and knowing that hour-counting began at midnight and noon.[57] Dials augmented bells rather than superseding them, as thousands of grandfather clocks attest. The considerable physical stress on a turret clock's hand(s), necessarily large and visible in an exposed situation, was a major obstacle to displaying the time compared with the sheltered environment and lightweight hand(s) on a domestic clock.

The visual equivalent of the hour-bell were dials with a single hand, from which telling the time was a relatively simple and accessible skill compared with later clocks with two or three hands, only one of which indicated the number at which it pointed, or with dial-rings indicating both hours and minutes. (Readers with young children can explore this by teaching them the time with just an hour-hand, and judging whole, half- and quarter-hours from that alone, replacing the minute hand once this has become familiar.) To reiterate, time-telling from clocks was intrinsically more intelligible in early modern England than later, and demanded comparatively unsophisticated counting skills. The widespread handling of clock times suggests these were readily acquired.

This point is vital because most discussions make anachronistic assumptions about time-telling. We link this to the assumption that before nineteenth-century schooling clock-time skills were acquired in the pressure of disciplinary contexts, like Thompson's early factories. Thompson certainly emphasized the importance of schools in training children in clock-time skills, habituating them in readiness for industrial work, but saw them as relatively unimportant until factory industrialization.

We regard early modern schools as significant sites of (broad sense) temporal learning, though in a different sense from Thompson's stress on schools during industrialization. Schools played significant socializing roles by arranging their everyday operation within a clock-time framework, although this was often taken for granted and evidence is patchy. Historians of education usually focus on endowed schools, whose formal foundations, income, and buildings created institutional continuity, and better documentation than parish elementary schools (Orme, 1989: 35).

'Schools and universities probably had their customary daily times of starting, changing and finishing their work as early as the twelfth century, although little or nothing of these patterns has been recorded' (Orme, 1973: 179), which were rephrased using clock time. But there was 'clear evidence of school routines timed by the clock...from the fourteen-forties' (Orme, 1973: 124), mainly through school statutes. Around that time, Oxford University statutes begin to fix colleges' hours, and the times of lectures. Specific hours are first mentioned in 1443, for All Souls, and for several other colleges before the early sixteenth century.

Such stipulations were not only found where schools were embedded within other institutions, but where they were established by community elites, as in Solihull parish school in 1669. That contract exemplifies specific and routine uses of clock time in school life.

Table 6.10. Feoffees' instructions to the master of Solihull Free School, 1669

1669, 19 April. Solihull.
At a publick meeting of the feoffees it was ordered in reference to ye Government of the free Schoole upon the entry of the new upper Schoolm^r Mr Jonathan Core & agreed unto by him as follth.

constantly and daily to keep the schoole from seaven of the clock in the morning at the Latest till eleaven in ye morninge & from one in the afternoone till five from Candellmas till St Andrewes Day save on ye days & p(ar)ts of days afore excepted & except all Thursdayes in ye year when the schollars are p(er)mitted to be dismissed at three of the clock in the afternoone; & except the two days before Easter Day & all Easter week, & two days before Whit Sunday, & all Whitsun week & the nine days p(re)ceding X^{t}mas day in wch number w^{t}ever Sundays happen are to be reckoned & wch ever days days following Christmas day in any yeare till the Monday follth ye feast of Epiphany for all which s^d schoolm^r is not obliged to teach schoole; & further except that from and after St Andrewes day till ye s^d nine days before Christmas he shall not be obliged to begin to teach till 9 of ye clock in the morning nor longer than till 3 in the afternoon; but then to suffer them to play but one houre at noone vizt from eleaven till twelve: & as from St Andrewes Day till Candlemass day.

Source: Warwickshire Record Office. DR.64/63 p. 277.

Table 6.11. Teaching hours at boys' grammar schools

		summer		winter	
Norwich	1566	6.00–11.00	13.00–17.00	6.30–13.00–	
Botisdale	C16th	6.00–11.00	13.00–17.00	7.00–11.00	13.00–17.00
Shawell, Leic	1608	6.00–	–18.00	7.00–	–17.00
Carlisle	c.1690	6.00–12.00	13.00–18.00	7.00–12.00	13.00–18.00
Tamworth	1693	7.00–11.00	13.00–17.00	7.00–11.00	13.00–17.00
Hours for senior masters					
Norwich	1566	7.00–11.00	13.30–17.00	7.30–11.00	13.30–17.00
Solihull	1669	7.00–11.00	13.00–17.00	9.00–11.00	12.00–15.00
Norwich					
Headmaster		7.00–11.00	13.30–17.00	7.30–11.00	13.30–17.00
Sub-master/usher		6.30–11.00	13.00–17.00	7.00–11.00	13.00–17.00
Boys		6.00–11.00	13.00–17.00	6.30–11.00	13.00–17.00

The sub-master taught the three lowest of the six forms

Sources: Norwich: Norfolk Record Office; Botisdale: Saunders, 1932: 141–3 n. 2, citing 'Tanner, pp. 1735–43'; Shawell: Leicestershire Record Office D/E 734/6, 'orders for the free school', 1608, at the end of this account book. Barnett: Tripp, 1935: 224–6. Carlisle: Cumbia Record Office, Carlisle. Uncalendared transcript in searchroom library, shelf for Carlisle local history. Solihull: Warwickshire Record Office, DR.64/63 p. 277; Tamworth: Bickley, 1930: II, 224–6.

Other seventeenth-century schools set out similar regulations, evincing concern with time and the school day (Table 6.11). Similar specifications were made in rural Shawell, Leicestershire, in 1608 with the foundation of a free school and almshouse by John Elkington's bequest of 1604. In 1608 John's brother Edward, the remaining executor, drew up detailed Orders governing the school, which with John Elkington's will were copied into the Shawell churchwardens' book. Fifteen Orders covered both academic matters (Grammar scholars must speak Latin) and day-to-day matters (scheduling punishments for breaking school windows, or infringements in prayer or punctuality), beginning with school-hours:

Imprimis it is ordered of every Schollar... shall be every morning at the school before 6 of the clock in the sum'er & 7 of the clock in the winter & There shall reverently upon their knees use such prayers as are agreeable to the word of god & the Lawes of this Relm befor the School begins as the Schoolm'tr shall approve & prayers likewise evry evening at 5 of the clock in the winter & 6 in sum'er before their departure.[58]

Late seventeenth-century Carlisle Cathedral's Free Grammar School had over forty detailed regulations.[59] The 16th rule required every scholar residing within Carlisle to:

come diligently and constantly to school at 6 a clock in the morning, when the bell rings during the whole summer... from Shrovetide till Michaelmas; and at 7 a clock from Michaelmas till Shrovetide. And not depart from or leave school till 6 a clock at night, either summer or winter, except the master shall think fit.

Rule 17 gave the master discretion over school-hours for boys from outside Carlisle, before rule 18 returned to the daily schedule:

every scholar having said his Morning Part, shall have leave to go to Breakfast at 8 a clock and to stay till 9. And that from 9 they continue till 12 a clock exactly, or near to it, when they are to be dismissed to Dinner. That at one a clock all again shall repair to school and remain there studiously and orderly till six a clock.

Further rules spelled out refinements (attendance at school was not required until 8 o'clock on Ash Wednesday), defined term dates, and required the usher to 'shut up the school at night punctually and constantly' besides teaching arithmetic, geography, surveying, gauging, and navigation. These subjects were often an usher's particular responsibility. Some ushers were clearly extremely capable—Samuel Foster rose from school usher in his home town of Coventry in the 1640s to become Gresham College Professor of Mathematics (Barbor, 1982: 114).

Descriptive accounts of schools closely echo the normative regulations, suggesting that pupils broadly experienced the school-days described in regulations. Tamworth Grammar School was described in 1693 by the tutor who accompanied George Hastings, heir to the Earl of Huntingdon, from 1690 to 1692 at Eton, and then at Tamworth for one year, before proceeding to Wadham College, Oxford. The move to Tamworth apparently agreed with the fifteen-year-old Hastings rather more than it did his tutor, who sent disparaging reports back to the Earl, including one describing hours.

The method of the school here is to go exactly at 7 and return at 11 and from 1 till 5 in the afternoon. They say parts in the Grammar and the books my Lord's form read are the Greek Testament and Juvenal only (for they read but few books at a time). There is too a small system of rhetoric. Their exercises at night are either... Latin or English verse, as their master pleases. They give over at 3 on Thursdays and play all Saturday in the afternoon. (Bickley, 1930: II, 224–6)

Had schools not actually operated such hours there would have been no cause in 1612 for a Puritan schoolmaster in Ashby-de-la-Zouche to write proposing quarter-hour intervals at 9 a.m. and 3 p.m., with school running on until 5.30 p.m. He argued that this would produce better work and less confusion from constant 'running out to the field ["*campo*"]', a debate almost as old as schools. For example, the thirteenth-century manuscript *De Disciplina Scolarium* had discussed the best parts of the day for learning and teaching (Weijers, 1976: 126–8). Similarly, a group of parents at Norwich petitioned against one Mazey, 'who kept the pupils until noon and 6 p.m.', beyond the hours at which morning and afternoon school should have finished, and indicates that stipulated hours were normally observed (Saunders, 1932: 142–3).

Whether schools maintained their own clocks or relied on signals from church clocks is rarely recorded. Where parish schoolrooms adjoined parish churches, a striking church clock would have been audible, but even schools close by church clocks had their own. In the 1680s the Cheshire village of Stoke funded both a church clock and a school clock, respectively overseen by the sexton and schoolmaster, each receiving twenty shillings annually.[60] At Cockermouth, ringing the school bell was written into the parish clerk's duties, alongside keeping the church clock and chimes, and ringing the supper bell (Bradbury, 1981: 87).

One might expect the schools whose statutes spelled out days organized with clock times would teach pupils to tell the time, but there is little evidence of this. There is little evidence of classroom teaching and learning of 'telling the time' through models or images of clock-faces, play situations, or books. Neither was 'telling the time' prominent in instructional literature for children, a major growth area of eighteenth-century publishing.[61]

There are three possible explanations for this invisibility: one, the teaching occurred, but was thoroughly taken for granted; two, that clock-time skills were acquired outside the classroom in preparation for work, rather than during school; and three, that there was no need for formal instruction because children had *already* acquired clock-time skills in everyday socialization, through the routines of their elders. The last possibility is strengthened by schools at various times relying on their pupils' already grasping clock times, in prescribed school-hours, and in their teaching materials.

Grammar schools' teaching of Latin, rhetoric, and classical authors, all involved English–Latin translation exercises, based on ideas and situations familiar to pupils, termed *vulgaria*. Several sets of *vulgaria*[62] survive containing several hundred sentences, as 'treasure houses of typical English scenes, proverbs, quotations and colloquial dialogue which preserve many details, both of life in general and schools in particular, which are not easily to be found elsewhere' (Nelson, 1956; Orme, 1973: 98–100, 134, 123). Like village by-laws, like school-hours, clock times first appear in *vulgaria* as modifications to older phrases.

For example, *vulgaria* variously dated to 1512–14 or 1522–27 describe life and work at Magdalen College School, Oxford, where a master and an usher oversaw boys aged between ten and eighteen (Orme, 1973: 126–8).[63] These are fuller and more polished than Thomas Schortt's *vulgaria* from near Bristol in the 1420s, befitting their advanced school context, with more material specifically about school life. Alongside general admonitions about time-wasting, 'how ungraciously and how wretchedly I have spent my time' (Orme, 1989: 135), are specific references, *c.*1500, to starting at 5 o'clock (Nelson, 1956: 2), and 6 o'clock starts were common (Orme, 1973: 124). Oxford boys arriving at seven are too late:

> He that hath the rule of me will not be content that I should go forth before seven of the clock striking, therefore I can not tell what I may do, whether I should after him or else the school master, which will forbid me the school except I use to come sooner in the morning. (Orme, 1989: 135).

I waked a great part of this night and about day I began to fall fast and heavy asleep, and so thought the hour that was appointed me to go to school in was come, yet my mother commanded that there should be no noise made and would not let me be called up. Therefore my mother is the cause of my late coming and not I, for if I had wakened myself or else any other body had wakened me, I would have been glad to a coming hither at the hour that was appointed to me.   (Orme, 1989: 139)

a scholar in our hall that had a little business to do in the town . . . [was robbed] . . . about eight of the clock as he was coming home again.   (Orme, 1989: 145)

In the very different context of Barlinch Priory, near Dulverton, Somerset, 'one of the poorest, remotest monasteries', a *vulgaria* from *c*.1500 has the pupil-writer at school from 5.30, expressed as 'half an hour before six o'clock'[64] (Orme, 1989: 119).

Vulgaria in general portray children experiencing time-discipline relatively early, as where one translation of *c*.1500 contrasts the slack daily patterns of younger children's lives with a twelve-year-old, for whom:

now the world runneth upon another wheel. For now at five of the clock by the moonlight I must go to my book and let sleep and sloth alone, and if our master lath to wake us, he brings a rod instead of a candle.   (Nelson, 1956: 1–2).

That schools presumed that boys were already able to tell the time, both in attending, and for *vulgaria* to make sense to those doing the translating, suggests clock-time skills already made general through everyday socialization, and not requiring special teaching. On this reading, formal instruction only became necessary when time-indication on clocks became more complex.

Even labourer-poet John Clare envisaged children reading clocks in his *Shepherd's Calendar* (1827): 'viewing with jealous eyes the clock' (May, line 22), and during the harvest, the 'driving boy with eager eye / Watches the church clock passing by— / Whose gilt hands glisten in the sun— / To see how far the hours have run' (September, lines 71–4).[65]

6.8 THE PLACES OF EVERYDAY CLOCK TIMES: EVERYDAY TEMPORAL ENVIRONMENTS

Despite the highly selective natures of both documentation and preservation of evidence, it is clear that clock times appeared in the writings and utterances of early modern Englishmen and women in a great variety of circumstances, and for many different purposes. Clock times, widely signalled through bells and, increasingly, dials, were everyday resources and skills constitutive of all sorts of social practices. According to time and place they involved varying combinations of purposeful 'foraging' or simple heuristics, usually the latter because of a considerable trade-off between the effort involved and the adequacy of the information obtained, for different purposes. Publicly struck hours were a routine descriptive resource, but the extensive 'foraging' required to tell the time really accurately ensured that, for everyday purposes, fast-and-frugal algorithms based on small amounts of robust temporal information were preferred, and adequation was relatively loose most of the time.

The successive 'revolutionary' changes of clock-time's uniform metric, increasing subdivision of that metric, and growing numbers of more specialized clock-time practices, were all grounded in long-run shifts in everyday temporal environments that transformed the trade-off between effort and adequacy. As greater numbers of better instruments circulated a richer array of information, they cumulatively produced environments that functioned more and more effectively as timekeeping devices, becoming increasingly self-referential and confirmatory.

The eighteenth-century density of instruments was much higher; they performed much better, and provided richer and more refined temporal information for particular uses associated with new communities of practice (Chapters 6 and 9). Particular temporal communities sought better adequation, and were willing to bear the (now-lower) costs of meeting their interpretation of the appropriate 'right time'. In the long-run, what Gigerenzer might refer to as 'more complicated foraging behaviour' was offloaded into the better-performing devices. By $c.1770$, cheap single-hand clocks cost around 20 shillings, less if second hand, and finely subdivided hour rings enabled close measurement. Many clocks even at modest levels of the market had a minutehand. For many everyday uses, clocks had a significant surplus of precision over what their owners usually used, and the readily available precise times were used in many circumstances when not specifically required.

Many practices changed in degree rather than kind, because pre-'horological revolution' practices rested on greater familiarity with clock time than has been supposed. Revolutions in practices rested on an environment much richer in material objects containing temporal information, and a complex of secular processes producing everyday environments whose temporalities were increasingly self-referential and confirmatory. That everyday temporal practices and everyday temporal environments were mutually reinforcing was an intrinsic feature of early modern everyday life. Their roots were multiple: in production and work; in consumption (with times imported into corporeality via clocks or publications); in changing patterns of informal sociality.

Assessing the relative importance of different factors in reshaping temporal environments requires sustained substantive exploration, rather than theoretical pre-judgement. So too do questions of how, how much, and when clock times mattered in particular temporal communities, how the dynamics of relations among specialized communities of practice unfolded, and how the clock-time practices of specialist communities influenced everyday practices.[66] Close timing was undeniably valuable for regulatory purposes and particular tasks, from coordinating factory work to timing eclipses to maintaining public order, but this chapter has demonstrated that very many clock times among diarists and deponents arose in everyday sociality, which we emphasize as a prime site for spontaneous uses of clock times and generating new timing practices. The 'everyday' does not equate with 'inconsequential': rather it was a site where different practices and notions of timekeeping circulated, were negotiated, and practically resolved. Frequently clock times were not recorded because timing was a specific objective of recording or regulation, but because clock times were increasingly integral to the fabric of everyday life, and part of 'common understanding' of living.

In that 'common understanding', clock-time skills were in some ways analogous to proverbial wisdom. Like axioms taking particular meanings according to context, clock times were flexible in line with people's goals and priorities in different situations. And they were acquired in ordinary socialization.

It was scarcely necessary to attend a school or church in order to be saturated with a wealth of proverbial imagery and aphoristic wisdom at an early age and to have it always at the tip of the tongue, ready to meet any situation. (Fox, 2000: 144–5)

Many general principles of practical conduct were absorbed through everyday experience by children (or migrants) from the regimes of adults around them, and through the distillation of experience, biblical and classical literature into 'proverbial wisdom'. The proverbs encountered in speech, almanacs, and depictions, were central to everyday oral cultures, embodying the voice and the values of the people (Fox, 2000, 112–72).

Prescriptive adages constituted a kind of oral textbook [that could] extend beyond moral and emotional issues to the practical tasks of daily life. For it was in this form that people carried much of their knowledge of meteorology, medicine, and the law; in this way that they transmitted and retained the tried and trusted methods of agriculture and husbandry, the intimate techniques of art and craft, the useful tips of manufacture and trade. (Fox, 2000: 151)

Proverbial wisdom was understood as a general resource for everyday life—Aubrey thought boys' judgement would be developed if proverbs 'be sometimes read a quarter of an hour' (quoted by Fox, 2000: 124).

The creation of everyday clock-time practices through socialization was much facilitated by the public nature of much time-signalling from the increasing numbers of public clocks, with hour-striking audible to a substantial population. That public time-signals were 'broadcast' often led early modern Englishmen and women to regard clock time as a public good, rather than perceived as disciplinary. That seems an entirely appropriate response, recognizing that changes in everyday temporal environments affected both sides of the trade-off between the 'foraging' required to find the time, and the adequation appropriate for different uses. The issue was a perennial one, but over time the limitations imposed by shortage of specific devices, or by narrowly distributed skills, were being relaxed. Notions of adequacy for general purposes were becoming tighter. In (say) 1550, complex and robust time-telling had been mutually exclusive, but by 1750 better timekeeping devices, and more self-referential environments had diminished the need for complicated foraging. The devices have received far more attention than the environments hitherto, whereas we see increasingly time-rich environments as central to the generation of everyday clock times.

Within more self-confirming temporal environments, once-novel clock-time practices were knitted into the fabric of everyday being, becoming 'second nature' to people in extended distributions of various embodied skills. These included the handling of routines, as more complex actions had to be dovetailed in time into complex schedules through coordination, controlled visualization, and the sequencing of activities and encounters. Repeated 'fitting-in' with times created senses of right and appropriate actions through the familiarity of unanalysed embodied experience. The counterpart of building routines were skills of improvisation, through which people could 'decide as they went', and could envisage alternative configurations of actions, materials and people. Better able to juggle pressures, people could respond rapidly to changing circumstances or goals. Capacities to handle routines and to improvise can both be seen as primarily urban sensibilities. However, as urban life and conditions became more familiar improvisation itself became routinized, as people built repertoires of experiences and skills that 'domesticated' improvisation as familiar and ordinary. Underlying many of these skills were improved abilities to estimate, allowing practical actions to be gauged.

To modern sensibilities these skills may appear so commonplace as to be hardly worth remark (though that is itself revealing of contemporary socialization), but in early modern England people were becoming more adept at them, they were being done better, and that made differences to lives, to social interactions, and to places. They reinforced clock times as integral elements of everyday life, distributed according to people's situations within diverse communities of practice and their social skills, and within everyday clock times and calculative practices. These environmental impulses were potent influences even where disciplinary clock times were strong. The cumulative significance of clock times grew substantially across the population, independently of whether or not particular people were subject to disciplinary impulses to explicit clock timekeeping.

6.9 CONCLUSIONS

Through various fragments and vignettes, we have tried to show that a large portion of the English population was familiar with everyday clock times from an early date, apparently handling them comfortably. Everyday uses of clock times and clock-time metaphors were comparatively widespread, much earlier than usually recognized. In certain respects, their social distribution narrowed as they became more complex over time, or new practices arose within the confines of particular temporal communities. However, the rise of specialist and technical clock times has led many commentators to underestimate the significance of everyday (and often only loosely 'disciplinary') circumstances. Forms of time-signalling technology were important enablers of routine uses and recording of clock times, but clock-time practices were also heavily influenced by the social and personal contexts of activities and recording.

Questions of how clock time was used in everyday sociality, and of how contemporaries understood the signals that communicated clock time are of very considerable interest, but are remarkably difficult to answer. The majority of the non-elite population have left no documentary evidence and, in any case, we have argued that the very taken–for–grantedness of clock time militates strongly against it being explicitly addressed, except in unusual circumstances.

Nevertheless, we think we have established several key arguments. First, clock times were familiar parts of the early modern everyday. Clock time and time-signals were widely used resources for conducting many everyday activities, that could go far beyond the particular trading or religious functions that may have provided the initial impulse for the public provision of clock time. 'If timepieces, by their visible and audible precisions, suggested possibilities of discipline, they intertwined them with promises of autonomy, self-empowerment, and liberation' (Sherman, 1996: 21). People's capacities to grasp and use clock times were widely distributed as capabilities within everyday practice, rather than confined to particular contexts or the formal exercise of power by authorities, employers, institutions. Orme (1978: 178) acutely observed the fifteenth-century 'growth of a more precise system of measuring time and a more sophisticated concept of using it', and this inexorably extended in early modern England. Clock times provided a language in which many social relations could be coordinated, impinging on many parts of the population, in both town and country. Very diverse sources convey the capacity, on occasion, of low-status men and women to use clock times. Early modern England was a much more clock-time-literate environment than has generally been assumed: few sections of society were unaware of clock times, and neither time-telling nor temporal sensibilities were such that people could not tell time beyond hours, as in some recent claims.[67]

Second, clock time is appropriately characterized as a web of practices linked to a particular metric, rather than purely a technological device. Telling the time in medieval and early modern England was very largely acquired in everyday life—the locality, the community, the home—rather than through formal teaching in educational institutions, and did not derive from clock or watch ownership. Those who could not find the time for themselves could still participate in temporal cultures through informal interaction and imitation. The breadth of temporal skills, and the growing saturation of everyday environments with temporal cues, mean that no presumption of a rigid divide between 'clock-time culture' and 'popular culture' is sustainable. We stress clock-time skills as adaptable, generic skills, rooted in mundane activities, and separable from, rather than yoked to, the specific disciplining contexts of work, church, and trade, that have received most scholarly attention. People operated at various, and varying, levels of temporal competence and sophistication, depending on *why* they were telling the time, and on the temporalities of their contexts. Their clock-time practices and their everyday dispositions toward time were mutually constitutive.

Third, early modern Englishmen and women worked with different clock times from those now 'obvious': their practices involved different ranges of skills, devices, and practical skills, from those 'natural' today. The barriers to telling the time were, in some respects, lower then than now. The experiencing and acquisition of temporal practices, in general, and clock-time skills in particular, were embedded in everyday life experiences, as practical knowledges acquired by observation and doing, rather than intellectual knowledges acquired through formal instruction, which more often characterized specialized knowledge of precisions (Chapters 7 and 9). Here again, the deployment of clock-time skills was strongly context-dependent. Here, context-dependent means not just 'who is regulating who for what', but much wider senses of the content of particular interactions, the relations among parties to those interactions, the moral weight attaching to precision or punctuality, besides more utilitarian and functional purposes.

Fourth, historians and social scientists should not presume that everyday practices were the lowest level of a hierarchy, down which temporal practices diffuse from high-order specialist centres. Such a sense is extremely common. That several people at papers where we discussed clock times in the Stonor correspondence commented that clock times must have been very pervasive of society to appear as a metaphor in love letters instances a massive assumption that such uses must have been derivative. However, the everyday could be a very strong impulse to clock time use in its own right, not merely a recipient of spillover from more important stimuli. Indeed, proximity and face-to-face arrangements were precisely the circumstances in which precision at the scale of clock time mattered. For an international merchant coordinating shipments or transactions, precision to hours or minutes was comparatively unimportant compared with calendar and rides, whereas the same person arranging to meet their fiancée's family faced different acute pressures of convention and manners.

That some contemporaries, at least, would have recognized our sketch of clock times as widespread in early modern England is suggested by William Harrison's *Description of England* of 1577, where, in discussing 'our account of time and her parts', he wrote:

Our common order . . . is to begin at the minute, . . . one-sixtieth part of an hour, as at the smallest part of time known unto the people, notwithstanding that in most places they descend no lower than the half-quarter[68] or quarter of an hour. . . . the common and natural day [being] observed continually by clocks, dials and astronomical instruments of all kinds. (Edelen, 1968)

Harrison is careful to distinguish people's *capability*—the minute as 'the smallest part of time known unto the people'—and *everyday practice*—in which 'they descend no lower than the half-quarter or quarter of an hour'. Much incidental documentation of clock times records the latter, but offers a poor basis for inferring the former, a task to which we turn in the next chapter.

NOTES

1. Ordinances for John Mowbray, Duke of Norfolk, *c*.1435, The National Archives, Kew, C115/K2/6682 fo. 251. Ordinances for Edward, Prince of Wales, 27 September 1473, *Lord Steward's Department, Misc. Book*: The National Archives, Kew, LS.13/280 fos. 277–9v.

2. On Lady Margaret Beaufort see Jones and Underwood (1992).

3. The original langauge of the ordinances was English: *A Collection of Ordinances and Regulations for the Government of the Royal Household*, (London, Society of Antiquaries, 1790) pp. 37–9. The spelling and punctuation has been modernized; Jennifer Ward (1995), 217–18. See also Armstrong (1983): 135–56.

4. Thus, for York, Humphrey (2001) stresses the importance of independent community control, rather than simplification in itself.

5. Accessed via the Medieval Towns online resource at <www.trytel.com/~tristan/towns/maldon6.html>.

6. This is a general instance of the sort of thing not recorded anywhere except by accident, prior to twentieth-century interest in oral history—the main source for Hussey's paper on gleaning—and in local customs.

7. Deedes and Walters (1907) collected details from elderly villagers of childhood memories of gleaning bells in nearly twenty parishes. Of these, eight rang at 8 a.m., one at 8.30 and six at 9 a.m., with two unspecified. The close of the gleaning day came at 5 p.m. at eight places and 6 p.m. at six. (Note however that this was not 8–5 and 9–6: all the places where gleaning went on until 6 p.m. were among those where it started at 8 a.m., so the contrast is between gleaning days of two different lengths,

not a 'standard' length gleaning day whose time was offset in some places relative to others.) Further Essex instances also started at 8 a.m. (Hussey, 1997: 63–4). Several participants recalled that the rationale for this was to enable mothers with young children to participate: the late start enabled them to provide for their families first.

8. The North Walsham account is nonetheless a striking example of the routine character of such uses in 1600 (Glennie and Thrift, 2005: 170).
9. Norfolk Record Office, MF/RO 461/4. The description may have reached a wider audience if the parish sought a 'brief'—parish collections throughout a deanery, diocese, or beyond, to benefit people whose possessions were destroyed. Circulated briefs contained formulaic statements about damage and loss to a settlement's inhabitants, but applications for grants, which rarely survive, may have included fuller testimony.
10. The edition used here is that edited by Richard Barbor (1982).
11. I.e. the square root of 666. A clergyman, Potter (1594–1678) was also a Fellow of the Royal Society.
12. I.e. 42 minutes 56 seconds past the hour.
13. This may hint that clocks did not pose the such an unfamiliar experience as watches, since most were turret clocks. The 'tick' Aubrey mentions as an element of fright may not have been very different from a household clock. It may well be that exactly how public clocks worked was not generally appreciated, and was not commonly apprehended as involving ticking.
14. Watchmaking, like dialling, was seen as entailing mathematical skill. The mathematician William Oughtred (1575–1660) wrote 'a little treatise of watchmaking' for his son Benjamin (Barbor, 1982: 231).
15. Aubrey's wider interest in quadrants and other observing instruments is evident from his notes on Edward Gunter (1581–1626). Francis Potter (1594–1678), gave Aubrey 'copper and silver quadrants, of his own projection, which serves for all latitudes' (1982: 256), observations on beam compasses, for precisely dividing scales, and Halley's improved backstaff design.
16. 'Four-miles' was the name of the course, not its length.
17. Grigson (1984): 177. Interestingly, Tusser presumes the availability of light in even poor households at both ends of the winter day.
18. Ward (1995), quoting Hunisett (1961). Similar use of canonical hours in murder and misadventure cases: ibid: 164–5. Midday was an exception: ibid: 170–1.
19. Quoting Hunisett, (1985). Similarly 10 a.m. was used in inquest cases of accident and misadventure (Goldberg, 1995, 171).
20. For McKay (2005: 205) 'it was the mind which was educated to think for itself which conceived the diary'.
21. The phrase 'point-and-stretch-time' first arose, to our recollection, during discussion following a paper we gave to the 1995 Social History Society annual conference at Lancaster. We are grateful to participants there, especially Robert Poole, Vic Morgan, Andy Wood, Paul Griffiths, and Anne Laurence, for discussions.

22. All dates in the following sections are given with 'New Style' years. Pepys used 'old style' dates: until 1751, the year number changed on Lady Day (25 March) rather than at 1 January. Thus in the old style, the day after 24 March 1661 was 25 March 1662.

23. For Pepys, explicit precision in the diary is more evident with regard to the calendar than to the clock, especially with regard to personal and national anniversaries. Pepys notes his own birthday (23 February) nine times in ten years, but birthdays of close family pass without comment, as do almost all those of colleagues and acquaintances—he records 'my Lord's' birthday twice (July 1661, November 1665), one was probably his patron Lord Sandwich. In several years, he records King Charles's birthday (29 May), but this was a national day of thanksgiving and bell-ringing, and need imply no particular remembrance on the part of Pepys. He several times mentions wedding anniversaries, both his own and others.

    Pepys specifically comments on the growing number of national anniversary days, especially 'Gunpowder Treason' day on 5 November, which by the 1660s was well established as the most powerfully evocative day in England's Protestant calendar, with its celebratory bonfires, fireworks, and church services (Cressy, 1992, 1993). Another major national day of celebration to intrude into Pepys' diary entries was 23 April, marking both St George's Day and the Coronation Day of Charles II. Occasionally Pepys refers to more traditional annual festivals, as on Valentine's Day in 1662, but such references seem incidental to accounts of events, rather than being thought worth recording in their own right.

    The outstanding example of an anniversary in Pepys diary, however, is something quite different. It concerns an anniversary which was very much of his own making, celebrated in a form of his own devising, which can be seen as, albeit on a micro-scale, an 'invented tradition'. This was what Pepys refers to as his 'stone feast'. On 26 March 1658 Pepys had undergone an operation for the removal of a kidney stone, at that time a hazardous procedure. The date immediately became a key anniversary in Pepys's year, a status it retains throughout the decade, the day marked by a dinner party and/or other entertainments as an ongoing reminder of Pepys gratitude for successful surgery and his continuing active life. Via the 'stone feast' Pepys constructs an intensely personal celebratory occasion, which rapidly occupied a central position in his annual round. It is striking that the stone feast evokes far more comment, sustained over more than ten years, than do official national anniversaries, and that no comparable meaning came to surround more general deliverances from danger, such as the passing of the very serious plague epidemic of 1665. The diarist was clearly capable of using calendrical precision in creating highly meaningful events.

24. Clock times are slightly more frequent in 1667 than previous years, though the entries tend to be longer. In 1662, for example, three years before Pepys got his 'minute-watch' there are 133 clock times, all phrased in hours only, and none with minutes.

25. Based on a sample of months. For example, between February and April 1662, Pepys made thirty-six mentions of clock times or clock durations in thirty-one days' entries of eighty-nine days in all (Latham and Mathews, 1970, III).

26. For example, see MacLean (1990: 263) on the 1660 Restoration as 'a large-scale media event . . . [T]he production of political poetry had become an integral part and necessary form of a public discourse largely managed by commercially minded printers and

professional writers.' Sherman himself noted parallels in how periodic news created a narrative indeterminacy resembling that used to build narrative tension in early novels.

27. Even if Pepys' actual writing-up was less frequent—he occasionally refers to completing several days entries in one sitting, based on earlier rough notes.

28. Diary entries that were marginal annotations on almanacs may include just a few brief phrases, but some of Dee's accounts of what he believed were conversations with angels, through Edward Kelley or Batholomew Hickman, reach well over a thousand words.

29. We shall develop this point in further papers, currently in preparation.

30. The 'hour' category includes one ambiguous entry: Fenton's version of the entry for the marriage of King Philip (of Spain) and Queen Mary (of England) quotes Dee's time of the event as '11h a.m.—20 ascending'. This is not clear as it stands, but possibly there has been a misreading of Dee using the astrological symbol for Jupiter, which resembles the number 2 in Arabic numerals (see the table of Dee's signs given by Fenton, 1998: 310). Similar use of an ascending zodiacal sign related to time appears on 12 February 1585, describing the birth of his son Michael.

31. The attention given in books about dialling to the depiction of equal clock-hours on sundials makes clear both that sundials were nor restricted to showing unequal hours, and the dominance of clock-hours, even though sundials remained important in setting clocks and watches, and as timekeeping instruments in their own right.

32. Mid-eighteenth-century population of about 350.

33. Vaizey (1985, e.g. 329–31). The following analysis is based on published sections of the diary, running to about 130,000 words of an estimated total of some 350,000 (Vaizey, 1985: v–vi). While the analysis is not comprehensive, therefore, it is based on a large volume of material, and in terms of its use of clock time, there is no reason to suppose the published sections to be unrepresentative. If anything, the elimination of repetitive elements from the published edition is likely to underplay Turner's habitual use of time recording. The original diary is now in the Sterling Library, Yale University, USA.

34. Turner also had explicit temporal interests on longer timescales. On 30 June 1757 he noted an article in *The London Magazine or gentleman's monthly intelligencer* for May (he routinely notes the issues) which discussed the expected reappearance of Halley's comet in 1758. He discusses the apparent alternation of seventy-five and seventy-six year periods in the comet's history, with recorded appearances in 1305, 1380, 1456, 1531, 1607, and 1682 (without supposing a single period of intermediate length), and worries about 'a dangerous situation as the denser part of the blazing tail should envelop the earth' (Vaizey, 1984: 103–4). The comet in fact appeared on Christmas Day 1758, but Turner did not refer to seeing it then. Turner's comments on the comet instance his astronomical interests, as when he sees and describes 'the modern microcosm', a clockwork model of the solar system, built in London by Henry Bridges *c.*1734 and widely exhibited in mid-eighteenth-century England (Vaizey, 1985: 160–1); Milburn and King (1978). His interest in topics such as birthdays and anniversaries is hard to gauge from the published edition. He appears less interested in days of public or national celebrations than did Samuel Pepys in the 1660s. There appears to be

considerable variation in Turner's noting or reflecting on his own birthday, birthdays of his wife or relations, or his wedding anniversary, but this may be an artefact of editing.
35. Published sections of Anne Lister's diaries have appeared in several works, including Whitbread (1988, 1992); Liddington (1995, 1998); Clark (1996).
36. Surviving portions cover 1635, 1643–9, 1652–3, 1657.
37. Dee's record of the birth includes conventional astrological various symbols, here spelled out and italicized.
38. Napier was the son of Robert Napier, a London merchant. Educated at Wadham College, Oxford, he became rector at Great Linford, Buckinghamshire, for several decades, practising astrological medicine. Forman and Napier corresponded regularly, visited one another, and met at Cambridge and elsewhere. In London, Forman purchased medicines, drugs, and herbs for Napier, receiving cash and 'good, country food' in return (Cook, 2002: 111–13).
39. Dee does not mention sexual intercourse from some seven months before the births of each of his children. After the birth of his daughter Katharine on 10 June, Dee does not record sexual intercourse with his wife until 5 July, after his wife had been churched earlier in the day.
40. Dee switched smoothly between the Julian calendar still used in England, and the Gregorian calendar, which many countries adopted in March 1583: 'on Saturday, the 14 of July by the Gregorian calendar, and the 4 day of July by the old calendar, Rowland my child (who was born anno 1583, January 28, by the old calendar) was extremely sick about noon, or midday, and by 1 of the clock was ready to give up the ghost, or rather lay for dead, and his eyes set and sunk into his head' (14 July 1584, Fenton, 1998: 133).
41. William Honeywell's diary remains in private hands, but sections are available at <www.express.demon.co.uk/diary/diary.html>.
42. Froissart reported that Sir Jean de Carrouges's wife was raped in 1386 by Jacques Le Gris, while Carrouges was abroad. Le Gris threatened his victim with dishonour should she not keep quiet. 'But she fixed firmly in her memory the day and the time when Jacques Le Gris had come to the castle' and after returning, her husband killed Le Gris in a trial of combat (Herren, 1999: 146–7).
43. Only two other family archives of similar scope survive, also from parts of the fifteenth century: the Pastons in Norfolk and London, and the Plumpton family in London, were both lawyerly families to a greater extent than the Stonors.
44. Note that Elmes was following an already familiar practice in reckoning the year as starting at 1 January, even though AD reckoning involved changing the year at 25 March. For official purposes, 25 March represented New Year's Day until 1752.
45. Such a comment raises two immediate questions: did Betson have his own (portable) clock? And were the specific times referred to those of the equal-hour day, reckoned as today with twelve o'clock at midday and midnight, or were they canonical hours of unequal length and counted from one to twelve in the course of each day or night?

46. Likewise, in late seventeenth-century Barrow-on-Humber, Lincolnshire, instructions for the parish neat-herd about moving and grazing parish livestock likewise presumed his understanding of clock-time instructions: Lincolnshire Archives, Barrow 10/1, see Chapter 10 below.

47. Employers and clerks varied in their willingness to record exact details of work, as opposed to lump sums for piece-work and materials. Payments cannot necessarily be taken at face value, especially during wage-regulation, when employers paying above the legal rate to get work done might disguise this fact in their accounts. Such practices may be concentrated in certain work, leaving other types of work (perhaps more compliant to regulation) in detailed form. However, our concern here is narrower: with the degree of temporal detail recorded in the most detailed instances.

48. General shortage of early evidence outside London makes local contexts difficult to discern.

49. Obviously, since survival of registers and details of recording vary greatly, these indicate the practice, rather than identifying its onset. Norfolk Record Office, Yarmouth (many from 1570s on), Essex Records Office, Chelmsford (from 1620s); Gloucestershire Record Office, Tewskesbury (from 1640s).

50. The earliest instance so far located among the London churchwardens accounts is in the Guildhall Archives.

51. Essex Record Office, D/P_94/5/1.

52. *The form of the proceeding to the funeral of Her late Majesty Queen Mary II. Of blessed memory, from the royal palace of Whitehall to the Collegiate Church at Westminster; the 5th day of this instant March, 1694/5* . . . The Savoy (London): Edward Jones, Accessed via Early English Books <eebo.chadwyck.com/search>.

53. Edward Millington, *A curious collection of prints and dravvings, by the best engravers and greatest masters in the world. Fit only for persons of quality and gentlemen . . . . London, 1690.* Accessed via Early English Books <eebo.chadwyck.com/search>.

54. This lengthy statement is part of the title of an entertainment by John Rastell published in London in 1520. The spelling has been modernized.
Rastell, J. (1520) *A New Interlude and a Mery of the Nature of the iiii Elements, declarynge many proper poynt of physlosophy naturall, and of dyuers straunge landys and of dyuers straunge effects [and] causis, whiche interlude yf ye hole matter be playd wyl conteyne the space of an hour and a halfe, but yf le lyst ye may leue out muche of the sad mater as the messengers p[ar]te, and some of experyens p[ar]te [and] yet the matter wyl depend conuenyently, and than it wyll not be paste thre quarters of an hour of length*, London: J. Rastell. (spelling modernized).

55. It would be interesting to know how closely timed were experiments at that time—although John Dee recorded several alchemical experiments in 1581, his use of clock times was limited to recording when he had done them rather than their duration, or how quickly changes were produced (Fenton, 1998: 308–9).

56. Artisans were sometimes recruited to conduct experiments, though this tended to complicate the identification of men such as Robert Hooke and George Graham as artisans.

57. Likewise, quarter-striking required hearers to count to three.
58. Leicestershire Record Office. DE.734/6, at the back of the volume.
59. Cumbria Record Office, Carlisle. Uncalendared transcript in searchroom library, shelf for Carlisle local history. Spelling modernized.
60. Cheshire Record Office, P31/3/1, microfilm 153/1, especially 1688 and 1690.
61. For example, telling the time does not appear among the 'Tom Telescope' series of childrens' primers in the 1740s. We owe this point to Neil McKendrick, former Master of Gonville and Caius College, Cambridge. See also Secord (1985), Fyfe (2003).
62. Whether particular *vulgaria* are model exercises for boys, or working notes in a schoolmaster's notebook, is discussed by Orme, but does not affect our argument. *Vulgaria* may have circulated among schoolmasters, especially when a uniform Latin grammar for England was promoted in the 1530s, to aid scholars forced to move schools during outbreaks of disease (Orme, 1973: 254–7).
63. Founded 1478–80. In 1501 the master was Thomas Wolsey. MCS boys were a mixture of Oxford natives and boarders from elsewhere who lodged in Oxford with tutors or 'creansers', who regulated behaviour, and oversaw their learning.
64. *Ego hic in scola hodie dimedia* (deleted) *semi* (*inserted*) *hora ante horam sextam fui.*
65. Clare explicitly envisaged both hour and minute hands.
66. People commonly participated in several communities involving different activities and timings—as witnessed by the varied technical, political, and spiritual times of John Dee. Specialized communities of practice varied greatly in their size, composition, stability, and interaction with other temporal communities.
67. For example, 'Technically Huygens accomplished . . . a sixtyfold improvement in accuracy (precisely the proportion of the old hours to the new minutes, and of the newly measured minutes to the newly measured seconds)' (Sherman, 1996: 5).
68. That is, seven and a half minutes.

7

Precision in Everyday Lives

7.1 EVERYDAY PRECISIONS

The previous chapter stressed that clock times were an integral part of everyday life for a very wide range of people in early modern England. It also suggested both the importance of, and the evidential problems arising from, the taken-for-grantedness of clock times in everyday life. To reiterate, it is very revealing that clock time was something that 'didn't need to be explained'. Precisely because everyday time senses were 'everyday', they were rarely remarked upon in normal practice except when they failed, and therefore had to be accounted for, or where they were invoked in justifying an account of behaviour or a belief. This same problem is certainly present when discussing everyday precisions, though in lesser degree because the more precise uses of clock times tended to be more explicit. Often—though not always—people were more aware that they were thinking (or doing) temporally.[1] They were more aware, too, that seeking greater precision could entail thinking about matters hitherto taken for granted, and they were sometimes therefore concerned to specify how the time was known. Indeed, sometimes, the action of being precise, or of planning precision, itself generated documentation. There was, in other words, a bit more bureaucracy involved in being precise than in just using or telling clock times, and this could generate recording and documentation.

Although precision has modern connotations, especially in scientific attitudes, our focus here is on precision as rather broader than self-consciously rational analysis. We focus on 'the everyday' community of practice, and the fineness of people's identification and distinction of exact times of day in connection with activities that were 'everyday' for most people. In areas like astrology, lay understandings could become quite technical, but in the main we focus on activities where precision was not in itself the intellectual purpose of action, and leave activities like astronomy, ocean navigation, surveying, or cartography until the following chapter. Obviously these were everyday activities for their few full-time practitioners, but they were hardly the norm. By comparison our concern is with the apparently mundane: in what circumstances did various types of non-elite people use relatively precise clock times? While not all the activities discussed here were literally everyday occurrences, commonplace experience they certainly were. We explore how detailed were the temporal capacities and imaginations called on in everyday life; which practices circulated and how generally in different communities of practice; what significance they had; and how they related (or did not) to the presence of certain sorts of time-signalling.

Our approach involves three main strands. First, we ask various questions about temporal referencing, especially in diaries, letters, and depositions where, even if indirectly, a subject's own words or utterances are recorded. What are the patterns of mentioning specific clock times for different people in particular times and places? How roughly or how exactly do people locate happenings or periods? For which happenings does exact specification seem to have mattered? Second, we explore the impulses to precise timing: how much precision was demanded of people by authorities, by their peers, or by themselves? Where was clock time merely the vehicle for expressing an intrinsic precision (e.g. observing an astronomical event, casting an alloy successfully, baking a cake), and where was urgency imposed by human choice (e.g. a market, a doctor's appointment, arranging a date)? And third, we ask, where did precision have positive or negative moral connotations?

Precision was always a relative matter, of course. And clocks were not necessarily a prerequisite, much as Thompson's task-orientation / time-orientation dichotomy may suggest this; 'senses of time deriving from nature were often associated with disciplined work, and fierce attention to saving time' (Glennie and Thrift, 1996: 285; O'Malley, 1992; Smith, 1986). Quite precise temporal identifications or sequencings were handled with whatever time-cues were available. What constituted a high degree of precision in earlier times might entail a distinction of detail that would not today count as precise at all. For the Stonor family and their correspondents in the 1470s and 1480s, as shown in Chapter 6, the very use of clock time at all constituted relative precision in timing events within the day. Their surviving letters do not include non-clock constructions of precision in timing so—for them—all their uses of clock time represent uses of—for them—relatively precise timing.

In analysing patterns of temporal referencing we need, in principle at least, to distinguish *incidental precision* from *purposeful precision*. Apparently greater precision may be *incidental* when it stems from increased availability of more refined systems of indicating the time. People may appear to be more precise, without doing anything fundamentally different if their world is now full of clocks. In that sense, incidental precision is misleading, but much interest attaches to incidental precision where we have supplementary information about the temporal information available. It may then be possible to say which of the time indications available in particular contexts were important in people knowing the time. *Purposeful precision* denotes the deliberate use of more precise temporal referencing to achieve specific goals. Who, where and when, was most likely to deploy more precise time-recording? How did people's use of precision vary with the circumstances of everyday life? Here, our approach is essentially comparative, uncovering activities or events where precision seems to have been a particular priority.

Purposeful and incidental precisions differ because the former explicitly entails more demanding temporal standards. Being purposefully precise could entail a considerable effort to determine exactly what the time was. Although there were growing numbers of public clocks and other timepieces in the mid-seventeenth finding, the exact time could still require considerable 'foraging behaviour', because they were unevenly distributed, and provided limited information. A good example comes from the autobiography of Samuel Jeake, a non-conformist merchant from Rye:

I was born at Rye in Sussex July 4th 1652 on the Lord's Day, ¼ of an hour past 6 a Clock in the morning, according to the aestimate time taken by my Father from an Horizontal Dial, the Sun then shining.   (Hunter and Gregory, 1988: 85)

Jeake's father rushing out to read a reliable sundial, and both father and son thinking it relevant (in 1652 and 1694 respectively) to record the type of sundial, show the importance attached to knowing an exact time for Samuel's birth.

Where precision is an important objective, the difference between times that are understood to be 'exactly right' rather than 'nearly right' can justify considerable extra effort. Other things being equal, ease of time-determination is preferable to laboriousness, of course, but where 'precision matters', rough-and-ready timings will fall short of more exacting standards. So considerable effort went into exact time-telling when people judged the situation called for it.

Temporal descriptions can be precise in various ways. First, through specification: 'seven minutes past eleven' is more precise than 'just after eleven o'clock'. Tracking such precision, among diarists for example, depends on identifying their use of particular times rather than rounding, and we can ask, 'how particular are the distinctions made in the patterns of their timings?' Second, precision could be conveyed by emphasizing the exactitude required in actions, as in 1669 when the feoffees of Solihull School (Warwickshire) required that the master should 'constantly and daily to keep the schoole from seaven of the clock in the morning *at the Latest* till eleaven in ye morninge',[2] or when George Hastings' tutor reported from Tamworth that 'the method of the [grammar] school here is to go *exactly* at 7' (Bickley, 1930: 224–6, our emphasis). A third facet of precision in dealing with time involves calculation based on explicit expectations of the rate at which things happened. This facet of precision is exemplified by Henry Best's instructing his son in 1642 about shepherd's rates of work in particular tasks:

one that is ready att it will easily gelde an hundrethe lambes in 3 houres.   (Woodward, 1984: 19–25)

Best's statement exemplifies that some types of precision may not necessarily involve access to very precise time measurement, so much as a relative sophistication in notions of speed or rate. Precise expectations were not impossible if only coarse time-indications were available, but they needed to be couched in comprehensible terms. With only a single-hand clock with hourly striking, available, for example, the difference between 2 hours 50 minutes and 2 hours 55 minutes might be difficult to determine, but numbers of whole hours signalled would be clear, and a difference of three lambs relatively obvious.

7.2 ENCOUNTERING PRECISION: ALMANACS AND INTERACTIONS

Almanacs were probably the most widely circulating publications in England between the late sixteenth and early eighteenth centuries (Capp, 1979). Costing somewhere between one penny and sixpence, they were beyond the reach of many plebeian readers, but their common presence in alehouses and other quasi-public spaces meant that their readership was not limited to those who could afford to buy their own (Capp, 1979; Raymond, 2005: ch. 7).

Early seventeenth-century almanacs typically mix sensitivity to precision in certain times with indifference to, and/or ignorance of, precision in others, as can be shown by comparing several almanacs published for 1642.

At least a dozen almanacs were published in that year (Capp, 1977), but we will focus on five, respectively compiled by John Poole, Nathaniel Nye, George Naworth, Samuel Ashwell, and Vincent Wing. Nye's was compiled for London, Naworth's for Durham, Wing's for Stamford, while neither Poole nor Ashwell specified calculation for a specific meridian. Poole's title page specified 51 degrees

Table 7.1. Contexts for precise clock times in five almanacs for 1642

	Ever gives times or durations using					When recording most exact
	¼-hour	10 min	5 min	1 min	seconds	
John Poole, 1642	yes	yes	yes	yes	occasional	positions of moon, sun, stars, planets
Nathaniel Nye, 1642	yes	yes	yes	yes	no	solar, lunar, planetary positions
George Naworth, 1642	yes	yes	yes	yes	yes	equinoxes and defining the seasons
Samuel Ashwell, 1642	yes	yes	yes	yes	no	tides; lunar positions
Vincent Wing, 1642	yes	yes	yes	yes	no	positions of moon, sun, stars, planet

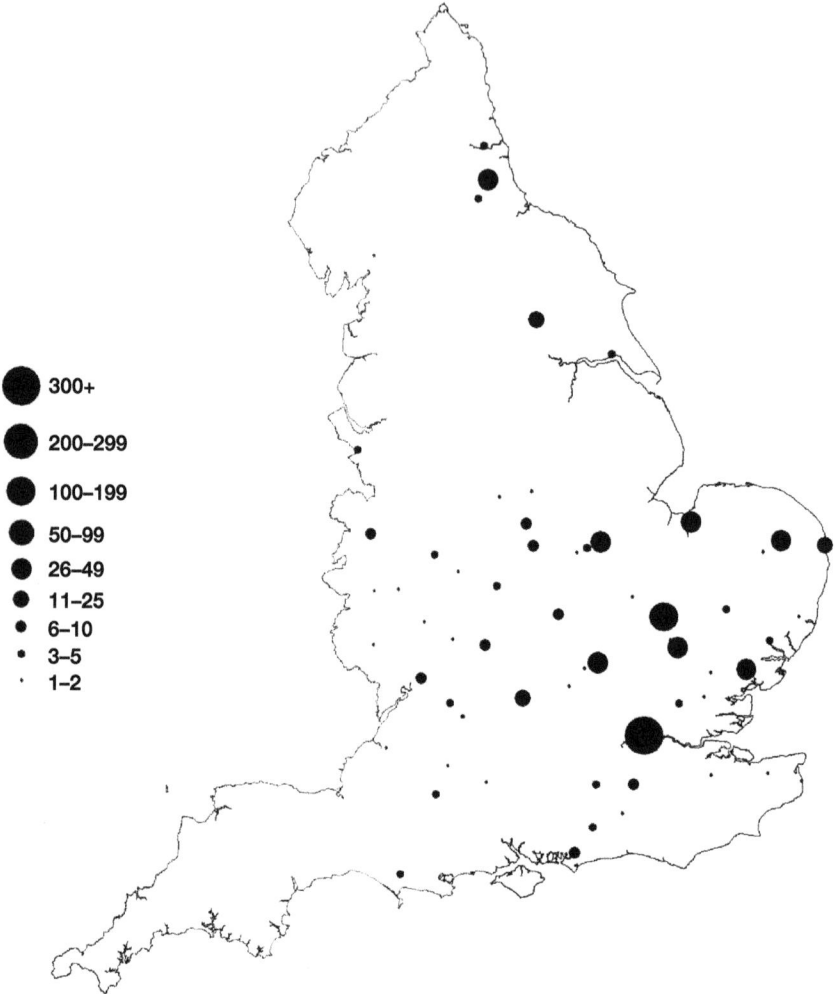

Fig. 7.1. Specific places for which almanacs were published.

35 minutes, the figure routinely given for London, but claimed his almanac would 'serve indifferently for any part of great Brittaine'.

Almanacs were the main type of printed material through which early modern English men and women were likely to encounter minutes and seconds. Their compilers presumed the familiarity of readers with these finer divisions of daily time, which they gave for various purposes. Minutes (and occasionally seconds) appeared in diverse places in Poole's almanac: rising and setting times of sun and moon (sometimes for every day of the year); the times that various heavenly bodies 'southed' (lay exactly south); times of full, new and quarter moons; tide tables giving times of 'full sea' in hours and minutes for several dozen British and overseas ports. Though proclaimed as 'An exact and perfect Tide-table', the timings given were not precise: rather they were given as 00, 15, 30, or 45 minutes. But there are also signs of comparative indifference: Poole claimed that sunset times 'being doubled is the length of the day' showing he assumed that 12 o'clock occurred exactly mid-way between sunrise and sunset, and elsewhere similarly overlooks the distinction between apparent solar time and mean time. Nye similarly gave longitude and latitude for various places around the world, with linked information, including the duration of the longest day in hours and minutes, and gave sunrise times. Ashwell's almanac was particularly detailed about times of 'the moon's coming to south' given to the nearest minute, and in giving highwater times for several ports both in hours and minutes locally, and the difference from high-water at Southampton or Portsmouth. Naworth's almanac gave daily day-length for Durham in hours and minutes, and paid close attention to times when each season started. Thus summer was described using the time at which the star Phoebus reached its greatest meridian altitude:

beginning this yeare on Saturday the 11 day of June, at 1 houre, 42 minutes, 53 seconds, after high noone, according to the mean or equall time, but at 1 houre, 49 minutes, 53 sec the true or apparent time.

Likewise, the beginning of autumn was specified as

when the sunne toughteth the first minute of the other Equinoctall... imballancing his raye with equall poyse to all terrestrial inhabitants, which this yeare falleth out the 13 day of September 16 minutes 38 seconds past 4 in the morning according to the mean or equall time, but at 32 minutes 18 seconds past 4 the true or apparent time... (Naworth, 1642: Unpaginated; quotes from the 17th and 18th pages of text)

Naworth shows a sophisticated grasp of variable dimensions of self-evidently 'natural' time, making explicit the difference between 'true or apparent time' and 'mean or equall time'. One noteworthy dimension is the making of this distinction, recognizing that noon-to-noon intervals were longer or shorter than exactly twenty-four hours at different times of year. This realization is often claimed to have resulted from the much more accurate timekeepers available after the 'horological revolution', but its appearance in popular almanacs decades earlier shows that the realization was less technologically driven than often assumed. Also noteworthy is Naworth's revealing description of what counted as authoritative and 'natural' in telling the time of day. That 'apparent' and 'true' were synonymous shows that the astronomical position of the sun had authoritative status, while the 'mean or equal' time of horological devices was a convention.

The distinction was important for keeping clocks effective as markers of celestial time. Within a decade or two, by comparison, a clear distinction was being made between 'true' and 'apparent' time, where solar time was apparent, but slightly variable, and no longer true time. 'True' time was now horological—a time more constant than the slightly variable solar time.

7.3 SHOWING PRECISION: CONVEYING AND DEPICTING PRECISE TIMES

In Chapter 1 we described the diversity of early clocks and hour-reckoning systems, stressing the contingency of the extended selection processes through which they emerged, which nowadays seem obvious and natural. The same applies to the display and signalling of precision, at least until the late twentieth-century spread of digital timepieces, when the system of hour, minute and seconds hand, all centrally mounted on a single dial, respectively rotating every twelve hours, one hour, and one minute, came to seem inevitable.

However, many different ways of displaying precision were being tried out either side of the decisive technological advances of the horological revolution period. Insofar as these are discussed in the horological literature, they tend to be treated anecdotally, and as curiosities or experiments. But they are more than that. In their schemes for marking and communicating precise times, they provide considerable insight into everyday senses of temporal precision. In other words, although these ways of showing precision were 'unsuccessful', in the sense that they did not develop into standard forms that have survived into the twenty-first century, the principles and techniques they incorporate are very revealing about interpretations and skills of precision existing at the time at which they were devised. They were, after all, devised not as abstract experiments, but as ways to engage more precise, and denser, temporal cueing into everyday life. They can thus shed light on practices that may have been more important then than now.

7.3.1 Single-Handed Clocks

Horology's supply-side approach to the history of timekeeping, centred on the horological revolution, has led the accuracy of pre-pendulum clocks to be widely denigrated. The literature is replete with claims that sixteenth- and early seventeenth-century clocks (never mind their medieval forebears) were hopelessly inaccurate; they lost or gained half an hour a day (that is, if they kept going for a whole day); they needed constant repair; they were mainly useful as a kind of joke about how bad they were. All too often, the same claims or anecdotes circuit around and around.

It would certainly be perverse to claim other than that pendulum clocks were much better than their verge-and-foliot driven predecessors. And it was not simply the pendulum, but the linked improvements in escapements. These clocks kept better time, with fewer problems, and often ran for longer. Much horological

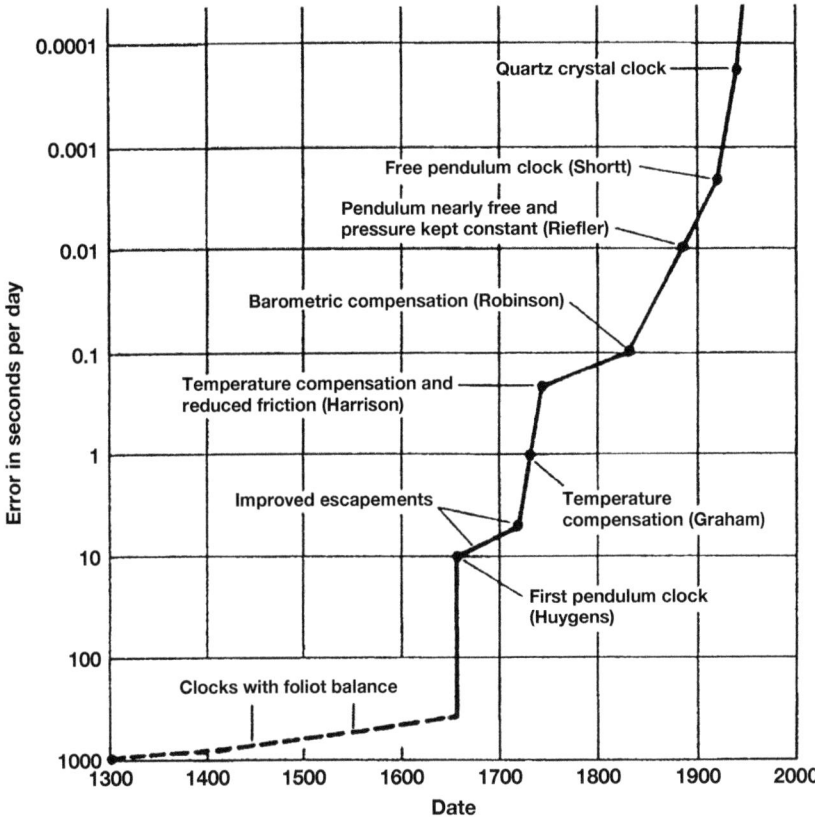

Fig. 7.2. Progressive improvements in the accuracy of mechanical clocks.
Source: Whitrow (1988).

literature goes further, suggesting that the more accurate pendulum-regulated clocks led to the introduction of the minute hand, in turn leading to minutes being used in timekeeping. At last, people were not limited to telling the time in hours. Such a neat contrast, however, overlooks several important factors.

For a start, single-handed clocks were already commonly interpreted to shorter intervals. Hour hands moved continuously, not in hourly jumps, so the hand's progress between numbered hour-marks provided continuous intermediate signals absent from aural signalling. Seventeenth-century clockmakers commonly engraved half- and quarter-hour marks on clocks' hour-rings, even on clocks aimed at the bottom of the market. Some marked the half-quarters as well, making seven intermediate marks between hours. Depending on the delicacy of the hand's point, its position relative to marks and gaps enabled reading of time to within a few minutes (Figure 7.2). People reliant on one-handed timepieces were clearly not restricted to describing the time in whole hours.[3]

In considering the precision with which one-handed clocks could be read, we pursued three enquiries. First, how far did clockmakers attempt to achieve distinctions in fractions of an hour with just a single hand? As we shall shortly show, the answer is that they attempted to show much finer distinctions. Second, since such finely divided hour rings were features of only a minority of clocks, we need to consider to what extent the simpler hour rings and broader hands of more typical clocks, both domestic and public, could be read to make distinctions below the full hour. And third, we look at the recording of time by people reliant on one-handed timepieces. We will show that they were not restricted to describing the time in whole hours.

The formal depiction of smaller time-divisions on a single-handed clock rested on the fineness of divisions on the hour ring. There were two important contexts for so doing. The first, and earlier, predated the pendulum and minute hands of clocks from the end of the 1650s. Very commonly, domestic clocks at that time incorporated at least a half hour and two quarter marks on the hour ring. Some examples are known with half-quarter marks (that is, eighths of the hour or $7\frac{1}{2}$ minutes). Whether these marks represented the finest possible reading of the clock depended on how wide the hand was relative to the marks. Fine-pointed hands offered the potential to interpolate times between the marks on the dial.

The second context was in the continuing production of single-handed clocks after the introduction of minute-indicating clocks. To be sure, one reason for the continued production of single-hand clocks through the late seventeenth and eighteenth centuries was in keeping the cost of basic clocks down to a level affordable for the bottom end of the market.

One strategy to make one hand subdivide time in a very sophisticated way was to dispense with the convention of accommodating twelve hours within one revolution of a hand on a dial. Obviously, the shorter the period in which a single hand rotated once, the longer the arc of the circle that it traversed in an hour. One such clock was constructed by one of the Porthouse family of clockmakers at Penrith, Cumbria, in the late eighteenth century (Loomes, 1997: 204, fig. 7/190). This clock used Benjamin Franklin's published suggestion, in which the hour hand rotated in four hours, with four, eight and twelve o'clock at the top of the dial, and the ninety degrees traversed each hour subdivided by five-minute marks.

Much earlier, in the 1720s, Andrew Knowles at Bolton, Lancashire, produced single-hand thirty-hour clocks with a twelve-hour rotation, in which the quarter-hour units on an external hour ring were further subdivided into three, to give markings at five minute intervals. These small divisions were still several times wider than the hand tip, so further interpolation was readily available. Such fine hour-ring divisions were unusual, but few single-hand clocks after about 1670 are without subdivided hour-rings, even the very few that survive from the lowest levels of the market.

Single-hand clocks usually existed in environments that included other timekeepers, and visual and acoustic time-signals, not in complete isolation. They may have been 'read' differently where quarter-striking bells were audible. It is nonetheless very clear that sub-hour distinctions, describing fractions of an hour, or numbers of minutes, were made by many people in diverse locations, well before clocks with minute-indication were at all common.[4]

7.3.2 Showing Minutes and Seconds

However, none of this is to argue that minute hands were completely unknown before the horological revolution. Clockmakers had been attempting to time and display minutes for more than a century. The earliest known reference to a minute hand comes from a monk, Paulus Almanus, in 1477 (Andrewes, 1985: 83 n. 3), and around 1500 Leonardo da Vinci made a detailed drawing of a clock mechanism with four separate dials. One turned once a year, and showed the month, the date, and the length of the day; the second showed the phases of the moon, and turned once every $27\frac{2}{3}$ days; the third turned daily, showing 24 hours; and the last, marked with 60 minutes, turned hourly (Baillie, 1951: 5). Clockmakers explored various means of showing minutes, with the familiar two-hands-on-one-dial eventually becoming standard. By mid-century minutes-showing watches were being produced in the major European clockmaking centres, both as masterpieces demonstrating apprentices' their skills, and for sale (Baillie, 1951: 32–40). While minute-indication remained unusual, it is significant that it was regarded as a key test of skill in the late sixteenth century. In England, the physician William Harvey owned a minute watch, mentioned in John Aubrey's account of his death from a stroke, in 1657:

> As soon as he saw he was attacked, he at once sent for his brother and nephews and gave one a watch . . . as remembrances of him. . . . *It was a minute watch with which he made his experiments.* (Barbor, 1982: 132–4, our emphasis)

There was no unanimous solution as to how clockmakers should indicate the precision made feasible by the pendulum. It was far from obvious how minutes might be displayed most conveniently (for clockmakers) and most clearly (for users). During the decades before and after 1700, especially, clockmakers produced clocks indicating precise clock time in several different ways. These are of much more than curiosity value: they are not just eccentric ways of indicating precision. Rather, they reveal something of how precision was conceptualized. Not all 'non-modern' means of showing minutes (and seconds) can be read as easy technical solutions, requiring lesser skills, cheaper materials or more familiar techniques than 'modern' depiction. More important than these supply-side factors were clockmakers' understandings of how clock-users read (or at least as often heard) the time from clocks.

Early attempts to show precision exhibit considerable variety. Initially, aural signalling was important, especially through clocks and watches that struck every seven and a half minutes, that is eight times an hour on the hour, the half, the quarters, and midway between each quarter hour. Such devices marked a greater number of specific temporal points and, obviously, reduced the periods between successive strikes. They provided more benchmarks for 'point-and-stretch' time, but there were practical limits to the frequency of aural signalling. Seventeenth-century watches that struck every minute are known (Baillie, 1951; Andrewes, 1985: 80–1), but as test pieces or to facilitate timing of astronomical observations rather than as everyday objects.

Visual time-display on a dial with a minute hand provided continuous display, with the display altering perceptibly over comparatively short periods. Whereas a one-handed clock in practice enabled the distinction of times to within, say, five to ten minutes, a minute hand (depending on its design, and the smoothness of its movement) could be distinguished down to a quarter of a minute or less.

Early minute-indicating clocks took various forms besides two hands on one dial. Indicating hours and minutes on separate dials was commonplace, usually with moving hands and a fixed dial, but with moving dials and fixed pointers, or dials appearing in windows. In the early decades of minute-indication, many combinations of hands, dial(s) and window(s) were tried before the single-dial-multiple-hands configuration became universal.

However achieved, separately indicated hours and minutes made for simpler-looking dials. The combination of hour and minute hands on the same dial necessitated two different numbering schemes, one to twelve and one to sixty, around the dial ring, and made reading the clock a significantly more complicated operation. Besides the differences in numbering, the hour ring was usually divided by three or seven marks between successive numbers, whereas the minute ring bore four intermediate marks. Since the minute ring and numbers were displayed to the outside of the hour ring, the minute hand needed to be longer than the hour hand, further complicating the interpretative task. Not that numbering minutes one to sixty was immediately adopted as standard. Several examples survive of clocks in which the minute hand rotated twice, or four times, per hour, necessitating more complex numbering schemes still. There were obvious advantages in rationalizing this diversity, but no inevitability about which scheme was chosen.

Similar considerations applied to the provision of seconds hands, though these remained rarer than minute hands. Early seconds-display clocks often showed seconds on a separate smaller dial perhaps two inches across, within the main hour dial (illustrated in Loomes, 1997: 24–75). Early examples include various periods of rotation. In cases where all three hands revolved on the same dial some economy in labelling was possible by having the second hand completing as many revolutions per minute as the minute hand performed each hour, but this was not the standard layout until later. For specialist uses, an hour hand could be dispensed with altogether. Clocks with just minute and second hands were used by astronomers to closely time short periods during observations. Some of these struck every sixty seconds and, with a pendulum beating seconds, provided an audible count of seconds and minutes (Andrewes, 1985: 80–1). Early seconds clocks were produced by makers in several parts of the country, though it is often difficult to distinguish original work from later alterations (Loomes, 1997: 45–72). Pocket watches with seconds hands only became more common in the late eighteenth century, though earlier examples survive, including one by John Bushman of London from $c.$1695 (Andrewes, 1985: 82).

Even before the invention of pendulum clocks, though, when clock times in seconds were derived more readily from calculation than from mechanical devices, their use is attested to outside the rarified spaces of astronomical observatories. For example, John Aubrey recounts being told by Sir William Petty that he had been born 'on Monday, the twenty-sixth of May 1623, eleven hours $42'56''$ after noon' (Barbor, 1982: 241–2). Aubrey's account establishes a concern to refine precision, even if its achievement remained beyond reach.

We emphasized in Chapter 1 that clocks and other devices influenced one another, and this was clearly the case regarding seventeenth-century changes in precision. The much larger numbers of domestic clocks and watches in circulation during the late seventeenth century, and their increasingly precise depiction of time created strong pressure for sundials to provide more precise time-indication than hitherto, especially to fulfil their long-standing role in the setting of clocks—recall that for Aubrey dialling ability was emblematic of precocious mathematical expertise. As precisely setting clocks received greater attention, it became more pressing for sundials to indicate precise times, and this awareness spilled over into other uses of sundials. This occurred whether people set their clock or watch directly from a dial, or indirectly from the hour-striking of a public clock, because increasingly accurate uses of timepieces required their coordination with one another.

More accurate clocks necessitated more accurate sundials from which to set them. As the disparity between apparent and mean solar time, long evident to specialists, became more familiar, the orientation of sundials shifted from their familiar vertical position, carved or scratched on walls, to horizontal instruments. On these more intricate horizontal instruments, the gnomon angled to be parallel to the earth's axis of rotation (and so varied with latitude), while dial plates were much more carefully divided and engraved so as to be interpreted to a few minutes or less (Turner, 2002: 16–19).

Minute indication on sundials and other devices had been growing during the sixteenth and seventeenth centuries. In most cases, the key obstacles to be overcome lay not in devising appropriately moving parts, but in the accuracy of geometrical calculation, the devising of appropriately laid-out instruments to display results, and their accurate engraving (Chapman, 1995a). In the sixteenth century this was usually achieved—especially for astronomical star-sighting instruments—by making them extremely large, but very precisely laid-out small instruments slowly eclipsed those giants. Minute markings on sundials, quadrants, nocturnals, and hybrid instruments slowly appeared (Turner, 2002), as the more precise times in ephemeredes and other astrological tables stimulated the production of instruments which at least gestured towards more precise indication. Equation-of-time tables were added to sundials as well as clocks, the earliest extant example being one by Henry Wynne of London, made in 1675 (Turner, 2002: 17).

At the humbler end of the range, sixteenth-century nocturnals were, to judge from surviving examples, commonly marked in quarter-hours in France, Italy, and England (Higton, 2002: 390–401). Humphrey Cole, one of the most versatile and prolific London instrument makers of the second half of the sixteenth century (Turner, 2000, especially 38–9) indicated half-hours even on the small dials within portable astronomical compendiums (Higton, 2002: 301–5, and plate 40). Among the instrument collection at Greenwich, in the National Maritime Museum, are an analemmatic dial by Thomas Tuttrell, made just before 1700, with times marked every two minutes, and Richard Howard's horary quadrant of just after 1700, on which times were marked in single minutes (Higton, 2002: 352–3).

Some eighteenth-century sundials illustrate the concern for highly accurate time-indication through their indication of individual minutes. Aimed at the top end of the provincial market, to judge by surviving examples, such as the garden sundial made *c*.1760 by Henry Hindley of York, accuracy here was aesthetic and rhetorical as well as practical. The brass dial of Hindley's piece was both very precisely engraved, and of a carefully composed symmetrical design. Hindley was primarily a maker of relatively sophisticated long-case clocks, rather than of sundials. His minute-showing sundial was a counterpart to the accurate mechanisms and complex displays. For example, a number of his clocks were year-duration, and include complicated clockwork to take account of leap years without a need for resetting. Both Hindley and his patrons were well aware of the equation of time, and the varying fit between apparent solar time and mean time, but the minute-showing sundial evidently still had a considerable appeal (Loomes, 1997: 143–6).

7.3.3 The 'Equation of Time'

Time from dawn to dawn on successive days obviously varies with seasonal day-lengths but that the intervals from one noon to the next vary during the year was little appreciated outside astronomical circles, and their comparison of solar and sidereal time. Because of the tilt of the earth's axis relative to its path, and the earth's elliptical orbit around the sun, apparent solar time and mean solar time vary by up to just over fifteen minutes. Whilst this was made increasingly evident by pendulum-driven clocks after the horological revolution, almanacs had been explicitly noting the discrepancy since the sixteenth century, and the implication that a perfectly constant clock would rarely agree with a perfectly laid-out sundial, because the former ran more consistently than the earth. The clock would appear to be anywhere between about a quarter-hour fast and a quarter-hour slow, depending on the date.[5] While several London and provincial makers constructed clocks showing both solar time and 'mean time', a complex but feasible task, which was essentially an extension of date-display mechanisms, mean time quickly became the standard display. Precision no longer meant being *as good as* nature, precision meant being *more regular* than nature.

The variations between solar time and mean time were summarized in what was termed 'the equation of time'. These were tables on a weekly or daily basis showing the difference between solar and mean times, showing which was ahead of the other, and by how much. Following Huygens' first pendulum clock came his first equation table in 1657, after which they became a relatively commonplace part of the running of timepieces. Interestingly, they were not just of utilitarian value. They were much more important than that. Through tabulations of the equation of time, precision possessed a further aesthetic dimension. For example, equation of time details were inscribed on the relatively expensive silver cases of some pocket watches, not merely printed on papers carried inside the case (e.g. Lippincott, 1999: 142) [though Lippincott refers to gold, not silver watches]. Plate 10 shows a long-case clock built in 1728 by John Harrison, the same John Harrison who later laboured so long in perfecting the marine chronometer. His equation-of-time table is displayed in a glass-fronted carved frame on the case door, the centrepiece of the clock as a piece of high-class joinery and veneering. The demonstration of precision was central to the clock's aesthetic, and similar equation-of-time tables were pasted in the doors of long-case clocks, by makers ranging from elite London makers like Thomas Tompion and Danial Quare, to humbler metropolitan makers like George Neale (whose 1733 table is illustrated by Turner, 2002: 18), and in Harrison's own clock of 1717 (his two tables are illustrated by Sobel and Andrewes, 1998: 80, 91).

7.4 THE AVAILABILITY OF PRECISE TIME INFORMATION

Even if clocks showing minutes were hardly known in the sixteenth century, contemporaries nonetheless believed that minutes were widely known. In 1577, for example, and as we have noted, chapter XIV of William Harrison's *Description of England* included 'Of our account of time and her parts', in which he commented on

> the common and natural day [being] observed continually by clocks, dials and astronomical instruments of all kinds...

Our common order [of time] is to begin at the minute, . . . one-sixtieth part of an hour, as at the smallest part of time known unto the people, notwithstanding that in most places they descend no lower than the half-quarter or quarter of an hour. . . . the common and natural day [being] observed continually by clocks, dials and astronomical instruments of all kinds. (Edelen, 1968: 379)

Harrison was writing long before the provision of clocks with minute hands, but minutes were already familiar from almanacs. Almanacs like Humfrey Lloyd's *Almanack and Kalendar containing the Day, Hour and Minute of the Change of the Moon for Ever*, published from 1563, and similar publications sold in very large numbers, and were directed at humble readers. William Bourne's much-reprinted *A Regiment for the Sea* of 1574, which we discuss in Chapter 9, was not atypical in distancing itself from the 'learned sort' of men, explicitly targeting 'meaner men' or 'the simplest sort of readers'. Such almanacs offer a striking contrast between an explicit concern to keep language basic and arithmetic simple, and their presumption that readers need no explanation of times given in minutes in eclipse predictions or moon-risings. Bourne did, though, explain seconds, and their further subdivision into 'thirds' ($1/60$ of a second) and 'fourths', as unfamiliar.

In time the spread of minute hands embraced many large public clocks as well as domestic clocks and pocket watches. Minutes were, though, particularly associated with private timepieces, because of the massively increased numbers of domestic clocks and pocket watches in circulation. The appearance, following soon after, of second hands was almost exclusively a feature of domestic clocks and watches.[6] Seconds-indicating clocks were being made by several clockmakers well before 1700, and by increasing numbers afterwards, but we cannot specifically estimate how widely owned clocks with seconds hands may have been.

One brief but tantalizing glimpse arises from a very specific set of circumstances, and an attempt to coordinate seconds timing across a particular temporal community, based on the Royal Society, in 1714. The pivotal figure was Edmund Halley, then Savilian Professor of Geometry at Cambridge, who planned the gathering of observations across southern England during the total solar eclipse of 22 April. Halley's central concern was to exactly measure the width of the moon's shadow on the earth's surface, as a step in calculating the earth's distances from the sun and moon. The Royal Society had collected information about the total eclipse in 1666, which had been published in the first volume of its *Philosophical Transactions*,[7] but the timings in seconds that Halley now realized were crucial had only been recorded by the Swedish astronomer Hevelius, and at the Paris observatory.

Not without difficulty (such as the withholding of information from the Observatory at Greenwich by the Astronomer Royal, John Flamsteed) Halley established a network of reporting observers. The most vital piece of information they were to observe and report was the precise duration of totality. Mapping and interpolating these data enabled Halley to determine the precise width of the moon's shadow at the earth's surface, and led to unprecedentedly accurate

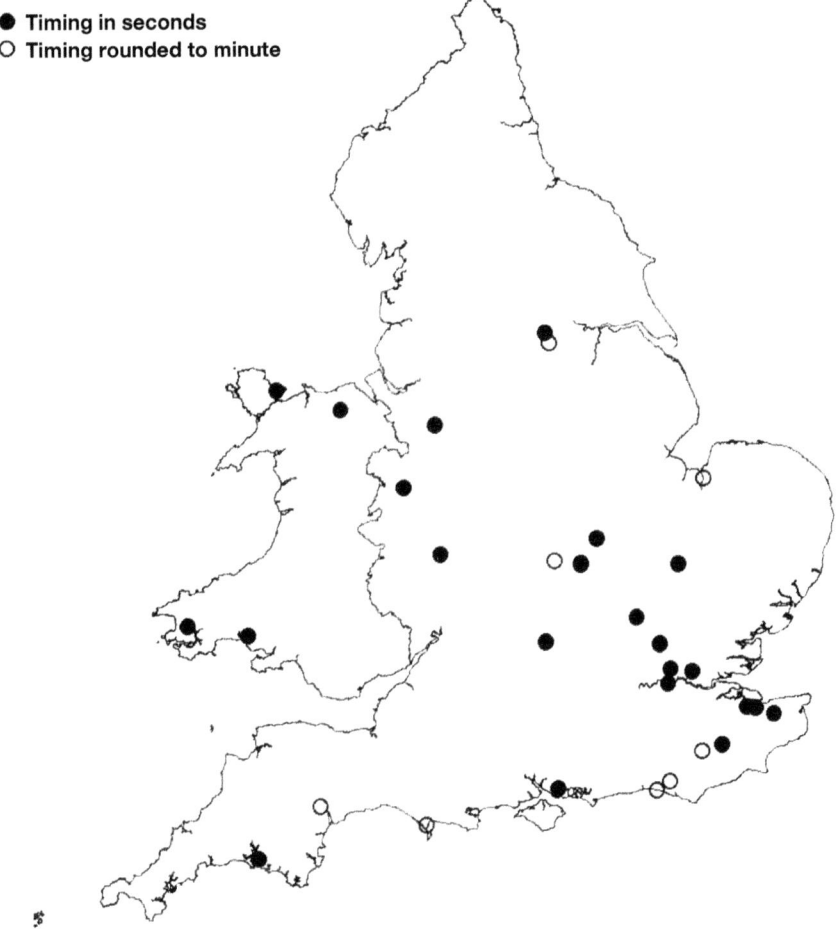

Fig. 7.3. Places from where durations of totality were reported to Edmund Halley in 1715.

solar and lunar dimensions and positions being incorporated into Newtonian mechanical modelling.

Halley intended to collect and map timings of totality made in different places, telling the Royal Society,

I caused a... request to the curious... especially to note the time of continuance of total darkness, as requiring no other instrument than a Pendulum clock with which most Persons are furnish'd, and as being determinable with the utmost exactness, by reason of the momentaneous occultation and emersion of the luminous edge of the Sun. (Halley, 1715: 245–6)

Halley's claim about widespread clock-ownership was hyperbole, but the dispersal of his correspondents shows that seconds timing was geographically widespread (though socially biased towards an educated elite of clerics and scientifically interested laymen). It is also noteworthy that the coherent pattern of the reported times shows them to be reasonably accurate. Halley specifies pendulum clocks because it was often feasible use them to count seconds, even if they lacked a seconds hand. Pendulums were commonly constructed to beat seconds or half seconds (respectively believed to be of lengths 39.41 and 9.81 inches, although exact length varied with temperature and barometric pressure). Hence an observer could hear, and count, the 'ticks' of seconds, or half-seconds.

However, Halley regretted that:

[several] observers give us no account how they measured this time, and therefore it may well be supposed... [some]... took it in a round number, and perhaps from pocket minute watches. (Halley, 1715: 254)

The observers whose casualness about measuring or recording seconds meant that their timings were less precise than Halley thought essential, illustrate how 'precision' translated into differing understandings of what the situation required. In enlarging an astronomical community of practice, Halley encountered looser definitions of 'exact' amongst the more diffuse community of practice comprising clock-owners.

This was not a passing problem, for a similar exercise in 1737, when a total eclipse was visible from much of northern Britain, found a similarly inconsistent response. Again, many observers lacked clocks showing seconds. Some observers judged seconds from a minute hand: 'the duration was six minutes as near as could be judged by a watch that did not shew the seconds' (Hopetoun House, outside Edinburgh), and the annular appearance at Montrose continued seven minutes 'as near as he could judge by an ordinary watch'. Several observers counted pendulum swings, or escapement 'ticks'. At Crosby, near Ayr, 'a distinct annulus... continued exactly seven minutes, measured by a pendulum vibrating seconds'; at Frazerburgh, 'from the time of the Ring's beginning to appear upon the lower and western part of the sun's disk, till it began to break on the east and upper part, there were 300 vibrations of a pendulum, or five minutes'; and at Longframlington, 'the duration [was] 40 or 41 half-seconds, measured by a pendulum 9.81 inches long' (Graham *et al.*, 1737: 175–201).

Eclipse-timings show the swelling of a dispersed network of precise timing practice, briefly mobilized and coordinated for a specific purpose. People's subsequent timing practices might or might not pay greater attention to seconds than hitherto, but documentation problems make it hard to gauge the impacts of precise timing on subsequent everyday practices. That a broader public awareness persisted between short-lived eclipse-timing activity is suggested by the appearance through the eighteenth century of commercially published broadsheets and pamphlets about further eclipses and other astronomical phenomena, including planetary transits (Armitage, 1997; Walters, 1999). Whether seconds-timing and other specialized practices moved into everyday life depended on whether people identified uses for them in everyday activities. Useful (or apparently useful) practices were readily taken up.

We have already highlighted everyday sociality as a prime site for spontaneous uses of clock times. Particularly close timing was certainly valuable in particular tasks, from timing eclipses, to coordinating factory work, to military coordination maintaining public order. But we reiterate the importance of much broader communities of everyday practice in generating new timing practices, and not just in copying specialized timing practices.

That seconds time was relatively widely available is not unexpected for horologists, since clocks showing seconds hands, on either the main dial or a small, secondary dial, were becoming much more common by the early eighteenth century. But it is less expected for much of the historical literature, which turns the (presumed) lack of everyday purposes for which measurement in seconds could have mattered, into an absence of the capacity to do so.

Our central point here is to emphasize once again how relative precision in timekeepers could be a valued attribute in its own right, not just by clock or watch owners but also (and perhaps more especially) by clockmakers and watchmakers. In other words, a significant degree of 'excess capacity' for precise indication was itself part of the aesthetics of timekeepers as objects.

7.5 DIARISTS' PRECISION IN RELATION TO TIME SIGNALS

Notwithstanding the explicit drive to precision in clock-use for certain specific purposes, where unusual pains were taken to obtain an exact reading, the form of the timings that were more routinely attached to specific events did depend on the timekeeping and signalling material to hand. To illustrate the contribution of time-signals, both audible and visual, Figure 6.3 compares the distribution of recorded minute times in two comparatively late diaries used in Chapter 6: Thomas Turner in the 1750s and Anne Lister in the 1820s, with those of Samuel Pepys in the years either side of his acquiring his minute watch (Latham and Matthews, 1980–91, V and VII; cf. Sherman, 1996). Both the later patterns are quite different from Pepys's recording. Turner's main time-cues came from the hourly striking of the Cuckfield church clock, and a handful of domestic clocks in local houses and inns. Anne Lister moved among a greater density of clocks, in which two-handed and quarter-striking clocks were relatively common.

Across the whole range of the diaries used in the previous chapter, there is a wide range of practices and habits regarding the precision of references to times of day (Table 7.2). Equally striking in the diaries, though less so in the table, is that these practices were deployed highly selectively. Just as with the use of clock-time

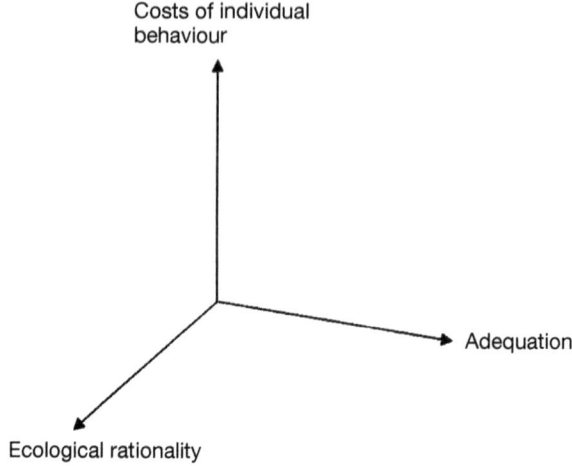

Fig. 7.4. Considerations in the demand for precision time-use.

references, there are striking variations in the contexts in which more precise times were given. It is unnecessarily naive to see such patterns as other than multi-dimensional, but the frequency with which births, deaths, travel, betting, and certain other activities recur conveys topics on when diarists felt greater need for temporal specificity, if it was feasible to do so. Obviously, though, there were signalling and adequation constraints as well as positive or negative motivations, which affected their dispositions or capacities to be precise, and much variation among individuals as to whether they were motivated to record the specialized activities, or facets of life, in which practices of precision were most involved.

Clearly, we cannot say how accurate their more detailed timings were, but their accuracy is not really the point, so much as their apparent perception that recording more detailed timings served some purpose. In other words, precision in references was being either actively recognized, or passively accepted, as useful. What, then, do the times recorded suggest about the precision of available clock-time information? In 1757, when Thomas Turner was in his late twenties, the published diary entries for that year include just under a hundred clock times. None of these are cited with ostensible precision beyond five minutes. That is to say, they are not more detailed than, for example, twenty-five minutes past, or five minutes to, the hour. Although every five-minute time point was stated at least once, their frequency of appearance was uneven (Figure 6.3). Times in the first half-hour after the hour were stated very much more often than times thirty-five to fifty-five minutes after the hour. Besides whole and half hours, he commonly used times of ten and twenty minutes past the hour. However, quarter hours were used comparatively sparingly. This last feature became more marked over time: in 1764 over 90 per cent of the specific times cited (including

Precision in Everyday Lives 263

Table 7.2. Precision and its uses in early modern English diaries and memoranda

	Ever gives times or durations using					When recording most exact
	¼-hour	10 min	5 min	1 min	seconds	
Robert Reynes	no	no	no	no	no	–
William Honeywell		no	no	no	no	–
John Dee	yes	yes	yes	yes	no	astrology, births, deaths, sex
Lady Margaret Hoby	no	no	no	no	no	devotional
Simon Forman	yes	yes	yes	yes	no	astrology, appointments
Richard Napier	yes	yes	yes		no	astrology, appointments
Lady Anne Clifford	no	no	no	no	no	–
William Brereton	no	no	no	no	no	–
William Whiteway	yes		no	no	no	births
Nehemiah Wallington	yes	no	no	no	no	
Roger Lowe	yes		no	no	no	
John Greene	no	yes	no	no	no	births
Elias Ashmole	yes	yes	no	yes	no	births, marriage, legal
Samuel Jeake	yes	yes	yes	yes	no	astrological
Samuel Pepys	no	no	no	no	no	–
John Aubrey	yes	yes	yes	yes	yes	births, deaths
Claver Morris	yes	yes			no	
James Warne	yes	no	no	no	no	births, funerals
Nicholas Blundell	no	no	no	no	no	–
Jeffrey Whitaker	yes	yes	no	yes	no	observing a solar eclipse
Thomas Turner	yes	yes	no	no	no	betting
Anne Lister	yes	yes	yes	yes	no	deaths, travel
Early C17th almanacs	yes	yes	yes	yes	no	positions of moon, sun, stars, planet

whole hours) are to a multiple of ten minutes. Taken together, these patterns suggest that quarter-striking clocks were largely absent among the timepieces that Turner read. The scarcity of times between half past and the hour suggests some difficulty in keeping track of time between hours. It also suggests a reliance on aural rather than visual information, since it is hard to see why such an imbalance should exist if Turner's main time-cue was visual. His preference for multiples of ten minutes may indicate that he often relied on single-handed timepieces, where inferences to a precision of five minutes required greater effort.

The facility with which Anne Lister used precise clock times to locate events in her daily narrative, to emphasize departures from expectations, and to make forward arrangements is considerable, as in her recording of times in 1824, illustrated in Figure 6.3. The preponderance of whole hours and half hours partly reflects Anne's recording of events which had either been arranged in advance, by herself or others, or were social events arranged at fixed times, and advertised accordingly. Nevertheless, her pattern of recorded time differs from that of Thomas Turner. In both cases, all five-minute times after the hour are recorded but there are some striking contrasts. While Anne Lister's lifestyle involved more formal public events and specific rendezvous than Turner's, the greater prominence of half and quarter hours in Anne Lister's account also owes something to the surrounding time-signals: minute-hand clocks that struck the quarters as well as the hours. It is also striking that she is precise about times between the half-hour and hour about as often as she is precise about times between the hour and half-hour, whereas Turner recorded the latter much more often than the former. Again, part of the explanation lies in his lack of, and her access to, quarter-striking timepieces, but less as a direct influence from the clocks themselves than because the temporal metric of everyday life was that much denser.

Likewise, in the mid-seventeenth century, the horological revolution had a very limited direct impact on recording by contemporary diarists. Samuel Pepys has been singled out by Sherman as pioneering a new sense of temporality and the self, yet, among the earlier diarists, Pepys seems surprisingly silent in recording clock times (which he used only very occasionally by comparison with vaguer formations such as 'in the morning' or 'late in the afternoon'); or detailed timings within hours; or time-cues. Pepys certainly was interested in precision, in a variety of ways but, though his days were often busy, this was not manifest in his recording of, or planning with, clock times. As we have pointed out even by the standards of his much terser predecessors he is reticent, even before the destruction of the majority of City churches in the 1666 Great Fire temporarily reduced the number of public time-signals in the capital. The diary conveys strong senses of sequence, and the texture of a multi-stranded life, but to say that the diary shows clocks as 'profoundly compelling [companions] . . . in the diary's fundamental engagement with the measured passage of time' (Sherman, 1996: 100), exaggerates both the novelty and distinctiveness of Pepys' diary-keeping. Pepys might be a key figure in shaping a new subjectivity of time, narrative, self, and property, that was 'especially responsive to Huygensian temporality in its early stages' (Sherman, 1996: 77), but Pepys's everyday precisions in timekeeping look far from distinctive.

7.6 THE SELECTIVE DEPLOYMENT OF PRECISION

William Harrison's statement, quoted earlier in this chapter, makes an important distinction between *capability for precision*, in which he identifies the minute as 'the smallest part of time known unto the people', and *everyday practice* in which 'they descend no lower than the half-quarter or quarter of an hour'. Precision in temporal reckoning and timekeeping was particularly manifest in connection with certain activities. This section outlines those circumstances when precision in clock time assumed particular importance in everyday practice and in recording. We do so via several vignettes stemming from a variety of sources and, equally deliberately, diverse places and periods in order to establish pattern of usage.

As in John Aubrey's biographical notes on his contemporaries, rich and poor diarists alike used relatively precise recordings of times associated with the death of relatives or close friends. In the sixteenth century John Dee recorded precise minutes for the births of his children, in the 1580s (Fenton, 2000: 13, 169):

[22nd February 1585] 'Michael born, Prague, 3 hours 28 minutes after noon . . .'

Approximation was, however, acceptable for other people's children:[8]

[18th February 1595] 'Mr Laward's son, Thomas, born at noon or a little after, ¼ or perhaps ½ an hour' (Fenton, 2000: 273),

Michael's first birthday was marked with equal concern for minutes, but in a painfully roundabout way (Fenton, 2000: 185):

[22nd February 1586] 'Michael Dee revolutus 9 hours 23 minutes'

The (obsolete) adjective 'revolute' meant 'that has completed a full revolution' (Fenton, 2001: 192 n. 3). Here the earth has completed a full orbit around the sun, returning to precisely its relative position at Michael's birth: 365 days, 5 hours and 55 minutes was the prevailing late sixteenth-century figure for the year.

Precise timing of significant births was no novelty. Its occurrence many decades earlier is shown by Dee's recording of the King of Poland's death in 1586:

[11th December 1586] King Stephen of Poland died 2 hours after midnight, in Grodno. He was born on the 13th day of January 1530, at 4 hours and 25 minutes in the morning in Transylvania, in Scholnio.   (Fenton, 2001: 204)

Apart from the recording of minutes in 1530, the entry indicates transmission of that information over the king's lifetime, and a considerable distance, to be accessible to foreign observers half-a-century later. Earlier still, some early sixteenth-century portraits record the (late fifteenth-century) birth-times of their sitters.[9]

Diarists commonly manifested similar concern with the precise times at which loved ones had died, or been born. The London physician Simon Forman recorded his first daughter's birth on 10 July 1605 at 'forty minutes after 4 of the clock in the morning', and the death of his servant Joshua during the 1603 London plague epidemic 'aged eighteen years, six months, twelve days and seven hours' (Cook, 2002: 178, 175). The phenomenon appears to have been quite widespread. We have already quoted the opening of Samuel Jeake's autobiography, about his father taking the time of Samuel's birth from a sundial in 1652, and William Petty's use of seconds in describing his birth-time to John Aubrey. Diarists who do not note such detail are the exception rather than the rule.[10] There is very little plebeian autobiographical comment from which similar uses could be sought, but Anne Lister records, on 9 November 1817, a verbatim account of her uncle's death the previous day, told her by a female servant in his household. Anne wrote down the servant's story 'on a scrap of paper, from which I am now copying it'. The account finishes:

> 9th November 1817 'These were the last words my master ever said & he spoke them very plainly. This was about 10 minutes after 12. He was ever after quite composed & went off without a groan—one would have thought he was only falling asleep—at exactly 5 minutes past 1 by his own watch that was hanging up in the room.' (Whitbread, 1999: 21)

Noteworthy here a previously pointed out, is that a maidservant was able to report a precise time to an absent relative. Even if the maid's knowledge of the precise time of death was second hand, the time of the last words spoken to her are more likely to have been timed by her own ability to read a clock or watch.

The concern cannot be specifically linked to Puritanism. Just as work during the last twenty years has dismantled the notion of a literate laity in Protestant Europe and an illiterate Catholic laity, so too with regard to time. Consider the following statement on time specification as a facet of personal morality:

> As soon as the child is born one should note in the family records and secret books the hour, the day, the month and the year as well as the place of birth. These records should be kept with our dearest treasures. There are many reasons for this, but, all else aside, it shows the conscientiousness of a father.

This comes, not from any Puritan text, but from the *I Libri Della Famiglia* of Leon Battista Alberti. This wide-ranging text of 1434 also included discussions of the economical use of time; of techniques for assigning a given time among many tasks; and which advocated a nightly accounting of how time had been used. What we have here appears similar to a 'Puritan' temporal agenda, but in the context of early fifteenth-century urban Italy. It provides a clear example of how similar positions regarding moral dimensions of time could arise in cultural environments that in many ways were quite different.

The association of relatively precise timekeeping with transport and communications was likewise a recurrent theme. This is no surprise. Long before the appearance of set timetables, times were central to the administration of communications, and important sources of indirect time-cues. In the late sixteenth century, scores of postmasters were appointed in towns on major roads (Figure 7.5), to oversee and monitor letters and packages, recording the volume, times, and destinations of the mail passing through their hands (Brayshay, 1991; Brayshay *et al.*, 1998).

Anne Lister's diary most clearly articulates a strong sense of time well beyond the utilitarian as fundamental to her experience of travel, exercise, and exertion.

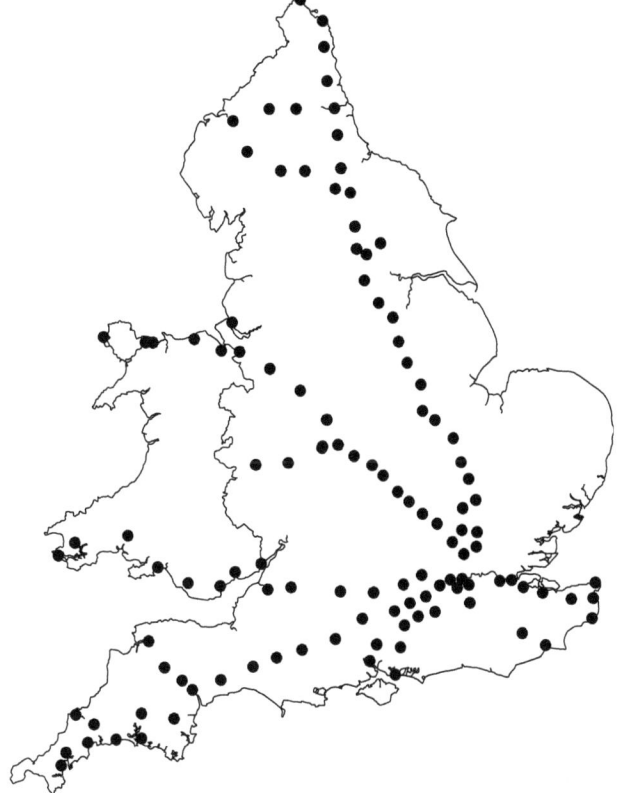

Fig. 7.5. Late sixteenth-century English post-towns.

The time taken was no neutral, objective property of activities, but woven into the very activities, and the meanings they held. Take this series of entries relating to coach and horse travel in December 1817:

December 18th 1817 . . . booked and paid a shilling for an inside place in the True Blue heavy coach that set off from the Black Swan in Coney Street every day at 2p.m. for the Golden Lion, Leeds, where it should arrive at, or a little before, 6.

December 19th 1817 . . . Just got the Black Swan in time. Took my seat in the True Blue heavy coach & drove off for Leeds as the Minster clock struck 2. . . . a rainy afternoon. We reached Leeds and stopped at the Golden Lion at ½ past 6. . . . to the Rose & Crown in Briggate . . . I could not secure a place in the Mail till 4 in the morning.

December 20th 1817 After a few hours disturbed sleep, Boots called me ¼ before 4 . . . got into the mail at 4 . . . Got to Halifax a little before 8. Walked . . . reached to Shibden a little before 9. My uncle & aunt not having expected me till one (by the Highflier) were agreeably pleased to see me . . .

December 26th 1817... met M____ ... very civil drivers from the Rose & Crown (Leeds) & which were in readiness according to my orders at 1, exactly the time she reached Leeds. M____ arrived here at 5 by the kitchen clock, ¾ past 4 by the Halifax, & ¼ past 4 by Leeds & York. (Whitbread, 1988: 27–31)

Anne Lister did not only use time-distance relationships to focus precise ideas or calculations about time and space with regard to stagecoaches and other transport technology, but with reference to her own body. Her recording of movement and time extended to walking: she times some of her walks, and records details of her particularly rapid pedestrian exploits. On holiday at Keswick in the Lake District in July 1824, she recorded how she and her companions 'had walked very fast up the mountain & averaged 3 miles an hour the whole way'. Time also provided a basis for reassessing distance: 'they called it ¾ mile, but we were 20 minutes walking it'. And in describing a particularly dramatic incident when Anne walked some fifteen miles onto Blackstone Edge in the west Pennines, in order to intercept her one-time lover Marianne's coach, to be spectacularly rebuffed by an alarmed and angry Marianne, she still finds space in her diary entry to record her pride in the excellent pace she maintained on her ascent, despite heavy rain and driving wind.

Precision was also valuable when close time management was necessary, and appointment systems had appeared by the early seventeenth century. The Court painter van Dyck worked, reported one observer,

on several portraits in the same day, at extraordinary speed. [He] never worked for more than one hour at a time on any one portrait... When the clock struck the hour, he got up and bowed to his sitter, to indicate that that was enough for that day, and arranged another appointment: after which his valet came to clean his brushes and give him another palette while he received another sitter who had the next appointment. (Cited in Campbell, 1999: 190)

Repeated appointments at the same time each day could enable sitters to be in similar lighting conditions each visit, weather permitting, but the centrality of time is strongly emphasized with reference to Wenceslaus Hollar a few years later:

Francis Place noted that he [Hollar] *worked 'by the hour, in which he was very exact*, for if any body came in that kept him from his business he always laid ye *hour glass* on one side, til they were gone, *he always recond 12d an hour'*. (Tindall, 2002: 104, our emphasis)

Appointment systems at this time, though, were not restricted to London, to painters, or to those dealing with elite clients. In the Chilterns, the healer-doctor/astrologer Richard Napier likewise used an appointment system to maximize the number of patients seen (MacDonald, 1981, 1996). Napier had been taught by Simon Forman who also appears to have operated an appointment system, frequently seeing five or more patients a day, and using minutes in astrological prognostications, besides to arrange social meetings, and—like Dee,

Hooke, and several others—record the times at which he had sexual intercourse with his wife, or other partners. Many diarists, though, were like Pepys, in that while they clearly handled complex sequences of meetings and engagements, there is little explicit sign of systematic appointment schedules.

Astrology was a powerful stimulus to precision for many people, at least as powerful as any we have identified. Here was an area where there was little perceived limit to the degree of precision sought. Francis Bacon had suggested in his *De Argumentis Scientiarum* (1623) that a scientific basis could be established for astrology by correlating astrological positions with details of the weather and of human events (Geneva, 1995: 77–8). Both lines were pursued, but we focus here on the latter, via the diary of Samuel Jeake.

Samuel Jeake (1652–1699), from the Sussex port of Rye, was a prominent non-conformist merchant. Light is shed on Jeake's life and views by an unusually diverse range of surviving documents, including many of his business records; several volumes of diary and autobiography; some correspondence; many appearances in legal, town, and church records; and the records of several national institutions with which Jeake was involved, including the Royal Society and the Bank of England.

In many respects, Jeake was a polymath of a type relatively familiar in late seventeenth-century England: he was active in the worlds of trade, science, and religion; and his reading ranged extremely widely, from Puritan and Anabaptist theology to Descartes and Bacon, from Machiavelli to Cervantes. Like other early members of the Royal Society he saw no contradiction in holding simultaneously a belief in God, an interest in science, and an interest in alchemy. Where Jeake stands almost alone among well-documented seventeenth-century English people is in the extent and forms of his practical work in astrology alongside his other interests. (Here too he was continuing an interest of his father's.)

Overall, Jeake's interests take in many activities which involved an ability to handle times, but that handling of time took extremely diverse forms. Time-consciousness came from trade (the scale and extent of credit in imports and exports), investment (as where Jeake compares the returns from different forms of share schemes), science (e.g. the return period of Halley's comet), as well as astrology. The overwhelming majority of Jeake's references to personal, commercial, or civic events do not go far below the level of the hour in specifying times of day (Table 7.3). When they do, they involve the use of half- and quarter-hours. Only very infrequently is a time given more precisely than, for example, 'a little after a quarter past ten o'clock'. The same is true of many astrological references, although not of his own time of birth, which he records as having been timed to within less than five minutes, by his father leaving the house to consult the nearest *reliable* sundial (Jeake's own emphasis).

Astrology provided by far Jeake's strongest impulse towards temporal precision. Certainly it accounts for the overwhelming bulk of the most precise references

Table 7.3. Selective precision in Samuel Jeake's diary entries, 1670–2

His *Critical Register of Several Paroxysms* (*CRSP*) compared with other entries.

Time relative to hour	CRSP	(%)	Other	(%)
On the hour	8	(5.6)	98	(61.6)
Half-hour	11	(7.7)	21	(13.2)
Quarter-hour	7	(4.9)	15	(9.4)
10, 20, 40, 50 minutes	4	(2.8)	9	(5.7)
5, 25, 35, 55 minutes	6	(4.2)	0	(–)
Finer distinctions	106	(74.6)	16	(10.1)
Total entries	142		159	

Source: Calculated from the published edition of the diary: Hunter and Gregory (1988): 85–220.

to the time of day, and to the duration of events or intervals. Jeake was preoccupied with the establishment of an empirical base for astrology, as a scientific enterprise. Whereas others, such as John Goad, were trying to establish an astrological analysis of the weather, Jeake's focus was to a considerable extent, personal. When he came to transcribe and edit more than thirty years of his diaries in 1694, his chosen title was *A Diary of the Actions & Accidents of my life: tending partly to observe & memorize the Providences therein manifested; & partly to investigate the Measure of Time in Astronomical Directions, and to determine the Astrall Causes, & tc.*

The core of Jeake's interpretations lay in the interaction of astrological causation with divine intervention. The position of the planets was not treated as itself an explanation of events: rather they formed potential dangers which would have become actual had not God intervened through, for example, the illumination by a lightning flash of an unseen hazard. Demonstrating God's power to affect human events was thus grounded in identifying potential dangers from the position of the planets, for which precise timing of the relevant events or situations was therefore essential, as it was, indeed, for refining interpretations of what particular planetary influences were. With precision the central element in Jeake's astrological analysis, he was extremely concerned to record precise times in relation to astrology, and to elaborate in giving numerical quantities to the good or bad aspects of planetary positions. This precision in timing, and—especially—in knowledge of times of birth for casting natal horoscopes, was essential for detecting both influences and their evasion.

The most striking case in Jeake's diaries and autobiographical notes occurs between 31 August 1670 and 2 May 1671, when Jeake sought to investigate an astrological basis to the incidence and severity of a series of nearly 150 ague attacks that afflicted him over that period. The times of these are recorded to a particular minute. There is very little tendency to approximation or rounding. Times involving whole hours, half- or quarter-hours account for only just over

10 per cent of times, not much more than what would be expected if episodes occurred randomly. The contrasts with Jeake's still explicit, but comparatively imprecise, precision in specifying times in relation to business, religious and social life, are very striking.

For Samuel Jeake's recording of business activities, social arrangements, or journeys, his adequation was what Gigerenzer calls 'fast and frugal': entailing rough-and-ready heuristics and limited-but-robust temporal information. The rounded times he recorded reflect this loose adequation, not an incapability of precision. He did not choose to be specific on these occasions. By contrast, Jeake's analytical approach to the exact onset-time of his ague attacks required much more exacting adequation than did anything else he did. We are not claiming that Jeake's times were exact, either as regards his reading of a clock, or as regards the accuracy of those timekeepers. But we do emphasize that the possibility of precision was real enough to Jeake for him to record times in fine detail. And it is that he thought the attempt at precision worthwhile that is our main point. The reasons for his making, or not making, the effort to be precise lay in the practices and meanings of the activities themselves.

Another important site of precision-use was gambling, which seems to have been endemic virtually throughout (male) English society. Two prime sites for betting were horse- and foot-racing, where match-racing and timing provided the two accessible ingredients of wagers. A stopwatch had been invented about 1690 by Samuel Watson, as a 'pulse watch' for use by physicians (Macey, 1994: 295) and by astronomers (Andrewes, 1985: 82), but racing and gambling were important motors for stopwatch making. Stopwatches timing to $1/5$ second were being used to time races before 1731, by which time they were in commercial production (Guttmann, 1978), seven years before the earliest use of 'stop-watch' known to the *OED*.[11] By 1757, times for some horse races were recorded to the half-second (Macey, 1994: 524). The idea of record times swiftly spread. For example, a ten-mile foot-race time of 54 minutes 30 seconds by Woolley Morris in 1753 was hailed by contemporary press reports as the fastest time yet recorded (Krise and Squires, 1982). If such claims often reflected lack of information—a man called Pinwire having been narrowly beaten despite running 52 minutes 3 seconds twenty years previously (Radford and Ward-Smith, 2003: 432)—the important point is less the accuracy of the claim, than the attention given to seconds. John Aubrey's biographical notes described Philip Herbert, 4th Earl of Pembroke (1584–1650), as a famed breeder of Arab horses, including the famous Peacock and Delavill: 'It is certain that Peacock used to run the four mile course in five minutes and a little more; and Delavill came since but little short of him' (Barbor, 1982: 142). But before the infeasible speed implied by Aubrey's claim is taken as evidence that he had no realistic conception of speed, recall that 'the four mile course' was the name of King's Plate course at Newmarket, not a measure of distance. While the 1665 rules did not specify race-timing, this clearly occurred for betting, and the availability of good timekeepers meant that it was easier to specify a distance and measure the time than vice versa.[12]

This is not to suggest that all uses of time in wagers were so exact. For example, on 12 June 1760 Thomas Turner entered into a bet on the precise time in which a horse could, at a steady pace, make a particular journey, 'between nine and ten miles and very uneven ground... within the hour' (Vaizey, 1985: 206). This illustrates well the low-key ordinariness of the incorporation of clock time into Turner's everyday outlook. Around 1750 a Sussex village shopkeeper, like Thomas Turner, or his Lancashire counterpart, were perhaps more likely to encounter 'seconds that mattered' when gambling on a footrace or a horse-race, than anywhere else (Vaizey, 1985: 104, 109; Sachse, 1938).

The printing of race timings in eighteenth-century press reports fostered a concern with notable performances and 'record' times for certain courses, or distances (Radford and Ward-Smith, 2003). Parry (2006: 214) identifies three factors—'socially produced focus on excellence, the landmark achievement, and the serendipitous description'—as stimuli towards the recording of, and regard for, measured performance. While Parry's focus is essentially modern, going back only as far as the four-minute mile record (1953), the factors identified also apply to early modern concern for precise timing, using new stopwatch technologies, though gambling gave a particular cast to a 'socially produced focus on excellence'.

How directly such uses connect to either interests in precision springing from astrology, or from a broader enthusiasm for technologies that measure, remains an open question. Rather, it is their commonplace character as indicating wider ideas about 'something that can be done with timing' that is striking. The treatment of stopwatches as general knowledge that underlies Lawrence Sterne's extended derogatory comments on stopwatches in *Tristram Shandy* (1759–69) is telling: these are not a few throwaway lines about something exotic, but a discussion of something generally familiar. So, while stopwatches were clearly being used in some factories in the late eighteenth century, workspaces were not the site of their earliest use.

7.7 AESTHETICS OF PRECISION

Not that accuracy was solely a utilitarian issue. Several other aesthetics, often in combination, bore on people's stances towards, and practices of, close timekeeping. These aesthetics mattered to different communities of practice for different reasons, and hence had their own geographies, sociologies, and histories. Together, they created multiple motives for seeking precise and/or accurate timekeeping. Some aesthetics were part of much more general fascinations or imaginations. Consider just seven aesthetic sources.

First, enthusiasm for time-measurement related to more general notions that measurement was a useful way of comprehending, understanding, or controlling, the temporality of things in general. Timing events or measuring durations were parts of approaching a problem, even without a specific objective. Time-measurement in early factories provides the classic—but not unprecedented—case (Thompson, 1967). Second, as Thompson also notes, making good use of one's (God's) time was a recurrent theme among post-Reformation writers. Within discourses hostile to waste and inefficiency, seconds-indication was an important rhetorical point, whether or not such close timings were actually used. Third, a fascination with gadgetry and things mechanical was also endemic in early modern England. Diarists like Samuel Pepys, Claver Morris, and Anne Lister record watching watchmakers at work, and their interest in machines and gadgetry (Latham and Matthews, 1972: VI, 337; IX, 109; Reeves and Morrison, 1989: entry for 20 March 1739; Whitbread 1992: 321).[13]

Fourth, novelty and newness were valued as interesting in themselves, especially in discourses of science and of consumption. The category 'new and exciting' was produced and contingent, rather than a given, but 'new kinds of newness'—including precision timing—had an appeal in themselves. Fifth, and remaining with consumption, timepieces sold on facets of craft and design, satisfying their owners as sensory objects, not just on technical grounds. Cases, finishes, and dials were important elements to the experiencing of clocks-as-objects revolution (Roberts, 1990; Loomes, 1995; Robey, 2001). Private clocks' performances as consumer goods were central to the early modern consumer attitudes towards them. Sixth, clock times quickly became an index of people's commitment to emergent forms of politeness and civility, to which changing attitudes to punctuality were central. Conventions about promptness formalized responses to an informal quantification of impatience. Anne Lister's diary, for example, records testing social relationships by deliberately forcing issues of inclusion/exclusion that centred on the punctuality of routines (Whitbread, 1992: 335).

Seventh, new precision in timekeeping was at once drawn in to explorations of timing and the human body. We have already cited Anne Lister's recording and commenting upon her walking speed in 1817, but a similar aesthetic is among those detectable 150 years earlier in Samuel Pepys's enthusiastic response to his new minute watch, on 13 September 1665:

Up and walked to Greenwich, taking pleasure to walk with my minute watch in my hand, by which I am now come to see the distances of my way from Woolwich to Greenwich. And do find myself to come within two minutes constantly to the same place at the end of each quarter of an hour.   (Latham and Matthews, 1972: VI, 221–2)

Recognizing these diverse aesthetics of timekeeping does much to explain why first public, and then private, timekeeping were much more widely distributed than can be explained by an exclusively technical drive to using clock times and owning clocks.

Concern with speeds of bodily movements has a particular manifestation in notions of 'record times' for particular distances, routes, or races, and with the associated technological development of the stopwatch. As already noted, stopwatches began to be made in the early eighteenth century, distinguished as the name implies by having a display that could be instantly stopped while the watch continued to keep time, even in the early days recording times to ¼ or ⅕ of a second.[14] Early stopwatches had diverse uses: to accurately time the pulse of a patient; to time astronomical observations; and to time the races or time-trials of athletes ('pedestrians') or horses. Once familiar, stopwatches figured a certain psychology, as where the *OED* cites an 1817 characterization of 'uncle . . . being a stop-watch person, always in a hurry' and an 1806 definition of 'automata' as 'people who regulate all their thoughts, words, and actions, by the stop-watch'.

Current debates among athletics historians about trends in running speeds achieved by humans over the last three centuries are in part possible precisely because various timings were documented in diaries, broadsheets, newspapers, and pamphlets (Radford and Ward-Smith, 2003). The seemingly obvious presumption of progressive improvement in times is simplistic, given the long-run impacts of diet, urbanization, and industrialization on human stature and performance. The case of racehorses differs, since central issues here concern the cumulative effect of three centuries of breeding among a narrow selection of thoroughbred bloodlines, but discussion also draws on long-run recording of pedigrees, records, and times.[15]

7.8 CONCLUDING COMMENTS

A number of broader points arise from these various vignettes, of which we want particularly to stress three. First, relative precision was used selectively. What counted as precision at any given time was clearly relative, among other things to be discussed here, to the technology of time indication. When clocks were relatively novel, indication of the hours was a more precise indication of time than had hitherto been available. Two or three hundred years later, discrimination to the hour was old hat, and everyday precision was a matter of quarter hours or below. Whether we look at diaries, letters, biographies, or depositions, people sometimes chose to be precise. But they were not automatically precise, and they were not always precise.

Over time, precise coordinations were increasingly a matter of inter-referencing within increasingly dense networks of clocks and other timepieces. Environments were undoubtedly patchy in the provision of clock times, but the main factor that explains the pattern of more precise uses is motivation rather than information. Over time, the increasing density of timepieces gave everyday environments a greater ecological rationality, and more precise indications became available with significantly less effort. But there were always differentials in how precise people chose to be on different topics. The documentation of Samuel Jeake's differential application of precision in timing provides an unusually well-documented manifestation of clearly widespread practices of selective precision in the timing of different activities. We have identified five subjects as likely—but not guaranteed—to stimulate particular precision: namely births and deaths, travel and communications, appointments, astrology, and gambling; but these were not necessarily the only ones. Impulses to precision could arise from quite 'non-modern' world views (like Jeake's astrology); and precision could be important as an aesthetic in itself, even if in hindsight the underlying logic appears misguided, or primarily rhetorical, rather than well founded.

Second, it follows that one must be cautious in inferring temporal indifference when people were vague about times in diaries, letters, and the like. Much depended on their motives in each context. Jeake's precise timing of the onset of ague attacks exemplifies that precision of references was *not* simply dependent on available time-cues, but on his perceptions of the value(s) of precision in different circumstances, and whether the required effort was worthwhile. For most diarists, whether in the 1580s or the 1820s, notions of adequation were relatively loose most of the time. An hour usually sufficed, and greater accuracy than a half-hour was not worth the effort involved. Early instruments relied on narrowly distributed skills, and the restricted distribution had major implications for whether devices worked at all.[16] The availability (or not) of heuristics facilitating certain calculations or uses of information was also important, with a trade-off between simple and robust forms of estimation, and complex but non-robust time-telling. This stark dichotomy was usually the extent of the choice available.

This point once again underlines William Harrison's careful distinction of a century earlier, between what people knew about time-divisions, and their habitual uses. But note that Harrison's mention of minutes predates the availability of pendulum clocks with minute hands by nearly a century. Samuel Jeake's distinction of single minutes in timing his ague attacks in the 1670s postdated the production of pendulum clocks, but can essentially be seen as continuous with his father's attempt to distinguish single minutes from a sundial, half a century earlier. Likewise, William Petty's—possibly playful—use of specific seconds in describing the time of his own birth in 1623 predated pendulum clocks, but was reported to John Aubrey some time after it was feasible to construct more accurate clocks.

Third, the findings here also have consequences for the general question of the relationship of the technology of timekeeping and the social (broadly defined) impulses to timekeeping. The tendency of earlier historians to write the history of time-consciousness from the technology of timekeepers—by no means extinct among writers for either academic or general audiences—has duly been severely criticized, and justifiably so. Technologies in this area clearly do *not* dictate the sophistication of time awareness. However, there has been quite a strong tendency to simply reverse the direction of the causal arrow; that is, to see the technology of timekeeping as a quite well-calibrated response to the social impulses to timekeeping.

Both these views are mistaken in presuming that the performance of clocks, and people's expectations of clocks, moved together. In this chapter we have shown they were substantially out of step. There may be considerable pent-up demand for precise timekeeping that clocks are unable to fulfil. Or clocks may keep time to levels of accuracy and precision going way beyond most people's notions of what may be done with them. Very broadly, the late seventeenth-century horological revolution marked a relatively rapid shift from the former situation to the latter. The rapid but selective take up of minutes and seconds indicates a pent-up demand for accuracy, that for most purposes was exceeded when seconds indication became commonplace, hence the very limited use made of seconds indication, even when it was common on long-case clocks.

The selective use of precision indicates that public timekeeping technology was well ahead of many of the uses to which it was put. For certain purposes, the possibility of specifying times of day more precisely *was* seized, but that possibility often exceeded what people felt they needed. People could take precision or leave it, and often they left it. But when they left it—when Thomas Betson merely dated a letter, or when Samuel Jeake wrote 'about three o'clock'—they were not being imprecise because they lacked understanding of the possibility of greater temporal precision. As yet, the exploration of prosaic, everyday time-usages has only just begun, and much about the form and sophistication of everyday temporal behaviour awaits the historian ingenious enough to tease it out from the records in which such actions and knowledges occasionally break the surface of documentation largely preoccupied with more specific concerns.

NOTES

1. We leave discussion of the distinction between 'precision' and 'accuracy' to Chapter 9.
2. Warwickshire Record Office, DR.64/63 p. 277 (emphasis added).
3. Production of single-handed clocks continued into the nineteenth century, providing basic clocks cheaply for the lower end of the market. For example, in the 1720s Andrew Knowles at Bolton (Lancashire) produced single-hand clocks, in which quarter-hour units on the hour ring were subdivided into markings at five minute intervals (Loomes, 1997: 162–3). These small divisions were still several times wider than the hand tip, so further interpolation was readily available.

   Increasingly accurate clock time also stimulated changes in the design of sundials (as in the Loomes example with minutes of a Vernier scale), following on from the dial transition between unequal and equal hours. As we have pointed out, the notion that there were few changes in use and construction of sundials over time can be dismissed—not only were there new changes relating to dials, but adequation of sundials also changed substantially. A dial had to perform to a considerably higher standard to suffice in 1700 than two or three hundred years earlier.

4. Those who want to reject our argument for the fine reading of single-hand clocks because of the prevalence of quarter-striking signals from public clocks can only do so by invoking another part of our argument: that public timekeeping and signalling ran ahead of the apparent crudity of single-handed clock dials. The objection thereby serves only to strengthen another part of our overall argument!

5. The two times coincide in mid-April, mid-June, early-September, and late December.

6. Few public clocks were ever fitted with seconds-indication, not least because of problems of wear and of ensuring regular running where gearing revolved relatively quickly. But even some parish church clocks do have a seconds hand, e.g. St Nicholas, Bristol.

7. *Phil. Trans.* 1: 296.

8. Dee used Julian calendar dates in England, but Gregorian calendar dates when in Bohemia, and other countries that had shifted from the Julian to the Gregorian calendar in 1582.

9. For example, a precise birth-time from the late 1490s is given in a portrait of Matthaus Schwarz of Augsburg painted by Christoph Amberger in 1542. Schwarz's horoscope was included in the picture, the time of the painting was given as 4.15 p.m. on 22 March 1542, and his exact age was recorded as 45 years, 30 days, and 21¾ hours. This Matthew Schwarz wrote the first treatise on double-entry bookkeeping published in Germany, and he also wrote a pictorial chronicle of his life and clothes. He had already had himself painted on his 29th birthday, with watch and hourglass hanging at his neck. His wife Barbara was painted (also by Amberger) on her 35th birthday in 1542 (Campbell, 1999: 191).

10. One who does not—though he had no children of his own—was Samuel Pepys, which suggests caution about Sherman's speculations about Pepys as the first modern diarist.

11. To measure the pulse of a horse, Henry Bracken recommended the use of 'a stop-watch which runs seconds, or a Minute Sand-Glass, as there are enough of them, especially in the Maritime Towns; I say either of these (in a good hand) will do, where a Person is not provided with a proper pendulum for the purpose' (Bracken, 1738: 184–5). Other references in the early 1740s come in contexts of medicine, astronomy (Long, 1742: II: 332), jewellers' stock and auction lists (Langford, 1750: 8).

12. There continue to be records in both running and cycling for the maximum distance covered in one hour, twelve or twenty-four hours, or six days, but such events are of low profile compared with competitions over set distances.

13. Anne Lister recording 'Got a new watch-glass at Pearson's & asked him to let me see a clock taken to pieces. I am to call next Tuesday for this purpose' (Whitbread, 1999: 321, 9 January 1824).

14. This is reflected in the usages cited previously. Samuel Johnson, though, did not include stopwatch in his *Dictionary* of 1763.

15. 'Pedigree Online' and other online thoroughbred databases trace the bloodlines of thoroughbred racehorses back to the 'foundation' sires and dames of the seventeenth and early eighteenth centuries.
16. For example, in the 1380s, a clock-keeper travelled in the King of France's entourage (Dohrn-van Rossum, 1996: 120).

8
'Posted Within Shot of the Grave':[1] Seafaring Times

8.1 INTRODUCTION: COMMUNITIES OF PRACTICE

New or modified temporal practices could arise in the specific circumstances of particular communities of practice. No one of these communities had complete access to all the mysteries of clocks and clock times, or a monopoly of them. Rather, gradually changing sets of communities had access to particular temporal practices in which clocks and other time-measurement devices were embedded. Some communities *were* tightly bound to specific practices, as where clockmaking practices and knowledges developed through several generations, only intermittently touching the formal, codified knowledges of primers and manuals.[2] There were similarly self-contained communities using clocks and clock time in quite specific ways, but here their temporal practices were never entirely isolated, and overlapped with those of others.

That most communities of temporal practice were more open begs the question of the degree to which certain temporal practices leaked across boundaries from one community to another, becoming gradually general. For example, notions of mean time were predominantly the domain of astronomers and other specialists, making their way into everyday life only later.[3] Further, as temporal practices moved among communities and became general, their meanings were negotiated and could change quite substantially so that, although they might appear generalized, they would still have their own 'signatures' in particular communities.

To recapitulate from Chapter 3, we can characterize the conduct of these communities of practice in four different but clearly interrelated ways. First, such communities provide particular norms of embodiment which span familiar bodily gestures, clothing, particular ornamentations and characteristic technical devices—like watches. Second, these norms of embodiment are set within the well-worn grooves of specific inter-corporeal routines, bodies keeping together in time and space (McNeill, 1995). Third, there are devices which both support and drive these inter-corporeal routines. A whole 'ecology of tools' is bound up with even the simplest of practices (Hutchins, 1995). These tools may be more or less robust (so requiring more or less work to maintain), may duplicate one another to some degree, and so on. Then, fourth, communities of practice depend both upon diverse knowledges, usually unevenly distributed within communities, and upon specific conventions for recognizing the 'enunciative authority' of particular people, not least through norms of embodiment (Shapin, 1994). In turn, this knowledge combines with norms and devices to produce a *practical epistemology* or sense of 'correct' procedure. This epistemology comprises, not formal rules, but a general stance and sense of acting into the world. It is not acquired by 'satisfying myself of its correctness... [but]... is the inherited *background* against which I distinguish between true and false' (Wittgenstein, 1958, section 94).

In this chapter, we substantiate these general points through a consideration of a set of communities which had very specific temporal practices: the practices relating to seafaring. Some of the ways in which shipboard communities operated differently from communities on land are particularly revealing about the broader workings of temporal practices.

8.2 NAVIGATIONAL DEVICES

Unsurprisingly, given their existence in the highly uncertain and sometimes dangerous[4] context of the sea, members of seafaring communities produced many novel temporal practices, often utilizing quite specific devices. The numerous and diverse navigational instruments excavated from shipwrecks by marine archaeologists have rewritten a seafaring literature previously inclined to stress the sparsity of instruments on board. However, recovered objects do not reveal exactly who owned them, or how they were regarded. Devices were clearly valued by writers of astronomical and navigational treatises, and (later) used by the Royal Navy, the East India Company, and other large institutions, but ownership and regard for them among the virtually undocumented bulk of merchant mariners still remains largely obscure.

Certainly, for ship-borne navigators, a sense of time is central. Navigators need a battery of temporal cues: the flow of tides, phases of the moon, the ship's speed through the water, and the like. Most of this temporal information was gathered together as embodied, practical knowledge, often termed 'by guess and by God', and was in regular use before 1300 (McGrail, 1998, 2003). Using this body of knowledge, experienced navigators built up considerable navigational skills, especially in familiar waters. It is also worth remembering that, at least on some voyages, margins for error were large: all that was needed was to strike a course that would intercept a particular shoreline, before working along it.

There were five chief components through which this practical maritime sense of time was constructed. One potent source of temporal knowledge, just as for direction-finding, was the sky, and the relative positions and movement of the sun, moon, and stars. Time of day for the medieval sailor was apparent solar time, with noon established by the sun reaching its highest elevation, and bearing due south. The period between successive noons was divided into twenty-four equal hours. In daylight, the hour could be estimated from the sun's elevation or its bearing; at night, from the position of the 'Guards' stars in the Plough as they circled the Pole Star, *Polaris* (Waters, 1958; Frake 1985). Later, these became known as the clock stars.

A second source of temporal knowledge came from natural rhythms. Tides provided a kind of clock:[5]

> Times of departure could be chosen to get the maximum benefit from the tide; rise and fall of water could greatly aid the emptying of dry docks; a high tide could lift a ship over an obstruction, and was often worth waiting for. A ship going aground at spring tide would be 'neaped', unable to lift off until the next suitable tide, which might be six months later. Tides, unlike North European weather, were almost completely predictable, and therefore could be used by the skilled navigator. (Lavery, 1989: 265)

A third source of temporal knowledge consisted of judgements based upon repeated experience of, for example, the feel of wind and swell, the ship's movement through the water shown by spray and turbulence, the appearance and sound of the ship's bow wave. These were (and remain) crucial elements in sailing, requiring the cultivation of an incarnate sense of time, allowing navigators to recognize 'the conditions under which a "day's sail" will be accomplished in a day' (Hutchins, 1995: 94). This standard was often expressed as simple folk measures of distance, such as the 'day's sail' or the course or the *kenning*, commonly understood as 'the distance at which the shore could first be seen from the offing when making a landfall' (Cotter, 1983b: 260); in usual circumstances, the equivalent of about 20 miles.[6]

A fourth source of temporal knowledge were those practical navigational indexes already incorporating a sense of time. For example, early warning of land could come from orographic cloud, seaweed and jellyfish in the water, reflected wave patterns, colour changes in the water, the sound of surf breaking in the shallows, the smell of sheep, and most importantly of all, the sight of seabirds like the tern, which sleep ashore and so can be seen heading out from and in to land at dawn and dusk respectively (Hutchins, 1995).

Finally, and probably most importantly, there was knowledge gleaned from simple familiarity with repeated voyages:

> masters . . . gained very considerable advantages by serving continuously on the same routes. Familiarity would soon reduce the navigational dangers of a particular route, . . . [M]asters desperate for a ship . . . would sail anywhere [but] the typical master was associated with a particular trade. . . . George Marlowes, a Yarmouth man, . . . reckoned to have 'made more than a hundred voyages to Newcastle' and Thomas Collyer made 'the voyage from this part of London to the Barbadoes for neare forty yeares'. (Earle, 1998: 47)

The pitfalls of lacking such local knowledge are made explicit in John Dee's diary entry for 28 Sept 1583, after a North Sea crossing:

> We fell on the Holland coast, and none of our mariners, master nor pilot knew the coast: and therefore to the main sea again with great fear and danger, by reason we could scarcely get off from that dangerous coast, the wind was so scarce for that purpose. (Fenton, 1998: 106)

Throughout 1300 to 1800 these kinds of practical knowledges were absolutely crucial to life at sea. By these five means, often lodged in the head by various routines, mnemonics, and 'arts of memory' (Yates, 1966), skilled navigators could have at least some idea of speed and position.

Even before 1300, though, navigators were using various ancillary means of navigation which all involved timing. The simplest were chants or standard phrases used to measure elapsed time. Most important were physical *devices*, used with appropriate routines, to produce new fixes on speed and position at sea.[7] In particular these devices were of three kinds. The first was the *lead-and-line* or *sounding lead*, already known to classical authors like Herodotus, now recovered from ancient shipwrecks (McGrail, 1998: 276), and clearly in general North European use before 1300. That the earliest English illustration, *c.*1500 in the Hastings Manuscript (Hutchinson, 1994), and description, in the 1620s, so long post-date such examples, underlines how the lead-and-line was thought too familiar to be worth recording.

Early modern sailors used two types of lead. In 'shoal water', up to about 25 fathoms (50m) deep, cone shaped hand leads weighing 7–12lbs (3–5kg) could be heaved from a coil of rope every few minutes if necessary. Deep-water leads were more laborious to use: weighing 14–24lbs and attached to a 150 fathom line, they might need a winch, and required that the ship was hove-to. Elizabethan seamen called their heavier sounding instrument the 'dipsie lead'.[8] The line-length approximately measured by the sailor's arm-spans, or fathoms, gave the water-depth. On dipsie lines, coloured rags were twisted into the line at specified intervals for quicker measurement: for example, some dipsie lines were marked at 2 and 3 fathoms with black leather, at 5 and 15 with white cloth, at 7 with red cloth, and at 10 with leather, enabling the leadsman to sing out 'by the mark seven', and so on (Taylor and Richey, 1962: 16). Leads also enabled mariners to collect sea-bed material, in cavities in the lead, or on tallow affixed to it. Sailors had 'remarkable familiarity with the nature of the seabed' (Earle, 1998: 74), and an ability to fix position by interpreting the colour, texture, smell, and even taste of the seabed was a key practical piloting skill. Depth soundings, seabed material, and seaweed information was often recorded on charts. An early sixteenth-century rutter warned that when approaching the English Channel from Spain, after coming into 'Soundings' (the western English Channel, between Ireland and Brittany),

> When ye be at lxxx fadome ye shall find small blacke sande and yee shalbe at the thwart of the lezarde

When ye be at lx or lxv ye shall find white sande, and white soft wormes. And ye shall be very nigh to Lezard. (Cited in Waters, 1958: 19)

The second device was the *compass*. Rudimentary magnetic compasses had long been known but were not widely used for some centuries. The earliest extant English description, by the monk Alexander Neckham in the 1180s, describes a reed pierced by a needle magnetized by a lodestone, and then floated on a bowl of water, to indicate the four cardinal points. As Hutchinson (1994) puts it, with considerable understatement, 'an instrument which had a bowl of water as one of its components would be of limited use at sea' (Hutchinson, 1994: 176). Neckham adds that the compass was chiefly used during cloudy weather when the sun and stars—his preferred means of direction-finding—were obscured.

During the thirteenth century, a dry compass was developed in which the compass needle was mounted on a vertical axis with a pivot at each end, and with the edge of the bowl fitted with a graduated ring. But, compasses with a bare needle and a compass card or wind rose, either underneath it or marked around the rim, were prone to errors of parallax. To overcome this problem, magnetized iron wires were glued beneath the compass card, so that the card itself pivoted.

By 1300, the compass was becoming a more familiar instrument of navigation; and more frequently documented and illustrated. Late medieval ship inventories are scarce, but 'lodestones appear . . . in the inventories of 1410–12 of the *Plenty* of Hull, which has "1 sailing piece", and the *George* for which "12 stones, called adamants", called sailstones, were bought for 6s in Flanders' (Waters, 1958: 22).[9] Compasses drawn as decorative elements of fifteenth-century maps usually show 'a round box and a fly similar in form and layout to those in common use in the sixteenth [century]' (Waters, 1958: 26).

The compass was a timekeeping device in its own right, not just a direction-finder. 'It hath beene an ancient custom among Mariners', explained John Davis's *Seamans Secrets* (1599), 'to devide the compasse into 24 equall parts of howers, by which they have used to distinguish time' (Waters, 1958: 30). Bringing direction and time together, 'the compass rose [became] the clock face of the medieval sailor' (Frake, 1985: 265). From bearing south at noon each day, the sun was envisaged moving around the compass rose each twenty-four hours: bearing north at midnight, east at 6 a.m. and west at 6 p.m. Any time could be expressed as a compass bearing, with the 32 compass-rose points marking out periods of 45 minutes. Fortuitously, these approximate the 48 minutes by which the moon and tides lag behind solar time each day (Hutchins, 1995: 99–102), which enabled 'sailing directions, and presumably the memories of sailors . . . , [to] specify the tidal regime of a given place by stating the lunar time, named as a compass bearing, of a given state of the tide, usually high' (Frake, 1985: 265).

Lunar time was also represented using compass bearings. Outside the near tideless Mediterranean, tides were critical for sailing shallow water, especially around ports, and the moon's connection with tides made it highly significant for medieval sailors. From northern Europe, the moon bears due south as it reaches its greatest elevation, and 'moon bears south' was thought of as lunar noon (even though the moon might then be below the horizon). Similarly, 'moon bears east' named the lunar time of moonrise whether or not the moon actually bears due east at that time.[10] At full moon, when moon and sun are in opposition, 'moon bears north' names a lunar time that corresponds with solar noon, 'sun bears south' (Frake, 1985: 264–5). '[F]rom the thirteenth century to judge from the rhumb (direction) lines of the oldest surviving chart, the seaman's horizon has been divided into thirty two directions or . . . "rhumbs of the winds"' (Waters, 1958: 21).

Thus the compass rose—a cognitive map of directions—was used as a rough-and-ready timepiece, because of an entirely fortuitous relationship between the 32-point compass rose and the 24-hour day. The compass-timepiece came into its own when mariners needed to know the state of tides in relevant ports, and when tidal streams would change direction.

> They accordingly hit upon the practice of reading the times of high water according to the age and compass bearing of the moon. For simplicity and ease of memory they noted the times of high-water at the various ports on the day of full and new moon—at 'full' and 'change'—in terms of the compass bearing of the moon at the moment of high water. This, since the highest high-waters or spring tides were found to occur at about full and change, became the *establishment* of the port. (Waters, 1958: 30–1)[11]

So, for example, the establishment of Dover expressed as 'full sea at Dover, the moon south' meant Dover high-water at the full and new moon occurs with the moon south, that is midnight and noon. Similarly 'All havens are full between Start [Point] and Lizard at a West and South-West moon' meant 4.30 p.m. and 4.30 a.m.; and 'Dieppe is north-north-west and south-south-east' meant that Dieppe high water at the full and new moons occurred at 11:15 p.m. and 11:15 a.m.

Obviously, mariners arriving other than at full or change needed to know high water times. Since high tides occur 48 minutes or $4/5$ hour later each day[12] it was convenient for a seaman to use his 32-point compass rose, and

> round the daily retardation [to] $24/32 = 3/4$ hour or 45 minutes. . . . making each point worth 45 minutes of time. When he knew the establishment of a port all he had to do to find the time of high water on a particular day was to find from his almanac the age of the moon and add the daily retardation—one point for every day of the moon's age—[to the port's establishment]. (Waters, 1958: 31)

From medieval times, the compass was permanently mounted before the helmsman in a binnacle—deck-secured protective housing on the vessel's centre-line—although he might steer the ship by reference to horizon-objects like clouds, not just compass bearings.[13] At night, the binnacle was illuminated by a candle lantern placed within it (Waters, 1958: 24).

A third device, which offset some limitations of timekeeping using the stars and compass, was the *sand-glass*. Commonly timing hours or half-hours, sand-glasses were critical in gauging the distance sailed on a particular course, and in timing watches.[14] Appearing in the twelfth-century Mediterranean (McGrail, 1998: 285), sand-glasses moved rapidly northwards, appearing in late thirteenth-century English ship inventories as 'horloge de mer' or 'running glass' (Hutchison, 1994). At first, shipmasters continued to measure speed by eye, using the sand-glass as a rough check (Waters, 1958: 35–6), but as sand-glasses were linked with other instruments, and improved manufacturing produced glasses more accurately measuring short periods of time, both the device and its usage became more sophisticated. Elizabethan ships carried a range of glasses to measure periods from half-a-minute up to four hours.[15] All were relatively large and heavy. Watch-glasses 'might measure a foot in diameter and two feet in height, weigh several pounds, and need both hands for the turning. A half-hour glass would be about half the size, but a half-minute glass for use with the log line might well measure 5 or 6 inches in height and 2 or 3 inches in diameter' (Waters, 1958: 309).[16]

Manufacture of accurate sand-glasses faced a complex of problems. First, the filling, which was usually a finer material than sand, like iron filings, marble dust, or powdered egg-shell in order not to wear away the neck of the glass, did need to be heavy enough to run freely unaffected by clumping or condensation. Second, glass-blowing was a difficult task, producing glass halves of inconsistent thickness and bore-width until after the mid-sixteenth century. Third, the putty, wax, linen, or leather seals that joined the two halves could leak over time. Fourth, the glass was prone to breakage, despite its protection by an oak frame suspended by leather straps. To overcome such potential errors the prudent navigator used several glasses in concert:

[B]atteries of running glasses, as many as four glasses in one frame, were carried. This avoided any mistake about the simultaneous turning of several glasses and enabled a mean reckoning of several glasses to be taken easily. In other models the rate of flow of the four glasses was so regulated that they registered respectively quarter, half, three-quarters and one-hour intervals.   (Waters, 1958: 310)

The lead-and-line, compass, and sand-glasses, together with a manuscript *rutter* (handbook of sailing directions) and some small *cardes* (outline coastline maps), comprised the shipmaster's array of navigational devices. Though basic, they sufficed for the coastal passages that accounted for most voyages.

Indeed, until the 1550s English masters remained essentially coastal pilots, relying on

accumulated knowledge of the line and appearance of the coast, of tides and of the sea floor, and on dead reckoning by compass, sand-glass, log-line and traverse board. (Andrews, 1984: 29)[17]

But as voyages lengthened, and knowledge of new sailing conditions increased, so new devices began to appear. Some new computational devices were mechanical simulations of familiar routines, such as finding time by the stars. Detecting the exact time at night was complex because the sidereal day is roughly four minutes shorter than the solar day. The Guards appear to 'slip back' slightly in their circuit each night. Keen shipmasters could memorize rules which allowed for the slippage and, from the mid-sixteenth century, refer to almanacs containing rules for finding the time. Though approximate, these sufficed 'for the shipmaster rarely had need to measure time accurately in hours and minutes for purposes of calculation' (Waters, 1958: 35). The sixteenth-century development of the *nocturnal* provided a device to calculate the time directly from sighting the Guards, allowing for their retrogression. The business of finding the time of the tides via the compass and tables was similarly incorporated into a *tide-computer*, consisting, like the nocturnal, of a series of concentric circles.[18]

But many new devices—especially those concerned with deriving latitude and altitude—were far more sophisticated. Their invention opened up new apprehensions of the world which led to new computations and, in turn, the invention of even more sophisticated devices. In these devices, *calculation* of some kind—however simple—was assured.

Three kinds of devices were paramount, respectively measuring distance traversed; measuring altitude and thereby latitude; and plotting position. In use, the devices were interconnected, but for convenience we consider them separately.

Without direct means of measuring longitude, accurate computation of speed and of distances sailed was vital. Two devices to calculate distances gave some purchase on the position problem. Use of the *traverse board* was increasing through the sixteenth century, enabling shipmasters to keep a reckoning of the ship's position, based on mean speed and course, and the effects of wind and tide. This was a small circular board marked with radius lines on which were bored a series of equidistant holes. Pegs were put in the holes to mark the time-periods during which the ship sailed particular compass bearings during each watch. One captain explained:

Upon the binnacle is also the Traverse which is a little round board full of holes upon lines like the Compass upon which, by the removing of a little stick, they keep an account, how many glasses (which are but half-hours) they steer upon every point. (Waters, 1958: 36, citing John Smith's 1627 *Sea Grammar*)

Widespread use of the traverse board was almost certainly encouraged by another new device, the *log-and-line*, probably an English invention of the 1570s. William Bourne's *A Regiment for the Sea* (1574) mentions a new instrument 'to know the ship's way':

a minute of an hour glass, or else a known part of an hour by some number of words, or such other like, so that the line being vered out and stopped just with that time that the glass is out, or the number of words spoken, which done they . . . look how many fathoms the ship hath gone in the time: that being known, . . . they multiply the number of fathoms, by the portion of time or part of an hour.

Though ignored by some shipmasters' proud of judging their ship's speed without artificial aid, the log and line was common from the 1590s, and was gradually improved by a mixture of theoretical and practical additions (Hutchins, 1995: 103–4). For example, the original log was replaced by a wooden quadrant with a lead-weighted outer rim to make it float upright, providing more water resistance and so a more accurate record of distance (Kemp, 1976). Provided powder in the sand-glass flowed smoothly, and readings were taken regularly and often (say two-hourly), experienced seamen could obtain a good estimate of sailing speed, though the conversion of distances sailed (in leagues) and measured (in fathoms) was complicated, unpopular, and handicapped by sailing lore's understatement of how many fathoms made a league (Waters, 1958: 138).

The solution lay in coordinating measurement units to simplify calculations, so the log-and-line was usable by anyone who could add and subtract. With knots spaced seven fathoms (42 feet) apart on the log-line, a ship's speed was one mile-an-hour for each knot of line that ran out in 28 seconds.[19] Even so, the log-and-line took some time for its use to become regular—it is only in journals from between 1600 and 1625 that entries recording the routine use of the log-and-line are first noted down (Waters, 1958: 426). This contributed to the routine of keeping 'log-books' recording information from traverse board and log-and-line, often two-hourly, though some masters continued to estimate their speed by eye and rote (Waters, 1958: 283).[20]

A closely related method of measuring distance run was the so-called 'Dutchman's log', which involved timing a float over the distance between two marks on the ship's side. This device was described by Gunter (1623):

The way that the ship maketh may be known . . . for some small proportion of time . . . by the distance of two known marks on the Ship's side. The time in which [the ship] maketh this way, may be measured by a Watch, or by a Glass, or by the Pulse, or by repeating a certain number of words. Then as long as the wind continueth in the same stay, it followeth by proportion,
    As the time is given is to an hour
    So the way made, to an hours way.   (Quoted by Waters, 1958: 427)

Gunter's description echoes Bourne's earlier account of the seaman's pulse, or rhythmic sequences of words, as timekeepers, in addition to sand-glasses.

Even with such refinements the log was an approximate instrument, ideally requiring three people to be effective (Hutchins, 1995). Further error came from friction of the rope, shrinkage of the rope, the effects of currents, and swell. But errors could be allowed for, for example, by routinely overestimating distance travelled, or by simple recalibration using nails driven into the ship's deck at the correct knot-intervals which showed whether the line had shrunk, thus providing 'a memory of distance' (Hutchins, 1995: 106). But simplicity also provided advantages: the log-line could be used at night, and distinctive gradations of the line facilitated the use of touch.

By the eighteenth century, more sophisticated devices for estimating speed, or distance run, had been introduced, such as rotators towed by the ship (Hewson, 1983). These were, like the log, means of making analogue-to-digital conversions. But such automatic logs were not widely adopted until the nineteenth century. Instead, 'streaming' the log became an automatic procedure itself on seventeenth- and eighteenth-century merchant and Royal Navy ships alike.[21]

A second set of devices show the growing importance of calculation in the navigator's repertoire. These were seagoing versions of long-known and sophisticated instruments for finding positions on land, such as the astrolabe, often an expensively worked instrument representing the 'physical residue of generations of astronomical practice' (Hutchins, 1995: 97). Astronomers used astrolabes in both observing and calculating the altitudes and motions of heavenly bodies. Simplified, they became the *mariners'* or *sea-astrolabe* (Stimson, 1988). Contemporaries of Samuel Purches in seventeenth-century London thought the German Martin Behaim the first to have adapted the astrolabe to navigation in 1484, but mariners' astrolabes were apparently used by Mallorcan pilots as early as 1295. However, recently excavated wrecks suggest astrolabes were not in general use until the sixteenth century (Bennett, 1998). Compared with their land equivalents, marine astrolabes were smaller and less lavish, making some sacrifice in detailed scale markings (hence accuracy) for greater stability in making observations (Waters, 1958: 55–6). Yet the astrolabe was usually unsatisfactory for maritime use, since users found it difficult to take observations even within four or five degrees.

Most navigators preferred to go ashore for reliable latitude observations, but plainly this was impossible for oceanic voyages. The instruments carried on Columbus's first transatlantic voyage in 1492 included a quadrant, marine compasses, charts, sand-glasses, log-and-line, lead-and-line, portable sundials, and an astrolabe of some kind. The quadrant and astrolabe aside, this had formed standard navigator's equipment since *c*.1200. Regarding the quadrant and astrolabe, while the principal of fixing latitude from the time of day (not necessarily from a clock) and a measure of elevation had long been appreciated, its uptake among mariners was uneven:

> any astronomer, Hellenistic or later, Islamic or European, from the time of Ptolemy of Alexandria (about AD 150) could have indicated to a mariner some form of altitude navigation, but the application of this knowledge required not only developments in cartography and its associated concepts, but also a communication between astronomer and seaman that was rare in the early history of navigation. (Maddison, 1992: 514)

The critical difficulty, therefore, was to find accurate devices that were usable at sea, and fitted to the calculating skills of navigators or other seamen. Long after 1700 the popular *cross-staff* 'remained in use amongst unlettered sailors', because it was simple to use—and cheap to make (Taylor and Richey, 1962: 40; Bennett, 1998). Derived from a thirteenth-century astronomers' instrument, the

Jacob's staff, the cross-staff is referred to from *c*.1514 (Kemp, 1976), and described in print in Martin Cortes's *Arte de Navegear* (1551). As used by sixteenth-century English seamen, the cross-staff was a wooden staff 70–90cm long, with a shorter, moveable cross-piece. Moving the cross-piece such that its extremities exactly covered the distance between sun (or moon) and the horizon, or between two stars, measured the relevant angular separation (Taylor and Richey, 1962: 37; Turner, 1996, 2004).

The cross-staff had serious drawbacks, some intrinsic—seasonal limitations, the dangers of looking at the sun, the need for 'blinking' between the two objects, and parallax errors—and others practical—its use on the deck of a rolling ship (Waters, 1958: 205). These defects led the navigator-explorer John Davis to develop the *back staff c*.1594. This allowed mariners to observe the sun's altitude from shadows cast, reducing problems of parallax, glare, and simultaneously sighting widely sun and horizon.[22] The observer pointed the staff at the horizon, moving a cross-piece to-and-fro until the tip of its shadow fell across the horizon-sighting slit just as this lined up with the horizon, thereby observing the horizon and the sun's altitude simultaneously (Waters, 1958: 205–6).

Compared with astronomical instruments, back staffs traded off length (and accuracy) with compactness (and ease of handling). Modifications designed to enlarge scales without increasing the instrument's size or weight, and (in Davis's own improved version) to increase observation angles up to the 90° needed for use in equatorial waters, were rapidly introduced, after which the back staff was widely used for over 200 years, in place of astrolabe and cross-staff, becoming known as a *Davis quadrant*, or just plain *quadrant*.[23]

Only in 1731 was a successful alternative invented, namely John *Hadley's quadrant* (despite its name, actually an octant), which used two mirrors.[24] Hadley applied the principle that a doubly reflected object subtends half the angle than when observed directly, using a 45° octant to measure angles up to 90°. With the instrument held vertically, the observer used the two mirrors to align the horizon seen directly with its image reflected via the mirrors. Having exactly aligned the directly viewed and reflected horizons, the observer adjusted the instrument's arm so that the reflected image of the relevant heavenly body also appeared on the horizon. The actual elevation of the body could then be read off the scale.[25]

In 1757, John Campbell demonstrated a more refined *sextant* whose arc of slightly over 60° read angular distances up to 120°, and whose greater precision was advantageous for the lunar distances method of longitude determination (see Chapter 9, below).[26] This moved into general circulation relatively slowly, because its complicated use demanded greater technical knowledge, and its cost (six guineas upwards) was thrice that of Hadley's quadrant—beyond all but officers.

The final group of important devices were *texts*, through which position could be marked. The simplest was the *rutter*, a small pocket-book containing information including magnetic compass courses between ports and capes; distances between them (in kennings); the direction of flow of tidal streams; 'full' and 'change' high-water times or 'establishment' of selected ports, headlands, or channels; the soundings (water-depth) and seabed material in the approaches to ports; and all manner of coastal features, land-marks, and sea-marks (Waters, 1967). The oldest extant English rutters are early fifteenth-century, but clearly based on earlier lore. Originally handwritten on paper or vellum by pilots, and often handed down from father to son, rutters circulated more generally after the invention of printing. The first printed European rutter was *La Routier de la Mer*, printed at Rouen between 1502 and 1510. The first printed rutter in English was the *Rutter of the Sea* in 1528, probably translated from a French rutter brought back from Bordeaux (Waters, 1958: 12–13). Manuscript and printed rutters for North European waters were used side by side into the seventeenth century, but the existence and contents of manuscript rutters for foreign seas is lost, except where extracts appeared in printed voyage-narratives or journals.

The second form of text was the *chart*. English shipmasters, like most seamen in northern and Baltic waters, were slow to use 'cards' or portolan charts. Even in 1578, the third edition of William Bourne's *A Regiment for the Sea* makes clear resistance to the use of even the quite accurate charts then existing for southern England and the Channel:

I have known within this 20 years that ancient masters of ships have derided and mocked them that have occupied their cards and plans . . . saying that they care not for their sheep's shins, for he could keep a better account upon a [traverse] board . . . If they should come out of the Ocean Sea to seek our Channel to come into the River Thames; I am of the opinion that a number of them do but grope as a blind man does. (Quoted by Waters, 1958: 15–16)

The decisive advocacy of shipmasters charts' value came with the 1588 publication of *The Mariners Mirrour*, Anthony Ashley's translation of the Dutch pilot Lucan Janszoon Wagenaer's *Spieghel der Zeervaerdt* of 1584–5, which described navigation from the Baltic to Cadiz. It is no exaggeration to say that 'Waggoners', as English seamen termed them, revolutionized the charts of north-west European navigators. They

incorporate[d] all the important information contained in a rutter, . . . amplified it with diagrams, elevations and charts, and it contained in the simplest possible form the basic elements of position-finding by celestial observations. . . . a wealth of systematised and corroborated information. [I]t brought for the first time within the ken, and from a crucial point of view, within the pocket, of the small sea-trader numerous charts designed to assist him in position-finding in what were, broadly speaking, waters of pilotage. . . . Wagenaer's . . . were the first [charts] for popular use at sea. . . . they standardised hydrological knowledge and included only observed facts essential for good pilotage, . . . they placed pilotage on a firmer scientific basis than ever before. The English edition furthered this process . . . [with] the express purpose of facilitating amendment... [T]he scientific spirit of inquiry, the urge for recording methodically the results of accurate observation and then putting them to practical use, [were] now thoroughly present in English seamen's minds. (Waters, 1958: 175)

Although charts in Waggoners were still very variable, drawn by compass bearing and distance, lacking latitude and longitude, and with coastline of variable accuracy, it was inescapable that 'Dutch seamen knew more about the English coasts than British seamen did themselves' (Kemp, 1976: 157). Not until 1681 did the Admiralty, in the person of Samuel Pepys, initiate a new survey of British coasts and harbours, by Captain Greenville Collins. Collins's work, with measuring chain, compass and lead-line stretched over several years, eventually producing his *Great Britain's Coasting Pilot*, published in 1693 with forty-eight harbour and coastal charts, supplemented with 'excellent views to illustrate leading and clearing lines by which vessels could safely enter harbour' (Kemp, 1976: 158).

Through twelve eighteenth-century editions, these were the main charts used by British seamen, but other more accurate charts based on extensive surveying and Mercator's projection were also becoming available, especially for the growing Royal Navy, though the Navy neither undertook the production directly nor provided the charts for their navigators.

Until the late eighteenth century, production of charts was largely done by private enterprise... The availability of charts aboard ships largely depended upon the initiative of individual officers. The master... according to Admiralty regulations was 'to provide himself with such charts, naval books and instruments as are necessary for astronomical observations and all other matters of navigation'. (Lavery, 1989: 182)

The problem of reliable charts was a perennial one, anywhere that Naval officers lacked local knowledge, whether on the other side of the world, where 'the difficulty of fixing a ship's position with accuracy was compounded by the lack of trustworthy charts on which to plot it' (Rodger, 1986: 48), or in English waters unfamiliar to particular seamen.

Plane charts marked with latitudes became more common from 1500, but caused severe navigational problems, especially on long-distance ocean voyages where spherical distortion and changes in magnetic variation were considerable.

[O]nly when navigators really began to navigate by celestial observation [were] conflicting hydrographical clams of bearing and distance, and latitude and longitude, thrashed out. Gradually charts for navigation by latitude-finding carried the day. This meant that the land had to be drawn in with bearings between places corrected for variation, so that they were true, and with distances which were, if possible, accurate. (Waters, 1958: 70)[27]

Only slowly was it generally appreciated that rhumb lines or loxodromes (bearings making the same angle with all meridians) were, latitudes excepted, lines curving around a sphere. Practical means of representing this situation were few, though *globes* were produced in increasing numbers from the late fifteenth century, after the discovery of the New World, (Crane, 2002). Sixteenth-century globes used printed paper 'gores' to cover spheres with mathematically accurate land outlines, with major advances in accuracy and coordination of geographical information made by Gerard Mercator's products from 1541.[28] The extent to which globes became practical navigational instruments is contested. Even the most accurate encountered practical scale-problems when very fine readings were attempted, but there is 'some evidence that a pair of globes, celestial and terrestrial, with perhaps an armillary sphere, were part of the normal feature of the "great cabin" of a ship' (Taylor and Richey, 1962: 93), and Waters went so far as to argue that 'with terrestrial and celestial globes, an almanac, a compass and astrolabe, the competent navigator could fix his position, tell the time and find the compass variation with considerable accuracy' (1958: 74).

Charts on Mercator's projection, published in 1569, widely superseded globes, because of their lower costs and greater durability, though at first few pilots 'could grasp that on such a chart a time bearing could not be laid out with a ruler'. Mercator recognized navigators' conservatism by devising 'a projection on which noted rhumb-lines were true, while the meridians remained parallel straight lines'. Even so, 'it was actually a century or more before the plane chart was restricted to the small areas for which it was, in fact, suitable' (Taylor and Richey, 1962: 29–30).[29] As Narbrough explained in the seventeenth century:

> Most of our navigators in this age sail by the plane chart, and keep their accounts of the ship's way accordingly, although they sail near the Poles; which is the greatest error that can be committed, for they can not tell how to find their way home again, by reason of their mistake; as I have some in the ship with me now that are in the same error, for want of understanding the time differences of the meridian, according to their miles of longitude, in the several latitudes. I could wish all seamen would give over sailing by the plane chart and sail by Mercator's chart, which is according to the truth of navigation; but it is a hard matter to convince any of the old navigators, from their method of sailing by the plane chart; shew most of them the globe, yet they will talk in their wonted road. (Quoted in Rodger, 2004: 125)

All changed with the development of a truly reliable chronometer in 1762,[30] though their high price limited their use. In the Royal Navy, only seven per cent of British warships had a chronometer in 1802 (Rodger, 2004) and they were only made general issue in 1825, so that sufficiently efficient cheap timekeepers were not affordable to navigators until then (Kemp, 1976: 167). And, obviously, if a chronometer broke down,[31] 'the navigator had to fall back on sailing down his latitude and dead reckoning his longitude, just as in the old days' (Lundy, 2002: 165). Though all non-naval captains could do sun sights with a sextant, many could not take stellar sights, so the fall-back could be risky.

Effective use of more sophisticated instruments, measurements, and charts was intimately connected with greater use of *tables* that summarized various numerical information. Tables proliferated as mathematics increasingly became the language of navigational problem-solving, and seamen's manuals had more and more tables attached to them, as essential adjuncts to the navigator's craft. Many involved detailed information on time, and were potent sources of time sense. There were navigational tables, including tide-tables, tables giving the establishment of various ports, and those setting out the leagues required to move a degree of longitude of latitude on different rhumb-lines. There were ready-reckoners for arithmetic, and trignometrical tables to assist geometrical calculations about courses and bearings. There were astronomical tables giving forecast positions and timings for the moon and/or certain stars; and supplementary tables for correcting observed celestial positions for atmospheric refraction, possible parallax effects, the earth's curvature not being that of an exact sphere, and the like. Tables of logarithmic or trigonometric functions meant navigators could solve astronomical problems quickly and efficiently, bypassing laborious long division and multiplication. All of these underwent progressive refinements, depending on state-sponsored initiatives, like the astronomical tables in the annual *Nautical Almanac* produced at the Greenwich Observatory (founded 1675) from 1763, and the circulation of information in numerous commercial almanacs.[32]

Charts, globes, and tables were also served by other new devices, alongside such traditionally indispensable tools as the *pairs of compasses* central to time-honoured practices of 'pricking off' charts; this procedure gradually gave way to '*laying off*' a position with pencil and ruler, with older 'rhumb and compass' course-finding methods superseded by use of ruler and *protractor*—an English invention of the 1580s. *Plotting* charts allowed the transfer of position on to a smaller sea chart by bearing and distance from a given point. In long-run perspective, *parallel rulers* in the 1580s, and *sectors* in the 1590s were precursors of seventeenth-century slide rules. The likes of Edmund Gunter, subsequently Professor of Astronomy at Gresham College in London, were important in adapting new methods and tools for use at sea, such that soon after 1600 the *sector* and *traverse tables* were the basic means of plotting position at sea. Around 1620 came *Gunter's Scale*, a set of logarithmic scales of natural numbers and trigonometric functions set on a two-foot long staff, exploiting Napier's recently invented logarithms: 'except that distances were measured by a pair of compasses and not by another rule, it was a slide-rule' (Waters, 1958: 419). By c.1650 Gunter's scale had evolved to become a true *slide-rule*, the nautical version becoming known as the *Sliding Gunter* (Waters, 1958: 73; Taylor and Richey, 1962: 75). Once again, there was some resistance to each of these instruments. Older practices, such as using Gunter's scale and dividers, rather than the slide rule, or the continued use of the Sliding Gunter by those lacking the arithmetic skills to use log tables carried on (Taylor and Richey, 1962: 74). But, overall, the spread of techniques linked to Gunter's 1620 *Tables* shaped 'logarithmical navigation' as a significantly new representation of what navigation was.

Another set of texts perhaps most acutely shows off the growing awareness of time measurement at sea: namely the *logbook* or the *journal*. Through the period under study, under the pressure of more frequent and longer voyages, what had been diverse and haphazard observations became much more standardized in both timing and content. Logbooks and journals were both indicative of the standardization of a whole set of practices, and an instrument in that standardization. The likely exemplars were John Davis's ship's journals from his 1587 voyage, which circulated in manuscript—eventually printed in Hakluyt's 1600 *Principal Voyages*, and his 1593 voyage, printed in *Seamans Secrets* (1593). Davis recorded the course(s) followed, and the reckoning of distance run, from noon to noon, calculated from the information recorded on the helmsman's traverse board, estimates of leeway made and currents encountered, and daily (where possible) observations of latitude (and, later, longitude). Along with other notes on conditions, these became the formal record of the voyage: the logbook being retained as the rough working book of progress, and the journal as the corrected 'minutes' of the voyage (Waters, 1958: 284). Given the rapid standardization in logbook-keeping (at least up until the wider availability of marine chronometers in the nineteenth century), it is perhaps surprising that pre-printed blank logs seem only to have been commercially available from until 1702, produced by the East India Company, and the Royal Navy persisted with hand-ruled logbooks until much later (Waters, 1958; Wheeler, 1999: 41).

Logbooks' format shows strong awareness of clock time long before printed blanks were available, arising from both normal shipboard routine, and the shipboard use of clocks and watches. A good example is provided by the sinking of the *Association* and other ships in Sir Cloudesley Shovell's ill-fated squadron off the Isles of Scilly in 1707, due to navigational error—an episode often cited as the stimulus to an organized drive for navigational precision. The logbook entries of the surviving ships all use clock time to describe daily events, including those on the day of the sinking itself.[33]

Another set of texts were the numerous navigational books published alongside Davis's *The Seaman's Secret* (1599) and *The World's Hydrographical Description* (1595), which the navigator might well possess in some number. Authors clearly perceived their readerships, actual and potential, as substantial. Thus William Bourne's oft-reprinted *A Regiment for the Sea* first published in 1574, and his astronomical and navigational almanacs, were explicitly popularizing works, distanced from the 'learned sort' of men in universities or observatories. Bourne explicitly targetted non-elite readers whom he called 'meaner men' or 'the simplest sort of readers' (Taylor, 1963). Addressing a readership lacking technical vocabulary and all but basic arithmetic, Bourne used vernacular language, simple diagrams, ready-reckoner tables, and worked examples.

Bourne's explicit concern with language and arithmetic skills as barriers to understanding is in striking contrast to his presumption that readers require no explanation for times in minutes in his tide-tables, eclipse predictions, and star risings, although he does explain seconds, further subdivided for calculation into 'thirds' ($1/60$ of a second) and 'fourths'.[34] Bourne's audience was strongly focused on certain communities of practice in and around navigation and seafaring, to whose temporalities and objectives the book spoke. In this sense, his audience was much narrower than 'the public' or 'the reading public'. Nonetheless, twelve reprints, translations, and overall sales of Bourne's *Regiment* all suggest that the extensive direct and indirect circulation of his material within the humbler levels of the seafaring community succeeded.

Bourne's book may well have been the 'regimento' bequeathed by one mariner at Leigh (Essex) to another in 1570 (Table 8.1).[35] Taking the chart, astrolabe and regimento together, Leonard Tele as an individual mariner possessed devices with which to tell the time, establish his latitude, and relate it to shoreline(s). At that time, he was probably unusual in using sun and stars to fix his position, but he was not alone:

John Borough, master of the *Michael* of Barnstaple in 1533 and a follower of Lord Lisle, owned some simple instruments including a cross-staff and pilot books in Spanish and Portuguese, which indicates where he must have learnt skills which must have been almost if not completely unique in England. (Rodger, 1997: 305)

Tele's executors naming of his instruments is highly significant, especially given the early date. Shipping from Leigh and other small North Sea ports is widely held to have been coastal trade with London, so it is particularly striking that instruments and texts associated with mathematical navigation were known and appreciated there.

Table 8.1. The will of Leonard Tele of Leigh, mariner, 20 April 1570

To my mother all my money; after her decease, to my grandam if she overlive my mother, i.e. £13 10s., whereof I lent to John Benn £3 which is to be paid within 2 years.
To my mother all my clothes.
To the poor of Leigh 6s. 4d. which John Myller oweth me.
To John Goodlad my card[1] and my escralabye[2]
To Edward Salmon my regemento[3]
To William Goodlad my writing books.
To my mother a crusado which Joan Goodlad hath of mine
Out of the £13 10s. my sister Joan's son shall have £5 when the money comes to my mother's hands, 20s. to be given to me grandam, and the rest to the discretion of my mother to use at her pleasure. I make her my executrix.
To Robert Camper my 'balestela'[4]
I make my mother executrix.

Source: Emmison (1994: 203); spelling modernized by Emmison.
Notes:
[1] Sea-chart
[2] Astrolabe
[3] *Regimento*: 'tables of the midday position of the sun (solar declination)' (Maddison, 1992: 512). Comparing the angle of ascension of the sun at local noon, measured with an astrolabe or quadrant, with those given in the table for the appropriate date, enabled the navigator to establish his latitude.
[4] *Balestela*: 'arbelast', a type of crossbow (Emmison).

Finally, there were *other navigational devices* which might easily be overlooked because they were not on board ship, namely *seamarks* (painted rocks, buoys, and the like), and *landmarks*, including church towers, a few beacons, and a very few lighthouses—like that built on St Catherine's Point on the Isle of Wight 1314 (Rodger, 1997: 161–2; Hutchinson, 1994). Sailors continued to rely on their 'very extensive collective knowledge of the appearance of islands, headlands and other aids to navigation' (Earle, 1998: 73–4), using visible landmarks like steeples and trees; beacons of various kinds (especially around ports); and floating wooden buoys, either barrel-shaped tuns or cone-shaped 'cam-buoys' (Hutchinson, 1994).

From the sixteenth century seamarks and landmarks were taken more seriously by the Crown, the Trinity Houses and other agencies that maintained and augmented them. Most significantly, the 1565 'Act concerning Sea-marks and Mariners' observed that

By the destroying and taking away of certain steeples, woods and other marks standing upon the main shores, adjoining the sea coasts of this realm of England and Wales, being as beacons and marks of ancient time accustomed for seafaring men, to save and to keep them and the ships in their charge from sundry dangers there to incident: Divers ships with their goods and merchandise, in sailing from foreign ports towards this realm of England and Wales, and especially to the port and river of Thames, have by the lack of such sea-marks of late years miscarried, perished, and lost in the sea, to the great detriment and hurt of the Common Weal, and the perishing of no small numbers of people. (Waters, 1958: 10)

The Act extended the authority of Trinity House to set up and maintain beacons, marks and other 'signs for the sea' as seemed necessary—both in the approaches to ports, and but also on the sea shores and heights. It prohibited, under threat of a £100 fine, the destruction of steeples or conspicuous trees used as recognized marks.

Over time, these marks proliferated and new marks became available, most notably lighthouses (although even by 1815 there were only thirty-six lighthouses in English waters) and light ships (which were rare; there were only three in 1793).[36] Many devices were simpler, like whitewashing Sandown Castle as a leading mark out of Sandown Bay, and laying a buoy on Bembridge Ledge. Significantly, seamarks themselves began to be formally represented as symbols on charts, starting with Wagenaer's *Mariners Mirrour* (1588), which showed buoys, beacons, landmarks, and hidden rocks using standard symbols. This new maritime language of indicating marks on representations of the seas rapidly became general.

Overall, then, what we see is a complex and interconnected set of temporal practices being continually carried out at sea. These practices involved not just one but whole families of devices interacting in such a way as to produce more effective movements, and greater and greater way-finding possibilities. Late eighteenth-century navigation involved a great deal of effort, both official and unofficial. It still mixed guesswork, art, and science, but the balance was shifting from the former to latter at an accelerating rate.[37] Navigation's landscape was becoming populated with new and improved instruments: astronomical and mathematical tables, charts, manuals and texts, records of earlier voyages, and larger numbers of people more or less trained in their use. All these elements combined to create navigational networks of people and objects that in manifold ways used timings in observations and calculations, and to which their interrelatedness was central. New instruments and techniques were being more widely used, but tended to be applied within existing specializations, especially the specialization of merchants, ships, masters, and mates, in particular routes, which built up relevant experience.

Thus, even at the end of our period, the method of 'lunar distances' was preferred, which called for advanced mathematics, but no instrument more expensive than the essential quadrant (costing about three guineas in the 1760s) necessary for all observations, or, better, the new sextant (twelve guineas) which could measure angles up to 120°. Lunars served either by themselves, or to rate the chronometer for those who had one. Even so, lunars were not a panacea. They could only be observed about twenty days in each lunar month, and of course all calculations are subject to error. Crossing the Atlantic in 1776, Home took a lunar which put them 300 miles west of their true longitude, they nearly ran on Nantucket Shoal when they thought they were off Long Island. Merchant ships were still crossing the Atlantic without chronometers, charts or sextants far into the nineteenth century, and all but the most daring navigators still made their landfalls by latitude sailing. (Rodger, 2004: 383)

8.3 PRINCIPLES OF USE

The preceding section shows different instruments, and uses of them, becoming widespread, or rarer, at different times. Until the sixteenth century most ships carried just a few devices: lead-and-line, compass, sand-glass(es), and a rutter. Thereafter, this basic 'ecology of devices', to use Hutchins's (1995) term, became progressively richer. Tables 8.2–8.3 show the greater sophistication of navigational and pilotage instruments used on a trans-Atlantic colonizing voyage in 1631 compared with those used in Frobisher's 1553 search for the Northwest passage, though some new devices were the technological cutting edge in 1631, and not yet in use by navigators at large. Mid-seventeenth-century London instrument-makers' lists show many new instruments coming into general circulation (Tables 8.4–8.5). By the late eighteenth century, a market driven by increasing numbers of naval officers was serviced by instrument-makers stocking very large numbers of different instruments—a burgeoning ecology of devices, reflecting in particular the rise of arithmetical navigation (Clifton, 1995). At the top of the market were London instrument-makers like Benjamin Cole and George Adams (Tables 8.6–8.7), whose prestigious products were priced far beyond the reach of ordinary seamen, who would have gone to a ship's chandler, one of the many smaller merchants, or one of the navigational warehouses.[38]

Table 8.2. Inventory of first Frobisher voyage in search of the North West Passage, 1553

Compasses	20
Hourglasses	18
Cross-staff	
Astrolabe	
Meridian compass	
Equinoctial/Universal Dial	
Astronomical Ring-dial	
Armillary Sphere	
Globe (metal)	
Standing level}	for charting
Plane table}	the coasts

Source: Waters (1958: 145).

Table 8.3. An early ship's navigator's instruments, *c.*1631

Instruments of Pilotage
Compass (mounted in gimbals in a square compass box, and fitted in a portable binnacle with a lantern inside)
Lodestone
½ Hourglass
Traverse Board
Telescope
Universal Ring-dial
Nocturnal
Two Pairs of Compasses
Pocket Dials
Log and Line (log chip, log line, stray line, log-line reel)

Instruments of Navigation
Mariner's Astrolabe (obsolescent, being superseded by cross staff)
Davis Quadrant
Cross-Staff
Brass Quadrant
Battery of four ½ hourglasses
Gunter's Scale
Charts (normally kept rolled) and Waggoners
Dial Watch
Mercator's Globes (Celestial and Terrestrial) – obsolescent but still often carried
(Two Pairs of Circular Compasses for use with Globe)

Source: Waters (1958).

Table 8.4. Instruments sold by John Seller at his house in Wapping or his city shop by the Royal Exchange, second half of seventeenth century

Azimuth and meridian compasses
Hanging and pocket compasses
Brass compasses [i.e. dividers] for Platts and Charts
Steel-pointed compasses [i.e. dividers] for scales
Cross Staves
Gunter's Scales
Sinical Quadrants
Astrolabes
Sliding Gunters
Gunter's Quadrants
All sorts of Running-glasses
One, two, three and four Hour Glasses
One Minute, Half minute and Quarter minute Glasses
Logs, Log-boards, Log Lines, Sounding Leads
Nocturnals
Plain Scales
Almicanter Staves
Height Rules
Davis's Quadrants
Tide Tables

Source: Taylor and Richey (1962: 109).

Table 8.5. Instruments sold by Thomas Heath and Tycho Wing from their shop near Exeter Exchange in the Strand, 1750 and 1765

Gunter's Scales
Sliding Gunters
Davis's Quadrant
Hadley's Reflecting Quadrant
Mr Smith's Reflecting Quadrant (by Caleb Smith)
Mr Smith, Captain (Christopher) Middleton and Mr (Joseph) Harrison improved Azimuth Compass (described to the Royal Society in 1738)
Common Azimuth Compass
Mariner's Compass either for the Binnacle or Cabin
Nocturnals
Plain Scales
General Quadrants
Telescopes of new contrivance with Six Glasses
Telescopes for Day and Night
Prospect and Spy-Glasses

Source: Taylor and Richey (1962: 109).

Table 8.6. Instruments sold by Benjamin Cole from his shop in Fleet Street, 1768

Pocket cases of drawing instruments in silver	3 gns. to 20 gns.
The same in brass	5s. to 5 gns.
Plain and plotting scales in brass, ivory and wood	8d. to 18s.
Gunter's 2 foot and 1 foot scales in brass and wood	2s. to 2 gns.
Protractor	1s. 6d. to £1. 16s.
Parallel Rules 6 inches to 36 inches	2s. 6d. to 18s.
Sectors in brass, ivory or wood	2s. 6d. to 4 gns.
Theodolites	3 gns. to 6 gns.
Theodolites with vertical arcs, spirit levels, telescopes, etc.	10 gns. to 20 gns.
Plain Tables	3 gns. to 5 gns.
Circumferentors, the principal instruments for surveying in the West Indies	£1.16s. to 3½ gns.
Gunter's four-pole chains	6s. to 12s.
Spirit levels	5s. to 12 gns.
Measuring wheels	42 gns. to 6 gns.
Hadley's Quadrant with Diagonal Divisions	£1. 14s.
Hadley's Quadrant with a nonius	2 gns. to 31 gns.
Hadley's Quadrant all in brass	3 gns. to 6 gns.
Davis's Quadrant	12s. to 1 gn.
Cole's Quadrant	18s. to 25s.
Sutton's Quadrant	6s. 6d.
Gunter's Quadrant	3s. 6d. to 1 gn.
Azimuth Compasses	5 gns. to 10 gns.
Amplitude Compasses	£1. 7s. to 5 gns.
Mariner's Compasses either for the Cabin or for the Binnacle	7s. 6d. to 3½ gns.
Pocket Compasses	1s. to 1½ gns.
Armillary Spheres	£12 to £50
17 inch Globes	6 gns.
15 inch Globes	5 gns.
12 inch Globes	3 gns.
9 inch Globes	2 gns.
6 inch Globes	£1.16s.
3 inch Globes in case	8s. to 10s.
Speaking Trumpets	10s. to 1½ gns.
Reflecting Telescopes	£1. 16s. to £50
Reflecting Telescopes with 4 or 6 glasses	7s. 6d. to 6 gns.
Achromatic Opera or Prospect Glasses	1 gn. to £1. 16s.
Achromatic Telescope of any length	1 gn. each foot

Source: Taylor and Richie (1962).

Turning to how navigators actually used the expanding ecology of devices, nine inter-related principles of use seem particularly pertinent. Seven concern the devices themselves, and two their patterns of use. First, most devices had to be *simplified* to be of service. Though many navigational instruments were used mainly by masters and other navigational professionals, they still had to be sufficiently robust and practicable to work on most occasions for a variety of users, remembering that there can be sophisticated use of rudimentary devices,

Table 8.7. List of instruments issued by the younger George Adams from his shop in Fleet Street, 1789

Cases of instruments and telescopes of different kinds, sizes and prices	
Night telescopes	1½ gns. to 2 gns.
An opera glass for the same purpose	1½ gns.
A telescope with an eye-glass micrometer for determining the distance of a ship at sea	[no price given]
Hadley's Quadrant in mahogany frame	2 gns.
Hadley's Quadrant in black ebony	3 gns.
Hadley's Quadrant of best construction	4½ gns.
Hadley's Sextant in wood	6½ gns.
Hadley's Sextant in brass on the most approved plan	11 gns. to 15 gns.
Knight's steering compass with improvements	22 gns.
Knight's steering compass on friction wheels	10 gns.
Circular instruments, to answer the purpose of a sextant	14 gns.
Dipping Needles	12 gns. to 30 gns.
An instrument to use instead of the minute glass, but much more accurate	[no price given]

Source: Taylor and Richie (1962).

and rudimentary use of sophisticated devices. Second, these devices required *redundancy*. If one device went wrong, was broken, or lost overboard, other devices or older procedures had to be available, so navigators were usually trained in several procedures. It was also better to have readings from several devices as cross-checks, where possible. Third, devices required routines to become effective—by themselves instruments were powerless:

The mere fact of sighting a heavenly body through protocols of an alidade had nothing per se to do with navigation. Their sighting, or the reading that corresponded to it, had to undergo a number of complex transformations before it could be converted into a latitude. The construction of a network of artefacts and skills, for converting the stars from irrelevant points of light in the night sky into formidable allies in the struggle to master the Atlantic is a good example of heterogeneous engineering. (Law, 1986: 124)

Fourth, the making visible of accuracy through instrument-use changes the whole frame of measurement itself. That how the world is measured out changes as the world is measured in new ways is nowhere clearer than in measures of distance and speed. Take distance first. Fifteenth-century English rutters show the seaman did not work in distances: he sailed from landmark to landmark, with little idea of the intervening distances. When out of sight of land he guessed his relative progress by soundings, knowledge of the sea-bed, and use of a running glass. When, after 1500, distances began to be recorded, it was as variable units like 'days of sail' or 'kennings' (the rough distance at which a coastline was discernable). Speed, meanwhile, was usually estimated by eye, usually from foam slipping past the ship.

However as new instruments appeared, so did new units. Increasing use of traverse boards in the sixteenth century promoted a switch to a more exact distance-unit, the *league*. After the mid-sixteenth century, references to leagues replace kennings as the favoured unit of distance. 'A league in northern waters was 3 miles, and this . . . could be gauged, for ships commonly sailed 3 to 4 miles in the hour' (Waters, 1958: 37). Similarly for speed. The addition of regularly spaced knots made the log line a direct-reading instrument. Aided by the practical imperative of regularly checking the distance run during long-distance voyages, and by the dissemination of books such as Gunter's *De Sectore et Ratio* (1623) and Richard Norwood's *The Seaman's Practice* (1637), the knot quite rapidly became the measure of speed. Where time was measured with a 'half-minute' sand-glass,

knots tied in [the line] every 7 fathoms did away with the need for arithmetic. [Counting] the knots gave the way of the ship in miles an hour, and it was accepted that 60 of such miles made a degree of the meridian. Thus a mile was a minute of the arc, and sailors have clung to this useful relationship ever since. Only very slowly and reluctantly did they re-knot the logline so that the nautical mile matched as many fathoms and feet as did the minute of arc as accurately measured by the astronomers. This proved nearer 50 feet than 42 feet (7 fathoms) in English units. (Taylor and Richey, 1962: 35–6)

Fifth, the role of particular devices changed as temporal practices changed. Some devices fell by the wayside; others were revived and inserted into new positions in networks of practices. Thus Hutchins notes how astrolabes were successively replaced as observational instruments by quadrants, cross-staffs, and sextants, but survive as the modern star finder (1995: 113). Sixth, at any point the complex 'ecology' of temporal devices (and devices with temporal connotations) needs to be considered as an overall assemblage:

tools of navigation share . . . a rich network of mutual computational and representational dependencies. Each plays a role in the computational community of the others, providing the raw materials of computation or carrying the products of it. In the ecology of tools, based on the flow of computational products, each tool creates the environment for the other. (Hutchins, 1995: 113–14)

Seventh, it follows that devices are transformational. They transform the task a person has to carry out 'by representing it in a domain where the answer or the path of the solution is apparent' (Hutchins, 1995: 155). Most of the navigational devices we have mentioned were bent to minimizing arithmetical reasoning by striking up concordances with other domains of representation.

Two further principles underline that the processes of device-selection were thoroughly cultural: 'a way of thinking comes with these techniques and tools' (Hutchins, 1995: 115). So, eighth, the selection of devices involves an increasing cultural preoccupation with, even passion for, *measuring* as an activity in its own right. Increasingly, there is a 'devotion to breaking down things and energies and practices and perceptions into uniform parts and counting them' (Crosby, 1997: 11). Ninthly, new cultures of measurement lead to an increasing tendency to take 'the representation more seriously than the theory represented'

(Hutchins, 1995: 115). The outcome of measurement becomes the world. Thus, increasingly, the stuff of reality is visualized as aggregates of units, through new means of visualization (Porathe and Svensson, 2003).

These nine principles can be summarized as follows;

a good deal of the expertise in the system is in the artefacts (both the external implements and the internal strategies)—not in the sense that the artefacts are themselves intelligent or expert agents, or because the act of getting into coordination with artefacts constitutes an expert performance by the person; rather, the system of person-in-interaction-with-technology exhibits expertise. (Hutchins, 1995: 155)

But three important caveats need emphasis here: expertise is differentially distributed, even in expert communities, and formalisms are often used quite informally. In practice, navigational instruments and skills were often hesitantly taken up by navigators, even though seafaring remained very dangerous, and mistakes were often fatal. New methods were sometimes resisted over long periods, as when simplified mathematical forms of navigation were regarded as suspect by older seamen. Norwood's *The Seaman's Practice* (1637) deplored opposition to the log-and-line. Norwood believed that estimating a ship's speed by experience was 'mere conjecture', but older seamen drew status and pride from their experience and feared of being 'accounted young [i.e. inexperienced] seamen' if observed using log-and-line (Waters, 1958: 433). Similar frictions arose later between experienced seamen using basic instruments on familiar routes, and the new navigators going to sea with numerous instruments, often custom-made, chests of books, and mathematical understanding.

Second, not all instruments were the same. Some were well made, others not. For example, in 1605 Richard Potter recommended 'Master Mullineux of Lambeth' as a supplier of running glasses because ordinary makers and vendors cared 'but little what error more or less was delivered in those glasses in 24 hours, nay in half an hour' (Waters, 1958: 305). Later, there are clear differences in quality between the expensive instruments supplied by professional instrument-makers and those supplied by chandlers, yet alone those made from templates by the ship's smith.

And, third, the skills of navigators and seamen were not all equal. Some navigators were far better than others; some seamen more 'seamanlike' than others. As the second mate of the *Temperance*, on a voyage to Archangel in 1701, lambasted poor Edward Knapman:

He was not an *able* seaman, or capable of taking upon him the office of foremast men or doing his duty therein . . . He could not take his trick at the helm for he was often tryed, and knew not his compass, for being often asked how they winded he would answeare to a quite contrary point of the compass as the wind then blew . . . and in general knew not the ropes or where to find the ropes he was bid to handle. (Earle, 1998: 44)

Navigation history thus echoes a constant theme in the sociology of science literature that good results can only be obtained from many scientific instruments by particularly adept operators. Good results did not always require the most complex devices, but a good 'fit' between device and the user's skills. Such a fit was not, by any means, always available.

8.4 COMMUNITIES OF PRACTICE IN PRACTICE

Seafaring communities clearly had cause to be obsessed by time, and logbooks and journals abundantly attest the routine familiarity of clock times on ships. The obsession was shaped by several factors: the routines and drills followed, especially on board; the time-keeping and time-finding devices that were either peculiar to seafaring communities, or distinct nautical variants, and the fabric of the day. Ships in sight of land generally kept civil time, but when land was lost from view ships kept nautical time, reckoned from noon to noon (since the ship's position was plotted from each noon-sight), rather than the midnight boundaries of the civil day. We show below how shipboard life consisted of a continuous set of embodied routines and drills, to sail the ship in a coordinated way and to respond rapidly to problems and emergencies. Devices were but a part of this temporal tapestry and yet, at the same time, they were also a means of representing and doing it.

However, it is vital to note that the shipboard temporal tapestry was based on clock time, rather than on clocks in the strictest sense. While many eighteenth-century sailors owned watches—an easily portable store of wealth—use of clocks and watches on ships was limited by their susceptibility to changes in temperature, humidity, and position.[39] Until accurate ways of systematically determining time at sea could cover the oceans, longitude remained an enduring technological challenge and obsession (see Chapter 9). The lack of precision chronometers until the late eighteenth century held up the development of devices and documents, including the sextant and accurate navigational tables.

Table 8.8. Timekeeping equipment on three naval ships c.1660

	Royal James	Royal Katherine	Harwich
Watch (4 hours)	1	1	1
Half-watch (2 hours)	0	1	1
Half-hours	18	16	10
Half-minute (for use with log)	8	6	4
Compasses	18	12	10

Source: Waters (1958: 580), citing *A Survey Book containing all the Rigging...Furniture and Stores belonging to his Majesty's ships the Royal James, the Royal Katherine and Harwich. No. 2265 Pepysian Library.*

Clock time nevertheless pervaded shipboard routines by proxy, through devices other than clocks. The best instance of this surrogacy is the *hourglass*.⁴⁰ Ships had glasses ranging from small 14- or 28-second glasses (so-called quarter minute or half minute glasses) for timing the log, through the half-hour 'watch glasses' even up to massive four-hour glasses for timing the whole watch, which had been calibrated ashore by clocks. Glasses were so taken for granted, so 'in the background', that this crucial nautical timekeeping technology is documented only in chance mentions—in Hakluyt recalling 'we lay six glasses (i.e. six half-hours) a hull [hove-to] tarrying for the pinnace' (1599: II, 126); in a 1660 Royal Naval inventory of glasses (Table 8.8); in a sailor being docked pay in 1700 'for an hour glass he broke' (Earle, 1998: 36), and in the evidence of bygone naval terms. Thus Smythe (1867/2005: 341) cites terms such as: '*to flog or sweat the half-hour glass*; to turn the sand-glass before the sand has quite run out, and thus gaining a few minutes in each half hour, make the watch too short' or to denote the length of a naval action' as in 'they fought yard-arm and yard-arm three glasses ie. three half-hours'. In the last instance, we see how the half-hour glass itself signalled a unit of time.

In looking at how timekeeping manifested itself in practice we consider in turn the temporal worlds of two parts of the maritime community: the merchant navy and the growing Royal Navy.⁴¹

8.4.1 The Merchant Navy

The medieval merchant marine was a key component of both the English trading economy, and of total capital equipment—indeed, the 'seagoing ship was probably the most expensive single piece of capital equipment' (Rodger, 1997: 119). Over time, ships grew steadily larger, more diverse, and became more technically sophisticated, with design innovations in their masts, rigs, rudders, superstructures, even pumps (Hutchinson, 1994; McGrail, 2003). Estimates of total numbers are necessarily approximate, but probably somewhere between 1,000 and 2,000 vessels just before 1400, of which some 80 to 90 per cent were under 100 tons.

Numbers (although not sizes) may have declined after 1400, but by 1582 the most extensive Elizabethan nautical shipping survey listed 1,630 vessels of 20 tons and upwards (Friel, 1995). Thereafter, the numbers and sizes of ships increased until by the early 1770s there were about 6,000 ships with a combined tonnage of just over 600,000 tons (Earle, 1998). The average ship by this time was still therefore a very modest 100 tons, but the range ran from 20-ton coasters to some East Indiamen weighing nearly 1,000 tons. Crew sizes, which could be just three (master, man, and boy), were correspondingly diverse. Ordinary seagoing vessels—the larger coasters such as colliers—had crews ranging from eight to forty men, specialized roles (carpenters, boatswains, gunners, cooks, cooper, sailmaker) and correspondingly complex command structures of mates and foremen (Rediker, 1987: 83–4; Earle 1998: 42). The largest East Indiamen could have crews of over 100.

In 1660 there were probably under 30,000 English sailors, of whom nearly 20,000 worked in fishing and the coastal trades (Harper, 1939; Earle, 1998). By 1770 there were over 50,000, 20,000 in fishing and coastal trades and 30,000 on ships sailing overseas. Of the latter about 5,000 were in crews working on slave ships, and 6000 on East Indiamen. Employment of foreigners on English ships was restricted by law to a maximum of a quarter of the crew (Earle, 1998), and most British sailors came from seafaring communities:

> unless they were Londoners [sailors] were overwhelmingly drawn from the maritime counties. Well over 90 per cent of provincial sailors were born in maritime counties and the great majority of these within sight of the sea and ships, in port towns, coastal villages or on navigational rivers. ... some areas were particularly important as 'nurseries' of mariners... Dorset, Devon and Cornwall... Norfolk and Kent. (Earle, 1998: 19)

Both on shore and at sea sailors lived largely in a male world of their own—though there were some female sailors (Stark, 1998)—often cut off from the wider community, and increasingly likely, through our period, to know no other life but the sea; 'once a sailor, the chances were... always a sailor' (Kemp, 1972: 92). Seamen were 'genuinely a peculiar class isolated by their profession from the bulk of their fellow countrymen, living in particular quarters of seaside towns, speaking a language of their own, "a people of a distinct nature in themselves, for the most part divest of common knowledge of things ashore"' (Rodger, 1988: 118).

On shore (and in a seasonal and casual industry this might be often), sailors were distinctive not only in their spatial concentration in coastal ports[42] (Earle 1998; Press, 1986; Gill, 1961), but in their bodily comportment and work rhythms. Contemporary comment on sailors' appearance emphasizes an amalgam of distinctive clothing and materials (like canvas, tar, and cotton), weather-beaten faces, and distinctive, rolling gait (Earle, 1998: 34). Their work day comprised either inactivity, or gruelling work in maintaining and loading ships ready for sea: 'ships fitting out and loading normally gave their crews Sundays ashore but otherwise the work was continuous, a typical twelve hour stint with breaks for meals with a crescendo in effort as sailing day approached' (Earle, 1998: 70). Ships at sea were their own wooden worlds, with the crew pursuing the three crucial tasks of steering the ship, managing the rigging, and working the sails.

Voyages varied in length from a day's coasting to trips lasting several weeks [Earle, 1998: 8]. Ocean trips took much longer: several months for round trips across the Atlantic; a year for the slave trade's triangular journeys. 'A typical voyage from London to the east and back to London lasted between eighteen months and two years but trading with Asia could extend such times to three years or more' (Earle, 1998: 8–9). Such extended voyages were broken up into 'passages' between intervening ports. Even so, trans-Atlantic passage times were typically between seven and ten weeks (and could be much longer), while passages to the East Indies could be as long as sixteen or even twenty weeks (Hayter, 2002).

However long the voyage, a completely enclosed temporal bubble came into existence for its duration. At sea, time was marked in two ways. First, there was the watch system. With the crew divided into two groups,[43] working alternate watches, the ship's day comprised seven watches—five four-hour watches, starting at 8 p.m., and two dog-watches of two hours, starting at 4 p.m. and 6 p.m.—so that each group would not work the same hours each day (Rediker, 1987). Each watch was marked by

> an hour glass or, as was customary by the seventeenth century, a half-hour glass, hung in the binnacle under the eye of the helmsman, who turned the glass each time the sand ran out and marked the passage of time by ringing the 'Watche belle' at each half-hour, sounding one stroke for every half hour that had passed of the four hour watch. (Rogers, 1958: 36)

Immediately a ship weighed anchor, its work organization was set up in half-hour parcels of time rung out by the ship's bell, from one until eight bells for each watch (except for the dog-watches which only reached four bells). Normally, a two-watch system was employed, dividing the company into starboard and larboard (port) watches serving four hours on and four hours off, although a few captains used a three or four watch system (Stark, 1998).

> Though broken up by breakfast, the midday grog issue, and dinner, this was an arduous timetable, made the more so by all... crucial chores that filled the day apart from the actual business of sailing (and especially all kinds of unremitting routine maintenance and repair)...
> And, of course, any emergency would mean all hands were called on deck whatever the time. Even so, relative to many other occupations, 'the seaman's workday remained "porous", marked by periods of intensity and inactivity that could not easily be filled with steady toil'. (Rediker, 1987: 114)

Time was also marked for navigational purposes, but these practices developed quite slowly, especially in the coastal pilotage that accounted for most medieval voyages. At the time Chaucer wrote his treatise on the astrolabe, in whose use he was clearly skilled (North, 1975)[44] some European sailors used charts, soundings, and position-reckoning, but Chaucer's 'Dartmouth shipman'[45] was a coastal pilot, not a deep-sea navigator. He would have made short passages out of sight of land, especially in good weather, but typically 'caped' along a succession of landmarks or seamarks. Fifteenth-century English merchant ships more often carried cards, compasses, and the like, but the details of their spread are unclear, though 'their knowledge (and with it the length of the sailing season) was gradually improving' (Rodger, 1997: 161–2). By 1500, better ships and navigation techniques made open-sea voyages somewhat less hazardous, though dead-reckoning was uncertain even in good conditions, being referred to by Castillian navigators as *navigacion de fantasia* (Rodger, 1997: 161).

Navigation on most ships around 1700 involved noon sun-sighting to find latitude, and dead-reckoning based on the traverse board and log-and-line to estimate longitude. Logbooks often estimated day's runs in miles, though the positions deduced from these were often guesswork, even in the absence of such behaviour as that of John Brown, mate of the *Glasgow* who (in 1735) was 'very drunk... could give no account of the ship's way from 12 noon till 8 at night and told me I might stick the log board in my arse etc' (Earle, 1998: 73).

Opportunities to compare knowledge of location were valued:

When ships passed and spoke they often asked each other what longitude they were in, while fishing boats were quizzed on the same subject. [eighteenth-century] seas were... quite well populated, especially close to land, and some exchanges of information saved many a master from an embarrassing landfall. (Earle, 1998: 73–4)

Importantly, as merchant seamen became increasingly knowledgeable about navigation, this knowledge was not confined to specialists but was distributed about the ship.

Masters and mates... had to know the principles of navigation, whereas the rest of the crew did not. Yet this separation of mental and manual labour was never complete... [T]he perils of life at sea placed grave limits upon the advisability of keeping trade secrets.... Much could be and was learned about navigation through observation of the daily work routine. Consequently older and experienced seamen... minimized the differential in knowledge... [in] the ship's labour hierarchy. (Rediker, 1987: 87–8)

Rediker may overstate his case, given the many contemporary criticisms of older seamen's reluctance to adopt new methods, but observation and experience certainly mattered.

The growing importance of new navigational devices and procedures meant that navigational knowledge increasingly resulted from formal training. Most seventeenth-century boys, going to sea in their early teens, had a least an elementary education in reading, writing, and possibly basic arithmetic. A minority would have had much more formal instruction in a grammar school. Signature-literacy rates among eighteenth-century sailors were typical of the better-off working class:

Some two-thirds of ordinary foremastmen and over 90 per cent of men who held any type of office in a ship could sign their names... The Londoners [were most literate], followed by the Scots and the sailors from the south-western counties, while the least literate were those born on the east coast. (Earle, 1998: 20)

Numeracy skills were increasingly necessary for ambitious sailors, being needed for many mundane tasks: reading numerical instructions, checking manifests, issuing receipts.

For those aiming to be mates, and chief mates in particular, numeracy was essential: they were expected to be competent at navigation and to be able to take over command if the master became ill or died... [By 1700] mates were increasingly required to keep logbooks... Navigation was crude, especially on small vessels, but certainly required numeracy, to calculate the ship's speed, the day's run and the latitude, blank pages in many log books being covered with the rough workings needed for those calculations. (Earle, 1998: 21)

Specific education in navigation was of two main kinds. The most important was on-the-job training whether informal, where boys learned the trade from older ship-mates and practised while off-watch, or formal, especially through apprenticeship to a ship's master. We learn something of the instruction expected from court cases brought by dissatisfied customers, as in a 1687 suit alleging

Captain John Ely of the *Coast Friggott*... 'did never teach or instruct [Gerrard Monger] in the art of navigation', but compounded his negligence by taking away for his own use the 'bookes and mapps touching navigation' given to the apprentice by his friends before he went to sea. (Earle, 1998: 23).

Increasingly, however, formal navigational training for officers was available in the seventeenth century, due to the growth of mathematical schools, usually staffed by men with practical skills, such as former ships captains or accountants (Taylor, 1954, 1966). Many boys about to go to sea were pupils here, receiving several months of relevant nautical education: writing a clear hand, mathematics, navigation, account-keeping:

William Dampier... was removed in 1673 'from the... [grammar] school to learn writing and arithmetick [and] soon after placed with a master of a ship at Weymouth', while Nathaniel Uring was sent to school in London for six months to be 'taught the first rudiments of navigation' before going on his first voyage. (Earle, 1998: 21)

Several early modern grammar schools had statutes specifically tasking Ushers (assistant masters) with teaching arithmetic and navigation (see section 6.4).

Thus merchant ships were enclosed temporal worlds, which perforce were highly time-conscious: from the constant turning of glasses and ringing of the watch bells, through methodical log book entries, to routine casting of the log-and-line. These pervasive routines help to explain the routine tagging of clock time to the recording of all events, as well as watch ownership amongst ordinary seamen, not just officers. Yet, if the merchant marine seems highly conscious of clock time—albeit often without the aid of clocks[46]—this was nothing compared to time-consciousness in the growing Royal Navy.

8.4.2 The Royal Navy

Hand in hand with the unparalleled expansion of English/British oceanic shipping and trade went that of the navy. War at sea was transformed from predominantly small-scale, local and coastal activity into the potentially worldwide province of a major arm of the state (Andrews, 1984; Harding, 1999; Padfield, 1999, 2003; de Souza, 2001). Caught up in various wars and the expansion of empire, the navy underwent continuous processes of technological change, bureaucratic systematization, and logistical and tactical refinement, in order to project power with greater force and precision. This already complex task was made much more complex by 'the wind, the primitive art of signals, the poor manoeuvrability of battleships, and the variable discipline and competence of officers' (Harding, 1999: 147), all of which made the precise and coordinated movement of ships difficult to achieve.

Until the sixteenth century neither the concept, nor the equivalent, of a modern, coherently organized standing navy existed (Rodger, 1997: xiv).[47] Ships were assembled for particular operations, usually by chartering or requisitioning from the merchant marine, rendering present-day distinctions between warships and merchantmen misleading. 'Not only were the two hardly distant in design, but "peaceful trade" was almost a contradiction in terms' (Rodger, 1997: 115). From the late fourteenth century, attempts were made to support a proper navy, but the ships were still likely to be merchant ships on war service (or king's ships on merchant service). Not only was the naval–merchant shipping boundary fluid, but there were few true sea battles. Most ship-to-ship engagements occurred in harbours, or close inshore, rather than between fleets at sea, unless ships encountered one another by chance (Rodger, 1997: 105–6).

A 'real' navy began to build up under Henry VIII, as a permanent fleet of ships intended (and increasingly specifically designed) for war, based on galleons as heavy-gun platforms,[48] and a corresponding naval administration. Continuing growth under Elizabeth meant that, by her death in 1603 'the English Royal Navy was in most respects stronger and more capable than... twenty years before, while Spanish sea power... was decaying fast' (Rodger, 1997: 243). The shift towards a standing fleet accompanied an expansion in English seagoing experience, beyond a secondary role in European trades largely within the relatively confined seas of the English Channel, North Sea, Baltic, and Biscay, to a much more ambitious independent role of, to use Ralegh's phrase, 'forcible trade', to both West and East. In this expansion, the English 'merchant fleet' was effectively a euphemism for privately owned men-of-war.

The uniformity of motives should not be exaggerated:

[An Atlantic] enterprise... was coming into being,... an aggressive drive by armed traders bent on breaking into the Portuguese and Spanish Atlantic trades, an unofficial war of trade and reprisals in the course of which emerged ambitions to colonise.... On the other hand the Londoners who led the quest for eastern trade seem to have taken a different view of England's oceanic role. The Muscovy Company was not interested in confronting the Iberian powers but in finding an alternative route to the East to avoid conflict... John Dee's... *General and Rare Memorials pertayning to the Perfect Arte of Navigation* (1577)... [envisaged] a peaceful empire of maritime commerce.

But it is easy to overstate the distinction... Power, whether concealed or overt, was an integral factor in the struggle for the rich trades of East and West alike. (Andrews, 1984: 9–10)

Whatever the impulses, the upshots were clear. First, wealth: Andrews (1984) estimates that prize goods brought into England during the war with Spain equated to 10 to 15 per cent of all legal imports. Second, a major increase in the number of merchant ships. In 1582 there were 20 English merchant ships above 200 tons, and 250 above 80 tons. By 1598, another 72 merchant ships above 200 tons had been built, with another 133 of above 100 tons. Third, the new kinds of nautical expertise first entailed in voyages to the Iceland or Newfoundland fisheries, became more general, in part through Franco-Scottish Huguenot charts and navigators from Dieppe and Le Havre in the 1530s and 1540s (Rodger, 1997: 305).

The 1550s search for the north-east passage brought together the talents of seafaring men—Sebastian Cabot, Richard Chancellor, Stephen, and William Borough—and three men of learning, Robert Record, Richard Eden, and John Dee. In 1558 Stephen Borough, a protégé of Chancellor and his successor as Grand Pilot of the Muscovy Company, from the *Casa de la Contratación*, brought back Martin Cortes's Spanish text *Arte de Navegear* (1551), from the navigation school in Seville, and persuaded the Company to have it translated. While William Borough's scheme for the formal training of pilots fell through, he managed to promote navigational skills through the Trinity House of Deptford, and as one of four Masters of the Queen's Ships (Waters, 1958).

Effective *oceanic* navigation was brought closer by convergences between theory and practice. Mathematical practitioners and practical seamen collaborated in improving navigational skills, through texts and devices (more and better navigational manuals, quadrants, cross-staffs); educational institutions (ranging from Gresham College in London, with its lectures on commerce-related subjects like astronomy, mathematics, and navigation, to several new 'mathematical schools');[49] and an emergent English School of cartography (evident in Hakluyt's reissued *Principal Navigations*), which

effectively announced the arrival of scientific geography in England. ... Sir Francis Bacon... thought it was part of God's plan that 'the opening of the world by navigation and commerce, and the further discovery of knowledge should meet in one time'. .... Science and seamanship finally came together.... (Andrews, 1984: 30)

Navigation may have became more purposeful as practices such as finding and holding courses along target latitudes became more common, but navigation remained closer to an art than a science over much of the period. Without reliable methods of determining longitudes, they had to be estimated by dead reckoning, defined by Dr Johnson in 1755 as

That estimation or conjecture which the seamen make of the place where a ship is, by keeping an account of her way by the log, by knowing the course they have steered by the compass, and by rectifying all with an allowance for drift or lee-way; so that this reckoning is without any observation of the sun, moon and stars, and is to be rectified as often as good observation can be had. (Johnson, 1755: 22–3)

Successful dead reckoning required a complex synthesis of close attention to impressions and measurements of speed and course, and the mental analysis of diverse measurements in the 'feel' of a changing situation. 'Experience and a talent for the feel of... the ship and the sea as they moved in uneasy conjunction—were what nurtured good dead reckoning' (Lundy, 2002: 162). But, in the end, dead reckoning could only do so much and, especially in the absence of visual confirmation, 'the accumulation of errors, big or small, makes dead reckoning perilously inaccurate, and eventually, it breaks down completely' (Lundy, 2002: 163), as many naval wrecks attest. Recent histories of longitude tend to gloss over the practical problems of determining latitude, but stories of gross error were legion.[50]

Within a few decades, certainly by the 1630s, the navy 'gives the impression, for the first time, of a settled service developing a sense of itself as distinct from merchant shipping and privateers' (Rodger, 1997: 405). The process was accelerated both by the greater size of naval ships—up to about 1,000 tons—and by their developing divisions of labour. Newly introduced tasks, ranks, and procedures contributed to a growing sense of discipline and order. Even so, in 1649 the navy numbered only some fifty warships plus some light sloops and ketches (Capp, 1989). But during the Commonwealth of the 1650s the navy was central to the new republic's survival, seeing near-continuous action, including England's first major war at sea, against the Dutch in 1652–4. It grew to about 266 ships by 1660, about half of them prizes, and the use of hired merchant ships was gradually dropped. Though many ships were destroyed or sold, Pepys's Restoration naval survey in 1660 still found 161 state-owned ships, 135 of fifth- or sixth-rated warships—the strongest naval force in Europe (Wilcox, 1966).

The navy now required very considerable manpower. Thus,

the summer guard planned for 1652, with the distant convoys already at sea, required a total of 10,024 seamen. A fleet survey of September 1653, in the midst of the Dutch war, recorded a figure of 19,524 men. The plans laid for the summer campaign of 1654 called for 30,000 men. ... The Navy tried to improve conditions to attract volunteers, but there were never enough: only pressing on a large scale could supply such huge numbers. (Capp, 1989: 9)

Though the Restoration meant a return to the gentlemen officers, the post-Commonwealth navy was clearly becoming a professional service in the modern sense (Wilcox, 1966; Davies, 1991), especially

as war became a permanent condition and large fleets had to be sent out each year. Increasingly appointments were filled by promoting men already in the service, not by recruiting outsiders. This process continued after the Restoration, with a largely separate caste of naval officers gradually evolving, and by 1688 the navy displayed a far more professional structure and mentality. (Capp, 1989: 401)

By the 1750s the Royal Navy was therefore a large tried-and-tested fighting machine. In 1754, 10,149 men were borne on ships books (including marines) either as complement or as supernumaries, while 9,797 were actually 'mustered' on board. During the Seven Years' War (1755–62) this number reached 84,464 (72,265 mustered) in 1757, dropping again to 17,424 (17,415 mustered) by 1764 (Rodger, 1988: 145–52). Numbers rose again with the onset of the Napoleonic Wars, to 45,000 in 1793; 85,000 (72,855 seamen, 12,115 marines) in 1794; 120,000 (100,000 seamen, 20,000 marines) in 1799; and to 130,000 (110,000 seamen, 20,000 marines) in 1800 (Pope, 1981: 93).

Both numbers and sizes of ships increased. In 1793, the Royal Navy had 493 ships and vessels divided into six rates, plus various unrated ships mostly under 500 tons (Lavery, 1983, 1989, 1992). All first- and second-rate ships were three-deckers; third- and fourth-rates were two-deckers (usually displacing about 2,500 tons fully loaded), and nearly all fifth- and sixth-rates were frigates with a single gun-deck. The largest ships had correspondingly large crews. A large two-decker might have a crew of over 600, while the crew of a common frigate might number about 280 (Lavery, 1989).

Accordingly, the organization of crews became more elaborate, and was far from haphazard even on the smaller ships. Watches were set by officers who were likely to have considerable nautical knowledge, since many had come from the merchant marine. The Commonwealth placed particular emphasis on practical skill and experience, choosing escorts ships and their captains for tasks around coasts and seas with which they were familiar: Capp cites examples involving Yarmouth and Scarborough captains in relevant areas of the North Sea, captains from Rye and Arundel protecting the Sussex coast, and James Coppin of Margate patrolling the Thames estuary. 'The same stress on experience [wa]s evident in... more distant operations' when convoy and/or defence roles around Newfoundland, Gibraltar, and Virginia were assigned to men experienced in trading voyages to those waters (Capp, 1989: 162).

But even the commonwealth's predominantly 'tarpaulin captains' were not unlettered or uneducated, notwithstanding their characterization by one Jacobean administrator as 'mechanical men that have been bred up from scrubbers' (Capp, 1989: 178). Tarpaulins were very varied both in their economic and social backgrounds, and in their behaviour and manners. If few were of gentle status, most were educated and literate—several lieutenants and captains published on navigation, or other topics—and appreciative that skilled knowledge, ability to learn, social and linguistic skills were all important for their careers (Capp, 1989: 178–9).

From the 1750s, the skills and work of the crew were increasingly regulated. As the size of ships increased further, the crew had to be divided up for different tasks and roles: campaigns were longer, and so were colonial voyages; captains... became [more] concerned with their welfare and discipline. Naval warfare was more intense, and a well-trained crew was a great advantage. (Lavery, 1989: 194)

Like the merchant marine, the late eighteenth-century navy was an enclosed world—nearly as much on shore as off. First, sailors were chiefly concentrated into only six home naval bases: Deptford, Woolwich, Chatham, Sheerness, Portsmouth and, from 1690, Plymouth. Second, sailors' overall bodily comportment was distinctive. Many impressed men and volunteers came from merchant marine backgrounds—any other origin was legally and practically difficult—so they already had the distinctive seaman's gait.[51] Third, was their distinctive clothing. Command officers and midshipmen had uniforms from 1748, but seamen had no official uniform until well after 1800. Nonetheless,

seamen had their own distinctive style of dress [short jackets and loose trousers], totally different from [the] landsmen's... long coats, breeches and stockings.... when the seaman Robert Wilson dressed in the 'long clothes' of a landsman, it was to disguise himself from the press gang. (Lavery, 1989: 209)

Their distinctive trousers, canvas 'petticoats', divided 'petticoat breeches' or checked shirts 'made seamen instantly recognisable'. Indeed, seamen 'scorned to wear landmen's clothes, and their best clothes were more elaborate and fancy versions of their working rig' (Rodger, 1988: 64).

Finally, there was the workday, which towards 1800, for which knowledge is fullest, had become quite exact, both daily and weekly. Eighteenth-century captain's logbooks show ships at sea for an average of 43 per cent of 'ship-days', varying between 23 and 50 per cent, with the largest ships and crews at sea least often (Rodger, 1988). Daily routines in port were less structured than those at sea, but still had strong temporal patterns. Once hammocks were slung, lashed, and stored—sometimes beginning as early as 4 a.m.[52] (Lavery, 1989) there came;

the order at 6.15 'watch below clean lower deck', followed by 'lower boats, wash round ship's side and coil down ropes', when any dirty marks on the hull were cleared off, 'breakfast' (half an hour from 7.15 a.m.), 'divisions' at 9.30 (when the men paraded under their officers), 'clear decks and up spirits' at 11.30, when the first half of the grog was issued, dinner which lasted an hour from noon, 'pump water and serve out' at 2pm, 'clear decks' at 4.15 p.m., which marked the end of the working day, supper at 5 p.m. and 'coil up ropes and sweep decks', the last task of the day, the time of which depended on sunset. Down hammocks was usually piped at 8 p.m. followed almost immediately by 'ship's company fire and lights out'. The officers had another two hours use of the purser's candles; the last order of the day, at 10 p.m., was 'gunroom lights out'. (Dove, 1981: 166)

The remaining ship's activities depended upon the day of the week. Routines generally depended upon the captain, but in home port[53] might include cleaning duties: scrubbing the upper deck each morning or holystoning[54] the upper deck on Thursdays and Sundays, and scrubbing the lower decks on Wednesdays and holystoning them on Sundays. It might include washing duties: 'wash clothes' was usually piped on Monday mornings immediately after hammocks were stored, with an hour allowed before the order was given for hanging the clothes up to dry and 'down wash clothes' being piped at 1 p.m. (whether or not they were dry), followed by 'make up clothes lines and sweep decks', and on Sundays, divine service.

However, timekeeping in the home port was relatively slack. Thus, 'at single anchor some ships kept three watches, but when securely moored most seem to have abandoned watches altogether' (Rodger, 1988: 43). Discipline was relaxed; Sundays could be taken off, women were allowed on board (Stark, 1998), shore leave might be granted to seamen (and was a right for officers), and so on. In other words, 'to weigh or drop anchor made a revolution in the internal affairs of every ship, suddenly altering the rhythms of everyday life' (Rodger, 1988: 37).

At sea, though, timekeeping became an all-enveloping grid, and time was more intricate, more concentrated, and more complex. On board at sea, time was marked in two main ways. To begin with, there were the exigencies of routine, with generally more activities to be carried out and coordinated at sea, both daily and weekly. Some captains made up very detailed lists of these routines: the early nineteenth-century captain of the *Bellerophon* listed activities for all parts of every day, including the drawing of clean hammocks and wearing of best clothes for divine service on Sundays (Lavery, 1989). Then, there was an exact watch system, usually, like the merchant marine, consisting of two watches (Figure 8.1). Most ships used two-watch systems that split the crew into two parts, though three- or four-watch systems were sometimes used, especially in port (Lavery, 1989: 194). Like the merchant marine, the two watches were known as larboard (port) and starboard (mainly because when, rarely, both watches were on deck, the larboard watch operated ropes on the port side). In addition, there were some men not assigned to watches—so-called 'idlers', usually men with relatively specialized occupations—who did not keep the night watch.

The watch system's intricate allocation of shipboard time and space, was absolutely necessary because:

> timekeeping was just as important as the allocation of duties. A sufficient number of men had to be available on deck to deal with any emergency, at any hour of the day or night. Duties had to be carried out at specific times, . . . Certain men in posts which required high concentration, such as helmsmen, sentries and lookouts, were relieved after quite short periods on duty;[55] those at more gruelling tasks, such as pumping, also had to be replaced regularly. Some activities, such as cooking, had to be begun well in advance, with the raising of food casks from the hold and the issue of provisions from cooks. (Lavery, 1989: 200)

As in the merchant marine watches were split into six four-hourly periods and two hourly periods. The watch from 8 p.m. until 12 midnight was called the *last watch* because it was the last watch set in night-time hours. The *middle watch* ran from midnight until 4 a.m.; the *morning watch* from 4 a.m. to 8 a.m.;

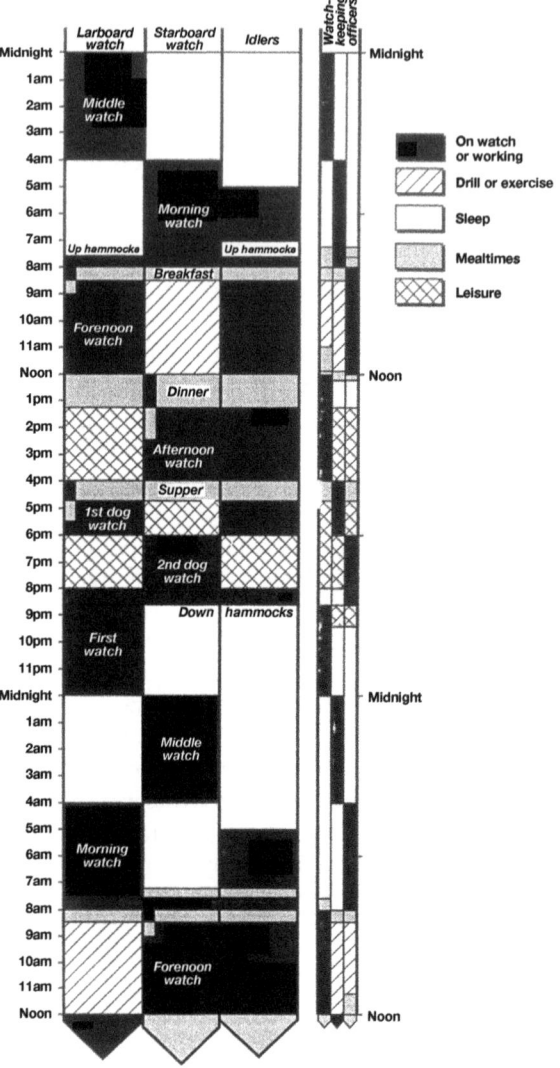

Fig. 8.1. Time–space arrangement of the naval watch system.
Source: Lavery (1989).

the *forenoon watch* from 8 a.m. until noon, the *afternoon watch* from noon until 4 p.m.; and two *dog watches*, 4 p.m. until 6 p.m., and 6 p.m. until 8 p.m., to ensure that men were not on duty at the same times every day.

Watches were announced by strokes on the ship's bell, usually housed in a prominent belfry. Ringing the bell formed a shipboard time system available to all; sounded every half-hour, the strokes counting the half-hours elapsed into each watch. So one bell meant half-an-hour of the watch had elapsed, two bells an hour, until eight bells ended the watch. Some larger ships had a petty officer ('quartermaster of the glass') specifically allotted this task. On smaller ships, it was carried out by the sentry to the captain's door. To keep time, these men usually used two sandglasses, a four-hour glass to signify when the watch had ended, and a half-hour glass to time the bell ringing. Piping was also used to coordinate many activities, such as dinner or 'down hammocks'. This temporal coordination was also a spatial coordination, when each watch was divided into 'parts of ship' allocating men to particular areas, with each man allocated a 'station' for all important ship's manoeuvres, and for 'quarters' in action.

The distinctiveness of shipboard time and space is nowhere better illustrated than by the fact that until 1805, 'the ship's day officially began at noon, half-a-day behind Greenwich time, and was so recorded in the log books' (Lavery, 1989: 202).[56] This was strictly an 'at sea' peculiarity; 'ships in port, freed from the requirements of navigation, reverted to the civil calendar' (Rodger, 1988: 39).

The second way in which time was marked at sea in the Royal Navy originated from navigation, the province of the master. Not surprisingly, the criteria employed in choosing masters were primarily technical skill, experience, and steadiness of character.[57]

> As well as setting courses and finding the ship's position, [the Master] was to supervise the pilotage, and represent to the captain every possible danger in or near to the ship's course, and the way to avoid it. He was to supervise the midshipmen and mates in taking the noon sights of the sun and look after the maintenance of the ship's compasses. Like all craftsmen of the ages, he was partly responsible for supplying his own books and was 'to provide himself with such charts, nautical books and instruments as are necessary for astronomical observations and all other purposes of navigation'. (Lavery, 1989: 101)

Although eighteenth-century naval navigational standards were improving, and probably superior to those on merchant ships, they remained inconsistent. Ships were repeatedly surprised by either the unexpected absence of expected ports, or the unexpected presence of land, finding themselves scores of miles (occasionally more) out in their reckoning. Not for nothing was it 'proverbial amongst seamen that it was better to trust a good lookout than a bad reckoning' (Rodger, 1988: 47–8).

Navigation's crucial importance on board ship was enshrined in routine, especially the noon sun-sight, using a quadrant (weather permitting) to obtain the ship's latitude. The captain and especially the master took the authoritative sights, but sighting was often no solitary affair as lieutenants and midshipmen took sights for practice (Lavery, 1989). The master entered his sights in his own log, along with the courses steered, wind direction, and strength, sail carried, and half-hourly speed, recorded by a mate or watch-officer on the log board. On warships, details in the master's log were copied in the captain's journal (Pope, 1989: 202).

318   *Seafaring Times*

Navigational expertise had improved considerably, partly due to generally increasing educational standards, several observers identifying improvements in both skills and manners (Rodger, 1988: 261). Near-illiteracy effectively barred a commission or even a warrant, and numeracy also grew in importance.[58] Naval officers like John Campbell, part-inventor of the sextant, and his namesake Alexander Campbell, experimenting with mathematical and other instruments, were among the leading practical specialists of their day. As Rodger (1988: 262) notes, 'both Campbells were of fairly humble birth, John the son of a Scots country minister, and Alexander the son of a purser, which suggests that a good education may have been among the qualities which raised them, respectively, to vice-admiral and commander'.

Although formal training was becoming more common, it supplemented on-the-job seamanship experience rather than replacing it. New naval academies to produce officers, like the Royal Naval Academy at Portsmouth (founded 1730), attracted criticism that 'the scholars... for all their theoretical studies were not learning seamanship, and the two years sea-time credited to them on graduation was no substitute for real experience' (Rodger, 1988: 265).[59] After all, most officers and sailors had previously been at sea since an early age, sometimes five or six, and almost always from ten or twelve: 'seamen had to learn their trade from boyhood if they were to attain the skill born of experience before they lost the vigour and agility of young manhood' (Rodger, 1988: 114). Ships' musters show that the peacetime navy found almost all its men from existing seamen or skilled craftsmen (Rodger, 1998: 114). Officers had invariably begun their career as ratings, often as young boys.[60] Since 'the regulations required a candidate for a commission to pass an oral examination in seamanship, to have served six years at sea, and two of those in the Navy in the ratings of midshipmen or master's mate... long sea service was indispensable' (Rodger, 1988: 263). Throughout, 'the quality which above all determined an officer's future in the navy, and marked it out from other professions, was his practical ability as a seaman. The capacity not merely to command and to navigate, but to hand, reef and steer, was the basic requirement for an officer' (Rodger, 1988: 259–60).[61]

Finally, time was marked at sea by the body itself. These were disciplined bodies, used to quite exact demands of numerous routines. Each man had his watch and mess numbers, his assigned position, and was often subject to the influence of drill books.[62] Other time-signatures intervened every day on ship, including the rhythms of sea-shanties and the coordination of the capstan.

8.5 CONCLUSION

This chapter has shown how a particular community of practice—seafarers—had a clear allegiance to temporal measurement, and that it evolved various specialized means of carrying out that measurement. In turn, temporal measurement became deep-rooted in embodied practice, a metric rarely discussed because it was so deeply engrained in every sensory register from sound (bells, pipes, commands) through sight (activities, lights, glasses), to smell (at mess times), taste (of the air), and even touch (of the lead).

It is remarkable that this time-obsessed community has been so little remarked upon, even though time-discipline on board ship was arguably fiercer than in the sometime coeval—and much-discussed—factories. We suspect there are two reasons for this. One is that seafaring communities *were* narrowly drawn; living in an enclosed world save for press-ganged landsmen.[63] In particular, the shipboard community was remote. Maritime time did not (much) travel to other communities of practice.[64] Secondly, there were very obvious reasons for much greater time-discipline and coordination being accepted at sea than on land, because of clear dangers: 'even weak or unpopular officers seem to have had no trouble obtaining obedience, for any seaman realised that disobedience was dangerous. This discipline was purely functional, and existed only to preserve the ship's company in safety' (Rodger, 1988: 207–8).

Nonetheless, attitudes to time-discipline were not solely utilitarian. For example, in 1755, one Captain Cuming was thought 'well beyond tolerable limits' in

> flogging men who took more than five minutes to answer the pipe for 'all hands'. This is a revealing attitude, for all hands were only called when they were needed, frequently in an emergency in which the safety of all on board might depend on the promptness with which the command was carried out. It should have been possible to get on deck from any corner of a small ship like *Blandford* in a couple of minutes at most; in the circumstances a modern captain might well regard five minutes as an excessively generous allowance. (Rodger, 1988: 214)

More generally, time-discipline was exacting without always being exact: witness the many activities on naval ships that were timed to the half-hour, rather than the half-hour timing them.[65] Such hints suggest many further questions about the history of times, and the wider cultures of timing within the seafaring community of practice, that currently await exploration.[66]

NOTES

1. The title phrase comes from Daniel Defoe's characterization of seamen in his *Les Enfans Perdue or the Forlorn Hope of the World*: 'They are fellows that bid defiance to terror, and maintain a constant war with the elements; who by the magic of their art, trade in the very confines of death, and are always posted within shot, as I may say, of the grave' (Defoe, 1697: 124).
2. Certain of these skills, especially relating to dialling, were subjects of a large literature, now lost as a discipline (Bennett, 1998).

3. See Chapter 9.
4. The peacetime life expectancy of sailors in European waters, on voyages to the North American colonies, or to the Arctic to hunt whales, was actually marginally better than that of young men working ashore in the ports from which they sailed. But in wartime, or on voyages to Africa, the West Indies, or the South American colonies it was very low. Earle (1998) suggests that 10 per cent of crews sailing East Indiamen lost their lives or drowned during the long voyages, and 20–25 per cent of crews on slave ships.
5. Though the variation in tides was itself a spur to the development of an understanding of clock time since it was necessary to know the relative times of tides in different places and the state of the water in particular ports was not just a simple consequence of tide time. Not surprisingly, special tide clocks were developed and tide times are a piece of information found on many early clocks (see Cartwright, 1998). Thus:

> sophisticated methods for predicting tide times [date] from the 17th century onwards... but the old simple rules, as embodied in clockwork mechanisms, continued to be used... for several centuries. As clock making developed during the 17th and 18th centuries, some of the more elaborate clocks were fitted with a lunar dial, driven by a gearing ratio of 59:2, and indicated the age or phase of the moon. With a simple adjustment, such a mechanism could easily be adapted to show the time of High Water at a chosen port, usually but not always London. (Cartwright, 1998: 20)

Cartwright goes on to cite the twentieth-century restoration of the original 1681 clock on the southwest tower of St Margaret's church at King's Lynn in Norfolk. He also cites examples of tide clocks mounted in wooden cases for interior use.

6. See also Forte, 1998. The nautical mile (6,076 feet, 1,852 metres) was not introduced until 1929, but such a measure had been in use since it was proposed by Norwood in 1637. Distance units like 'cables' were also used.
7. If circumstances allowed, seamen could put ashore, or anchor in sheltered water, and await clear skies, to observe the altitude of the sun or pole-star as a measure of latitude, even if only by crude measures of elevation or shadow-length (McGrail, 1998: 279).
8. Possibly signifying 'deep-sea' lead.
9. The importance of lodestones at this time can be gauged by the fact that they were often carried in small containers hung around the captain's neck.
10. The rising and setting bearings of both moon and sun vary through considerable arcs, being due east and west only when moon/sun is on the celestial equator.
11. The earliest surviving tide table, simply consisting of a column of thirty times in hours and minutes for London Bridge, is thought to be largely the work of John of Wallingford who died in 1213 (Cartwright, 1998)—not to be confused with the horological work of Richard of Wallingford, also an Abbot of St Albans (North, 2005).
12. This is approximate. The actual lunar cycle is just under $29\frac{1}{2}$ days, but $24/29\frac{1}{2}$ has no simple solution, hence the preference for an easy-to-execute approximation.
13. First mentioned in English ship inventories of 1410–12. The binnacle gradually transmuted into a permanent feature of a ship's construction and by the

eighteenth century, ships usually carried two compasses in a binnacle—with a lantern for night work—one an azimuth compass designed to detect magnetic variation (Harland, 1984).

14. To time watches the helmsman kept an hourglass or, customarily by the seventeenth century, a half-hour glass, hung in the binnacle. He turned the glass each time the sand ran out and marked the passage of time by ringing the 'Watch bell' every half hour, sounding one strike for every half-hour that had passed of the four-hour watch (Waters, 1958: 36).

15. William Bourne's *Regiment of the Sea* (1574) envisaged that ordinary ships carried various timekeeping equipment, including an accurate two-hour watch-glass, ideally reset each noon. He advocated the use of accurate shorter timings when using the log-and-line to measure speed (Taylor, 1963).

16. See, for example, Turner (1998); Estacio dos Reis (1997).

17. Pilots, aiding ships finding their way into ports, are a constant presence from 1300 to 1800. Pilot ships, or 'lodships', though often much resented, were an essential part of the maritime landscape. 'Pioneer explorers of the fourteenth and fifteenth centuries had been essentially... masters of the art of pilotage, finding their way at sea principally by observation of terrestrial objects while in sight of land, and by observation of the ship's course as indicated by the compass for the short periods when they were out of sight of land' (Waters, 1958: 41).

18. To focus narrowly on the development of the marine chronometer obscures the other reasons for which it was necessary, or certainly useful, to have clock times available on board. Nautical information could best be used, and sometimes could only be used, if clock time was available. Various sailing instructions presumed that clock time was at hand. For example, during the mid-seventeenth century Commonwealth, instructions about the complex running of tides around the Isles of Scilly were compiled by Abraham Tovey. Navigation among the islands hinged on knowing where and how fast, local currents were running relative to the turn of tides, and these were routinely expressed in hours:

A True Description of the Setting of the Tides, As Also the Time of High Water in the Islands of Scilly, Taken by Abraham Tovey, Master Gunner of the said Islands, c.1740

Between the Lands End and Scilly the best time of tide for to make a short passage from the Lands End let it be half-ebb, for at that time the tide sets W.S.W., right on the Islands. But if you run to the northward of the Island you will have the first of the flood run strong to the northward but if you keep to the southward of the island you will carry the W.S.W. stream for six hours untill it is half flood by the shore, by which time this W.S.W. stream breaks and runs in towards the islands, but not very strong. The reader must understand that this tide means about a league or two of the Island to the southward, and as far to the westward as the outermost rocks.

The second is Crow Sound where the tide sets out S.E. at three-quarters flood and continues till three-quarters ebb and then turns and runs back N.W. from the entrance of the Sound to the Bar, where you will meet with the tide, which sets into St Mary's Road W.S.W., and when you are there you will see a round rock called Nut Rock. Let that bear W.N.W. about two cables lengths from you, and then you may come to an anchor in clean, good, holding ground, where you will have at low water five or six fathom water at spring tides.

The third is St Mary's Sound where the tide sets out E.S.E. half southerly and continues till two hours ebb. And then it turns and sets in W.N.W. northerly. Peninnis Head and St Agnes Island will make the Sound and as you keep St Mary's shore aboard, the first point you come to will be the Woolpack Point, on which there is a battery of guns. Then give that a good berth for two or three ship's length in your sailing and the length of a cable distance from the shore. Import your helm and keep the starboard shore close aboard as far as The Steval, which is {illegible} and as you run N.N.E. to the Road, you will meet the Broad Sound tide setting E.N.E. and out W.S.W. Take no notice of the Mills for that mark is pulled down and altered, but mind these directions and Captain Collins' Draughts of these Islands and you will come in safe.

The fourth is Smith Sound. It is very good and the water deep but it is narrow, and as you come from the southward you must mind to keep St Agnes Island on your starboard side and all the Western scragged rocks and islands on your larboard side. Now the tide sets out S.E. at four hours flood and continues till two hours ebb and sets pretty nigh the same all through the western necks and rocks as far as the Bishop. But to the northward of that you will meet the Harbour tide mentioned in the next Sound.

The fifth is Broad Sound which ought to be nicely handled by reason the ships that fall in with it are very much puzzled, not knowing the setting of the tides. For suppose a ship falls in with the Bishop and Clerks at low water for about half an hour and no more the tide sets N.W. And if you stand to the northward the tide trims round until you leave all the islands on your starboard side, and then you will have, about a mile from the shore, a true tide setting E.N.E. as on the other side of the islands away to the eastward. I shall now return to the Bishop. At this half hour's flood you will have an ENE stream all the way to St. Mary's Road and also as far as St Martin's Head but then you must mind not to go too far to the southwards amongst the Necks for as the tide flows it parts and turns away to the southward and if you run too far to the northward it does the same. Therefore you must take short trips to get out but if the weather is so bad that you cannot carry sail, or a contrary tide, then steer in east by north from the Clerks or Crim and fear nothing. But when you come the length of a net head haul off N.E. for the Road, to avoid a ledge of rocks called the Old Wrack if the ship is large and water low, but if the waters are high don't fear him, for there is 12 foot at low water on him.

The sixth is New Grimsby. It is an excellent harbour but the tide runs as followeth. From the northward at low water the tide sets in an hour and a half and then it sets out three hours and then it turns and sets in four hours and a half until it is half ebb and then turns and runs out the other three hours until low water. But between Samson and Bryher Island the tide sets in eight hours from the westward from low water until two hours ebb, and then it runs out to the westward till low water.

The seventh is Old Grimsby, St Helen's Gap and Tean Sound, the tide runs all alike. For at low water the tide sets in from the northward and runs away S.E. nine hours until it is half ebb by the shore, and then it turns back for the other three hours till low water. Old Grimsby and St Helen's Pool are both very good places for small coasting vessels or such as will lie with the ground for where the vessels lie the ground is clean sand, and about six foot at low water at spring tides. The entrance from the northward is deep and when you are in you are all land-locked and you may go to sea with all winds.

The eighth is St Martin's Head being the northernmost part of all the islands. There is a steeple built for a landmark and between that and the eastern islands there is a Sound where small vessels may come in. But few come that way, not being acquainted with the marks nor setting of the tides, which is as followeth, viz. that tide as comes in from Broad Sound the same runs through St Mary's Road and so runs out E.N.E. through the above Sound. But as soon as you open the sea to the northward of the headland you will meet with the tide at four hours flood come round into the E.N.E. stream, and that makes the race of Hingeg [part illegible] and it gets the victory and sets away south by west as far as Minnywether Island, and there it meets with the Crow Sound tide which sets out south-east. And that makes a very great race at spring tides, and so both tides go off to the southward together.

As I have already given the reader an account of the true setting of the tide I come now to let him know the true flowing, as I myself have duly observed for this fourteen years past. It is as followeth. With moderate weather, in and about the islands, it flows E.N.E. and at neap tides with a northerly wind N.E. But when it has or does blow hard, and the wind out to the southward it shall flow above an hour longer and a northerly wind puts it as much back.

19. '. . . By [1633] it had become general practice to mark the log line so as to facilitate the calculation of speed. This was done in the following way. If a half-minute glass was used

then the length of time necessary to indicate a speed of one mile (of 5000 feet) per hour was (30 × 500 ft)/(60 × 60) or 41⅔ feet. In other words, at one mile per hour the ship would advance, and the line would run out 41⅔ feet in 30 seconds. The line was then divided as follows: from 10 to 20 fathoms, depending on the size of the ship, were allowed as 'stray line' next to the log chip, to ensure it being clear of the effect of the wake. The end of this stray line was marked by a knot of a piece of white or red rag, and then from these the line was divided into sections of 41⅔ feet or 42 feet, each section being marked by a knot in the line. Thus came into being the term known as the measure of a ship's speed in nautical miles per hour.' (Hewson, 1983: 610)

As another example, when Sir Henry Norwood wrote a personal dictionary of sea terms for the newly appointed Lord High Admiral, the Marquis of Buckingham, probably in 1620, and no later than 1623, he wrote of the log and log-line in ways which, though sceptical, already indicate a change in the method of calculation, based upon the number of fathoms sailed in a minute which is proportional to a distance of one mile per hour (since one mile or 5,000 feet per hour = approximately 84 feet per minute = 14 fathoms per minute).

> A *log-line*. Some call this a minute line. It is a small line, with a little piece of a board at the end, with a little lead to keep it edge long in the water. The use of it is that by judging how many fathoms this runs out in a minute, to give a judgment how many leagues the ship will run in a watch; for if in a minute there run out 14 fathom of line, then they conclude that the ship doth run a mile in an hour, for 60 [the number of minutes in an hour] being multiplied by 14 [the number of fathoms run out in a minute] make just so many paces [fathoms] as there are in a mile; so accordingly as in a minute there runs out more or less, they do by judgment allow for the ship's way. But this is a way of no certainty unless the wind and the seas and the course would continue all one, beside the error of turning the glass and stopping the line, both at instant; so that it is rather to be esteemed as a trick for a conclusion than any solid way to ground upon. The manner of doing it is: one standing by with a minute glass, whilst another out of the gallery lets fall the log; just as the log falls into the water the other turns that glass, and just when the glass is even out he cries 'Stop'; then he stops and reckons how many fathoms are run out, so gives his judgement.

20. 'The log-board used by English navigators measured, [Champlain] stated, 3 feet high by 15 inches wide, and was marked into four columns and thirteen lines. The top line contained the column headings—Hours—Knots—Fathoms—Courses and Rhumbs. The twelve lines in the Hours column were numbered at 2-hourly intervals from 2 to 12 and again from 2 to 12 to cover the 24 hours of the day. In the succeeding columns were entered the knots and fathoms logged every two hours and the mean course steered. Every 24 hours the knots and fathoms were added up, doubled to produce the sum of the hourly distance sailed and converted into leagues sailed in the 24 hours, on the basis that a knot equalled a mile, and three miles a league. The distance was then transferred to the log-book.' (Waters, 1958: 283)

21. On the Royal Navy, see Lavery (1989): 183.

22. In the late sixteenth century, it became customary to provide the cross-staff with three or four transoms (cross-pieces) to measure a range of altitudes. It could also be reversed and used as a back-staff to measure the angle of the shadow cast by the sun. So the two instruments were incremental developments of one another.

23. The Davis quadrant used two acute transoms, both forming integral parts of the device, an upper 'great arch' containing 65° and the lower 'lesser arch' 25°.

24. Hadley appears to have been unaware that his quadrant echoed seventeenth-century reflecting quadrants demonstrated by Robert Hooke and Isaac Newton.

25. Critically, once the image of 'the heavenly body had been brought . . . down to the horizon it did not leave it, no matter how much winds and waves tossed the ship,

the observer and the instrument about, provided always that he still held it vertical. The seemingly insoluble difficulty of precise observation at sea had been overcome. Shown to the Royal Society in 1731, and immediately tested by the Admiralty, it was put into commercial production at once and within ten years was in general use' (Taylor and Richey, 1962: 53–4). In practice readings needed to be corrected for atmospheric refraction, but these arguments need not be rehearsed here.

26. Three separate angles had to be taken simultaneously—the height of the moon, that of the star, and the angle between the two. These had to be carefully calculated using spherical trigonometry, and the time was calculated using tables which gave the moon's position for periods of three hours. All navigators were expected to understand lunars, even after chronometers came into use (Lavery, 1989: 186).

27. Waters continues:

   [Many mariners] preferred charts drawn by compass bearings, uncorrected for variation and distances... partly because [he]... felt far safer relying for latitude variation on his compass, and a chart drawn from bearings taken from it, than on changing sights and a chart drawn with compass bearings corrected for he-knew-not-what uncertain and varying quantities of variation.

28. Mercator's terrestrial globes showed the equator, with lines of latitude at 10° intervals, a prime meridian through the Canaries, and other longitude lines at 15° intervals.

29. Only after 1700 did marine charts approach the accuracy of land surveys, as marine surveyors adopted the geodetic methods of land surveying, and as more accurate measures of longitude were available (see Chapter 10, below).

30. A copy of Harrison's watch, by Larcum Kendall, was used by Captain Cook on his second voyage of exploration in the Pacific. When Cook made his final landfall at Plymouth in 1775, after circumnavigating the globe, it gave an error of less than eight miles in calculated longitude.

31. For this reason, it was usual to carry three chronometers, so that if one went wrong the error could be detected (Rodger, 2004: 363).

32. On almanacs, see Capp (1979).

33. Particularly the journals of *Torbay*, flag-ship of Vice-Admiral Sir George Byng, captain James Monypenny, log kept by Lieutenant Field; *St George*, Captain James, Lord Dursley, log kept by Lieutenant Wiscard; Journal of *Monmouth*, log kept by John Baker, Captain; *Swiftsure*, log kept by Richard Griffiths, Captain; Journal of *Isabella*, 'yacht', kept by Finch Reddall, Captain; and *Somerset*, Captain John Price, log kept by Joseph Line, Sailing Master.

34. Bourne took the principle of the subdivision of time way beyond levels of precision with practical utility. In a brief general discussion of space and time, Bourne explains that 'For as 60 minutes is a degree or an hour, so 60 seconds is a minute, and 60 thirds is a second, and 60 fourths is a fourth, etc.' So an hour contains 12,960,000 fourths (Taylor, 1963: 166). Bourne does not calculate anything with thirds or fourths, but thinks them worth explaining to readers.

35. The bequest of Tele's astrolabe and chart to the same man suggests they were intended for use by the recipient John Goodlad, who was prominent (with several other Goodlad family members, in Trinity House, an institution that, then as now,

undertook various projects and support for mariners and their dependents: (Fury, 1998). It hasn't been possible to find how the various items were acquired by Tele: neither he nor they are mentioned in any other wills from Elizabethan Essex.

36. A few beacons and lighthouses were constructed relatively early on in English maritime history (see Hutchinson, 1994) but it was not until Henry VIII's reign that the Crown became seriously involved.

37. According to Waters (1958: 476), a competent early seventeenth-century navigator knew how:

> 'to observe that Altitude, Latitude, Longitude, Amplitude, the variation of the Compasses, the Sun's Azimuth and Almicanter, to shift the Sunne and Moone, and know the tides, his Roones, pricke his Card, say his Compasses'. To be able to do this [the Captain] would have equipped his ship with leads and lines, and a dipsie lead and line; common sea compasses, a dark compass, and an azimuth compass; half-hour, hour and four-hour watch glasses, and a watch bell; a traverse board; a log and log-line, logboard and half-minute or one minute running glasses; and in his chest his instruments would include plane, circumpolar and Mercator charts; compasses and protractor; a terrestrial globe and compasses; a plain-scale and a sector or Gunter's Scale; an astrolabe, astrolabe quadrant or celestial globe; ring dial or pocket sundial; a nocturnal; a tide-computer; a lodestone; a sea-astrolabe, cross-staff or back-staff, and a Davis quadrant, a log book and a journal. He would probably prize a telescope and might well keep a six-inch dial watch in his cabin. He would be able to handle these instruments correctly because he would probably carry a copy, and would certainly be well acquainted with the contents, of the 1610 edition of Wright's *Certaine Errors of Navigation*; the 1625 edition of Tapp's *The Seaman's Kalendar*; the 1615 edition of Tapp's revised text of Eden's translation of Cortes's *The Art of Navigation*; the 1626 edition of Bourne's *A Regiment for the Sea*; the 1626 edition of Davis's *The Seamans Secrets*; of Blaeu's *The Sea Mirrror* (1625); of Gunter's *De Sectore et Ratio* (1625); of Aspley's *Speculum Nauticum* (1625); the 1614 edition of Norman's *The Newe Attractive* with Barolyn's *Of the Variation of the Compass*; of Wright's *Description and Use of the Sphere*, reprinted [in] . . . 1627; and of either Hue's *Tractatus de Globis* (Amsterdam, 1624), or his *Tractaet van Globe* (Amsterdam, 1623), or his *Traité des Globes* (Paris, 1618).

38. See also Turner (2004).

39. Earle (1988: 57–62) notes several instances of merchant sailors owning silver watches, and of officers on East Indiamen privately trading watches for themselves or others. Lavery's argument (1984: 200) that few navy men 'had time pieces of their own, and most would not have known how to use them' is highly unlikely, as we hope to have shown in this book. Note also that the use of almanacs and astrological guides is strong evidence of a need for detailed timekeeping. Capp (1989) argues that such texts were in use on ships in the 1650s, hardly surprisingly given their widespread use in England and Wales.

40. Remarkably little has been written about hourglasses given their crucial roles in shipboard life.

41. We have not discussed the fishing fleet separately, which was a significant source of ships and men

42. 'In the early 1770s the twelve largest ports account for over 70 per cent of all sailors' (Earle, 1988: 209).

43. Confusingly, the term 'watch' was used both for the parts of the crew, and the periods of time. For clarity here, we use 'group' when referring to the crew.

44. Skeat's 1870 edition, for the Early English Text Society, of Chaucer's treatise is available online via <www.hti.umich.edu/Chaucer>.
45.
>But of his craft to reckon well his tides,
>His streams and his dangers him besides,
>His harbouring (herberwe) and his moon, his lodemenage,
>There was more such from Hull to Carthage,
>Hardy he was and wise to undertake;
>With many a tempest had his beard been shake,
>He know all the havens, as they were,
>From Gotland to the Cape of Finisterre,
>And every creek in Brittany and in Spain.

46. 'Exact means of measuring times besides noon were difficult to use at sea because they relied upon the sun. Sun dials, or the more sophisticated ring dials popular on shore until... the eighteenth century' (Waters, 1958: 60), had fundamental defects at sea, needing to be adjusted accurately north–south, set accurately for latitude, and held stationary. However pocket dials were clearly carried by some mariners (Waters, 1958, Appendices 7 & 8), and a compass fly could be used as a sundial.
47. Although such a concept was becoming familiar in several southern European nations.
48. Galleons could have a crew of up to 180 men
49. On Gresham College, see <www.gresham.ac.uk/uploads/historygreshm_bk2.pdf>; on mathematical schools, Money (1993).
50. 'Latitude in principle was easily observed, but in 1596 Lord Howard found the masters of his fleet differed widely in their reckoning after only a few days at sea. Returning from the South Atlantic in 1583, Captain Luke Ward and the officers of the *Edward Bonaventure* could not agree whether they had sighted Ushant or the Pointe du Raz; it turned out to be the Scilly Isles, so they were 100 miles of longitude and nearly 120 miles of latitude out of their reckoning. Luck and judgement continued to play as large a part as science and mathematics in any successful voyage. The weather was forecast from the behaviour of porpoises and seabirds, to say nothing of pigs and cockroaches. Masters were often supplemented by pilots. Sometimes these were local or coasting pilots of the modern kind, familiar with a particular port or coast, but there were also deep-sea pilots, specialized ocean navigators who provided skills which many masters had not yet acquired' (Rodger, 1997: 306). In a later period, even a keen navigator like James Cook could get it wrong—when Cook sailed from the Society Islands to New Zealand in1769, he was four degrees (or more than 150 miles) out on his longitude by the time he arrived at the coast of New Zealand (Lundy, 2002).
51. 'A majority of those who served in the Royal Navy during wartime had served in merchantmen during peacetime, so that naval service was common in the career of the typical merchant seaman' (Earle, 1998: 185). This was not surprising, given the large demands in wartime. It has been calculated that an average of 25,000 extra men had to be found for the navy in 1739–48 and over twice as many men in 1756–63. In turn, this meant that the wartime merchant marine included many

foreigners and landsmen (Earle, 1998). Even on Royal Navy ships, foreigners were not necessarily a rarity (Pope, 1981).

52. 'No man in a ship of an anchored fleet had any doubt when it was day break: in every ship a drum began to beat, and all the drummers kept thumping away . . . until "a man could see a grey goose a mile", when the flagships fired a gun. Each ship hoisted her colours—ensign and jack—at sunrise' (Pope, 1981: 166).

53. 'In port' usually meant at anchor in a reasonably secure road, not actually in harbour.

54. Holystones were stone blocks used to scrub the decks, so-called because their size and shape resembled bibles.

55. For example, helmsmen usually served for a 'trick' of about half-an-hour.

56. In reality, the ship's day began at 4 a.m. with the off-watch men being called, the men at the wheel and the lookouts being changed, and the log being hoved (Pope, 1981).

57. 'The master's cabin or office, situated on the greater deck on the opposite side from the clerk's office, served as the chartroom and centre for navigation. It was fitted with a table for navigation, and the charts were stored there. Other navigational instruments, such as the log and lead lines were stored in drawers in the binnacle, while instruments like sextants were the property of the officers using them, and were kept in their cabins' (Lavery, 1989: 184).

58. Many of those serving on board would have needed literacy skills as a matter of course; for example the purser. Even in the 1650s, though some of the warrant officers on board might be now regarded as technically illiterate—that is, unable to sign their names—they were almost certainly able to read (Capp, 1989: 207). For example, the master's knowledge tended to be local until quite late on in the history of the navy. In the 1650s:

> 'a master needed to be familiar with the area where the ship was to serve. One captain was asked to check that his master knew the coasts and harbours of Ireland, where he was to be stationed. When Robert Dennis found his master knew nothing of Virginia, where he was bound, he asked at once for a replacement. Masters themselves sometimes asked to be released if they lacked appropriate expertise. One said he was totally ignorant of the Caribbean and begged to serve in the Channel instead. Another was dismayed to find his ship appointed to the French coast 'contrary to my expectations when I first came on board. I cannot speak myself able, my ability being altogether for the nor'ard, where I expected our station would have been.' (Capp, 1989: 202)

59. Which, in turn, was a modification of the old scheme of volunteers per order or 'King's letter boys' established by Samuel Pepys.

60. Aided by the use of schoolmasters at sea.

61. Though we should not overstate this: 'Senior officers were less than clear about the geography of the British Isles. A captain found it necessary to explain to the Admiral where Sunderland and Blyth were, while a Sussex man assumed that the flag officers in the Downs would not have heard of Newhaven' (Rodger, 1988: 49).

62. Especially for gunnery. Our discussion has generally ignored the presence of marines on many naval ships. Marines often ran to precise drills.

63. In their partial isolation, there may be some similarities between large ships and medieval monasteries.

64. Again, perhaps like medieval monastic time.
65. In other ways, though, naval time-discipline has retained distinctiveness—thus naval marching has remained a slower affair, quite clearly meant to distance it from the army.
66. In later years, time-discipline in the Royal Navy became much more exact. For example, ratings were expected to arrive five minutes before assigned time.

9

The Pursuit of Precision

9.1 DIMENSIONS OF TEMPORAL PRECISION

This chapter tightens our focus from the 'everyday precisions' to be found across large parts of the population, to the explicit search by certain specialized temporal communities for more and more precise timekeeping. This topic has become a familiar one in recent years, largely thanks to Dava Sobel's international bestseller *Longitude*. But concern with precise time-measurement long predated the availability of precise mechanical timekeepers. Indeed it long predated mechanical timekeeping of any description, as most clearly seen in many early societies' efforts to maintain a consistent, exact correspondence between their calendar and the celestial year. In those very different technological environments definitions of precision took forms sharply different from early modern Europe's focus on very precisely measuring very short periods of time. But in both cases, and much more generally, the search for greater precision was successively pushing against, and sometimes pushing forward, the limits of technology and calculation in measuring and keeping track of time.

However, temporal precision is not a singular thing. Without trying to be comprehensive, we can distinguish at least five components: (i) the precise definition of particular moments or times of day (such as noon or midnight); (ii) the precise measurement of set periods of time; sometimes between recurrent events or celestial positions (one year, one lunar month, one day), sometimes of conventional fractions of such a unit (one hour, one minute, one second); (iii) the capacity to precisely specify the time of a particular event (such as an infant's birth, or a specific star-transit); (iv) precise recording of a duration of something as it happened (such as a solar eclipse); (v) the coordination of different times, events, or actions (such as the operation of machinery). Thus precision entails several dimensions. Particular devices or practices are rarely capable of tackling them all at once, but commonly attempt to address a sub-set of these dimensions.

There are also some definitional complexities to unravel. For example, *precision* is not exactly the same as *accuracy*. The state of being accurate entails the correspondence of a measure to an external reference, whereas precision refers to the exactness of the measure whether or not the measure is accurate. For example, the statement 'the time at noon is four minutes and 36.579 seconds past eleven o'clock' is precise but not accurate. In both past and present, it has frequently been assumed that precision and accuracy are two sides of the same coin, but in the history of timekeeping the distinction is an important one. Apart from the point that clocks may be much more (or less) accurate than their signalling displays, precision and accuracy diverge because of the variability of the earth's rotation. The divergence of mean and apparent solar time means that a clock cannot be accurate with respect to both unless incorporating some kind of dual-display. A clock that keeps perfectly accurate solar time will diverge from mean solar time by up to about fifteen minutes, depending on the time of year.

Where the sun provides the ultimate authority as to time of day, timepieces are intermediaries to the true time: their accuracy is defined in terms of their correspondence to solar position. Both sundials and clocks might be more accurate or less accurate, but a true sundial is authoritative, whereas a clock is always an intermediary. However, once the divergence between apparent solar time and mean solar time had been recognized, and mean time became the 'real' (socially meaningful) time, authority derived from particular clocks recognized or defined as official. So that clock was no longer an intermediary to the true time, although other clocks or watches still were. In practice, the extent to which people regard other clocks as authoritative sources of precise time depends on the reliability of the performance of clocks in general; the particularities of devices; and the standards of adequation applying to timing for different purposes.

The most prominent form of temporal precision in early modern Europe involved greater refinements in accurately subdividing time, and in measuring very short periods of time. The two critical communities for precise time measurement and/or calculation in the Middle Ages were, first, astronomers and astrologers (between whom there was then no clear distinction), and second, computists. Computists were concerned with compiling and refining calendrical and astronomical calculations used in determining the future dates of Easter for Christian observance. The contemporary Latin term for such a treatise was *computus* (Borst, 1993). The *computus* was necessary because much of the Christian calendar tracks backwards from the date of Easter, so that Christian communities did not have the option of determining its date by direct observation of the equinox and full moon, as may be done for festivals such as Ramadan, in Islam. The central computistical techniques were the coordination of solar and lunar data, and calendars based on them, the interpretation of biblical texts, and maintaining the temporal relation of Easter and the Jewish Passover. Bede was the outstanding English contributor to the European computistical project, in his *De temporibus* ('On Times') of AD 704) and *De temporem ratione* ('The Reckoning of Time') of AD 725 (Bede, trans. Wallis, 1999).

Bede explicitly discussed 'the smallest intervals of time' in *De temporem ratione*, in which he made an important distinction between fine temporal distinctions useful in calculation, and those that were perceptible. Calculators reckoned down to 'the smallest time of all, and one which cannot be divided by any reckoning, they call by the Greek word "atom", that is, "indivisible" or "that which cannot be cut". Because of its tiny size, it is more readily apparent to grammarians than to computists' (Bede, trans. Wallis, 1999: 15). On the other hand, he suggested that the shortest perceptible time was that 'in which the lids of our eyes move when a blow is launched [against them]', while the shortest perceptible time measurement on a sundial at that time was about one-fifth of an hour.

Medieval Europeans were using minutes long before the development of mechanical clocks, as in the recording in 1181 of the length of a solar eclipse in minutes, which seems to reflect Islamic astronomical influence (Walker, 1996; Dohrn-van Rossum, 1996: 42, 360 n. 35). That early mechanical clocks, whether their users were aware of it or not, functioned as part of broader webs of temporal information, such as dawn, noon and other celestial time-cues, is occasionally explicit in contemporary records. So too is the way in which a clock's going train could be used directly to count subdivisions of time, as when the Nuremburg astronomer Bernhard Walther recorded on 16 January 1484 how:

I observed Mercury with a well-regulated clock, which gave the time precisely from noon the day before to noon. I saw Mercury in the morning in contact with the horizon, and simultaneously I put a weight on the clock, which had 56 teeth in the hour wheel. It made one complete revolution and 35 teeth more, by the time the Sun appeared on the horizon, whence it followed that Mercury on that day rose one hour 37 minutes before the Sun, which almost agrees with calculation. (Andrewes, 1985: 69, citing D. Beaver, 1970, 41)

But European attempts to time seconds in the late sixteenth century were a new development. The Danish astronomer Tycho Brahe had four different clocks which 'show with precision, not only the minutes of the hours, but even the second parts' in his observatory at Uraniborg in 1587 (Baillie, 1957: 28–9). Tycho commended their performance in his *Mechanics of the New Astronomy* of 1598, although he was subsequently more critical. He was not alone in his aspiration to time in seconds: it was also shared by John Dee (Fenton, 1998); one among several astrologer-astronomers aspiring to significantly greater precision than that sought by the layman Samuel Jeake, whom we discussed in a previous chapter.

In certain respects, various earlier non-European societies were concerned with very precise timings, of which the Maya are perhaps the best known. However, the Mayan approach to precision timing differed substantially from the European preoccupation with the exact measurement of short units of time, to which mechanical clocks were applied almost from their invention. Precise Mayan calculation of year-length, for example, or eclipse predictions, or tracking cycles in planetary movements, all centred on long-run record keeping and the progressively more refined understanding of the recurrence intervals of particular relative positions (for an introduction to the large literature on Mayan calendrics, see Aveni, 1992). The maintenance of records over long periods enabled certain parts of Maya societies to eventually attain an extraordinary calendrical accuracy regarding the lengths of the cycles, intervals, and repeated positions that were central to the whole of Mayan timekeeping. For example, the calculation of the exact intervals at which Mars, Venus, and the Moon repeated their own positions, and their positions relative to one another, was subject to corrections on the scale of fractions of a day in several decades: 'nowhere else . . . but among the Maya do we witness the mathematisation of nature's time cycles carried to the borders of obsession' (Aveni, 1996: 285). They illustrate how some past precisions took rather different forms from today's, and we should not be reluctant to acknowledge the sometimes very different precisions found in societies distant in space or time from our own.

9.2 COMMUNITIES OF PRECISION

In early modern Europe, precision timekeeping was a concern for several technical and scientific communities of practice that directly or indirectly handled astronomical information. It would be wrong, though, to treat concern for precision as solely a feature of small, mathematically skilled communities. For precise times served diverse 'non-modern' ends like astrology, as well as the 'modern' ends of navigators and instrument-makers. It was, after all, essential that precise navigational methods and techniques could be carried out at sea, not just in the observatory, so the practical use of precise navigation required that skills be accessible to at least some seamen and officers. As we show, important extensions of precision were less to do with theoretical precision in the abstract, than with making precision a practical proposition, realizable in the everyday world. A mixture of intrinsic technical difficulties, on the one hand, and the complications involved in making precision transferable from the ideal space of the dedicated spaces of the observatory, laboratory, and workshop, to the less-than-ideal circumstances of everyday life in busy multi-purpose spaces, explain why the pursuit of precision in timekeeping involved many different types of people over several centuries.

Several different stimuli towards longitudinal and temporal position co-existed in early modern Europe. These motives related to different contexts for using timing practices and instruments, and they involved different—though overlapping—'knowledge communities'. We want to highlight five of these specialist knowledge communities. The majority were primarily involved as users (or consumers) who applied precise times and/or made use of precision timing practices in their activities and social networks. Astronomical communities were additionally important as 'time (or technique) providers' for the various 'precision-applying' communities. The *astronomical* community, as the first of our knowledge communities, was itself diverse, with concerns ranging from cosmologies of the solar system and beyond, to mathematical calculation of positions and orbits. Astronomers also addressed broader groups, though, through tables and instruments showing high water for particular ports,[1] and reached out to European astronomers increasingly engaged at least indirectly with various strands of an Islamic astronomical community, as a net consumer of new instruments, observational conventions, texts, and data.

Although the definitional boundaries between astronomy and astrology had been laid out explicitly by Bishop Isidore of Seville around AD 630 in his *Twenty Books of Etymologies*, in practice the overlaps were considerable for around another millennium for a majority of non-monastic astronomers, and for some of those within monastic walls as well. The *astrological* community, which formed the second of our knowledge communities, was concerned with precision as a facet of accurate casting of natal horoscopes, including the specification of astronomical positions and aspects at particular times or events, and of their trends (Tester, 1990). The rhetorical appeal of science and precision was no less powerful for astrologers than for astronomers, despite the differences that hindsight places between them. Many of the most systematic timings, and some of the most precise astronomy, were carried out by and for astrological practitioners.

Third, various *mathematical* practitioners increasingly drew on and endorsed precise measurement of times, not least in debates in mechanics on topics such as rates of gravitational fall, acceleration, and the like (Drake, 1990), as well as in describing and predicting the behaviour of objects in the world, first geometrically and subsequently algebraically. Specialized practices of precision could spill over into mathematical practitioners' everyday usage. For example, as we have pointed out, while in Prague John Dee recorded his son Michael's first birthday with a painfully roundabout concern for minutes:

[22nd February 1586] ... 'Michael Dee revolutus 9 hours 23 minutes'

The adjective 'revolute' meant 'that has completed a full revolution' (Fenton, 1998: 192, n. 3). Dee is referring to the earth having completed a full orbit around the sun, returning to precisely its relative position at Michael's birth, which on 22 February 1585 Dee had recorded taking place 'hora. 3. min. 28, a meridie...' [i.e. 3:28 p.m.]. The calculation makes explicit that Dee was using the then-prevailing precise estimate of solar year length, of 365 days, 5 hours and 55 minutes.

Fourth, temporal precision had critical uses in connection with *cartography*, especially at national and international scales. Longitudinal differences ultimately measured through time-differences were critical to accurate world and continental mapping (see section 9.3). Rhetorics of precision are encountered in smaller-scale mapping projects, but in practice these could afford to disregard temporal differences at such a small scale, much as map-makers could afford to ignore the curvature of the earth in producing estate maps. Fifth, demand for temporal precision was apparent amongst *seafaring* communities involved in exploration, militarism, or long-distance trade, and especially those persons responsible for navigation. This impulse has also already been discussed in general terms in Chapter 8.

Though some of these specialized communities were small, even at an early date they could be strikingly international. For example, as we have pointed out in Chapter I, that Jupiter had four major moons was discovered by Galileo, in Florence, on the evening of 7 January 1610. Before the year's end, they had been observed from several other European centres, and Galileo and the Frenchman Claude Fabri de Peiresc were both compiling tables of their motions. Within another year, observations had spread considerably further afield.

Beyond communities of practice that used relatively precise times, and were concerned with making timing more accurate, were two considerably larger communities that we label 'lay' and 'public'.[2] We distinguish at least one important contrast between 'lay' and 'public' communities of precision. The former was a diffuse secular community of relatively educated people, though not necessarily of high social status (thus it included some traders and artisans as well as leisured people), with largely non-utilitarian interests in measurement and in calculative practices (Dohrn-Van Rossum, 1996; Crosby, 1997; Shapin, 1994). On the other hand, the 'public' community involved people exposed to one or more concerns with accurate timekeeping, and/or its results, but whose involvement was passive. All in all, then, the 'community of practice concerned with precise timekeeping through clocks' embraced several distinct but overlapping communities of practice, with very different motives and objectives. Their diverse concerns with, and practices of, precision were often quite specific and often implicit, so they do not necessarily show a 'modern' world view.

The first printed manifestations of the pursuit of precision came from astronomer-astrologers. Seeking to provide advance information about impending eclipses, for a broad audience, and specialized planetary positions for astrological interpretation, and more precise stellar positions for an international, but very small, community of astronomers, they published various tables, calendars, and *ephemerides*. For example, Regiomontanus's *Kalendarium*, published at Nuremburg in 1475, made forecasts for 1475–1530 giving calculated dates of predicted solar and lunar eclipses, their projected time of day in hours and minutes, and the projected obscuring of the sun in twelfths of a diameter (illustrated by Swerdlow, 1996: 193, and Gingerich, 2004: plate 5a, following p. 146. See also Gingerich's plate 1a). *Ephemerides* gave daily positions for the sun, moon, and planets, along with planetary positions. That this information was often of doubtful reliability stimulated ongoing attempts to refine previous work (Swerdlow, 1996). The earliest printed equation-of-time tables showing the difference between solar time and mean time through the year followed shortly, in the 1492 edition of the Alfonsine tables (Andrewes, 1985: 84, n. 18).

The astronomical observatory was not the only space in which very precise timekeeping was a priority, and from which precise temporal devices and knowledges circulated. Another, from the late thirteenth century, was the clockmaker's workshop. An integral part of the production of clocks or watches that kept good time was a reliable, accurate, reference clock, against which the performance of clocks could be checked before they left the workshop. Reference clocks were generally termed 'regulators,' since their purpose was as an authoritative device to regulate the going of other clocks. Clockmakers' regulators illustrate that temporal precision was not just pursued in astronomical observatories, but more widely, too. Many of the technical innovations that became important in chronometers were foreshadowed in the workshops of clockmakers. The eventual solver of 'the longitude problem', John Harrison, is a clear example. For some years Harrison's work on precision timekeeping concerned regulators in general, rather than sea-going clocks in particular.

Precision in a clockmaker's use of a regulator had at least three dimensions. First, a regulator needed to be consistently accurate as a timekeeper; second, comparisons between the regulator and the clock being tested had to be exact; and third, any difference had to be not just detectable, but measurable in a short time. Accuracy in regulators hinged on precision at microscopic levels, small fractions of a second. In dealing with the second and third issues, various means were deployed to use not just the dial-work, but the regulator's mechanism clock itself, to obtain finer temporal divisions. In effect the regulator was customized to show the exact correspondence or difference in the going of the two clocks.

One strategy was to use a very fast-beating pendulum, such as that which George Graham constructed for John Desaguliers in 1730. With a verge escapement and a short pendulum less than two-and-a-half inches long, it was fitted with a dial divided into quarter-seconds and with a quadrant to arrest the pendulum to within the nearest quarter of its vibration, thereby providing a measurement of time to one-sixteenth of a second. Desaguliers used this chronometer for astronomical observations, measuring the descent of falling bodies and velocities, and a slew of other experimental purposes (Andrewes, 1996: 192 n. 13).

Alternatively, a scale could be placed beneath the pendulum, showing fractions of its arc of swing. This seems to have been the method used by John Harrison in achieving an almost unbelievably consistent pendulum, by being able to measure tiny differences in mechanical performance, and make minute adjustments to delicate equipment (Burgess, 1996).

[T]he seconds hand only shows you what second you are in, but Harrison says he could read his clock to one-twentieth of a second or better. To do that, he would have had to look at the instant position of the pendulum...

In order to adjust, test, and regulate a [precision] timekeeper... observation of the time of day from the position of the seconds hand would not have been sufficient, because very small differences between the rates of the two timekeepers would have been impossible to determine in such a manner. Such an observation could only be made by determining the exact position of the pendulum at a particular moment, and, to help with this, Harrison might well have marked the scale beneath the pendulum in time rather than in degrees. The pendulum's wide arc of swing would greatly assist this precise observation. (Burgess, 1996: 259 n. 27)

Precision mattered more widely to the growing clockmaking community, though, quite apart from the technicalities of building regulators and comparing the going of clocks, because accuracy and precision came to have an important aesthetic standing, as important elements of clockmakers' sense of their own craft, skills, and knowledges. And, as discussed in Chapters 6 and 7, precision and accuracy were central to many customers' sense of clocks and watches as consumer goods, how they used them, and the pleasure derived from them.

9.3 TEMPORAL PRECISION AND SPATIAL POSITION

The relationship between temporal precision and the precise measurement of spatial positions was *the* key issue of temporal precision in early modern Europe. Above all, it related to the many ways in which ability to calculate longitude was vitally important within, and beyond, early modern Europe. When Robert Hooke outlined a research agenda for science, in his *Animadversions* of 1674, 'the determination of absolute coordinates for important geographical positions' was among the top priorities. The next year, John Flamsteed's warrant as director of the Royal Observatory spelled out his main task: the 'perfecting of navigation and astronomy' with special reference to 'finding the longitude by lunar distances'. Put at its simplest, determining longitude was a matter of knowing the time in two places at once. On land and on sea, the keys to knowing where you were, and the keys to knowing the shape of a territory, a route or a network, were both bound up with the precise determination of time. Seconds clocks were integral to accurate observation. When the Royal Observatory was established at Greenwich in 1676, high-precision clocks were as essential as high-quality telescopes.

Knowledge of position entailed measuring latitude and longitude. Latitude could be found relatively simply, by observing the angular height of the sun above the horizon (or rather, above the perpendicular to a plumb-line hanging vertical). If one were stationary, this was in principle an uncomplicated procedure, though one that could require some time to identify the sun's maximal elevation, assuming the sun to be visible. In finding longitude, it was necessary to know how far one was east or west of a set longitude, usually that of a home port, or a key astronomical observatory. That calculation required two key components: the local time in one's current location, and, simultaneously, local time at the reference longitude. The first was straightforward, the latter was the difficult part.

Finding longitude at sea is sometimes discussed as though finding longitude on land was unproblematic. It clearly was not, as the well-known apparent shrinkage of late seventeenth century France illustrates. The 1693 Picard-Cassini map of France, based exclusively on precisely timing eclipses of Jupiter's moons visible from different places, produced a considerably smaller France than the maps it superseded (Louis XIV supposedly remarked that France had lost more land to cartographers in a year than to foreign armies in the previous century). New longitudes for ports like Brest proved to be a whole degree different from their old longitudes (Figure 9.1). Not that the English had grounds for complacency—witness the inaccuracies of latitudes and longitudes mapped by Saxton in the 1580s and Adams almost a century later (Figure 9.2).

Accurate positions for particular places were only part of the solution. Their incorporation into more accurate maps depended on the cartographer being able to 'stretch' the hitherto-believed outline over the revised pattern of fixed point. This was among the problems faced by Joel Gascoigne in getting his (remarkably accurate) position for Lizard Point—some 30 miles different from that previously accepted—into the realm of cartographically depicted information (Ravenhill, 1976: 88–91).

On a much larger scale this was also the central problem for early modern world maps and globes, which incorporated massive errors, even in European navigators'

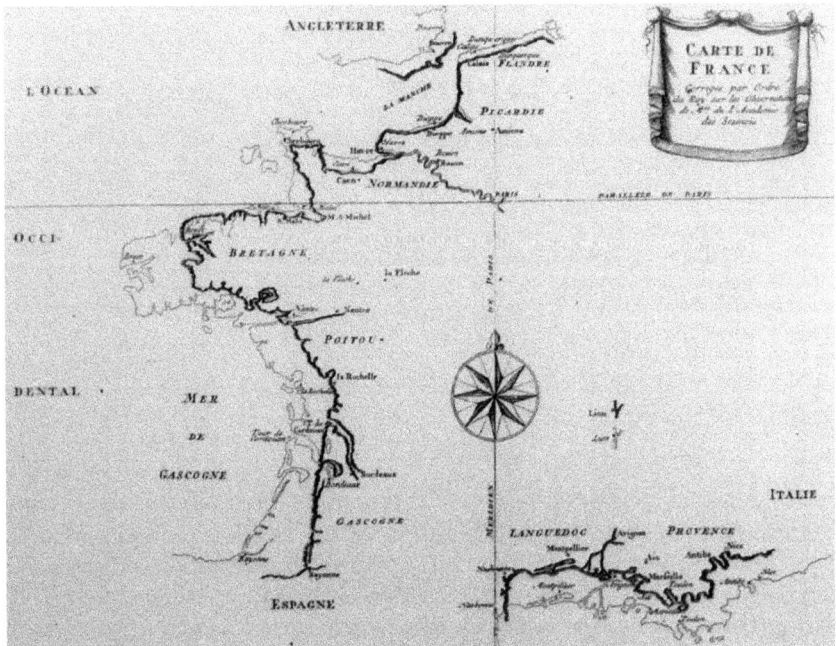

Fig. 9.1. The shrinkage of France, following the 1693 Picard-Cassini mapping of France with longitudes based on Jovian eclipses.

Source: Andrewes (1996).

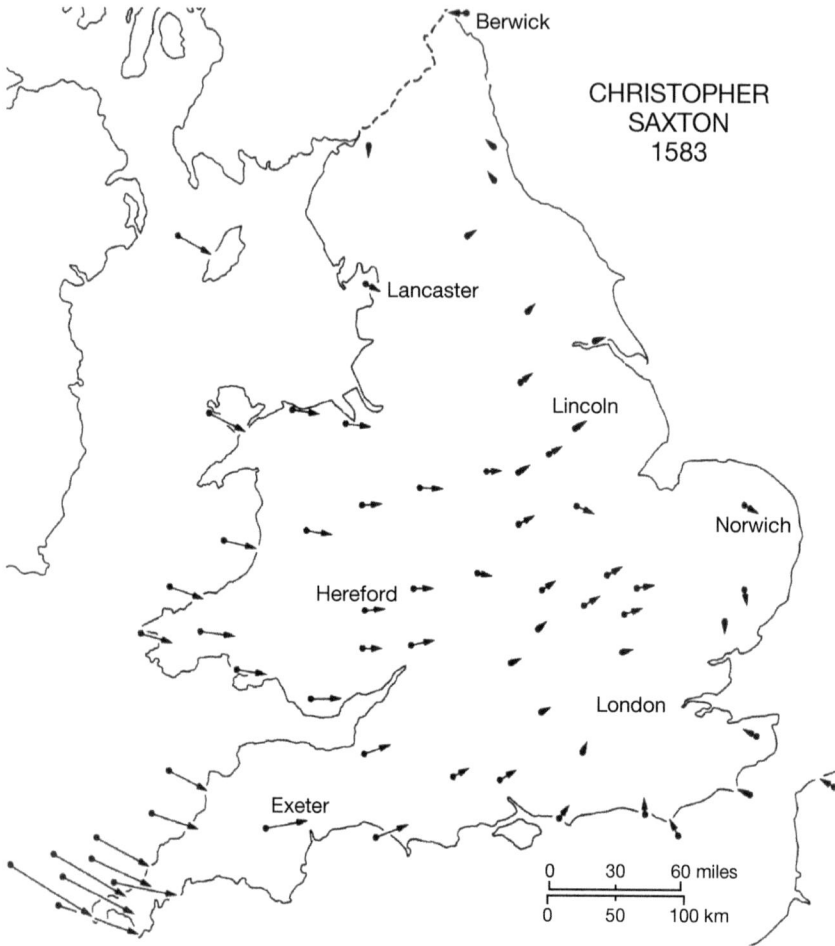

Fig. 9.2. Inaccuracies of latitudinal and longitudinal positions in maps of England by Saxton (1580s) and Adams (1670s).

own backyards. In its fifteenth-century Latin versions, Ptolemy's *Geography* gave longitudes for around 8,000 places, mostly determined by dead reckoning (that is, inferred from voyage times, Dilke 1985). Fifteenth-century editions of Ptolemy's *Geography* gave the length of the Mediterranean as 62 degrees. This was cut to 56 degrees by Oronce Fine in 1531, and progressively by Gerard Mercator from 58 degrees in 1541 to 53 degrees in his European map in 1554 (Crane, 2002: 80–1, 182). From the 1570s Abraham Ortelius's *Theatrum Orbis terrarum* (Antwerp 1570, London 1606), and the early seventeenth-century atlases of Mercator and others, put its length at between 51 and 53 degrees. All these were

Fig. 9.2. continued

still considerable overestimates: by 1700 its length was established at just over 41 degrees (van Helden, 1986: 88).

Unsurprisingly, errors outside Europe were generally much larger. For example, information about the coastline of Africa available from Portuguese explorations had, before 1500, produced maps that were fairly accurate with respect to latitude, but very crude with respect to longitude (for example, some of Abraham

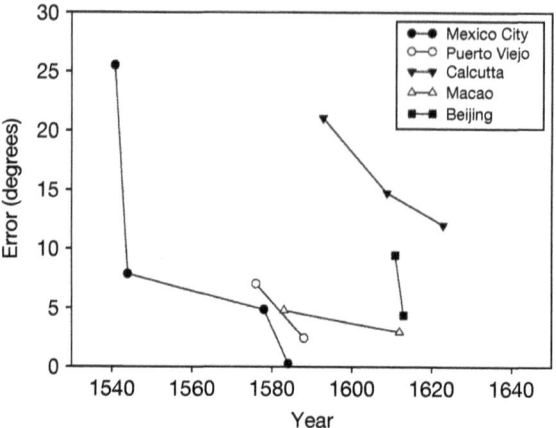

Fig. 9.3. Progressive reduction of longitude errors for selected major ports in Spanish and Portuguese territories.

Ortelius's maps of China and Japan are in error by more than 50 degrees). Mercator's Atlases contain a number of inconsistencies among different maps, even within the same edition, despite Mercator's strenuous efforts to reconcile divergent reported positions (Crane, 2002). World cartography spent decades in an awkward hybrid phase, as earlier maps were stretched over, or compressed into, coordinate frames of newly observed time differences.

It is perhaps somewhat surprising that longitude errors were reduced at very uneven rates, even for the Spanish and Portuguese (Figure 9.3). Spain and Portugal, after all, had especially powerful reasons for longitudinal precision, since the 1494 Treaty of Tordesillas divided colonial rights between them according to longitude. It is also striking just how long some inaccuracies persisted (as late as 1769, the charted longitude of Baja California was found to have been in error by about 5 degrees).

Finding longitude centred on being able to do to two particular things simultaneously. The first was to precisely identify particular times of the day, principally noon, for the resetting of clocks or watches to keep local (apparent) solar time. Having established that fixed point in time, reasonably accurate local timings could be made of a particular event or celestial position. The second was to know the local time in another place (typically an astronomical observatory) at the same moment as the observed event or position was timed locally. Predicted timings of those celestial events or positions for the observatory were calculated, and published in ephemerides or almanacs. So the longitude problem could be summarized as knowing the time in two places at once.

Early sixteenth-century commentators understood that longitude could in principle be found either astronomically (Werner, 1514; Apian, 1524) or

Table 9.1. Required elements for two methods of determining longitude

	Astronomical	Chronological
Accurate method for observing & correcting local time	X	X
Accurate astronomical tables for a fixed meridian	X	
Precise observing instrument (the sextant)	X	
Correction tables (atmospheric refraction, etc.)	X	
Calculation tables (trigonometrical, logarithmic, etc.)	X	
Reliable, precise timekeeper set for a fixed meridian		X
Accurate coastal charts	X	X

chronometrically (Frisius, 1530). The former involved measuring the relative positions of the moon, sun, planets, and/or stars, and comparing these observations with tables setting out the times at which particular conjunctions would be visible at an observatory, such as Greenwich. The difference between local times at the reference point and the observing position, if accurate, could be converted to their longitudinal separation. The latter, chronometer, method consisted of carrying a clock sufficiently accurate and reliable to keep running at (say) Greenwich time for at least the entire duration of an oceanic voyage. Longitude could then be calculated directly from the difference between Greenwich time and local time (making the necessary adjustments for the equation of time). Both methods, however, required much more accurate instruments than were available in the first half of the sixteenth century, or for a considerable period thereafter, and in the early eighteenth century solutions were still being sought down both routes.

For practical purposes, there were four main astronomical bases for determining positions from times: solar eclipses, lunar eclipses, Jupiter's eclipsing of its four visible moons (after their discovery by Galileo in 1610), and 'lunar distances' (which involved measuring the distances and angular separations of the moon from the sun or other stars). Assuming the meridian of the observatory was correctly known, and the predictions were accurate, the time difference between the actual local time of an event and its predicted observatory time measured the difference in longitude between observatory and observer. As already remarked, the key role of early Royal Observatories in England and France was to provide very accurate predictions for the future occurrence of celestial positions viewed from the particular observatory.

The successive application and refinement of new celestial timekeepers underlay revisions of world maps. For example, Phillipp Eckebrecht's world map published in Nuremburg in 1630 was based on time differences in the observation of lunar eclipses, using the star positions set out in Kepler's 1627 *Rudolphine Tables*, with longitudes measured with respect to Tycho Brahe's observatory at Uraniborg on the island of Hven between Denmark and Sweden. Later, potentially more precise representations of locations were based on timings of lunar distances, and of the visibility or disappearances of Jupiter's moons.

Almost immediately after discovering Jupiter's moons, Galileo had proposed that the moons were eclipsed by Jupiter often enough to act as a celestial timekeeper. Besides frequency (about 1,000 events a year visible from earth), their great advantage was their short duration (facilitating precise observation and timing). Realizing the potential of Jupiter's moons for timekeeping required precise models of the satellites' motions, accurate prediction tables, and improved telescopes, problems adequately handled only by the publication of new tables by Giovanni Domenico Cassini in 1668. Cassini's 1668 Jupiter tables were, within five years, deployed in longitude determinations for various parts of Europe, Africa, French Guiana, and southern and eastern Asia (Thrower, 1996: 55). The English Astronomer Royal John Flamsteed followed suit, publishing annual Jovian satellite tables from 1683 (Figure 9.4).

Astronomical measures continued as the mainstay of longitude determination through much of the eighteenth century. Although Captain James Cook is prominent in the history of exploration for using chronometer-watches during his second and third Pacific voyages, on his first voyage in 1769–70 he had produced several new maps, including the first detailed coastal map of New Zealand, which was

> based entirely on lunar distance calculations derived from observations taken aboard ship . . . As expected, latitudes were excellent but longitudes were not generally accurate, with the result that New Zealand was placed too far to the east by an average of about 25′ (about 20 miles at that latitude), with errors of up to 40′ (32 miles). (Thrower, 1996: 60)

By contemporary standards, though, these errors were remarkably small.

We shall discuss *finding* the precise time, first, and then *keeping* the precise time. It takes some mental effort to think of these as separate issues in a contemporary world full of clocks and watches, and saturated with precise and coordinated temporal infrastructures (standard times, precise time signals, international agreements and conventions and so on). But in the sixteenth century, such infrastructures did not exist. Most timepieces were rudimentary and/or imprecise, such as sundials, or weight-driven clocks. Finding and keeping temporal precision were approached in different ways: the former astronomically, the latter through attempts to develop precision timekeepers. While finding and keeping the time are facets of the same problem (since finding the time repeated frequently enough would amount to almost-continuous time), they were particularly separate problems.

9.4 FINDING AND KEEPING PRECISE TIME AT SEA

Although with hindsight, contemporaries' preferences for astronomical rather than chronometric methods of time determination can be portrayed as a

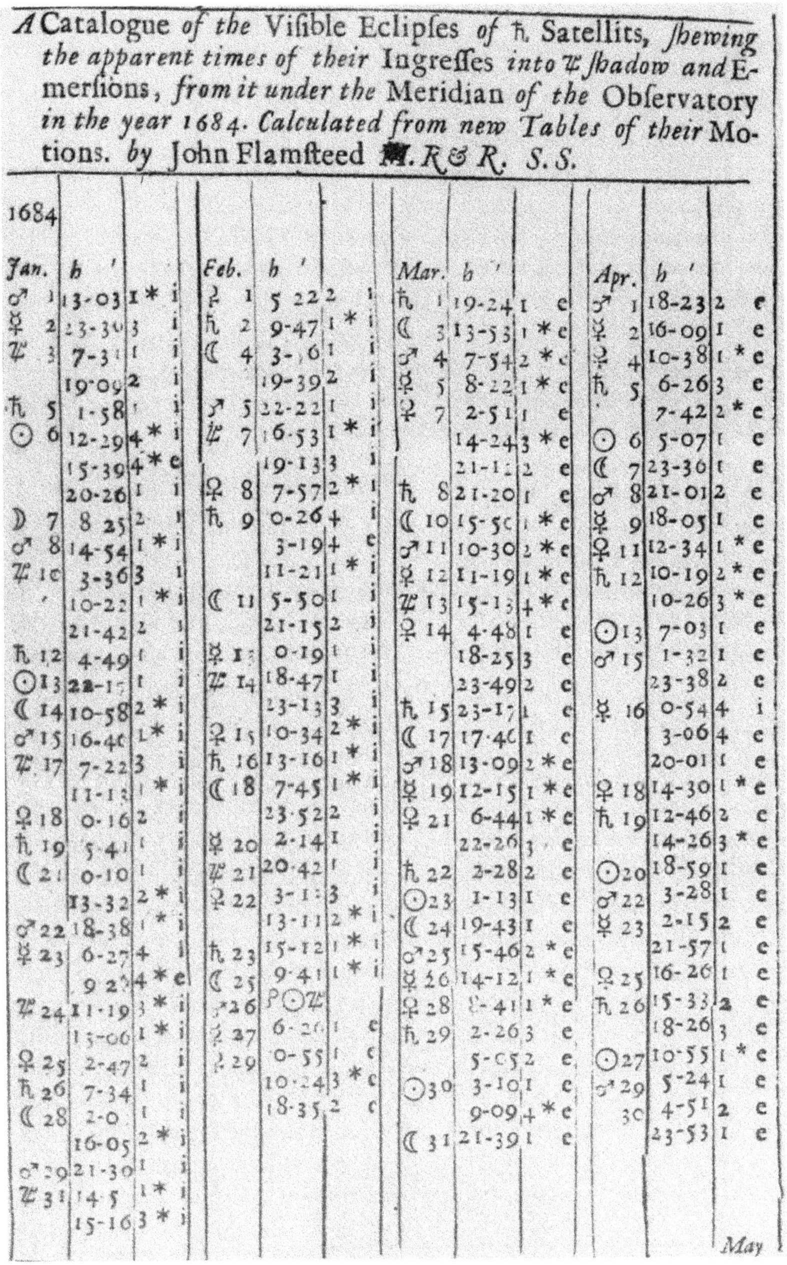

A Catalogue *of the* Visible Eclipses *of* ♄ Satellits, *shewing the apparent times of their* Ingresses *into* ♃ *shadow and* Emersions, *from it under the* Meridian *of the* Observatory *in the year 1684. Calculated from new Tables of their* Motions. *by* John Flamsteed M. *R.* & *R. S. S.*

1684

Jan.	h '		
♂ 1	13-03	1 *	i
☿ 2	23-30	3	i
♃ 3	7-31	1	i
	19-09	2	i
♄ 5	1-58	1	i
☉ 6	12-29	4 *	i
	15-39	4 *	e
	20-26	1	i
☽ 7	8 25	2	i
♂ 8	14-54	1 *	i
♃ 10	3-36	3	i
	10-22	1 *	i
	21-42	2	i
♄ 12	4-49	1	i
☉ 13	22-15	1	i
☽ 14	10-58	2 *	i
♂ 15	16-40	1 *	i
♃ 17	7-22	3	i
	11-13	1 *	i
♀ 18	0-16	2	i
♄ 19	5-41	1	i
☽ 21	0-10	1	i
	13-32	2 *	i
♂ 22	18-38	1 *	i
☿ 23	6-27	4	i
	9 23	4 *	e
♃ 24	11-19	3 *	i
	13-06	1 *	i
♀ 25	2-47	2	i
♄ 26	7-34	1	i
☽ 28	2-0	1	i
	16-05	2 *	i
♂ 29	21-30	1	i
♃ 31	14-5	1 *	i
	15-16	3 *	i

Feb.	h '		
♀ 1	5 22	2	i
♄ 2	9-47	1 *	i
☽ 4	3-16	1	i
	19-39	2	i
♂ 5	22-22	1	i
♃ 7	16-53	1 *	i
	19-13	3	i
♀ 8	7-57	2 *	i
♄ 9	0-26	4	i
	3-19	4	e
	11-21	1 *	i
☽ 11	5-50	1	i
	21-15	2	i
☿ 13	0-19	1	i
♃ 14	18-47	1	i
	23-13	3	i
♀ 15	10-34	2 *	i
♄ 16	13-16	1 *	i
☽ 18	7-45	1 *	i
	23-52	2	i
☿ 20	2-14	1	i
♃ 21	20-42	1	i
♀ 22	3-13	3	i
	13-11	2 *	i
♄ 23	15-12	1 *	i
☽ 25	9-41	1 *	i
♂ 26	P ☉ ♃		
☿ 27	6-20	1	e
♀ 29	0-55	1	e
	10-24	3 *	e
	18-35	2	e

Mar.	h '		
♄ 1	19-24	1	e
☽ 3	13-53	1 *	e
♂ 4	7-54	2 *	e
☿ 5	8-22	1 *	e
♀ 7	2-51	1	e
	14-24	3 *	e
	21-12	2	e
♄ 8	21-20	1	e
☽ 10	15-50	1 *	e
♂ 11	10-30	2 *	e
☿ 12	11-19	1 *	e
♃ 13	15-13	4 *	e
♀ 14	4-48	1	e
	18-25	3	e
	23-49	2	e
♄ 15	23-17	1	e
☽ 17	17-40	1	e
♂ 18	13-09	2 *	e
☿ 19	12-15	1 *	e
♀ 21	6-44	1 *	e
	22-26	3	e
♄ 22	2-28	2	e
☉ 23	1-13	1	e
☽ 24	19-43	1	e
♂ 25	15-46	2 *	e
☿ 26	14-12	1 *	e
♀ 28	8-41	1 *	e
♄ 29	2-26	3	e
	5-05	2	e
☉ 30	3-10	1	e
	9-09	4 *	e
☽ 31	21-39	1	e

Apr.	h '		
♂ 1	18-23	2	e
☿ 2	16-09	1	e
♀ 4	10-38	1 *	e
♄ 5	6-26	3	e
	7-42	2 *	e
☉ 6	5-07	1	e
☽ 7	23-30	1	e
♂ 8	21-01	2	e
☿ 9	18-05	1	e
♀ 11	12-34	1 *	e
♄ 12	10-19	2 *	e
	10-26	3 *	e
☉ 13	7-03	1	e
♂ 15	1-32	1	e
	23-38	2	e
☿ 16	0-54	4	i
	3-06	4	e
	20-01	1	e
♀ 18	14-30	1 *	e
♄ 19	12-46	2	e
	14-26	3 *	e
☉ 20	18-59	1	e
♂ 22	3-28	1	e
☿ 23	2-15	2	e
	21-57	1	e
♀ 25	16-26	1	e
♄ 26	15-33	2	e
	18-26	3	e
☉ 27	10-55	1 *	e
♂ 29	5-24	1	e
30	4-51	2	e
	23-53	1	e

May

Fig. 9.4. Table of predicted positions of Jovian satellites, the first in England, by the Astronomer Royal, John Flamsteed, 1683.

Source: Andrewes (1996).

combination of vested interests and conspiracy (Sobel, 1995; Barnett, 1997), there was far more to this bias (Andrewes, 1996). Astronomical methods had been successful for large-scale surveying on land, and ocean navigation can hardly be divorced from the global positions of land. Indeed, determining longitude at sea is only a meaningful problem in relation to longitude on land. As seventeenth-century commentators asked, what use was knowing the longitude of one's ship if you did not know where the Lizard was? Finding positions at sea took for granted that positions on land were known, even when this was manifestly not the case. But land surveys and cartography had established most of the methodological principles and practical techniques for oceanic navigation, and the obvious practical requirement of charts of coasts, inaccurate as these were, with which to compare one's computed position at sea.

Various astronomical positions and conjunctions provide potential temporal markers in the sky. For example, at any given place, the maximum daily solar altitude defined noon. But using the sky to establish longitude required the knowledge of local time in two places at once, and hence required both an event or position observable simultaneously at places of different longitude, and a means to establish local time.

In theory longitudes could be established from knowing the different local times at which an instantaneous event had occurred. But in practice such conditions were not fulfilled by events like solar eclipses, because they are not seen simultaneously in different places. The moon's shadow, and the track of the eclipse shadow, move across the earth's surface. A lunar eclipse (or rather, certain moments of it) is much more nearly simultaneous in different places.

Human societies all over the world had long been fascinated by, and often fearful of, some such events, such as solar or lunar eclipses. A widespread belief that they were portents helped the development of record-keeping in many parts of the world, and an interest in making predictions of eclipses in the future. But solar eclipses are rare, and lunar eclipses are relatively infrequent events. Lunar eclipses were therefore more or less restricted to measuring longitudes on land. Galileo's telescopic discovery of the four principal moons of Jupiter in 1610 was the first 'new' sort of eclipse to be visible. And with, he estimated, about 1,000 moments a year when one or other satellites was passing an identifiable momentary position, Galileo immediately appreciated the possibilities for longitude determination.

Already, there had been attempts to obtain a more continuous time-measuring system from the moon's position in the sky, first by the moon's occultation of individual stars, and second by the so-called 'lunar distances' method. The latter involved measuring the separation in the dimensions of altitude (height above the true horizon) and direction (compass bearing). It is striking how rapidly tables predicting astronomical events and positions, long circulating in manuscript in both Islamic and European societies, were brought into print—within twenty years of the first European printed books, in the case of Regiomontanus's *Kalendarium*, mentioned above.

Mariners' use of trigonometric tables for solving the 'nautical triangle' was one sign of a new phase of 'arithmetic navigation' in the years after *c*.1600. Such calculations were much helped by use of the logarithms first published by Napier in 1614 and incorporated by Edmond Gunter's in tables published in his *Canon Triangularum* in 1620, 'one of the fundamental books of arithmetical or modern navigation' (Waters, 1958: 408). Armed with the new logarithmic canon, 'a navigator who memorised the necessary rules could solve nautical astronomical problems with relative ease' (Cotter, 1983: 242).

Improvements in the precision of astronomical calculations of longitude came most obviously from better-quality predictions, based on more and better positional observations. As long as the exact longitudes of places that were the reference points for tables remained uncertain or in error, astronomical tables were bound to be inaccurate, as would be the longitudes derived from them. So one element of greater precision in tables was the more precise establishment of longitudes for reference observatories, as when the longitude for Greenwich was corrected in 1677. Flamsteed's first Greenwich-based tables were a significant step forward from earlier records of precise celestial positions, but were still relatively coarse. In the long run, though, successive improvements made both observing instruments and astronomical reference tables, fuller, more detailed, more accurate, and in some cases more tailored to the particular needs of their users.

The key advance in the accuracy possible with the lunar distances method came

through the work of the German astronomer Tobias Meyer of Göttingen which required advanced mathematics, but no special instruments. Meyer's lunar tables were first published in 1755, were tried at sea by Captain John Campbell in 1757 and 1758, tried again by the Reverend Neville Maskelyne on a voyage to St Helena in 1761, and were published to an English-speaking audience at the end of the war in his *British Mariner's Guide*. Well before this, however, the news had diffused through the navy. Less than a year after Meyer's original publication a young officer, Lieutenant John Elliott, wrote home to his father from sea that he had no news, only the discovery of the longitude by a Hanoverian. The observation is simple and easy but the calculation is extremely perplexed. (Rodger, 1988: 52)

Within a short period, this perplexing calculation was already being taught in the mathematical schools which specialized in preparing boys for the navy and merchant service.

The various elements needed to make the lunar distances method practical came together when Neville Maskelyne published his *New Mariner's Guide* in 1764, which consisted of a more accurate set of the necessary lunar tables along with an enthusiastic advocacy of the method. The tables had been compiled by Tobias Mayer of Göttingen, using Newtonian principles, and were the first to achieve real accuracy, and the first, therefore, that made a long-known theoretical method for the longitude practicable. The use of the vernier on the sextant was now indispensable, for readings had to be made within two or three seconds of arc. The index arm was tightly held by a screw when an approximate observation had been made, and the last fine adjustment of its position was carried out by means of a fine threaded side-screw. The scale was then read through an eyepiece carrying a lens. A telescopic sighting tube was substituted for the original simple aperture through which the fixed mirror was observed. (Taylor and Richey, 1962: 57)

As tables became more accurate, other complications emerged although they were not always recognized at first, until it became apparent that they were a significant source of variation or error. As knowledge increased it was found necessary to have additional correction tables—for dip of the horizon, for refraction, and for lunar and solar parallax—to correct for geometric and atmospheric factors causing apparent positions to be (slightly) misleading. In addition, the progressive improvement of telescopes involved in the pursuit of precision in the observatory led to small but increasing discrepancies in timing between telescopes and the naked eye. Observers with better telescopes 'saw' some phenomena earlier than could an unaided human eye. Similarly, telescopes varied. It was realized that Jupiter's moons did not appear or disappear at an absolute moment: the 'real' time of these events depended on the instrument used to view them. Prediction tables, therefore, could only be absolutely exact for a particular power and performance of telescope (van Helden, 1996: 99).

Besides observing instruments and astronomical tables, a third area of related precision lay in the tools with which to make comparative calculations, through auxiliary mathematical tables, set within precise performative routines. In speeding up calculations, and in reducing the scope for arithmetic errors, important auxiliary tables included sine and tangent tables for trigonometric calculations, and logarithmic tables to replace large-scale multiplication and division calculations with additions and subtractions. The use of auxiliary tables was one part of a more general process of constructing explicit performative routines. The central aim here was to restructure very complex and demanding procedures into a much more easily learned routine of small, clear stages.

Complexity and simplicity are relative of course, as may be gauged from the example of the framework for rendering lunar distance calculations into such a routine. The various sections of this sheet contain different stages, from the recording of preliminary observations in the top-left hand corner, to the eventual emergence of a precise location in the bottom-right. 'Speeding up' is relative too, and even professional astronomers are on record as struggling to perform this sort of task in less than three to four hours (though their critics accused them of taking too long being perfectionist about the observing stage). Even so, this clearly remained a highly skilled operation.

Maskelyne also oversaw the application of the lunar distances method amongst naval masters, breaking down the highly complex series of observations and adjustments required to produce an accurate longitude estimate into a series of dozens of small steps in which mariners could be drilled. The step-by-step 'Maskelyne's method' required an ability to follow complex sequences of instructions rather than to understand the mathematics being deployed, but was capable of relatively accurate results, though it was prodigiously time-consuming—a minimum of three to four hours.

With his lunar tables (reprinted with many other tables in the first *Nautical Almanac and Astronomical Ephemeris*, printed in 1765, but providing tables for the year 1767) the Astronomer Royal (as Maskeleyne now was) explained the method of using them to find the longitude, besides what were termed 'The Requisite Tables' for clearing the observations of errors, and computing the result. The process of working out the longitude from the fixed observation of lunar distances filled a whole foolscap page with mathematical calculations (Taylor and Richey, 1962: 106–7).

The intrinsic advantage of the astronomical approach to reference times was that it enabled the reference time to be found afresh. The reference tables provided a means of renewing the time for the reference observatory and, in combination with the determination of local time, obtained either astronomically or from an accurate clock, established a longitude. On the other hand, exclusive reliance on the horological method was precarious, being vulnerable to any disruption to the reference timekeeper. Of course, the two methods could be used in conjunction and even after the general use of chronometers the taking of celestial sightings with a sextant remained part of the training of naval officers, and of some seamen. But the chronological approach alone was mistrusted because of the impossibility of refinding the reference time without accurate reference tables, astronomical instruments, and good observing conditions.

The chronometric route to precise timekeeping is now known to far more people than seemed conceivable when we began this project in 1992, largely through the enormous attention generated by Dava Sobel's *Longitude*. For the ship-borne community, chronometric methods had proved particularly problematic to implement, given the variations in temperature and humidity during long voyages, and the unavoidable physical disturbance of ocean sailing. More than that, the chronometer method of longitude determination was hamstrung in the absence of devices with natural periods (i.e. with a repetitive, highly regular oscillation). No such devices existed prior to the mid-seventeenth century, although Galileo had formulated the idea of pendulum driven clocks in around 1620.[3]

On its own, the chronometer approach faced three main obstacles. One was the construction of a timepiece sufficiently accurate to keep the time for the known meridian. Rapid improvements in the accuracy of timepieces were being made hand over fist in the late seventeenth century, but not to a point that could provide precise longitudes, even in ideal conditions. Even small daily errors could accumulate over a few weeks, resulting in positional errors of scores of miles. The second obstacle was reliability in the face of the physical duress of long sea voyages: the physical buffeting, which might accelerate or retard the mechanism; the fluctuations in temperature which could cause components of the clocks to expand or contract, and affect the viscosity of lubricants used to reduce frictional wear on the mechanism. Third, knowledge of the time at a fixed point produced by this method was precarious. Once a clock was out of time, it would be impossible to correct, at least without making landfall at a place whose longitude was known. The last of these was a powerful argument in favour of the alternative—astronomical—approach.

Although identified in principle by Gemma Frisius in the 1520s, and notwithstanding the advance in both substantially advancing theoretical and practical understandings of mechanical timekeeping over the next two centuries, there was still no workable device in the early eighteenth century. The construction of a sufficiently accurate timepiece to keep accurate time for base meridian had proved an impossible task. Improvements in the accuracy of timepieces were being made hand-over-fist in the late seventeenth century, but not to a point that could provide precise longitudes, even in ideal conditions. Some contributors to seventeenth- and early eighteenth-century debates on longitude chronometers, such as Jeremy Thacker in 1714, thought themselves on the verge of a decisive breakthrough, but Thacker was merely the latest of a long line of people to hold such a belief. And none of them had succeeded. This was the context in which John Harrison made his decisive contributions from the 1720s to the 1770s.

9.5 INTER-DEPENDENT PRECISIONS

As we have hinted, temporal precision was not self-contained: it necessitated other precisions. At least five areas of precision were critical to temporal precision, and not until Maskeleyne's day were these five essential conditions fulfilled: there must be accurate lunar or other reference tables for some fixed meridian; there must be high accuracy in the use and construction of celestial observing instruments—Maskeleyne had the sextant (Chapman, 1995a); there must be sufficiently precise tables for really accurate calculations and trigonometry; there must be an accurate method for observing local time, which had become possible by 1763; and all this took for granted a high degree of precision.

Figure 9.5 makes clear the interdependence of various elements of precision, for both methods of determining longitude. Each element had its own developmental trajectory, with a larger or smaller number of applications (and hence greater or lesser breadth of demand for improvements), but, from the standpoint of navigation, each was entwined with the others. While individual components could be improved in isolation, the effort put into improving them was markedly affected by progress or problems with other, related precisions.

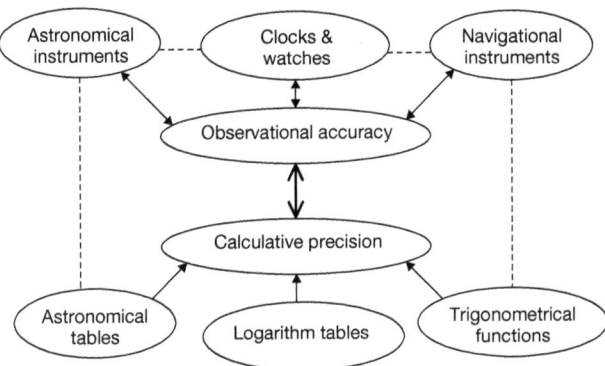

Fig. 9.5. Astronomical precision-timing as a network of interdependent precisions.

These inter-connections were often characterized by obstacles or bottlenecks that involved thresholds, rather than steady improvement. For example, astronomers' tables of the positions of the sun, moon, planets, and stars were (by and large) successively more accurate than their predecessors, but they were useless without accurate clock time. Likewise,

Before seagoing chronometers were perfected, there was little incentive to develop better sextants. . . . Since the earth turns on its axis one degree of arc in 4 minutes of time, there is no utility in having an instrument that can measure celestial angles even to the nearest degree unless it is coupled with a chronometer that is accurate to within 2 minutes. (Hutchins, 1995: 112)

Hutchins slightly over-states his case, since astronomical methods could function with a reasonably reliable watch or clock, rather than requiring a chronometer, provided that noon or some other daily benchmark could be reliably established, but the interdependence of facets of precision remains. Both sextants and precision tables were technological possibilities before the chronometer, but the utility of each was substantially increased by really accurate time-reckoning.

Greater instrumental precision was sought (and sometimes obtained) from many sources: larger scale instruments, better-constructed instruments, more precisely graduated scales, more precise sighting devices, more sophisticated trigonometry in the design of observations, telescopes, micrometer controls on instruments and sights, and—last but by no means least—the cultivation of great manual dexterity. Observational precision was also enhanced by more systematic observation, the precise integration of series of observations, more rigorous cross-checking of networks of mutually calibrating observations, and by the design of very specific observations in relation to particular cosmological questions.

Systematic clock times could not cover the oceans without an accurate way of determining time at sea, and hence longitude, one of the most enduring technological challenges and obsessions in early modern Europe. The lack of precision chronometers until the late eighteenth century held up the development of devices and documents, including the sextant and accurate navigational tables. Undoubtedly, enormous progress was achieved in observational precision, but there were larger and larger divergences between state-of-the-art astronomical instruments, and practical instruments for use in the field, or on the deck. We turn next to the balancing of these precision-obtaining and precision-applying preoccupations, and to the bridging of the gap between them.

9.6 TRANSFERRING PRECISION AMONG DIVERGENT COMMUNITIES OF PRACTICE

Over time, one community's progress was another's handicap. The sources of greater precision for astronomers involved a complex of increasingly precise observations, measurements, techniques, and calculations. These precisions, so valuable for the astronomical community, were positively unhelpful to the navigational community. Increasing observatory precision entailed more accurate instruments. Prior to the telescope, 'more accurate' was generally synonymous with 'larger'. After the spread of telescopes 'more accurate' generally meant 'more delicate'. Neither increasing size nor greater delicacy was a helpful development in trying to reproduce 'observatory-precision' at sea. The problem was to make extremely specialized and technically demanding astronomical skills—demanding in both observational skills, and abilities in calculation—'work' on board ships, and in the hands of non-astronomers.

The communities in different contexts handled components of temporal precisions in rather different ways: they demanded or imposed different degrees of precision, they set different criteria for what counted as precision, they endorsed and validated different techniques and instruments in pursuit of appropriate and achievable levels of precision. Precision was both an abstract and a practical issue, but the mix of abstraction and practicality differed markedly among the observatory, the horoscope, the plane table, the deck of a ship, and the study. For example, oceanic navigation imposed a need for (nearly) continuous observation in ways that land-based surveying and cartography did not. And acute precision was still extremely important to the navigator, since measurement or calculation errors could be disastrous, but the practicalities of obtaining measurements at sea at all was absolutely central.

What we can see here is a separation between two formerly highly similar communities. The gap between the instruments on which astronomers were observing and those on which mariners were measuring was widening. The needs of the two groups, as regards information about Jupiter moons, were diverging.

Cassini, Flamsteed, and others faced a growing tension between satisfying their astronomical and navigational audiences. On the one hand, they sought greater precision through increasing attention to instrumental adjustments, and the running of clocks, to accurately measure times of immersion, eclipse, and emersion.

But at the same time, they sought to tailor methods and tables for use by an overwhelmingly non-expert audience of mariners and solo explorers on land. This involved simplifying the complex procedure into a series of routine and unambiguous steps, using pre-printed forms, auxiliary tables, and the like. The goals here were to simplify and accelerate the transformation of raw observational data into a precise time.

The divergence of astronomical and navigational communities brought with it greater difficulties in translating between them. Tables that were improved by astronomical standards were less useful in relation to navigational observations. As observational technologies diverged, tensions increased between measures simplifying techniques for navigational use (tables, reckoners, and manuals, which made a highly skilled process more routine and mechanical), and tables requiring increasing adjustment to bridge the growing gap between astronomical and navigational instruments.

The considerable progress made between the mid-seventeenth and mid-eighteenth centuries in precision timekeeping was in part predicated on much greater control of conditions in increasingly specialized spaces: observatories, laboratories, or precision workshops. Both astronomical and chronometric approaches to obtaining precise time-indication then faced the problem of transferring such practices of precision from their particular dedicated spaces to the far-from-ideal space of the ship. Each had a largely distinctive strategy for doing so: astronomical timekeeping was made transferable from the mathematically expert astronomer to the ship's master through a 'routinization of skills' that broke complex calculations down into very many simpler stages; while chronometry was made transferable by an almost exactly opposite approach, which made the chronometer almost entirely independent of human activity by 'taking human functions into the machine'.

The first solution is easy to summarize: it was to break down the observational and calculative tasks into steps that were individually straightforward enough to be performed by non-specialists, and then to embed these individual tasks in a routine procedure so that each would be performed reliably, and in the correct order, with the results from some steps feeding into others, or being brought together in further calculations. The apogee of this approach came after the middle of the eighteenth century, as illustrated in Figure 9.6.

This figure shows the sort of master sheet used to handle the 'intermediate products' of each stage of the procedure. The observer made a series of observations, which were recorded in the appropriate places, and performed various averagings, calculations, and corrections. These made use of supplementary tables

Fig. 9.6. Recording sheet for lunar distance observations, and calculations of longitude based on them.

Source: Andrewes (1996).

of various kinds, both to speed up calculations (including logarithm and tangent tables) and to enable the derivation of the time at the base observatory (in this case Greenwich) at which that specific celestial configuration would have been seen. The latter took the form of predicted positions at three-hour intervals, with the navigator needing to interpolate intermediate positions.

Such calculations were fearsomely time-consuming. When the Astronomer Royal Nevil Maskelyne went through the full suite of tiny stages and calculations of the procedures he had designed to be taught to Royal Naval navigators in the mid-eighteenth century, he took nearly four hours, though it was admitted that he might have taken longer than necessary in trying out various observing situations. Presumably, though, he was likely to have performed calculations more swiftly and assuredly than the average ship's master. However, notwithstanding the unwieldy performance, and the large amount of time required, these methods were very widely performed, on ships ranging from admirals' flagships to sealing vessels. And they did work.

Several recent publications prompted by the 1999 total eclipse of the sun have retold the story of Cook's 1768–70 Pacific voyage, and its goal of reaching Tahiti to observe the transit of Venus across the face of the sun (e.g. Steel, 1999). Tahiti was an important location, as more or less the only site in a huge expanse of ocean, and observations of the transit were wanted from a wide spread of places around the earth. Like the early eighteenth-century attempts to measure the size of the sun's shadow on the earth during total solar eclipses, the transit was expected to help resolve major astronomical questions about the scale and geometry of the solar system, in this case, the distance between the earth and the sun. As had been suggested by Edmund Halley more than fifty years earlier, simultaneous observations from widely separated parts of the earth's surface, and close comparisons of the exact onset and ending of Venus's transit at different points, would provide the key data. First, though, it was necessary both to identify sites from which to observe, and to know exactly where they were.

In 1768 the naval vessel *Dolphin*, returning from a circumnavigation lasting a year and three-quarters, reported its discovery of the island of Tahiti. Its longitude was established using Maskelyne's method with impressive accuracy, given the circumstances, differing by less than 20 kilometres from present-day reckoning. In the log of the *Dolphin*, though, alongside a delightful name-shift, is the telling comment that longitude had been determined by:

taking the Distance of the Sun from the Moon and Working it according to Dr Masculine's Method which we did not understand. (Cited by Gurney, 1997: 10)

But in many ways that was precisely the point. In its own terms, this was Maskelyne's method in action; as appropriately drilled seafarers made observations and performed calculations that produced the right answers, despite the men's lack of understanding of why they did what they did. Maskelyne's method required only that men accurately performed a long routine of clearly structured tasks: it was irrelevant that they lacked a deeper grasp of what they were doing, and could not have produced a proper calculation or the correct result except through the long procedure in which they had been drilled.

Very much the same sort of approach—structuring a complex operation as a sequence of small, relatively straightforward steps, in which unskilled men could be drilled—had been taken by Sir William Petty in organizing his comprehensive survey of Ireland. Petty designed a framework of surveying procedures comprising sequences of very particular instructions and

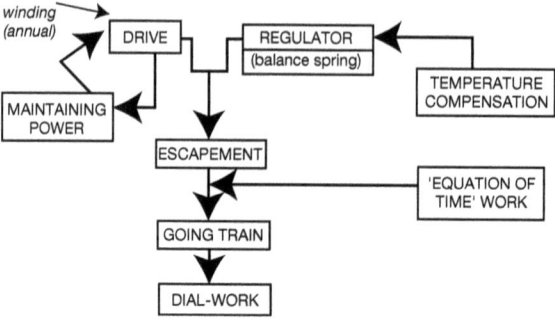

Fig. 9.7. The marine chronometer as a network of minimal human interference (compare with Figures 2.3 and 2.5).

measurements, in which his men were instructed, trained, and drilled. In Aubrey's words:

> these that he employed for the geometrical part were ordinary fellows, some (perhaps) foot-soldiers, that circumambulated with their box and needles, not knowing what they did, which Sir William knew very right well enough how to make use of. (Barbor, 1982: 243, original emphasis)

The second approach took almost completely the opposite strategy. The marine chronometer made very precise longitudinal positions knowable by extending the technological trajectory outlined in Figures 2.3 and 2.5 to its logical extreme, that is, by removing human activity almost completely (Figure 9.7). Rather than unpacking calculations to be individually drilled into non-specialist astronomers, marine chronometry involved removing the people from the process altogether, except in terms of reading the instrument. Formerly human functions were completely 'taken into the machine'.

In 1522 Gemma Frisius pointed out that an accurate timepiece carried by a traveller, and kept wound up, would enable the calculation of longitude from comparison of the timekeeper and local time. An English translation of Frisius's work was published in the 1550s, but more than two hundred years elapsed before a sufficiently accurate timepiece was available. Since an error of four minutes in time produced a positional error of one full degree of longitude, so accuracy within a few seconds over several weeks was required. In the mid-1650s, Christian Huygens had developed the concept of mean time, and discussed an intimidatingly long list of reasons why even his best-performing clocks were unreliable: the expansion and contraction of metal parts with temperature changes, variations in running speed with atmospheric pressure, corrosion of parts in moisture-laden conditions, clogging caused by lubricants, and more. Every occasion for humans to intervene in the running of the chronometer potentially raised problems in the stability of the timepiece's performance.

With hindsight, one reason for the failure of Huygens's sea-clocks lay in the whole tenor of his approach to the pendulum as a physical object. Huygens central interest in pendulums focused on 'the mathematization of nature', that is, on providing a mathematical understanding of the general problem of objects' behaviour when falling under gravity, especially their speed (Yoder, 1988). This had also been the root of Galileo's interest (Drake, 1990). Huygens's concern with mathematical rather than applied priorities is tellingly illustrated by the pendulums in his proposed clocks. He used a light pendulum, suspended by as fine a cord as possible, in order to conform with the mathematical case in which the weight of the cord was negligible, and the pendulum small and light (enabling its position to be plotted more exactly as a point). This also enabled Huygens to leave aside the effects of friction caused between the pendulum rod and the 'cheeks' used to force the bob to follow a cycloidal path (Yoder, 1988: 1–8, 11). Harrison later made a very similar criticism of George Graham's regulator clocks, while his own use of a very heavy pendulum, whose movement he sought to stabilize, clearly conveyed that his supreme priority lay in applied science, rather than in mathematics for its own sake.

This priority led to a quite different strategy in producing regulators and pendulum-regulated sea-clocks. They used much heavier pendulum bobs than Huygens envisaged, though this in turn necessitated heavier rods to hold the bob rather than a very light cord. In doing so, they increased the cheek-rod frictions that Huygens sought virtually to eliminate, and had to consider devices to overcome the consequent loss of speed in the pendulum, through new designs of escapement, like Harrison's own 'grasshopper' escapement. In short, the clockmaking community adopted a quite different strategy to that of Huygens. While Huygens thought to make clocks that resembled as closely as possible the mathematical system he had (in a broad sense) modelled, the makers sought a less elegant system that could prove robustly accurate in use.

Harrison not only solved the technical problems, but completed the removal of humans from direct involvement in the running of the chronometer. Harrison's development of a reliable chronometer, becoming available from the 1760s, solved the technical problems, but the early chronometers were expensive (very few ships' masters could have afforded them), and they therefore spread quite slowly. Royal Naval ships going on a foreign voyage were entitled to a chronometer, on the captain's application, but they were not made general issue until 1825 (Lavery, 1989: 186). Outside the navy, they spread rapidly only when a combination of improvements in mechanized manufacture, and significant reductions in production costs, and prices, brought chronometers within reach of many navigators (Kemp, 1976: 167). Even so, they still needed very careful handling on board ship and required periodic checking against lunar observations, which still provided a fall-back should a chronometer go wrong (Lundy, 2002: 165).

There is a recurrent theme here, of divergence among the priorities and practices of different temporal communities. Whether it is the contrasting settings of the observatory and the quarterdeck, or the contrasting mathematical and applied priorities among clockmakers, these cases involve communities of practice with different understandings of, and attitudes towards, temporal precisions The devices, and related instruments and texts, used by different communities also diverged, as features which had practical merits vis-à-vis the goals of one community often offered no attraction, and perhaps significant disadvantages, to others. What might seem like a fixed practice took different forms in different communities.

In turn, this affected the ability of precisions to 'travel' among different temporal communities, and whether 'leakages' or 'spillovers' in knowledges were readily transmitted to, or absorbed by, other specialized communities. Precisions could have specialized and popular forms, depending on how readily the specific material and contextual facets of certain precisions could be taken into general, 'everyday' practices and popular discourses. Sometimes, representations and aesthetics of precision likewise diverged, where different temporal communities appreciated or valued different facets of precision-related practices.

We earlier distinguished between 'lay' and 'public' communities of practice in popular attitudes towards precision. We identified several features of 'lay' community of precision with respect to the ephemeral community of practice of eclipse timing, created by Edmund Halley in 1715, and intermittently thereafter. Notwithstanding the misunderstandings or deficiencies in equipment that led to Halley's reservations about some reported timings, his community of reporters did work. While ephemeral, the community performed adequately enough to generate the information Halley needed. So the

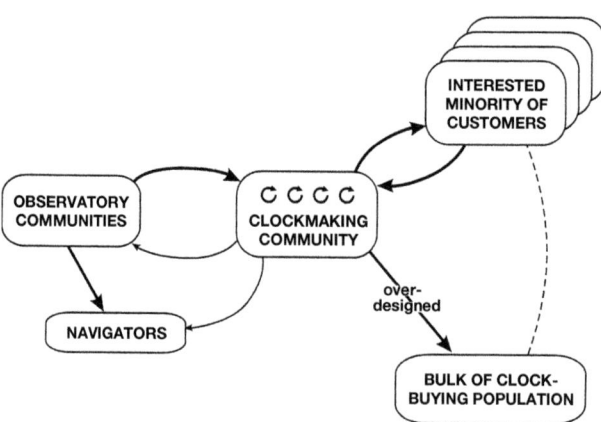

Fig. 9.8. Interacting communities of practice in clockmaking.

fact of divergent communities of practice did not necessarily preclude successful cooperation between them, or the transfer of practices into general use and circulation.

9.7 CONCLUSION: FROM ACTION TO WORLD INTELLIGIBILITY

Using a range of substantive evidence, we have explored a variety of precise times in early modern timekeeping, each with different stances (senses of appropriateness), conventions, and embodied knowledges, all played out performatively. (Relatively precise times were both an ingredient of activities, and a product of them.) Precise times were kept more by some communities than others, sometimes as a matter of circumstance, sometimes as a matter of policy, and specialized communities, contexts, and social relationships generated their own values of precision. The histories of 'precision-achieving' processes show that finding the precise time was a complex and difficult activity. Several strategies, sometimes overlapping, were developed by various different communities, with which they could achieve temporal and longitudinal precision in ways appropriate to their circumstances for observing and calculating.

Why, and how, the uses and values of precision spilled over and diffused among different temporal communities are important research questions. Usually, temporal precisions did not simply diffuse from donor-community to imitator-community. Transfers of precision were complex processes of negotiation and integration with other temporal values, understood as frameworks of action which have to be brought into rough correlation with these new values through practices, usually through the negotiation of specific 'ecological' situations (Gell, 1992). In these shifts the values associated with precision could change, therefore affecting some contexts and spaces more than others, for cultural and conjunctural reasons.

To varying degrees, specialized values of precision fed into other everyday lives, and the sorts of usages we examined in Chapter 8, but not necessarily in simple ways. In moving from the 'action intelligibility' associated with particular specialized communities to the 'world intelligibility' associated with the general fabric of everyday life (Schatzki, 2001), practices and values of precisions changed as they became something carried out quite generally across a number of different practices.

The piecemeal and opportunistic nature of emergent precisions merits emphasis, because present-day precisions are commonly experienced as integrated with one another. But this is largely a development of the last two centuries. There is a danger in portraying long-run change as (too) smooth and progressive. The general directions of change have certainly been towards more accurate devices and texts, more precise temporal usages, and more coherent understandings of precision as a general feature of 'modern' timing, but this has not necessarily been a unidirectional process.

There were grand concepts of precision in timing, but much about even these relatively intellectualized times was produced in piecemeal and opportunistic fashion in specialist communities and then made its way into the practical networks of the everyday—via intermediaries like manuals, rutters, and the like. In turn, these temporal practices were subject to further change, only gradually producing everyday time senses that were an amalgam of different practices, layered and constantly renegotiated. Important extensions of precision came not just from 'laboratory precision' in specialized spaces like observatories, or through refinements in abstract, 'theoretical precision', but also from the finding and accumulation of various ways to make precision a practical proposition, realizable in the heterogeneous spaces of 'the everyday world'.

NOTES

1. For example, a fourteenth-century tide-table from St Albans gives times for 'flod at London Brigge' in hours and minutes (Hughes, 2006: 446). It seems likely that this is in some way related to Richard of Wallingford's celebrated clock there, which indicated high-tide at London Bridge, among its several displays (North, 2006: 3).

2. The lower-case p in 'public' (as opposed to 'Public') is intended to highlight the contrast with the 'Lay' community. The latter had a more specific interest in, and enthusiasm about, timekeeping, whereas the 'public' community used and grasped clock time without much reflection upon the more abstract dimensions of timekeeping. We stress this ordering, of *use* preceding *grasp* of clock times.

3. Galileo's initial interest in pendulums was unpromising, although it is amusing now to visualize Galileo using his own pulse in order to conduct experiments as to the rates at which pendulums slowed! It did not take long for pendulums to be used to measure pulses, rather than vice versa, but, even so, Galileo had found a major problem, in that pendulums did oscillate more slowly over time.

10

Clocks from Nowhere? John Harrison in Context

> Harrison completed his first pendulum clock in 1713, before he was twenty years old. Why he chose to take on this project and how he excelled at it with no experience as a watchmaker's apprentice, remain mysteries. . . .
>
> Historians wonder which clocks, if any, Harrison might have dismantled and studied before fashioning his own . . . no one can guess where the boy might have gotten such a thing
>
> Dava Sobel, *Longitude: the True Story of a Lone Genius Who Solved the Greatest Scientific Problem of His Time*, 1995: 64.

10.1 A WONDER AND A MYSTERY

In this chapter, we want to put flesh on the bones of the last chapter by questioning and replacing two portrayals given very widespread currency in Dava Sobel's bestseller *Longitude*. These respectively concern the characterization of particular settlements in north Lincolnshire, where John Harrison lived between the ages of (about) five and thirty-seven (when he moved to London), and the influence of Harrison's own upbringing on his subsequent clockmaking. In particular, we present a range of new information about John Harrison's context and childhood, and use it to produce a new understanding of Harrison's key role in chronometry and his breathtaking technical achievements as a clockmaker.

In *Longitude*, Dava Sobel provides a lively account of the protracted attempts during the eighteenth century to develop a reliable method of establishing longitude at sea, that is, to surmount the difficulties in knowing precisely what the time was at some other location than one's own. The hero of Sobel's enthralling story of struggle, factional intrigue, and (eventually) goodness rewarded, is the English clockmaker John Harrison, whose extraordinarily precise constructions did most to ensure that mechanical timekeepers finally prevailed, and subsequently became the standard means of ensuring precise oceanic navigation. If we are ultimately dissatisfied with Sobel's explanation, we must say that *Longitude* does not distort the horological literature: it fairly summarizes relevant material (Quill, 1966; Hobden, 1988, 1999; King, 1993; and, the prime source, Andrewes, 1996).

Harrison was pursuing the less-favoured clock-based mechanical methods of establishing the time at a fixed meridian, whilst elsewhere, and comparing that 'base-line' time with local time. His extraordinary achievement was to produce a succession of timekeepers that out-performed anything previously made, and to refine their design and performance over a period of roughly fifty years, even though there were few British scientists competent to understand the mechanics of his devices.

Perhaps the most striking thing about John Harrison was his isolation from the traditional centres of English elite scientific understanding, in what King (1996: 168) calls 'a very remote corner of England'. Born at Wragby in the West Riding, he was brought up at Barrow-on-Humber in north Lincolnshire. His father, variously described as a joiner, a shopkeeper, a small farmer, and a surveyor, was of limited social standing. His formal education went no further than local schooling; he had not trained as a clockmaker's apprentice. One could hardly have predicted that a boy from this background would succeed where Europe's scientific and mechanical elites had failed.

John Harrison remains in certain respects an enigmatic figure. Typically for the period, his early life is poorly documented. It remains very difficult to answer some basic questions about that time, and relatively little has been added to Quill's account (Quill, 1966), drawing on various memoirs prepared by John Harrison himself, baptismal and burial registers, wills, and a smattering of local administrative records. He was born at Wragby, in the Yorkshire parish of Foulby, near Wakefield, on 24 March 1693, the eldest son of Henry and Elizabeth Harrison. Henry Harrison was a joiner by training, but also worked for Sir Rowland Winn as a surveyor, especially of timber, on his estate. The Harrison household remained in Foulby until at least 1697, when John's brother James was baptized there. Probably from 1697, and certainly by 1700, the Harrison family was living at Barrow-on-Humber in north Lincolnshire, east of the market town of Barton-upon-Humber (Figure 10.1).

It is unclear whether there were family or other contacts behind the move from Foulby to Barrow, but it may be more than coincidence that Sir Rowland Winn owned land at Barrow (Hobson, 2001: 17). There do not appear to have been any Barrow inhabitants named Harrison in the generation before Henry and family moved there, but other family links are of course possible. That, within a few years at most, Henry Harrison was the parish clerk at Barrow may strengthen the supposition of some connection or other—it would have been unusual for a hitherto unknown in-migrant to take up this office unless they were already known to the more prominent families of the parish. As adults, both brothers married and remained in or near Barrow. John and James were two of the four appraisers of Henry Harrison's property following his death in 1729. His inventory called him a carpenter, and included the stock of a fairly typical

Fig. 10.1. Locations of Barton-on-Humber and Barrow-upon-Humber.
John Speed's map of Lincolnshire shows Barton-upon-Humber and Barrow in the northeast corner of the county, from which they appear separated by a line of hills, though these only occasionally exceed 150 metres above sea-level. Brockelsby, where Harrison built a turret clock for Sir Charles Pelham, is shown to the south-east.
Source: Hawkyard (1988): 118–19

village grocer's shop (Table 10.2, below).[1] John Harrison moved to London in 1730, within a few months of his father's estate being settled.

The main contemporary biographical records available were prepared by Harrison himself, at various dates from the late 1720s, when he moved to London, until the 1780s.[2] Throughout, his prime concern was not to provide a comprehensive biography, but to furnish explanations and written instructions for the making and operation of his chronometers, to defend his experiments, methods, and innovations, and to secure recognition that his clocks fulfilled the performance criteria of the prize specified by the 1714 Longitude Act.[3] Harrison elaborated the sources of his interest, experience, and influences only where they bore on the technicalities of his timekeepers and their performance, as a result only sketchily outlining how technical and practical knowledges were acquired.

Besides his own much later writings, critical evidence for Harrison's early clockmaking comes from several surviving clocks made by Harrison early in his career. Four of his early clocks survive, dated 1713, 1715, 1717, and 1721, bearing John Harrison's name either alone, or alongside that of his younger brother, James. Also surviving is the turret clock Harrison made for Sir Charles Pelham's country house at Brocklesby Park, some ten miles from Barrow.

In Barrow, both brothers were active in experimenting and clockmaking. Their clocks are especially striking because the movements are wooden, using the naturally oily wood lignum vitae to minimize the problems caused by metal components' need for frequent oiling (King, 1996). Notwithstanding their wooden construction, these clocks show considerable technical sophistication and originality. Led by John, the two brothers systematically investigated alternative design elements for their clocks, and how performance varied with variations in conditions, especially temperature. They also incorporated an 'equation-of-time' table, showing the difference through the year between the apparent solar time that could be observed from a sundial, and the 'mean' time kept by a clock. Such equation-of-time tables had become much more common from around 1660, with the widening realization that the interval between successive noons varied—only a few times annually was the interval of exactly twenty-four hours. By 1700 these were common accessories to high-quality timepieces, and the brothers' 1717 long-case clock has a hand-written equation-of-time table displayed in a central glass panel in the clock-case door.

Sobel is struck forcibly by the observation that Harrison seems an unlikely figure to have revolutionized what could be done—and imagined—in precision clockmaking. Central to her narrative of the whole Harrison–longitude puzzle, is the question of how the hero, John Harrison, came, as it were, from nowhere to solve the problem. Not only was north Lincolnshire far from the intellectual currents of scientific and technological innovation, it was also peripheral to late seventeenth- and early eighteenth-century economic change, and marginal to emergent consumer society. Hence Harrison is unlikely to have been familiar with clocks:

Harrison completed his first pendulum clock in 1713, before he was twenty years old. Why he chose to take on this project and how he excelled at it with no experience as a watchmaker's apprentice, remain mysteries... Historians wonder which clocks, if any, Harrison might have dismantled and studied before fashioning his own... no one can guess where the boy might have gotten such a thing... Clocks and watches carried high price tags in Harrison's youth. Even if his family could have afforded to buy one, they could not have found a ready source. No known clockmaker, other than the self-taught Harrison himself, lived or worked anywhere around northern Lincolnshire in the early-eighteenth century. (Sobel, 1996: 64–5)

There were no clockmakers among Harrison's relatives. Clockmakers were thin on the ground in Lincolnshire, with London maintaining a virtual monopoly on high-quality work. Nonetheless, the ingenuity of Harrison's handling of woods and joinery, and his almost infinite capacity to take pains in improving devices' accuracy have attracted horologists' acclaim. Apart from its wooden components, though, Harrison's 1713 clock was of fairly standard design. Over the next several years, he developed and incorporated several innovations (Betts, 1993). By the mid-1720s, the cumulative impacts of gridiron pendulum, grasshopper escapement, better gear-wheels, reduced friction, near-elimination of lubrication, and sustained calibration-testing, gave the Harrisons' regulators unprecedented accuracy, notwithstanding that their creators were 'a couple of country bumpkins' (Sobel and Andrewes, 1998: 91).

Even before his marine chronometers, then, John Harrison was at the forefront of clockmaking technology. His temperature experiments, his temperature-invariant 'grid-iron' pendulum, and other components for his regulator, had no provincial precedent. The nearest comparable investigations were conducted in London by George Graham, England's premier clock- and instrument-maker in the 1720s and '30s. Graham was not a native Londoner, having been born in the Lake District, but spent his entire working career in London, and was central to debates about precision timekeeping in the Royal Society and other metropolitan circles. Yet even Graham had not progressed as far as Harrison in countering the disruptive effects of thermal expansion in his clocks.[4]

Sobel's solution to the question 'why could Harrison solve the longitude problem?' essentially mobilizes the narrative trope of 'the outsider'. Having little experience of clocks or watches, and lacking contacts with local clockmakers, Sobel's Harrison was able to solve the problems precisely because he was outside the orthodox clockmaking establishment. With his freedom to think outside the orthodox mind-set, and free of conventional assumptions about feasible technical performance, Harrison's 'outsider' situation was his critical advantage. His originality came—or rather, it survived to be realized in his clocks—because, unlike his formally trained peers, he was not restricted by a training confined within the accepted limits of techniques and imagination.

But reassuringly familiar as this narrative is, it is hardly an explanation at all. For how, then, was Harrison able to construct reliably performing clocks? His lack of training in clockmaking, which Sobel presents as an advantage, carried with it some considerable drawbacks. Early eighteenth-century clocks were not such crude devices that they were produced without instruction or training, by men 'good with their hands'. Precision clockmaking drew on several detailed and sophisticated knowledges, and a complex of refined technical skills, which were precisely what was instilled in young clockmakers during their apprenticeships.[5] Several features of Harrison's early clocks imply his familiarity with the conventional appearance of high-quality long-case clocks, including his wheel-trains, equation-of-time table, and case design.

The general argument that only an outsider to clockmaking could have produced a precision marine chronometer only applies insofar as the skills or attitudes of established clockmakers were actually obstacles to its construction. On the other hand, much of the knowledge, and many of the skills of skilled clockmakers were essential. With regard to the cutting of geared wheels, the calculation of gear ratios, the balancing of components, adjusting a pendulum, the provision of maintaining power, and numerous other facets of construction, the experience and skills of expert craftsmen were hardly optional extras. That they did not inhibit Harrison's work strongly implies he had had some significant instruction, training, and experience, even without a formal apprenticeship.

So Sobel's 'lone genius' narrative simply replaces one problem with another: if Harrison was unconnected with clocks or clockmakers, whence came his familiarity with contemporary techniques and designs, or his knowledge of approaches to technical clockmaking problems? How did the 'solver of the longitude problem' become interested in, and capable with, clocks in the first place? Or acquire the information and skills to teach himself? Apart from showing Harrison's access to material from Nicholas Saunderson's Cambridge University lectures on mechanics, Sobel emphasizes the 'lone genius' at the expense of exploring the sources of his knowledge about clocks.

We shall pursue these questions at various levels. First, we will consider the broader topics of scientific and technical activity relating to clockmaking in early modern northern England, including north Lincolnshire. Then, we look more specifically at the lower Humber estuary, and at clocks, and them at Barrow-upon-Humber and Barton-upon-Humber, before our focus narrows further to the Harrison family, and finally to John Harrison's clockmaking, knowledges, and experiments before his move to London.

10.2 SCIENTIFIC ACTIVITY, CLOCKS AND CLOCKMAKING IN NORTHERN ENGLAND

Much writing about scientific activity in seventeenth- and early eighteenth-century England centres on activity in London, and especially that occurring in and around the Royal Society in the century or so after its foundation in 1660, and through the Royal Society-centred networks of correspondence and publication that incorporated interested men elsewhere into scientific discussions (Schaffer, 1988; Shapin, 1994; Dear, 2001). Many key figures in the early Royal Society, though, had been born and brought up in the provinces.[6] For example, Robert Hooke (1635–1703) was the son of a minister at Freshwater in the Isle of Wight, and Isaac Newton (1642–1727) came from Woolsthorp in Lincolnshire: the former coming to Westminster School after his father's death, and returning to the capital after study at Oxford University, and the latter arriving in London after studying at Cambridge. Newton was not the only eminent mathematician to grow up in Lincolnshire: amongst others, so too did the astronomer Henry Andrews (1743–1820), born at Frieston near Grantham, and George Boole (1815–64), born at Lincoln.

Likewise, although 'Political Arithmetic', the numerical investigation of population, economy, and government, from the late seventeenth century, is frequently thought of as a metropolitan phenomenon, many of its leading figures were provincial men. Gregory King was a herald at Chester, for example, and Sir William Petty (1623–87) was the son of a clothier and dyer in the small Hampshire port of Romsey, where according to Aubrey the young Petty became fascinated with watchmaking. A key figure in the related area of cartography, John Ogilby (1600–76), was Scottish (Barbor, 1983: 224).

Oxford and Cambridge Universities could certainly be a key staging post for educated migrants between the provinces and London, but the proportion of each generation who studied there was relatively small. Much larger centripetal flows resulted from the movement of adolescents of (usually) lesser social rank to train as apprentices to London craftsmen and tradesmen. Provincial migration to early eighteenth-century London apprenticeships involved over 600 boys a year. This was under 10 per cent of total provincial migration to London, but these were the movers most likely to acquire the technical or calculative skills to encounter the capital's scientific communities (Kitch, 1985: 226; Wrigley, 1967: 44–70; Boulton, 2000).

Major figures in English political and cultural life had also come from modest provincial backgrounds. Cardinal Morton (1410–1500) was the son of a shoemaker at Blandford in Dorset, and Cardinal Wolsey's father was a butcher in Ipswich. Shakespeare provides the most celebrated instance among authors and poets, but was far from an isolated example. From Humberside came the highly regarded poet Andrew Marvell (1621–78), the son of a Hull clergyman. Actor-playwright John Lacy (died 1681) came from near Doncaster, and the composer William Byrd (1543–1623) came from Lincoln.

All these men gravitated to London, but an increasingly visible theme in recent work has been the variety of provincial scientific and technical investigation that existed beyond the metropolis. Some communities of mathematical, scientific, or technical practice existed far from London, and were only loosely connected to it. For example, several leading naked-eye astronomers located were in northern England, decades before the Royal Society or Greenwich Observatory were established (Howse, 1997). The most accurate and innovative observations and calculations of celestial measurements, such as the sun's size, and distance from the earth, were made by a small network of men, of whom only Jeremiah Horrocks (1618?–41) had attended university. Horrocks, at Toxteth Park near Liverpool, William Gascoigne (1612–44) at Middleton in Yorkshire; and William Crabtree (1610–44) across in Manchester, maintained links through their correspondence, accessed ideas elsewhere mainly through published books, and were supported by the contact networks of interested local gentry, including Jeremiah Shakerley, Richard Towneley, and Jonas Moore. This group operated right at the frontier of knowledge despite their isolation from the embryonic scientific establishment (Aughton, 2004; Chapman, 1990, 1995a, 1995b).

Such men were part of a distributed community of mathematical practice, engaged in varying mixes of investigation (e.g. trigonometry, astronomy), teaching (e.g. geometry) and applied work (e.g. land surveying). The latter activities, an important part of the work of seventeenth-century mathematical practitioners, provided points of contact between scientific and local, low-level mathematicians (Taylor, 1954; Cline-Cohen, 1982; McRae, 1995; Money, 1995). William Crabtree had made land surveys outside Manchester, for example (Chapman, 1996). Among Gascoigne's pupils was Jonas Moore from

Whitelee in Lancashire (1617–79), later knighted for his work in surveying and draining the Fens.

In the applied mathematical field of navigation, the coast of north-east England was notable for the production of skilled seamen, and for widespread expertise relating to sailing. The navigator and New England colonist John Smith (1579–1631) was born at Willoughby, Lincolnshire. And from the tiny fishing village of Staithes, fifty miles further north from Hull, came James Cook, perhaps the major navigational figure in the late eighteenth century (and an articulate enthusiast for chronometers as longitude instruments). Overall, then, there is considerable evidence for the involvement of provincial English men, including men from Lincolnshire, in activity and debates that were broadly mathematical or scientific.

The same was true of clockmaking. When John Harrison went to London in 1730, Halley sent him to see George Graham, who had been another 'outsider' to London clockmaking circles, having moved to London from Hethersgill in Cumberland. Graham was regarded by Royal Society and other elite investigators as not merely the best clockmaker, but the supreme astronomical instrument-maker in early eighteenth-century London. Graham was the capital's leading investigator of problems caused by thermal expansion of metal components in all types of precision instruments (Chapman, 1995a). Harrison and Graham's key common interest was temperature-compensated pendulums, not just clockmaking in general.

The metropolitan concentration of clockmakers is not in dispute, nor the capital's dominance of training and technological innovation. But while the capital dominated these activities, it did not monopolize them, and the provinces were not entirely passive and reactive. We have already shown (Chapter 5) that public clocks were familiar northern English church towers in both towns and non-market parishes. Turret clocks were as common in northern town halls, market buildings, or country houses, as they were further south. That there were fewer public clocks per square mile in the north reflected lower population densities and much larger northern parishes—their component townships were often larger than southern parishes. Neither did the north lag behind technologically, for several church clocks were converted to pendulum drive very early (Chapter 5, above).

Regarding domestic clocks, Loomes (1997) records more than forty men from northern England (here defined as Yorkshire, Lancashire, and the four northern counties) with surviving work, whether lantern, long-case, or bracket clocks, from before 1700, and numbers were much higher by 1730 (Figures 10.2a & b). The figures are minima, referring only to surviving clocks, excluding makers with no surviving work, no matter how securely documented in diaries, household accounts, court depositions, handbill advertisements, or newspapers, or as masters training other makers with extant work. Such men sometimes appear in passing in horological literature, like the York clockmaker Seth Agar (Loomes, 1997:

78), but horologists usually concentrate on makers with surviving work. Taken together and notwithstanding the patchiness of the sources from which they must be reconstructed, maps of both these clockmakers, and of pre-1700 church clocks, point to widespread clock-time provision in northern England. There are few provincial towns, even among the smallest urban centres, without a recorded clockmaker before 1750, even if some may have been repairers more than makers. By then, towns the size of Barton-upon-Humber usually had one or more clockmakers, which must qualify views of the southern Humber shore's 'remoteness' from clockmaking.

Section 5.4 showed that regional disparities in clock ownership between the 1660s and 1730s were more muted than for many new consumer goods, as well as the rapidly broadening social profile of clock and watch ownership by the early eighteenth century. The latter trend was clearly under way in northern Lincolnshire, and southern Yorkshire, either side of the Humber estuary, though

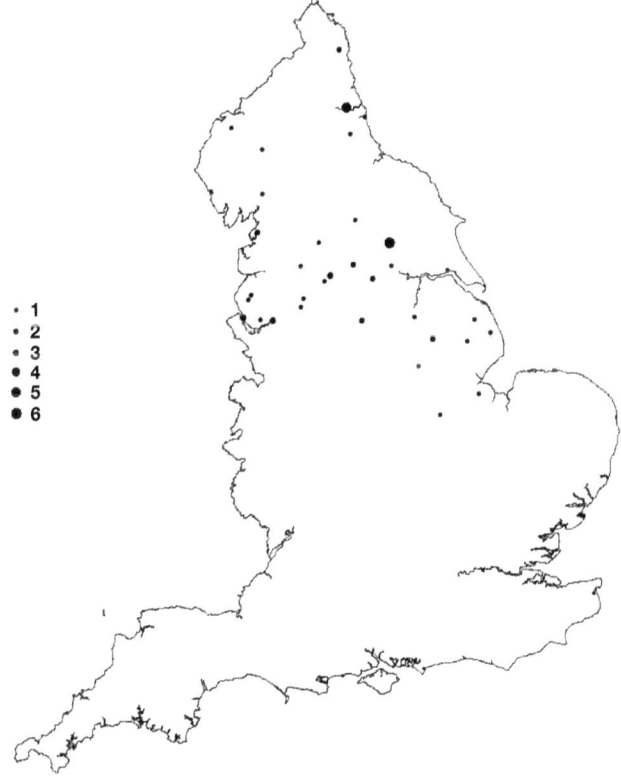

Fig. 10.2. Clockmakers in northern England with extant work: 10.2(a) before 1700; 10.2(b) 1700–30.

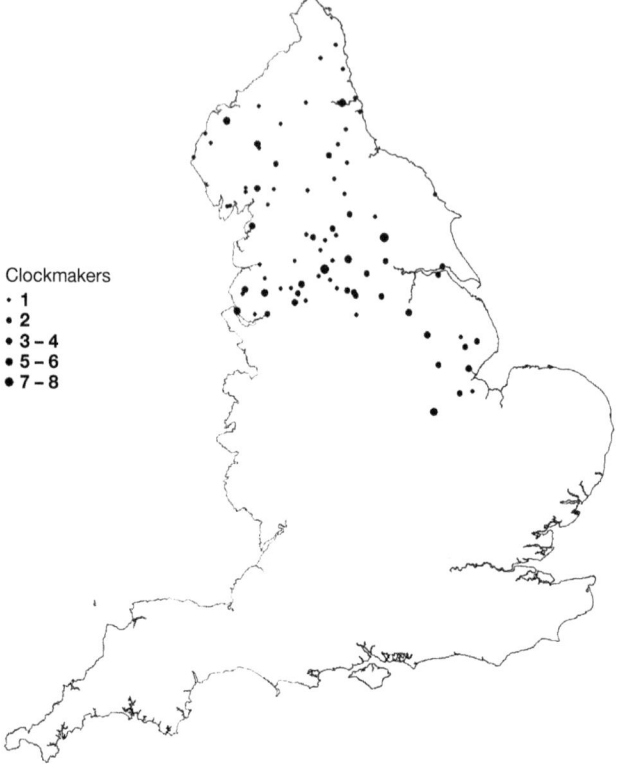

Fig. 10.2. continued

they were rare among poorer households at Clee, further along the north Lincolnshire coast (Watkinson and Ambler, 1987).

Further north, the shipbuilding port of Whitby, with some 3,000 people in 1700, was intermediate in size between Hull and Barton-upon-Humber. About 100 probate inventories survive for eighteenth-century Whitby, and although nearly half are terse summaries of wages due to mariners dying at sea several include clocks. In 1701 a wealthy shipowner's possessions included a clock, but clocks were not confined to the elite. In 1709, intriguingly given Henry Harrison's occupation, 'one old clock' belonging to George Attey, joiner, was appraised at £1 0s. 0d., of his movable estate of £10 6s. 8d. The next recorded Whitby clock-owner was a master mariner Christopher Hill, in June 1718, whose possessions included a clock and a case in his 'Fore House' (the usual term in Whitby inventories for a hall). Clocks were generally valued at between one and two pounds, although several clocks were valued together with other furniture items (Vickers, 1986: 38–85).

Clocks had been familiar items in the inventories of Yorkshire's social elite for decades. These included 'one Clocc' in a chamber at Denton Hall, owned by Ferdinando Fairfax, 2nd Baron Fairfax of Cameron, in 1647/8. Another wealthy Baronet, Sir Thomas Wentworth of Bretton Hall (west of Doncaster), also owned a brass clock at his death in 1675, valued at £2, with two more clocks in the kitchen. That two other rooms at Bretton Hall were called 'the Clockhouse' and 'the Clockhouse chamber' suggests the house also contained a turret clock (Stears, 1972: 146–50). But clocks were not the exclusive prerogative of the titled or very wealthy. Lower down the social scale, a substantial yeoman at Great Gristhorpe had a clock in his hall by 1682.

Richard Richardson of Bierley Hall, to the south of Bradford, in 1656, and his son William Richardson eleven years later, both described as 'Esquire', owned clocks, but their character changed significantly. The son's appraisers described 'one brasse Clocke... £.3 0s. 0d.' in 1667, whereas his father had owned 'A wood Clocke... 13s. 4d.' (Stears, 1972). Many writers have seen John Harrison's early clocks of wood as very unusual, but the 1656 Richardson inventory is among several, hinting that wooden clocks were quite familiar.[7]

Neither were wooden clocks merely early or 'primitive' timepieces. There were still men in the Humber area identified as 'wooden clockmaker' a century later. In the 1790s, Laurens Barringer of Hull was described as 'wooden clockmaker', and wooden clocks were being made at Hull a generation later by members of the Drescher family, who produced both wooden and brass clock movements. Usually these were 30-hour double-weight clocks for the cheaper end of the market, but the Drescher's wooden clock output included some of the more expensive longcase forms (Walker, 1982: 9). In Barton-on-Humber itself, Robert Sutton produced clocks with wooden movements, of which at least three survive, and others of brass, in the early nineteenth-century, based on what he had learned while repairing clocks built by Harrison (Walker, 1982: 31). In short, there is a range of evidence for local traditions of wooden clockmaking before the mid-seventeenth century, and men capable of making and repairing wooden clocks were still present in the early nineteenth century.

10.3 CLOCKS AND TIMEKEEPING IN BARROW AND BARTON-UPON-HUMBER

Sobel is quite right to describe north Lincolnshire as among the least urbanized areas of England in around 1700. The whole county of Lincolnshire contained no towns that were large by national standards, and its largest town—Lincoln itself—was a traditional ecclesiastical centre with a slowly growing population, rather than one of the new industrial or financial centres at the cutting edge of rapid urban growth (Langton, 2000). There was no Birmingham, Bristol, or Liverpool hereabouts. Local economies still revolved around agriculture (with some fishing on the coast) and local societies were still strongly oriented around relatively large aristocratic and gentry estates. But this is not synonymous with the area being 'remote' (Betts, 1993: 6), or 'very remote' (King, 1996: 168), or with the Harrisons being 'country bumpkins' (Sobel, 1995).

Places like Barton and Barrow, bordering the lower reaches of the River Humber, were less isolated and remote than Sobel implies. The Humber was both barrier and communications route, on which early medieval Barton was a major port. A ferry across the Humber there, according to *Domesday Book* (2003: 920), produced £4 a year for the Crown in 1086. Around 1300, boats from Barton-on-Humber were shipping grain to London (Campbell *et al.*, 1993: 62–9), and Barton was a major mid-fourteenth-century wool town (Miller and Hatcher, 1995: 244). Barton's fortunes declined as it lost ground to the rapidly growing new town of Kingston-upon-Hull, less than five miles away, which had been founded in 1293 (Beresford, 1967), but traders and mariners both in Barton and Barrow exploited opportunities produced by Hull's growth. Their late seventeenth-century populations included several sailors, mariners, and ferrymen, the latter plying people and goods to and from Hull itself (a function eventually replaced by the late twentieth-century construction of the Humber Bridge, whose southern end is at Barton). Hull ships and mariners were much involved with long-distance and even oceanic voyaging, besides the coastal trade that accounted for most activity in provincial ports. The longer-distance trips arose in connection with import and export trades, with north Atlantic fishing, and with whaling. The port at Barton-on-Humber had regular links to Hull, and in Harrison's own time some imported timber came directly to Barton from the Baltic, rather than via Hull, leading Hobhouse to suggest that access to good quality woods for joinery influenced Henry Harrison's decision to move from Foulby to Barrow (Hobden and Hobden, 1998).

Barton should be seen as market town and port, then, not an introspective village. The coastal region of John Harrison's upbringing was not 'the back of beyond'. With the international connections of Hull and the smaller Humber ports, this was an area within which the longitude problem was familiar. It may not be entirely coincidental that among the most perceptive proposals spilling from the presses after the 1714 Longitude Act was a pamphlet written by Jeremy Thacker: Thacker came from Beverley, half a dozen miles inland of Hull (see section 10.5, below).

North Lincolnshire had also shared in the late medieval spread of church clocks. The earliest documented clock in the Humber estuary area comes from another port, namely Grimsby, some dozen miles around the coast south-eastwards from Barrow. At least two of Grimsby's parish churches had clocks before 1411. Grimsby aside, few north Lincolnshire parishes have any surviving early churchwardens' accounts, but, slowly increasing numbers survive through the fifteenth and sixteenth centuries, and it becomes clear that there were also church clocks in several inland market towns in the county. There are clocks in the earliest churchwardens' accounts for Louth in 1500 (Dudding, 1941), and Kirton-in-Lindsey in 1486.[8] Clockmaking expertise could be found fairly locally: Louth's church clock was overhauled by a clocksmith from the small market centre of Kirkby-on-Bain, and Kirton's was refurbished by a craftsman from Gainsborough. The Grimsby parish church clocks were maintained by local smiths and metal craftsmen from at least the early fifteenth century, while the church clock at Heckington was mended by George Smith from Hull in 1572, and a man from Burton in 1585.[9]

Sobel's difficulty in identifying many clockmakers hereabouts mainly reflects the regional domination of the supply of domestic clocks and watches (which are usually signed with the maker's name and location) by clockmakers and watchmakers further afield, in London, York, Beverley, and Lincoln (Wilbourn and Ellis, 2001; Walker, 1982; Loomes, 1997).

The earliest clockmaker from Beverley whose work survives in numbers of securely attributed pieces is John Hall, who was working there from about 1730 (Loomes, 1997: 134–6). About twenty examples survive, from a long working life of almost seventy years (perhaps suggesting he had a son of the same name), most being relatively sophisticated long-case clocks with engraved brass or painted dials, two hands, and quarter striking. Since Hall did not number his output, we cannot say what proportion of his output is represented by extant pieces. The latter end of Hall's career in Beverley overlapped with that of Hudson Fox, who produced and sold eight-day, painted-dial, long-case clocks, for sale in shops in Beverley and Hull (Loomes, 1997: 119). Clockmaking had become established at Hull in the late seventeenth century, rather earlier than in Beverley, when John Rooksby moved there from after his apprenticeship in York. Thomas Cliff of Hull's surviving products include several long-case clocks, and a quarter-repeating bracket clock of *c.*1700 (Loomes, 1997: 24, 108). Around 1740 a maker named Joseph Squire was producing some highly sophisticated clocks in Hull (Robinson, 1981: 394–5), though only two are known to survive (Loomes, 1997: 247–8).

Probate inventories show clock-ownership becoming increasingly common, and not limited to the rich. The social deepening and broadening of timepiece ownership gathered pace after 1700, and Barton was among the places it touched. Wills survive for 115 residents of late seventeenth-century Barton-on-Humber among the records of the Consistory court of the Bishop of Lincoln, with sixty-one accompanied by probate inventories (Table 10.1). More than one-third of inventoried men were gentlemen or yeoman farmers, with eight labourers at the other end of the social range. Predictably, in an extensive parish around a market town, agricultural and non-agricultural designations each comprise about half of the men inventoried, most of the latter in Barton itself.

Overall, the majority of craftsmen's inventories date from the 1660s or early 1670s, a generation before the Harrisons arrived, and early in the diffusion of clocks and watches.[10] Only seventeen inventories record the estates between 1680

Table 10.1. Occupational profile of Barton and Barrow testators with probate inventories

	1660s	1670s	1680s	1690s	
Gentleman			1		1
Yeoman/Agric.	3	8	3	3	17
Labourer	2	4	1	1	8
Craft/Trade	7	4	3	3	17
Marine		2	1		3
Unspecified male			1		1
TOTAL MALE	12	18	10	7	47
Female	4	3	3	2	14

Notes: (1) 'Agricultural' comprise 16 yeomen, 1 husbandman (1698), 1 shepherd (1675). Labourers' main employment in all cases appears agricultural. (2) 'Craft/Trade' comprise 3 carpenters (1663 1665 1685); 3 weavers (1676 1685 1692); 2 mercers (1663 1664); 1 maltster (1669); 1 chandler (1679); 1 musician (1670); 1 blacksmith (1681); 1 whitesmith (1663); 1 bricklayer (1692); 1 wheelwright (1667); 1 ironmonger (1698). (3) 'Female' comprise 12 widows, 1 spinster (1698), 1 unspecified (1666). (4) 'Marine' comprise 1 ferryman (1671); 1 mariner (1674); 1 boatman (1681).

and 1700, but they include a clock in the inventory of Robert Trower senior, ironmonger, which on 27 January 1698 amounted to the considerable sum of £248 16s 8d. Trower had wider horizons than Barton, with further shops in Barton Brigg and Caistor. His Barton house had at least nine rooms, and appears recently extended since the downstairs rooms included both a 'new parlour' and a 'great parlour'.[11] Notwithstanding the two parlours, the 'house' [hall] was at least partly a room for social interaction and display, containing not only a clock, but a [dining?] table with chairs and stools, a dresser with pewter and brass, and other items, the room's goods together valued at £5 10s 0d.

In April 1711 a clock worth 13s 4d was among the inventoried possessions of Mr Ralph Feustall of Barton, and in June 1723 at Barrow, William Jackson—ascribed no occupation but dying possessed of horses, cattle, sheep, pigs, and poultry—had a clock in the 'house' [hall] of his five- room cottage. Together with a table, five chairs, one stool, a dresser and some 'small implements', his clock was valued at £1. Also at Barrow, six years later in June 1729 Robert Harding, a farmer with nearly 50 acres of arable land, and a considerably larger, but equally diverse array of livestock to Jackson's also had a clock in his 'house' [hall]. Thus domestic clocks were present among the middling ranks of farmers and craftsmen in late seventeenth and early eighteenth-century Barrow and Barton. Where exactly they had been produced, and where lived the craftsmen who could be called on to maintain them is not recorded, though, as we have already indicated, much routine maintenance and repair probably lay within the expertise of local craftsmen.

An eighteenth-century inventory from Barton-on-Humber without a clock is not of itself noteworthy, but in one case it most certainly is, and illustrates that an inventory without a clock cannot, indeed must not, be taken as evidence of a household unfamiliar with clock times. For there is no clock in the inventory compiled at Barrow on 4 June 1729, for Henry Harrison, joiner, John and James Harrison's father. According to John, there were several clocks in the house, but if they ever had been Henry's they were disposed of before his death, when his sons were among his appraisers (Table 10.2).

As discussed in Chapter 5, one important factor in the widening social ownership of clocks and watches was that they were less expensive than Sobel suggests. That horological attention has focused primarily on expensive pieces at the top end of the market, whose greater value and resilience means that more have survived compared with clocks and watches for plebeian customers, detracts from a key feature of eighteenth-century clock production: the many clockmakers who, even before $c.1730$, focused their production on relatively modest (though not necessarily plain) clocks.

Although no traceable churchwardens' accounts survive for Barton or Barrow until the mid-nineteenth century,[12] a good deal of evidence points towards both parishes having had clocks before 1700. Some is circumstantial and general: across England, settlements' status as central places is a fairly reliable indicator of public clock-time provision. The size of a settlement, its relative wealth, and the presence of a ring of bells or other investment in the church fabric, are all well correlated with the local presence of church clocks. Clocks were very commonly required for specific uses relating to markets in order to signal market opening and restricted periods. Almost all towns like Barton, and many settlements like Barrow, had clocks well before 1700 (Chapter 5).

Saying 'traceable' in the previous paragraph is deliberate, because volumes of accounts for both Barton parishes, St Mary's and St Peter's, were extant until at least the 1850s, perhaps the 1890s. They may still survive in private hands.[13] They clearly existed at the time Henry Ball was compiling his *The Social History and Antiquities of Barton upon Humber*, privately printed locally in 1856. Ball recorded seeing accounts for St Mary's covering 1640–1725, and for St Peter's from 1650 to 1750, though quoting only from the former. Items from St Mary's accounts, about bells being taken to Lincoln for recasting, were privately printed in 1896, though it is unclear whether they were cited directly from the accounts or from extracts copied by Ball.

Ball's quotations from the St Mary's accounts include five items of expenditure concerning the clock. That he does not mention the construction or purchase of a new clock, given his interests in church equipment and in contacts between Barton and the wider world, implies that the clock already existed by 1640. Ball's thought the clock 'had a propensity to go wrong, from the repeated items to its correction'. However, this seems based on a misapprehension: three of the

Table 10.2. Probate inventory of John Harrison's father, Henry Harrison

A true and p(er)fect inventorie of all the singuler the goods, chattels and credits of Henry Harrison of Barrow in the County of Lincoln Joyner Deceased died in possession of taken and apprised the 4th day of June by use whose names are hearunder written

	£	s	d
In primus his Purs and Apparrill	12	0	0
Item in the house three Chairs kupboards two tables	1	10	6
Item in the chamber one bedsted with bedding	3	0	0
Item in wheate and peais and one cheaste with other	2	6	0
Item one Cow and Calfe and hours and sheep	8	10	3
Item one karte & furnetyre with other things	4	3	2
Item in wood and deales and waneskote	16	15	0
Item in Linnin & Latts & Beales and Skops & other things	5	6	6
Item one acre of Bienes & one Acre of Barley	1	15	0
Item in olde iron Candles and Iron wair	8	14	0
In locks & Candelsticks Cotons Canvis fustan Dimathy	7	13	0
_____ and whole Butans hare and threed	4	15	0
buckels & Spurrs Silk galoune Riddins	3	0	0
Bodies & Stomachers wool cards & corks Coffin irons & nales	3	4	0
_____ lls and chisels Manchester small wair Bout Straping	3	0	0
Item powder & shott hardewair as knifes and Bookes	5	12	0
glooses & Crape Nalls and gimlits and Druggs and dry spices	10	10	0
_____ ttling Corrans and Rasens Starch and blew and Rice	4	0	0
tobackeo & hops Suger Resin and Rud	10	5	0
_____ pitch and tarr Treakell and Sope	4	3	0
ginger Fins and Glew Salt and otmill 0and Brandeyss	5	3	0
Item bees wax honey Black bear lofe Suger	1	6	0
Dishes and Laddles Rills washintubs	3	19	0
Pruners and pipse and soes	1	13	0
Charnes and buckets and Skepts	1	1	0
Spanish white trenchers and Lamblack	0	9	0
Item pattens and Clodds pepper and starch	1	12	0
Ropes and pools and shackels	3	13	0
Handes Sishes and ox bralls hand skops Latts	2	10	0
Pigg and Gun and some Brasses	2	5	0
Item the Saddle with some Bookes other things	1	5	0
In Wollen Shop Goods and other things	6	17	0
	151	15	2

  Valued by us
    John Harrison
     James Harrison
      John Colworth
       Benj'n Wilkins

Source: Lincolnshire Archives Office, Lincoln, INV.207/284.

five payments he quotes are each of 6s. 8d., respectively for 'ceping the clocke' and (twice) 'looking after the clock'. That form of words, and the consistent amount, strongly suggest that the payments are annual fees for the routine clock maintenance integral to keeping the clock working, like winding, setting, and oiling. A fourth entry is unhelpfully vague with 18d. paid for 'mending the clock', but the fifth, 'for mendinge the clocke and for wyer to the same ... 12s. 6d. [and] towards the clock strynges ... 2s. 3d.' suggests that the clock had chimes, since the distinction of 'wyer' and 'strings' is elsewhere symptomatic of churches with both clock and chime mechanisms.

Though the loss of churchwardens' accounts means that we cannot date the construction of St Mary's church clock, that it was in place long before 1640 can also be inferred from references to clock times in the 1676 manuscript *Barton-on-Humber Town Book*.[14] This comprises various 'Paines and Customs of Barton', with a commentary, and copies of wills and jury presentments bearing on parish resources and customs. Many 'paines and customs' may long pre-date the 1676 volume, since its preamble describes the book as revising 'an antient town book ... date 1 May 1600, then collected by the consent of the Grand Jury', following an earlier 'antient' book's destruction by fire. If that book was 'ancient' in 1600, its compiling was probably nearer 1500.

Several regulations describe customs and practices that incorporate clock time.

Item. that no householder within this Town do occupy any Baking or Brewing in the night season, neither for their own use nor for the victualling of marketts or any otherwise after 8 of the clock at night or before 5 a clock in the morning, in pain to be amerced for every default five shillings.[15]

Besides baking and brewing, clock-time restrictions also applied to gathering wool: 'None to gather wool before 9 of the clock.' Further time-based regulations in documents abstracted by Ball also specify hour times: for example, the swineherd was to blow his horn between 6 and 7 a.m. in summer and between 7 and 8 a.m. in winter to collect the town swine. Hours 'of the clock' clearly presume a timepiece of some kind, providing public time signals. All in all, then, and as in hundreds of other early modern English parishes, there was at least one church clock in Barton-on-Humber by the mid-seventeenth century (when John Harrison's grandfather Henry [born 1632] was still a boy), and possibly much earlier. Long before John Harrison ever thought about making clocks, clock times were incorporated into the conduct of everyday life in Barton-upon-Humber.

Neighbouring Barrow-upon-Humber also lacks churchwardens' accounts or vestry minutes until the mid-nineteenth century, and here there are no fortuitously surviving extracts made some 150 years ago. However, Barrow also has a manuscript town book like that for Barton: *The Towns Book of Barrow Containing the Dues and Customs Belonginge to the Said Town, One Thousand Seven Hundred & Nine*.[16] Like Barton's this book updated and systematized local customs, going back at least to agreements about the parish's obligations for mending flood banks along the Humber, made in 1553.

An early section of the book sets out 'The Office and Duties of the Parish Clerk': these begin with church attendance; the setting out of the communion cloth, cushions, books, and other items in readiness for church services; the sweeping of the church, chapel, and seats; before setting out a variety of more detailed responsibilities such as 'to pike grease or oyle & keep the bells in good order' and 'he must be cerfull that no boys or Idle p'sons janle the bells'. In Barrow, as in most English parishes, it was not the parish clerk's responsibility to keep or wind the clock, but the sexton's, augmented by local maintenance agreements with specialist craftsmen.

Unfortunately, two pages missing from the Barrow town book mean that the list of sexton's duties (including 'waking sleepy people in the Church in time of divine service') is incomplete.

Nonetheless, that a public clock signalled times of day is implied by several of the parish clerk's duties:

Item. He is to Ring a Bell att nine a Clock In the morning & at 4 at afternoon every working day from Munday in the first whole week of Lent untill East'r. Ex't such days as there are prayers in the Church.

Other peoples' duties explicitly used clock times, their locations implying aural signals, most likely from the church. For example, the neatherd tending Barrow's cattle was:

to keep the herd upon Hawk untill 6 of the clock in the morning & then to go into budforth untill 12 of the clock & then to drive them to the Lars untill 2 of the Clock in hott weath'r or else he must not come on the Lars . . . be at home by sunset.[17]

The longstanding presence of church clocks at Barton and Barrow throws new and suggestive light on John Harrison's awareness of clocks, particularly because Henry and John Harrison were both closely involved in local church life, as John recorded in his 1720s and 1770s essays (King, 1996: 168–70; Hobhouse, 2001). Both were involved in bell-ringing, and Henry was parish clerk. The close proximity of bells and clocks in church towers and steeples, and the parish clerk's duty of sustained supplementary ringing at set hours of the day, would have brought the young John Harrison face to face with several dimensions of timekeeping, both mechanical-cum-technological, and social.

10.4 JOHN HARRISON'S CLOCKMAKING

10.4.1 Harrison at Barrow

With some twenty years clockmaking behind him when he moved to London in 1730, aged thirty-seven, Harrison was an accomplished craftsman well into his clockmaking career, unlike those migrating to pursue apprenticeships. However, much uncertainty surrounds the likely scale of John and James Harrison's clock output at Barrow; horologists generally regard this as more or less restricted to the surviving examples (Betts, 1993; King, 1996). This assumption seems deeply problematic for several reasons.

First, Harrison would be unique among early modern clockmakers if all his clocks had survived. This is not true of George Graham, for example, or even for Thomas Tompion, the most prestigious clockmaker of all, and the one whose products were the most lavish, most expensive, and highly prized, so is much less plausible for frailer and less luxurious objects. For mid-market makers in northern England, the majority of work is lost, even for the best represented among surviving clocks.[18] There are, though, many makers whose securely documented long working lives are represented by just one or two surviving clocks. For example, only two known clocks bear the name of Joseph Squire of Hull, active in the 1740s, one of which shows not only hours and minutes, but also the date, day of the week, time of sunrise, phases of the moon, and lunar calendar (Loomes, 1997: 247–8 for Squire's output; Robinson 1981: 395, figure 11/16 shows the clock). John Birchenlee of Birchenlee, near Colne in Lancashire, was the first of a dynasty of clockmakers there, living from 1711 to 1794. However he is known by a single sundial, and one clock including a dial showing the time in several cities around the world (Loomes, 1997: 246–7). John Brewer of Rochdale (Lancashire) is known by one surviving clock, as are James Kirton of Alston, Cumbria, John Haythornthwaite of Lancaster, and James Langworth of York. Other clocks may survive as yet unknown to horologists, but the vast majority of these makers' output has clearly been lost over the intervening three centuries. In such cases horologists do not usually assume that a clockmaker produced only the output that survives. That most clocks have not survived is clear when the relatively few late seventeenth- and early eighteenth-century domestic clocks are compared with the scores of thousands indicated by clockmakers' serial numbers, account books, and inventories.

It therefore strains credulity to equate clocks of high technical quality with the lifetime output of their maker, to that date, being just a few pieces. Even if Harrison's clocks took longer to make than those whose components were filed from brass, an annual output of a dozen or more clocks looks quite feasible.

Second, if Harrison produced only the surviving clocks, how did he support himself and his family? As Sobel stresses, the family lacked the resources for John to spend much of his time on a hobby, yet Harrison describes large and sustained series of experiments on the performance of materials and designs. The shortage of documentation that he pursued another 'main' livelihood should not be overstressed because there is so little direct evidence for his activities, but it seems far more plausible to us that experimentation and the building of regulator clocks accompanied the making and selling of further clocks, than that Harrison combined a different line of work with the exhaustive experimenting, making, and checking that Harrison describes. Other high-quality provincial clockmakers were full-time specialists, and Harrison's clocks out-performed them all. We do not see him as a hobbyist.

Third, Harrison was not alone in becoming an accomplished clockmaker without training in London, even if he provides the most dramatic example. Regional horological monographs contain several examples of makers trained by their father, or another relative. The orthodox career-path from home to apprenticeship in London or a provincial centre, and back to the family workshop was just one possibility, and many trajectories were messier.[19] Similarly, the standard trajectory of provincial migrants to London being trained in current London practices, after which their practices stabilized, exaggerates the extent to which provincial makers were necessarily conservative, once away from the London clockmaking community. London clearly was very important, but other places also contributed. Innovations made by locally trained makers who had never visited London point to a potentially more dynamic atmosphere in provincial workshops than conventional portrayals allow.[20] Some innovations were developed by small-town or rural clockmakers, trained within family workshops. The work of successive family members could show a growing sophistication, in the quality of both mechanisms and cases, over both individual careers, and generations.

Fourth, there is evidence of later specialists in wooden clockmaking in Lincolnshire, whose clocks show familiarity with design features from Harrison's clocks, especially those of Robert Sutton of Barton-on-Humber (1774–1865), who produced clocks with movements of oak, lignum vitae, and boxwood, and radially cut wheels, inspired by John Harrison's clocks. Sutton had learned about Harrison's clocks through repairing them, and added his own modifications to Harrison's techniques in his own work, claiming an accuracy of one second a month (Walker, 1982: 31). In 1790s Hull, Laurens Barringer was also described as 'wooden clockmaker'.

The worlds of some provincial clockmakers, even in small towns and isolated villages, then, were not necessarily as static and remote from wider practices, knowledges, and aspirations as accounts of John Harrison have commonly emphasized.

10.4.2 Family Context

We now focus in closer still, on John Harrison's contact with timekeeping knowledges and clockmaking skills, and with ideas and people involved in early eighteenth-century longitude debates. This involves both material specific to Harrison and his immediate family, on the one hand, and the circulations of ideas in various local communities of practice around.

Harrison's immediate family descent is familiar from previous work (Figure 10.3), but researchers from Quill (1966) onwards have not looked more widely.

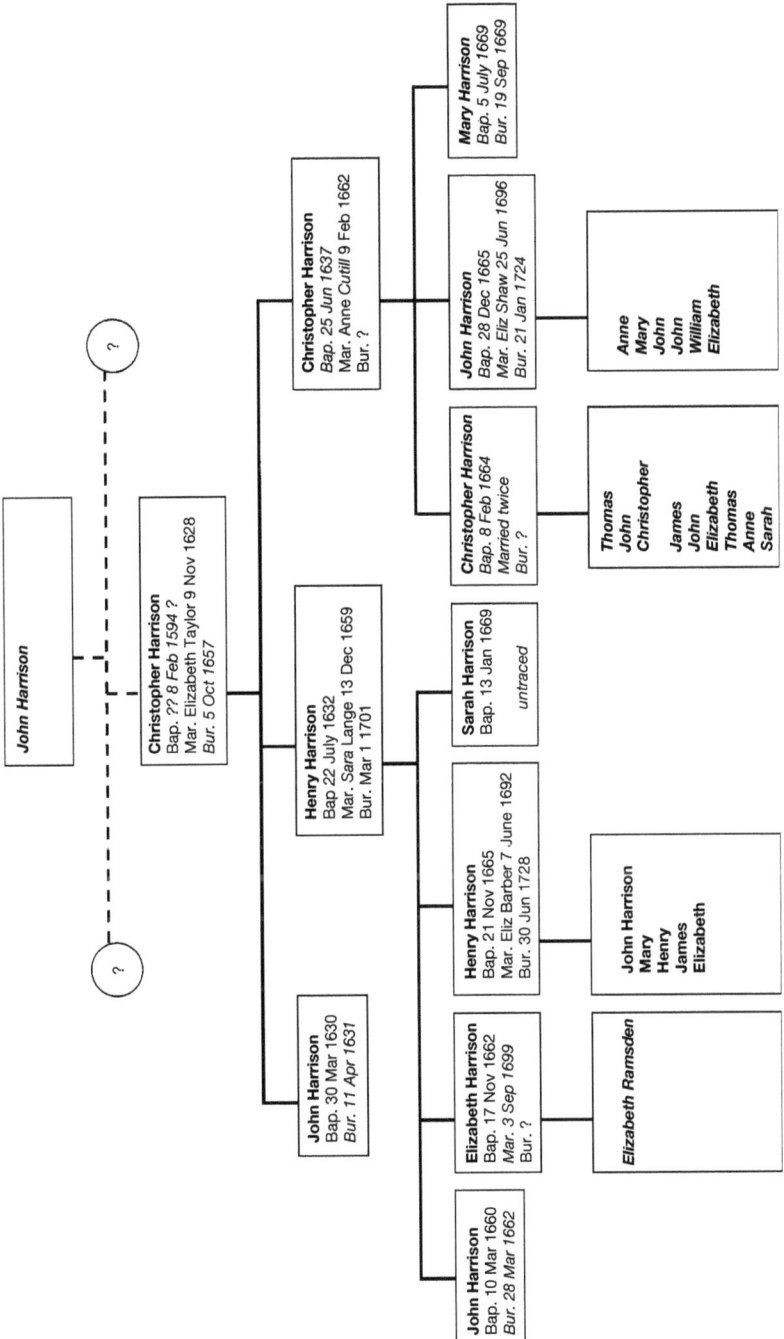

Fig. 10.3. Genealogy of John Harrison.

We enter unknown territory once we move outside the families of his father, grandfather, and great-grandfather. John Harrison himself is partly responsible, insofar as his writings make few references to background beyond his immediate family, and his father's involvement in parish life, especially bell-ringing.

Unfortunately, biographical research on John Harrison (Quill, 1966; King, 1996) has identified only these five male forebears: there is no information currently available on further members of these preceding generations. John, Henry, and Thomas figure particularly prominently as names used by the Harrison family, whereas no subsequent births before 1800 were christened Edward.

Accounts since Quill rightly emphasize Henry Harrison's rapid incorporation into parish life in Barrow, especially through the church, pointing to his long tenure as parish clerk from shortly after 1700, following the family's move to Barrow between 1698 and 1700, emphasizing the local reputation for reliability and trustworthiness that this implies (Quill, 1966; King, 1998; Sobel, 1996). Henry and (later) John Harrison's connections with churches, in Barrow and Barton, were both spiritual–civic, including bell-ringing, and utilitarian, working on the church fabric and furniture. Churchwardens across England paid craftsmen for their carpentry and joinery expertise in making and maintaining clock frames (usually oak), on which the going and striking trains were mounted, and for the clock housing, external dials, and chime barrels that were increasingly seen as integral parts of clocks (section 5.5, above; Jennings, 2000). Such components needed to be strong and accurately made, creating work for competent village craftsmen over the years.[21] So did bell-frames, and the Harrisons' involvement in bell-ringing at Barton, implies the presence of several bells, a feature commonly associated with church clocks in early modern England.[22] Both Barton's churches now possess full rings of eight bells, several of which are old enough to have been in use in Harrison's day (Johnston *et al.*, 1990).[23]

There were several reasons, then, why the young John Harrison would have seen church clocks at close quarters. We can improve on this circumstantial evidence by returning to Ball's extracts from the St Mary's churchwardens' accounts, because amongst the men paid for work on the clock was one Henry Harrison (Table 10.3).

It would be premature to assume that this Henry Harrison was John's father, since Henry was also the name of one of John's brothers (though he died aged twenty-six) and of John's grandfather. The clock-repairer may have come from outside the family. Other sources shed some light on these possibilities. There had been earlier Harrisons at Barton, including 'Anthony Harrison, gent. of Barton-upon-Humber' whose will was proved at Lincoln in 1625. Several Harrisons appear in Barton's parish registers between 1640 and 1730, including two men called Henry Harrison, one during the 1640s, another late in the century. Ball does not give dates for the payments he lists, but if he listed them in chronological order then one way of identifying the clock-repairing Henry

Table 10.3. Churchwardens' payments for work on the church clock of St Mary's, Barton-upon-Humber

to Robert Fairwedther for the ceping the clocke...	6s.8d.
to Robert Fairweather looking after the clock...	6s.8d.
to Charles Rowbotham for loucking after the clock...	6s.8d.
to Henry Harrison for mending the clock...	18d.
to Edw'd Page for mendinge the clocke & for wyer to the same...	12s.6d.
towards the clock strynges...	2s. 3d.

Source: Ball (1856): part two, p. 8.

Harrison may be to date each of the payments to Henry Harrison, by finding the possible date-ranges for the other names.

There were several Fairweather families in Barton-upon-Humber between 1640 and 1750. Robert Fairweather's name appears in five register entries for one parish or other. One born in September 1686 was dead by December; another born in April 1688; a third in September 1714 was a son of Robert and Edith Fairweather, but died by November; and the last in June 1725 was the son of the same parents. This father seems to be the man born in 1688 and so, because the others died in infancy, is the only candidate identifiable from baptisms or burials. He would have been a working adult from the mid-1700s. Seven of Charles Rowbotham's children appear in St Peter's parish registers: John (baptized December 1692), Joseph (baptized and buried in December 1694) and Charles (baptized November 1695, buried three months later), Mary (buried 1700, possibly born after Charles), Edward (baptized January 1700), Richard (baptized February 1702), and Elizabeth (baptized January 1704). The last three entries call Charles Rowbotham a cooper. His work on the church clock (as a cooper, probably on the chime-barrel) came between about 1690 and his death in the 1720s. Edward Page appears in the Barton registers later on, towards 1740.

So, if Ball did list the names in date order, the payment to Henry Harrison was made between perhaps 1710 and the 1730s, which fits well with the Henry Harrison who was John Harrison's father, who died in 1728. The other two Henry Harrisons in the Barton registers fit poorly: one had long been dead, while the other was older than the first two men in Ball's list, and a full generation older than John's father. He is therefore the strongest contender to be the man who mended the Barton church clock.

10.4.3 Career and Expertise

There is a tension, if not an outright contradiction, in the literature that simultaneously presents John Harrison as the uneducated son of a joiner, and as a technically well-informed experimental clockmaker. The realization of the profundity of Harrison's thinking about mechanics is perhaps the most important shift in horological understanding of Harrison in the last generation. Since the 1970s horologists have placed great stress on Harrison's exceptional understanding of mechanics (Burgess, 1996: 256), though this reassessment reached mainstream horology only in the 1990s (Andrews, 1996; Hobden and Hobden, 1999). The Harrison brothers, in Barrow, and John after his move to London, carried out extremely demanding experiments and systematic measurements into the effects of thermal and atmospheric variations on the performance of pendulums of different materials and designs, which are compelling evidence of technical understanding as exceptional as their skill and precision in clock construction. It is the combination of theoretical and practical skills that took Harrison so far beyond the achievements of even the highly skilled contemporary artisans.

For all our reservations about Sobel's characterization of north Lincolnshire as an almost clock-free environment, she clearly is right to highlight the practical issues that made Harrison's context far from ideal for this combination of theoretical insight and practical skills. Even if the technical knowledge came from Quill's hypothetical visiting clergyman, and his Cambridge lecture notes, that still leaves the practical skills to be acquired—the kinds of skill that were central to an orthodox craft apprencticeship. So how, while they were learning their father's trade(s), did they find the time and resources to acquire those skills, and to design, perform and repeat their experiments? The puzzle is deepened, moreover, by horologists' belief that the young Harrison brothers did not raise resources by selling some of their early clocks.

In this light, certain facets of prevailing accounts of John Harrison's early clockmaking seem dubious. King (1996: 172) suggests that by 1721 Harrison had produced four long-case clocks, one every two years, three of which survive. This strikes us as implausible. It is not just that Harrison needed some income during these years, but that he evidently acquired and practised the skills that enabled him to construct the clocks he had designed. He needed to build not just clocks, but also his local reputation. This mattered most obviously for the commission to build the Brocklesby Park turret clock, which King (1996: 172) dates to 1722. However, since the early clocks listed by King all remained in the ownership of the Harrison family, they could not have brought in an income from sales. If they were Harrison's sole output, his early income must have entirely derived from other activities, but it is then difficult to see how his skills and techniques could have developed to the level they evidently reached while he was still in his early twenties. Harrison's marriage in 1718 presumably implies his earning an income, greater than he would have needed as a bachelor. Yet apart from the Brocklesby Park clock, King suggests that Harrison did not sell clocks he had made until 1726 and 1727 (1996: 172).

We therefore suppose that John and James Harrison—like their contemporaries—made more timepieces than survive. We see clockmaking as the main way in which John Harrison generated the income necessary to maintain his family and fund his experiments, and to establish the reputation that led to the Brocklesby Park commission. Despite the absence of better records of Harrison's early products, and the loss of virtually all the churchwardens' accounts in north Lincolnshire for that period,[24] there are indications that Harrison had produced more clocks.

First, there were sufficient Harrison clocks in the locality around 1800 for a Barton clockmaker, Robert Sutton, to specialize in repairing them (Walker, 1982). Moreover, while some of Sutton's own clocks had conventional brass movements, others had wooden movements. Sutton's extant clocks blend of commonplace features in case layout and dial design, with others typical of Harrison's clocks, including use of lignum vitae for high-friction parts like roller bearings; wooden wheels assembled from radially cut sectors of oak, so that the wood's grain strengthened gear-teeth; and the grasshopper escapement. He made his own innovations, too, including double rows of teeth on wooden gear wheels, and a novel design for maintaining power. Sutton exemplifies a point of much wider importance, namely that clockmakers could learn from, and knowledge could circulate through, clocks themselves: 'having repaired a Harrison movement he copied it and was one of the few people capable of understanding and reproducing such work' (Walker, 1982: 31).

Second, the commissioning of the Brocklesby Park turret clock, some ten miles from Barrow, in itself implies a significant local reputation for skill in major maintenance tasks on turret clocks, and probably their construction. Horologists often stress that, even though they faced similar technical obstacles such as friction and wear, the activities of domestic clockmakers and turret clock makers differed significantly in their scale, materials, and skills. There are examples of men who excelled at both, but most makers specialized.[25] Turret clock commissions, whether for parish churches or country houses, rested largely on a maker's reputation specifically for turret clock work. As one of Lincolnshire's major landowners, it is implausible that Sir Charles Pelham would commission a large turret clock for his country seat from an unknown non-specialist with little experience in clockmaking, and none at all in crafting turret clocks.

The existence of a reputation implies satisfied customers, which in turn implies clocks, and therefore experience. The Brocklesby Park commission probably attests previous experience, most likely in local parish churches, though the disappointing sparseness of churchwardens' accounts hereabouts is problematic. If they had survived to show John Harrison's involvement in maintaining, repairing, or building turret clocks, the Brocklesby Park commission would have an altogether more convincing context. One might, though, turn around the horologists' question: given the reputation implied by that commission, is there a more likely candidate to have been maintaining church clocks in north Lincolnshire's churches than John Harrison?

Moreover, calling the commission 'a stable clock' understates the significance of timekeeping at Brocklesby Park, the largest single estate in north Lincolnshire, because of its importance as a stud in the breeding of both thoroughbred horses and foxhounds. Sir Charles Pelham was an enthusiastic horseman, and co-founder of the Brocklesby Hunt, which met at, and often hunted in, the Park (Collins, 1902). Pelham instituted one of the earliest recorded packs, and foxhound breeding projects, from at least as early as 1713. Interestingly, breeding and keeping foxhounds was a shared interest of Sir Charles Pelham and Rowland Winn of Nostel Priory, where Henry Harrison had worked before John's birth, perhaps bearing on the family's move from Foulby to Barrow.

Sir Charles Pelham's involvement in racehorse breeding and training also connects to his desire for a high-quality clock, given timing's potential importance in training, racing, and gambling. Several leading seventeenth- and eighteenth-century racehorses were bred at the Brocklesby Park stud, which was home to several thoroughbred bloodlines, with stallions brought considerable distances. Thus, Curwen's Barb, a stallion originally given by the King of Morocco to Louis XIV, was imported in 1698 to the Lowther stud in Cumbria and visited Brocklesby stud several times between 1703 and 1721.[26] Until *c.*1750 northern England, especially Yorkshire and Lincolnshire, were the pre-eminent centres of horse-breeding and racing (Figure 10.4). Although Hampton Court and Newmarket were among a handful of important breeding and racing centres in southern England, the centre of gravity of racing and gambling shifted decisively southwards only after mid-century. All three of the most celebrated imported stallions—the Byerley Turk (*c.*1680–?), the Darley Arabian (1700–30) and the Godolphin Arabian (1724–53)—and several lesser late seventeenth-century imports were northern-based (Wilkinson, 2003). The breeding, racing, and hunting uses of Brocklesby Park make the stable clock far more than 'just a clock for the servants'; rather it was integral to a more time-aware space. Such a clock was fundamentally unlikely to be commissioned from an unknown, on a whim.

The emphasis on Harrison as relatively uneducated creates a problem when it comes to understanding just why his clocks performed so well. The Harrison described in *Longitude* is a primarily empirical investigator and inventor, not a theorist, whereas in recent years Harrison specialists have strongly argued for his extraordinary theoretical understanding of oscillatory movement (Burgess, 1996). This was a topic that clearly defeated not just those who sat in judgement on Harrison at the Board of Admiralty—'poor fellows these to be my master'—he called them, but virtually all his clockmaking peers and successors. Only comparatively recently, and largely through analysing the performances of his restored clocks alongside his often-tangled writing, has Harrison's scientific originality begun to be appreciated by technical specialists (initiated by Laycock, 1976; developed by Burgess, 1996 and Hobden, 2001).

John Harrison's formal schooling seems limited, though there were local schools, notably Barton-on-Humber grammar school from the sixteenth century, and perhaps much earlier if it was connected to Barton's documented fourteenth-century schoolmaster (Hodgett, 1975: 141). Barton Grammar School had

Fig. 10.4. Centres of horse-breeding and racing before 1750.

Source: Andrewes (ed.), 1996: 98 (Original: J. Flamsteed (1683), Philosophical Transactions of the Royal Society, 13: 413)

potential links with Cambridge and Oxford colleges through endowed scholarships and exhibitions.[27]

The importance of informal knowledge networks beyond the schoolroom is recognized when Quill, Gould, and Andrewes stress Harrison's access, around 1713, to notes of lectures on mechanics given at Cambridge University by Nicholas Saunderson, the Lucasian Professor of Mathematics (Andrewes, 1996: 201). Quill speculated that they were lent by a clergyman (King, 1996: 169), but there may have been a closer, personal connection through members of the Saunderson family living in Barton. Nicholas Saunderson had been born (1682) and educated at Thurlstone, on the river Don, north of Sheffield (Yorkshire). The Don is a major tributary of the Humber, and Thurlstone lay about 10 miles south-west of Wragby, and some 35 miles west of Barton. A branch of the Saunderson family lived at Barton at that time, though we have as yet found no direct evidence that the Thurlstone and Barton Saunderson families were related.[28]

To summarize, we think it very likely that John Harrison had rather more clockmaking experience than has usually been described. It would be very unusual for an unknown and inexperienced maker to gain a major commission, like that for the Brocklesby Park clock, without significant experience in maintaining, and perhaps in constructing, turret clocks. These were most likely to have been in parish churches, but the loss of so many early churchwardens' accounts prevents us testing this idea.

10.5 JOHN HARRISON AND CONNECTIONS TO LONGITUDE DEBATES

Even the best-performing clocks in 1700 were far short of the accuracy required by the Longitude Act. Whether in London or in remote Barrow, makers required command of a range of relevant ideas, information, and practices, to solve the twin problems of building a sufficiently accurate timekeeper and of achieving the required precision in a shipboard environment. The sheer diversity of ideas suggested in print in the wake of the 1714 Act attests to the uncertainties surrounding both problems. Against that background, Sobel's 'outsider genius' narrative gains a greater impact emotional appeal from its equation of 'remote' with 'out-of-touch', but this presumption requires investigation: just how isolated was John Harrison from relevant knowledges?

Two possible connections worth exploring between John Harrison and longitude debates are respectively genealogical, with Edward Harrison (*Idea Longitudinis*, 1696), and geographical, with Jeremy Thacker of Beverley (*The Longitude Examin'd*, 1714). Within the available evidence, we explore both Harrison's specific connections, and the wider local networks and knowledge-circulations.

Edward Harrison was a naval lieutenant whose ambitious initiative in determining longitude led him to publish a pamphlet, *Idea longitudinis, being a brief definition of the best known axioms for finding the longitude* in 1696 (Turner 1996: 120). He claimed service on at least six Royal Naval ships and two merchant ships sailing 'to the East', probably at least ten years experience. Though disparaging navigational equipment and methods on naval and merchant vessels alike, he did not think reliable longitude-finding was a lost cause:

Some blockheads are apt to say, the Longitude cannot be found: no, no it cannot Accidentally... but by Care. Diligence and Industry; it may be found, without which it cannot be understood.   (Harrison, 1696: Dedication)

Harrison endorsed the experience of 'tarpaulins' over philosophers and scientists, and was scathing about English, Venetian, and Dutch longitude prizes.[29]

Nonetheless he was familiar with several Royal Society publications (raising questions about where he saw them), had attended Gresham College lectures in the late 1680s though not university,[30] and his reading Latin suggests grammar schooling. Harrison allied his shipboard experience and reading to a broadly conventional variant on longitude-determination by lunar distances.

Modern scholars have been unimpressed:

Scientifically and technically the book is poor, second-hand and ill-digested. It is shot through with aspersions on mathematicians, the manner is aggressive, and it shows a deep inferiority complex. (Cook, 1998: 264)

Contemporary views were apparently similar—Harrison presented a copy to the Royal Society, for whom Edmund Halley reviewed it unfavourably (Cook, 1998: 140).

It was ironic, then, that two years later Harrison was lieutenant and mate on Halley's ship *Paramore*, which tested finding longitude from magnetic declination, with Harrison blaming Halley for the Royal Society's disregard for his work, and bitter to be commanded by a much less experienced, perhaps younger, man.[31] Relations deteriorated, and back in England Harrison was court martialled and returned to merchant vessels (Cook, 1998: 275). Harrison's observations were seemingly competent; some were later used by Flamsteed. Halley and Harrison's independent observations agree on latitude and magnetic variations, but differ—sometimes considerably—on longitude (Cook, 1998: 264). Edward Harrison's subsequent career is unclear. He was not among post-1714 pamphleteers, whether because of death is unclear (no death has been traced by historians).

The possibility of links between the two men can be sought by working backwards from John, or forwards from Edward, but neither has proved fruitful. John Harrison's reticence about his background is unhelpful, though he would have had good reason not to mention any connection when seeing Halley in London in 1730!

Historians have focused their efforts to tracing John Harrison's immediate ancestral line (Figure 10.3). John Harrison was not directly descended from any Edward Harrison: both his father (1665–1728) and grandfather (1632–1701) were named Henry, and his great-grandfather was Christopher Harrison (died 1657). However, descent lines from any male siblings of these three forebears are unclear, and whether both John and Edward were descendants of one of them cannot be established. Trying to work forwards from Edward Harrison also faces acute difficulties. His description of his sailing experience to 1696 leaves disconcertingly elastic parameters for a demographic search: he was born somewhere in England, probably between 1640 and 1655, possibly dead before 1714. His name is quite common—many Edward Harrisons were baptized within that span, and with nominal-linkage problems exacerbated by our ignorance of his birthplace, by migration, by the International Genealogical Index (IGI)'s relative disinterest in mortality,[32] and by patchy record survival, it is difficult to identify him. Early eighteenth-century naval personnel records are very brief, and similarly unproductive. Consequently, we can neither prove nor dismiss a connection between Edward Harrison and John Harrison. The shared surname may be just coincidence. Even if related, their proposals were so different that John could have gained more than a general awareness of longitude issues—and wariness, even hostility, towards the Board of Longitude.

John Harrison's second potential link with longitude involves Jeremy Thacker of Beverley, on the Hull River, 6 miles inland of Hull (some 10 miles from Barton, 12 from Barrow). Thacker's *The Longitude Examin'd*, promoting shipboard chronometers, was among the most perceptive proposals spilling from the presses in 1714. Debates on finding longitude mattered to mariners from Hull and other Humber ports involved in trading timber, grain, and other commodities with ports around the North Sea and Baltic; or in north Atlantic fishing and whaling. The pamphlet demonstrates that knowledge about navigational problems and the Longitude Prize had reached the Humber during John Harrison's boyhood, notwithstanding the distance from London, even if of little concern to ordinary seamen.

A connection between Thacker and Harrison is plausible because of both proximity, and the striking parallels between Thacker's proposal, and Harrison's later work. Both used a spring-driven timekeeper, circumventing problems created by the thermal expansion of metal components, through sheltered housing, temperature-invariant components, and performance-correcting tables. They use all but identical methods to find sidereal time, and measure clocks' performance.

Discussions of the flurry of publications in 1714 usually take one of two mutually exclusive forms. Horologists have emphasized the slow development of marine chronometers, as diverse technical and practical obstacles were recognized and overcome, culminating in Harrison's eventual success. Historians of science have been at least as interested in the other extreme, with the proposals that led nowhere, whether based on longstanding folk beliefs, newly recognized phenomena, or personal eccentricity (Gingerich, 1996).[33]

The Longitude Examin'd, falls into two parts, respectively '*a Short Epistle to the Longitudinarians*', and '*the Description of a Smart, Pretty Machine Of my Own, Which I am (almost) sure will do for the Longitude, and procure me the Twenty Thousand Pounds*'. Clearly appearing late in 1714, since Thacker discusses several proposals published earlier that year. His commentary shows that these proposals reached the Humber relatively quickly, notwithstanding his opening observation that

The Books that are written about the Longitude are so acceptable to the public, that the whole Edition is commonly sold off before any of them can reach our Northern Booksellers. (Thacker, 1714: i)

Thacker avers a preference for experimentation over writing: only the conditions of the Longitude Prize, he claims, have caused him to write:

If it be ask'd why I wrote the Book at all, I'll frankly answer, *That I wanted Money*; and that if I had thought that the Commissioners would have been prevail'd upon to have given me some, to carry on Experiments, I had never set Pen to Paper. (Thacker, 1714: iii)

Thacker was nonetheless well informed, with shrewd insight on longitude problems. He characterized different approaches by the problems to be overcome, concluding that the practical difficulties of making sufficiently accurate telescopic observations, sufficiently often, rendered astronomical methods impossible. Hence solutions must be sought through precision timekeeping, and Thacker advances his own 'chronometer',[34] concluding that the *Phenometers, Pyrometers, Selenometers, Heliometers*, and all the *Meters* are not worthy to be compared with my *Chronometer* (Thacker, 1714: 23).

Thacker's proposal had several original elements, showing a sophisticated awareness of issues in constructing extremely reliable clocks. First, his chronometer had 'maintaining power', a secondary spring to drive the mechanism and keep the clock running during the rewinding of the main spring (1714: 18). Second, while long-case clocks kept accurate time without following Huygens' advice to ensure the pendulum's centre of gravity moved through a cycloid, since 'small Arcs of a Cycloid, and ... a large Circle, coincide' (1714: 10), on ship this equivalence broke down because the swing was unstable. Therefore Thacker abandoned pendulums, driving his timekeeper with a balance spring. Third, having shown the substantial effects of heat and moisture on his chronometer, 'by making a Steam in the Room, with a large vessel like one of Mr Savery's Boylers' (1714: 16),[35] he proposed partial vacuum housing for the clock to exclude changes in atmospheric pressure and humidity.[36] Fourth, Thacker realized that chronometers need not completely compensate for temperature changes, so long as they always responded in the same way to a given temperature (Thacker, 1714: 19). He claimed that exhaustive testing enabled him to produce a table of corrections at different temperatures, and conceived a weighted spiral spring to indicate corrections directly, that 'ingenious[ly] ... foreshadowed the later development of the bimetallic strip' (Andrewes, 1996: 192). Finally, for stability the whole device was suspended in gimbals, like a compass.

Thacker claimed that, after extensive experimentation and testing over more than four years, his chronometer consistently performed within six seconds a day against sidereal time (1714: 17). The Longitude Act's £20,000 prize required a clock to be accurate within three seconds in twenty-four hours (to find longitude to within half a degree over a six-week voyage), and the *The Longitude Examin'd* announced Thacker's intention to undertake sea-trials to achieve that performance: to, as he put it, 'give the Mariners such Ocular Demonstration of the Certainty of my Contrivance' (1714: ii). However, 'Its trial at sea ... was doubtless disappointing, because nothing more was heard of either the machine or its maker' (Andrewes, 1996: 192).

The 1714 pamphlet was Jeremy Thacker's only published work, and says little about its author, beyond describing his experiments and design. There is a little internal evidence, insofar as Beverley is consistent with his opening observation that 'Books... about the Longitude are so acceptable to the public, that the whole Edition is commonly sold off before any of them can reach our Northern Booksellers' (Thacker, 1714: i). Despite this inconvenience, he had obtained several, which he cites[37] and presumably travelled to acquire them, or had them sent. More generally, Thacker seems at ease with technical material, and in addressing specialist readers and elites. Though not a Fellow of the Royal Society, he was familiar with material presented there, like Savery's atmospheric engine. Some of the friends whose work he cites were Fellows, and his vacuum jar and air pump echo one of the Royal Society's earliest, and enduring, enthusiasms.[38] Aware of complications caused by the Equation of Time, Thacker uses sidereal time and fashions a home-made but effective transit instrument, sighting along a long alignment between sights attached to two buildings. Besides using this to exactly track sidereal time in calibrating his chronometer, Thacker mentions his own failure to 'find any sensible parallax of the fixed stars'. This is unsurprising: it would be another century before sufficiently precise instruments enabled the measurement of such small angular differences, but this demonstrates Thacker's awareness of broader astronomical issues and technological questions. He both reads Latin, and uses it himself. He signs himself 'Jeremy Thacker, Philomath', a familiar late seventeenth epithet, often translated on title pages as 'Well-wisher to the Mathematics', though Thacker with disarming directness described himself as 'Well-wisher to the Twenty Thousand Pounds'!

Thacker describes building and testing his chronometer over several years, and horologists routinely describe Thacker as a clockmaker, but there is no independent evidence to corroborate them. He does not label himself clockmaker, and no early clocks by any maker named Thacker are known to horologists (Loomes, 1997). This is not conclusive, since many well-attested clockmakers have no surviving work,[39] but neither is a clockmaker called Thacker documented before 1800 in English horological literature.[40] Both Beverley Minster and St Mary's had church clocks, but their clocksmiths are not named, and the earliest Beverley clockmakers named are Walter Clitheroe and Jonathan Craythorn, c.1710 (Walker, 1982).[41]

The name Jeremy Thacker is also absent from surviving Beverley corporation documents, and both Anglican and Congregationalist registers there (though the latter are partly illegible).[42] He may have lived in Beverley unmarried, without owning property, paying rates, or holding office. If a Catholic, he may have married or fathered children without appearing in surviving registers. He may have adhered to one of the other non-conformist communions in Beverley. Nonetheless, Jeremy Thacker's invisibility leaves open possibilities either that he was not from Beverley—but then why claim so?—or that *The Longitude Examin'd* was published under a pseudonym.

This is not pure speculation, because in 1745 *The Longitude Examin'd* reappeared as the final item in the two-volume *Collected Works of John Arbuthnott*.⁴³ Born in Kincardineshire in 1667, Arbuthnott trained as a medical doctor in Aberdeen, Oxford, and St Andrews, eventually serving as Physician-in-Ordinary to Queen Anne in London. Noted as an energetic essayist, both in his own right and in collaboration with Swift, Pope, and Gay, he had also, as a young man, taught mathematics in Doncaster.⁴⁴ Elected FRS in 1704 for his mathematical and scientific work, he was a council member for many years, including 1708–14, bringing close contact with longitude proposals.

The 1745 ascription of *The Longitude Examin'd* to Arbuthnott is presented without elaborating the grounds for supposing him the author. Neither his biographers (Aitken, 1892; Beattie, 1935), nor *The Cambridge History of English and American Literature* (Ward *et al.*, 1907–21: volume 9:), clarify the attribution. The latter lists Thacker's pamphlet among 'works attributed to John Arbuthnott' though tersely observing 'not all are by Arbuthnott'.⁴⁵ His authorship may have been privately known to contemporaries, who noted the disorganized production of his prolific writings. The foreword to *Collected Works* observed that comprehensively listing his work was difficult, given his large output, his many pseudonyms, and his household circumstances:

No adventure of any Consequence ever occurred, on which the Doctor did not write a pleasant Essay in a great folio Paperbook which used to lie in his Parlour; of these, however he was so negligent, that while he was writing in one end he suffered his Children to tear them out at the other, for their paper kites. (Editorial note in Arbuthnott (1745) 1770: xv)

However, much about Thacker's pamphlet seems strikingly un-spoof-like, to modern and contemporary eyes alike, compared with other schemes proposed in 1714. Several less plausible schemes had a hostile reception, but Thacker was not among those ridiculed (Gingerich, 1996). Modern historians of horology and technology regard Thacker as a serious contributor to early eighteenth-century longitude debates (Andrewes, 1996; Turner, 1996). None has thought his suggestions implausible, or doubted that Jeremy Thacker was a 'real' author. The practical tone of Thacker's emphatic preference for experiments sits oddly with Arbuthnot's verbal satire (unless this, too, is part of the joke).

But if intended as spoof, *The Longitude Examin'd* singularly failed to provoke satirical ripostes. It is telling that it was clearly seen as genuine by John Ward, who also proposed a timekeeper kept in a vacuum, and who resented Thacker beating him into print.⁴⁶ Ward's 'I had reason to suspect, as well from private Circumstances, . . . That the knowledge he pretends to have in this affair came by *Trap*' (Ward, 1715:1) is an accusation of plagiarism by a rival, not of hoaxing by a satirist. Unlike Thacker, John Ward is well documented: 'a pragmatic, highly competent professional mathematical practitioner' from Chester, teaching mathematics, writing textbooks, designing mathematical instruments[47] and employed in London as chief surveyor and gauger to the Excise (Turner, 1996: 123–7).[48] No sense of hoax had surfaced by 1716, when William Palmer treated both 'Mr Ward's and Mr Thaqueray's Books' as having solved problems of varying temperature, humidity, and pressure in a marine timekeeper (Palmer, 1716: 5, 11).

A further possibility of flagging satire lies in the text's Latin epigrams, but these seem not to signal satirical intent.[49] The opening couplet '*Quid non mortalia pectora cogis / Auri sacra fames*' or 'Accursed love of gold, what do you not compel mortal hearts to do', fitting comfortably with Thacker's sentiments about the Longitude Prize and pressure to write, is taken from Virgil's *Aeneid* (3: 57–8). '*Vobiscum certasse iuvabit*' or 'It will give me pleasure to have competed with you' appositely closes his 'Epistle to the Longitudinarians'. In concluding his proposal, Thacker quotes Virgil again, from the Fifth Eclogue: '*Lenta salix quantum pallenti cedit olivae*'—'As much as the pliant willow gives way to the pale olive tree',[50] to reiterate his proposal's superiority. The pamphlet signs off with '*Opus exegi*', 'I have erected a work', but eighteenth-century readers would readily have identified the whole of Horace's boast about the immortality of his poems: 'I have erected a work more enduring than bronze'. Thacker's usage of the phrases seems intended as seriously as those of the original poems, rather than ironically.

Our working hypothesis is that Jeremy Thacker was *not* a pseudonym, and that he was not necessarily a clockmaker. Accordingly, we searched much more broadly for any Jeremy Thacker in early eighteenth- century England. No Jeremy Thacker matriculated at Oxford or Cambridge Universities, and no son of his did so, though this may indicate non-Anglican beliefs rather than either lack of education, or poverty.[51] But whereas quoting Virgil and Horace nowadays might indicate specialized, high-level instruction, knowledge of the work of these authors was then a staple of grammar school curricula, typically tackled in the fourth or fifth year of formal Latin instruction. Citing them does less than might be supposed to narrow down Thacker's social background.[52]

'Jeremy Thacker of Beverley' has been surprisingly elusive despite lengthy archival searching. The name itself is less common than 'Edward Harrison', so much so that it is difficult to find any at all. If not common names in early modern England, neither Jeremy (Latin form Jeremiah) nor Thacker (variants Thaker, Theaker, Thaeker) was particularly unusual. However, the combination was rare, and the field of candidates is meagre. The IGI provides the largest compilation of early modern English genealogical information, notwithstanding several well-known deficiencies.[53] The IGI's coverage is extensive (over 60 million names) but far from comprehensive: IGI indices of baptisms, for example, can include only parishes for which registers or transcripts have survived, and not all such parishes have yet been incorporated into the IGI. Their incomplete character make such databases just one starting point for finding specific names, and primarily useful for revealing general distributions of names.[54]

The IGI contains only three baptisms of a Jeremy (Jeremiah) Thacker during the seventeenth century.[55] Two occur too early to have been the author, in 1609 (Grimsby, Lincolnshire) and 1636 (Keighley, Yorkshire), and the third too late (Sutton Coldfield, Warwickshire, 26 July 1695). However, this boy's father was another Jeremiah Thacker, whose own baptism, burial and, possible marriage escape the IGI. We therefore systematically searched the entire IGI for any other baptisms where the child's father was Jeremy or Jeremiah Thacker, a laborious task requiring a separate search for each possible forename. Searching all Thacker baptisms, 1690–1730 for over 250 forenames produced just four further instances, which clustered in space and time. Two sons of Jeremiah Thacker were baptized at Curdworth, William (16 October 1691) and John (14 January 1693[56]) and two daughters back in Sutton Coldfield, Elizabeth (18 July 1701) and Mary (4 May 1713). Curdworth and Sutton Coldfield are adjoining parishes in north Warwickshire, near Birmingham, the churches less than 4½ miles apart.[57]

This was close to where another Thacker was actively publishing mathematical works between the 1720s and the 1740s. He was Anthony Thacker, teacher of mathematics at Birmingham Free School, author of volumes of mathematical texts and problems, and author-compiler of *The Gentleman's Diary* and *The Ladies Diary* (both which regularly included mathematical problems) until his death in 1744.[58] Anthony Thacker was part of a growing community of mathematical writing in early eighteenth-century England which, though densest in and around London, included growing numbers of men from the West Midlands (Money, 1993: especially 335–46, 366–371 and figures 17.1–17.2).[59] A generation later, the Lunar Society of Birmingham was one of the key scientific-industrial spaces of the age (Uglow, 2002).

Two other men named Thacker or Theaker—both Robert—were also connected with the search for longitude. One was an architectural artist, serving as 'sketcher of fortifications' for the Ordnance in the 1670s. His set of topographical drawings of house and gardens of Longford Castle, near Salisbury, were possibly the earliest architectural drawings published in England (Harris, 1979: 89). His paintings inside and outside the new Greenwich Observatory are familiar as the engravings by Francis Place (Plates 10.1, a & b; Jardine, 1999: 211–12). Robert Theaker published *A Light to the Longitude* in London in 1665, describing a type of astrolabe to 'facilitate the affair' (Turner, 1996: 119, n. 26, who says Theaker died in 1687 but gives no source). It is not impossible that the artist and the author were the same Robert Thacker, but neither has a known date or place of birth, or clear family context, and no Robert Thacker matriculated at Oxford or Cambridge at this time.[60]

Table 10.4. Locations of male Thacker baptisms, 1650–1730

County	no.	%	% IGI data overall	% English pop'n in 1801
Lincolnshire	164	37.4	5.6	2.5
London	62	14.2	5.3	11.5
Yorkshire	52	11.9	8.3	10.3
Staffordshire	30	6.9	7.1	2.9
Derbyshire	28	6.4	7.6	1.9
Norfolk	27	6.2	13.6	3.3
Nottinghamshire	18	4.1	8.1	1.7
Oxfordshire	11	2.5	12.7	1.3
Warwickshire	11	2.5	5.3	2.5
Northumberland	9	2.1	7.2	2.0
Suffolk	9	2.1	9.9	2.9
Others	17	3.9		
Unstated	1	–		
Total	439			

Note: 'Others' comprise 3 each in Somerset and Surrey; 2 each in Buckinghamshire and Northamptonshire; 1 each in Bedfordshire, Berkshire, Cambridgeshire, Durham, Leicestershire, Sussex, and Wiltshire.

Source: Compiled from online searching of the International Genealogical Index Maintained by the Church of Jesus Christ of Latter Day Saints at: <www.familysearch.org/Eng/Search/framesel_search.asp>. The information remains indicative both because the index is incomplete, and because searching is available for only one forename at a time. Information in this table derives from searches on 245 male forenames.

The surname Thacker was relatively concentrated across England. Of over 400 male Thacker baptisms on the IGI for 1650 to 1730, more than one-third occurred in Lincolnshire (though none in Barton or Barrow), even though the county contained only 2.5 per cent of English population in 1801 (Table 10.4). Outside Lincolnshire, only in Staffordshire, Derbyshire, Norfolk, and Nottinghamshire did Thacker baptisms occurred disproportionately. London contained the second-largest cluster of Thacker baptisms, but not excessively so given its 11.5 per cent of English population. The Thackers in Yorkshire, Oxfordshire, Warwickshire, Northumberland, and Suffolk were not above the counties' population shares, but this conceals local concentrations.[61] The patterns are not artefacts of the IGI's incompleteness, since the counties with most Thacker baptisms are, if anything, under-represented in IGI. Only Norfolk and Oxfordshire of these counties are over-represented in the IGI compared with England as a whole, whereas Lincolnshire, London, and Warwickshire are particularly under-represented, and Staffordshire and Derbyshire also under-represented (Ecclestone, 1989).

To summarize, there are intriguing coincidences of surname among men involved with diverse facets of longitude, but we have not established definite connections. These may be clarified by future work, but the problems of lost and incomplete information may prove insuperable obstacles to a definitive account of knowledge and contacts.

As it stands, we regard the Thacker pamphlet as genuine, rather than a spoof. Internally, the proposal seems to show serious and original thinking, by someone generally well informed about contemporary longitude proposals, and having particular interest in overcoming the obstacles to precise timekeeping. That it was not produced later than 1714 is evident from John Ward's reaction to it in his essay of November 1714. We cannot prove conclusively that the Birmingham-area Jeremy Thacker was the author of *The Longitude Examin'd*, but have found no Jeremy Thacker in Beverley, and no other Jeremy Thacker of appropriate age and education anywhere in England.

Taking Thacker away from Beverley lessens the geographical link to north Lincolnshire, but it stretches credulity that the many parallels in ideas, design, methods, testing-procedures, and approach to experimentation, of Thacker and Harrison are purely coincidental. Likewise, the exact route by which Saunderson's lectures reached Harrison remain unclear. Whilst probably indirect, via the manuscript lecture notes of a former student, a more direct linkage cannot be ruled out, given the wider distributions of their families and knowledge communities in north Lincolnshire and adjacent parts of Yorkshire. More firmly established are that clockmaking knowledges were more widely distributed in provincial England, and around the Humber in particular, than widely supposed, and that specific knowledges about longitude problems, the 1714 Act and the Longitude Prize, and that chronometry was one of two routes to 'solving the longitude', were circulating locally. Together, these knowledge-circulations and their associated communities of practice constituted the zeitgeist of the young John Harrison.

10.6 CONCLUSIONS

10.6.1 The Maker(s) in Context

This chapter has presented new information about John Harrison's context and childhood, aiming towards a new understanding of Harrison's key role in chronometry and his breathtaking technical achievements as a clockmaker, and in particular responding to Sobel's questions about how and where Harrison acquired the technical knowledge and expertise that underlay his chronometers. Poor documentation limits the knowable details, but our answers span several themes and scales, of which three are particularly important.

First, the provincial circulations of relevant craft and (broadly) scientific knowledge were both broader and denser than in Sobel's account, and extended into north Lincolnshire and Humberside. Scientific and technical knowledges about timing, clocks, and clockmaking were clearly present within regional and local networks and temporalities. Works used by Sobel drastically overdraw the isolation of 'very remote' north Lincolnshire from early modern scientific and horological knowledges. Church and other public clocks were long established and, though the area contained few men explicitly described as clocksmith, sufficient local craft expertise was available to keep them running. The conversion of church clocks from foliot to pendulum drive, going on before and during John Harrison's boyhood, was the major mechanical issue relating to church clocks, even if Harrison did not himself see the changeover at Barton or Barrow. Here, as elsewhere in England, domestic clocks had become more widely owned over the preceding half-century, stimulated by a combination of dramatic improvements in clocks' performance, and falling prices. No longer a gentry prerogative, clocks were possessed by several local households, including some craftsmen and traders, and were liable to be seen by the likes of John Harrison.

Second, Harrison was relatively well located within timekeeping and numerical communities, through his father (joinery, estate surveying, parish administration, bell-ringing), through manuscripts and texts (Saunderson's lectures, longitude pamphlets), and through personal contacts and local experience (clockmaking and experimentation). Harrison's activities necessarily imply either quite specific local contacts and networks, or access to considerable temporal information and expertise within the Humberside area. The latter were clearly much more extensive than horologists have presumed. Clocks and clock times were familiar environmental features, with public clock times put to various uses, both for official, regulatory purposes, and in everyday sociality, well before 1700. A generalized capacity to treat with clocks and clock times was widely distributed well before John Harrison's time, and preceded the seventeenth-century boom in the ownership of domestic clocks and watches. Claims that a lack of local clocks meant that the young Harrison brothers were unfamiliar with clocks are clearly unsustainable. They began making clocks in a place where some households had clocks, and where public clock times were longstanding. Moreover, they had links with church clocks through their father's maintaining the church clock, and his duties as parish clerk.

Third, and overlapping with the last point, it seems likely that Harrison had local and/or family connections with specific contributors to late seventeenth- and early eighteenth-century longitude debates. His thinking blended the chronometry advocated by Jeremy Thacker with the mechanics of Nicholas Saunderson, and his own determination to translate these ideas into a clock of unprecedented performance. That Harrison was in some way connected with either of two earlier contributors to debates on longitude determination, namely Edward Harrison and Jeremy Thacker, remain a possibility, but has not been established here. At the same time, in both cases there are some suggestive links.

This argument offers an alternative to the 'untrained individual genius' explanation. It is not that Harrison's originality, innovations, and determination were not dimensions of a horological genius, but that they should not eclipse the concerns circulating in Harrison's time and place. The enormous skill, expertise, and imagination in Harrison's chronometers are indeed astonishing, but astonishment at the very existence of his interest in marine timekeeping is misplaced. His interest, at that time, in that place, was neither incomprehensible nor bizarre, and detracts from appreciation of his achievements. By the time his thoughts turned to marine timekeeping and he journeyed to London, he had some fifteen years' clockmaking experience, especially that directed to his regulator, and was well informed enough to be at the cutting-edges of mechanical theory, of precision design, and of construction.

Neither does shedding new light on some sources of Harrison's interest in clocks reduce our admiration for Sobel's masterly storytelling in *Longitude*, especially of Harrison's long struggle for recognition. But demystifying the absorption in clockmaking of a provincial small-town boy in early eighteenth-century England lets us see 'the clock that came from nowhere' in a more nuanced context. Outlining the local grounding of clock times enables us to refocus attention on Harrison's achievements. John Harrison's early interest in clockmaking was neither inexplicable nor freakish. He lived in a society much more familiar with clock times, and their associated technologies and debates, than many horologists have imagined.

10.6.2 Communities of Practice

One way to grasp the contexts of John Harrison's clockmaking is by thinking of communities of practice in action. Over time, Harrison intersected with several communities of temporal practice, defined in part by circulations of relevant knowledge among people pursuing certain tasks or interests. These ranged from clockmakers and clocksmiths, respectively specializing in the production of domestic timepieces, and turret clocks; communities of craft practice centred on woodworking and metalworking practices and knowledge; broadly scientific communities interested in experimentation, in astronomy, in mechanics, in thermal properties of materials; and calculative communities of mathematics and navigation. In some cases, Harrison's involvement came through detailed engagement and participation, in others it primarily entailed a grasp of what particular skills, expertise, or access to materials or contacts, could offer him.

But the communities of practice involved in clockmaking were considerably larger than usually thought not just because of the range of links just mentioned, and because horologists have often underestimated numbers of clockmakers, but because clockmaking knowledges circulated more widely around an actively interested lay community. This broader 'clockmaking community of practice' encompassed those customers taking an active role in deciding what clocks should do and how they should be, appear, or sound. Such customers, instanced in different times and places by Samuel Pepys, Claver Morris, and Anne Lister (Chapter 6), took active parts in the process of producing a clock, and were fascinated by seeing clocks or watches taken to pieces and reassembled, and by how timepieces performed. Their senses of clockmaking and time-telling as creative activities were, we think, not uncommon: if most common among provincial Royal Society members, especially clergymen, they were also present among artisans.

The active interest of a 'knowing audience' means that clockmaking was not simply 'maker-led', but was also 'pushed' by clients' timekeeping aspirations, and makers' own senses of craft accomplishment. Both the latter were important in shaping the standard forms of clocks. Very many clocks provided greater precision than most customers needed, but rudimentary clocks were more expensive when 'normal' clock construction involved divisions of labour and economies of scale in tasks like cutting gear-wheels and making dials.

For elaborate one-off commissions, the 'active consumer' is especially important, as was the case with the Brocklesby Park turret clock. Harrison's commission for this is implausible had he not some local reputation, based at the very least on significant experience in repairing or refurbishing turret clocks, even if this was more in terms of generic skills and a deep understanding of wood and woodworking, than solely and specifically as a clockmaker. It is strong evidence of reputation beyond the immediate locality, with all that implied for Harrison's previous activity and experience.

The expertise involved in constructing wooden clocks, the traces largely lost with the clocks themselves, hardly registered with later horologists. However, depicting the Harrisons' early wooden clocks as very unusual novelties, central to casting him as clockmaking's 'creative outsider', neglects the long tradition of wooden clocks in northern England. Harrison's wooden clocks were less a novelty than a once-extensive but now almost lost knowledge. Woodenness aside, Harrison's early clocks indicate his familiarity with 'orthodox' clocks: the calculation of trains and the equation-of-time, the internal layout, the external decoration, and makers' signing practices. His clocks incorporate many prevailing practices, rather than ignoring them; he appears familiar with up-to-date elements in both domestic and turret clockmaking, alongside his own important innovations.

The importance of craftsmen 'learning from objects' merits greater attention because it was a key way in which communities of practice 'progressed'. Objects were significant stores of information to those understanding the general features of their construction and working. Harrison seems to have done so (Betts, 1993, 2006), and, as we noted for Robert Sutton, others did so with Harrison's clocks, incorporating materials and design ideas into their own work (Walker, 1982: 31). More generally, indirect interactions mediated by handling clocks were one way in which communities of clockmaking practice sustained themselves. Another was by providing enough discourse to support critique and continued reflection, through circulations of critique, taking practical or theoretical orientations.

The continuation of learning after apprenticeship is an important point. Apprenticeship was not just about the transmission of specific techniques, but of a broader capacity to solve problems in particular types of objects or types of materials, as they arose, which was the essence of a master craftsman. Even if certain skills were used only very rarely, the capacity to adapt and extend skills was important expertise. Hence, in addition to the formal apprenticeship training that internalized rule-based activity, subsequent experience and learning through objects within a community of practice gave a forward 'push' to 'best practice' elements.

Where a community of practice possessed critical mass, this 'push' might over time cohere as 'momentum' through the community. While momentum varied with life-cycle and place, communities were rarely static. (Seeking a contemporary parallel, while it might be true that many general practitioners practise medicine in ways which may change little over several decades, one could hardly argue that 'medicine is static'.) There is no contradiction between a certain conservatism in much practice, alongside an ongoing sanctioning of inventiveness. From the late seventeenth century, indeed, technological changes in clockmaking were very general and rapid, under a combination of external and internal pressures, such that only by focusing on small scales and short periods can communities of practice be regarded as static. And even the spread of 'standard' time-indications from the plurality of forms taken by precise indications of clock time (Chapter 7) entailed change for individuals. In long-run perspective, any community of practice that did *not* mutate was thereby unusual, and that non-mutation is what requires explaining.

The innovations that worked most rapidly through clockmaking networks and communities of practice after about 1660 included equation of time tables, mediating between the precisions of the mechanism and the universe, the depiction of minutes on two-handed clocks, the pendulum, and the anchor escapement. Various circumstances or features can be adduced to account for the rapidity with which certain innovatory practcies spread. One possibility was where clockmaking communities of practice were particularly 'thickly drawn', that is, dense both in their content (numbers of personnel or the strengths of their interests) and their diffusion mechanisms (where there were established links between those sharing interests). Secondly, at a more abstract level, rapid innovations might be facilitated by specialist literatures acting as fast immutable 'mobiles' (Latour, 1997), accelerating the circulation of specific knowledges and orientations. Third, as we suggested above the presence of an already existing 'knowing audience' brought that interest directly to bear on clockmakers, influencing their ideas as to the features of clocks that customers would value, in terms of some combination or performance and aesthetics (Chapter 7).[62]

These considerations are useful pointers, even though we cannot precisely establish the personnel and connections in the network of John Harrison's acquisition of his technical skills, and theoretical-conceptual understanding. We regard the notion of an informal training in clockmaking within a network of interest in technology, calculation, longitude, and materials, as considerably more plausible than the notion that John Harrison existed in a virtual clockmaking vacuum.

We have argued that communities of practices, and networks of knowledge both about clocks, and about longitude, were deeper and more extensive than historians and horologists generally suppose. Only with some sense of what knowledges were 'common knowledge', can we sensibly infer how likely it is that Harrison's possession of particular ideas imply either contact (e.g. with 'Jeremy Thacker') or plagiarism (of Ward). We have not found clinching evidence either way. The closeness between, on the one hand, Thacker's and Ward's proposals, and, on the other, Harrison's practices, hints at a more direct connection, especially with Thacker's methods for heat trials, and timing for exactly 24 sidereal hours, which are those reported by Harrison to Halley in 1730. On the other hand, the density of circulating knowledges was such that we cannot dismiss the idea that these knowledges were sufficiently widespread for Harrison to have adopted them from a locally accessible common pool of knowledge and speculation. There is, in other words, no need to plump for a dichotomous choice between plagiarism and deceit, on the one hand, and inexplicable genius, on the other. Genius, yes, especially in Carlyle's transcendent capacity of taking trouble (1858: book 4, Ch. 3) or the proverbial 'infinite capacity to take pains', but not inexplicable; not 'from nowhere'.

NOTES

1. For village shops and the goods they sold, see N. Cox (2000) *The Complete Tradesman*.
2. John Harrison's autobiographical writings comprise various accounts of his clocks and experiments, written over nearly half a century: (i) an untitled manuscript dated 10 June 1730, Guildhall Library, Clockmakers Company records, 6026/1. This describes work over several years to bring a clock to performing to one-second-a-month; anticipates getting to 2–3 seconds/year, and not varying by more than 2–3 seconds/year from one year to next; (ii) a Royal Society memorandum, dated January 1742, reporting performance of a pendulum clock by Harrison to one second/month over 10 years, signed by President, Astronomer Royal, professors from Oxford and Cambridge, and 8 committee members; (iii) 'An explanation of my watch or timekeeper for the longitude...', 1763 manuscript, Guildhall Library, Clockmakers Company records, 3972/1; (iv) '*The case of Mr John Harrison*' (London, 1766, 2nd edition 1770).
3. Especially in his dispute with Maskelyne, see Harrison (1763, 1766).
4. Graham's specific interests in temperature compensation in scientific instruments (Chapman, 1985) provides an additional motive for Halley sending Harrison to Graham, over an above his specific expertise in clockmaking.

5. There were many printed manuals of clockmaking in the late seventeenth century, of which the most important were John Smith's (1675) *Horological Dialogues*, and Wiliam Derham's (1696) *The Artificial Clockmaker*.
6. The *Dictionary of National Biography* (2004) provides a convenient starting point.
7. There are also earlier references to boys building wooden clocks, as in the case of Robert Hooke, growing up at Freshwater in the Isle of Wight in the 1640s, so as to develop their design and manual skills (Cooper, 2005: 3).
8. Locations of cited documents are: Grimsby—E. Gillet (1956) *An Early Churchwardens' Account of St Mary's Grimsby*; Kirton-in-Lindsey—Lincolnshire Archives Office) Kirton in Lindsey, 7/1.
9. East Yorkshire Record Office, Heckington 7/1.
10. Intriguingly, given one of Henry Harrison's occupations, three inventories are for carpenters, two in the early 1660s and one in 1685.
11. Including, besides the parlours, a house [hall], a kitchen, three upstairs chambers and at least two cellars.
12. The earliest extant accounts are Lincolnshire Archives, Inventory 193/434.
13. Our enquiries in locating their whereabouts (or fate) through the twentieth century have so far proved unsuccessful.
14. Reference for Barton-upon-Humber town book, 1676: Lincolnshire Archives, Brears Mss. Vol. 7.
15. Lincolnshire Archives, Brears Mss. Vol. 7, p. 61, reiterated on p. 79.
16. Barrow Town Book Lincolnshire Archives, *The Towns Book of Barrow Containing the Dues and Customs Belonginge to the Said Town, One Thousand Seven Hundred & Nine*, Lincolnshire Archives, Barrow 10/1, pp. 9–10 have been lost. A copy is at Hull University Library, DX/1.
17. Barrow Town Book Lincolnshire Archives, *Barrow* 10/1, page 11.
18. Nevertheless, many early eighteenth-century clockmakers have twenty or more clocks extant, several more than fifty, and Samuel Roberts of Llanfair Caereinion—with his bottom-of-the-market customers—has well over one hundred surviving clocks (Pryce and Davies, 1985, 108–15). No northern clockmaker of the period has left comparable records, but the scale and range of their work is clear, especially for those who engraved serial numbers on their output, from surviving clocks.

    Jonas Barber of Bowland Bridge in Cumberland (now Cumbria) began to produce clocks there in 1717, moving the short distance to Winster in 1717. The business at Winster was taken over by his son, also Jonas, by 1748, and in turn by the younger Jonas Barber's former journeyman Henry Philipson at the century's end. The three men maintained a single sequence of numbers of their clocks, and the younger Jonas Barber's use of the numbers between the late 1740s and 1790s has been traced (Cave-Browne-Cave, 1979: Loomes, 1997: 81–5). The numbers in more than twenty surviving clocks, bearing the younger John Barber's signature, indicate an average production of over twenty clocks a year for more than forty years, from number 584 in 1761 through to 1435 in *c.*1800.

19. The case of Thomas Page's training in Chew Stoke and Bristol was discussed in Chapter 4. Loomes (1997) cites several northern English makers whose training took forms other than a full-term apprenticeship term, and Samuel Roberts (see Chapter 5) acquired his impressive range of technical skills without there being 'any record of Samuel Roberts ever having served and apprenticeship'; Pryce and Davies, 1985: 37–9.
20. Loomes (1997).
21. Moreover, many carpenters had some experience of working with wooden cogs and gearing, for example in connection with mills.
22. Based on linkage of databases of church bells (Johnston et al., 1990) and of church clocks and related payments in all surviving pre-1700 English churchwardens accounts (Glennie, <www.ggy.bris.ac.uk/clocks>).
23. At Barton-upon-Humber there were two churches, though strictly St Peter's was the parish church and St Mary's a chapel-of-ease (Bryant, 2003).

    With both parishes having invested heavily in equipping their churches with bells, there was possibly an element of competition between them—dual-parish towns often manifested such rivalry in facilities like bells (Sanderson, 1992) and clocks. Churchwardens need not have gone far to obtain bells, for there was a bell foundry at Barton (and James Harrison established a foundry at Barrow after John moved to London).

    Bell and clock maintenance commonly involved the same people, except where really major refurbishment was involved—and even here clockmakers and bell founders usually worked in conjunction with the local craftsmen who performed routine maintenance. In his later career James Harrison specialized in large bell frames, including those for York Minster and Beverley Minster.
24. Which might have documented Harrison's making or mending of church clocks. The only one of the thirty parishes within ten miles of Barrow and Barton for which churchwardens' accounts survive is Broughton, almost ten miles to the south-west: Lincolnshire Archives, Broughton-by-Brigg, 7/1.
25. Moore (1999); Bird and Byrd (1996); Ovens and Sleath (2002); McKenna (2000).
26. Online database of early pedigree thoroughbreds, at ww.turfnsport.com/pedigree.html.
27. It does not seem to have had any scholarships and/or exhibitions.
28. Details of Saunderson families from Barton parish register.
29. Axe's prize, precursor to the Longitude Act prize, was 'never to be paid, I think' (1696: 76). He was equally scathing about Venetian and Dutch prizes.
30. On Gresham College, Harrison 1696: 22–3. No Edward Harrison was at Oxford or Cambridge at the relevant time, Foster (1891–2), or Cambridge, Venn and Venn (1922–7).
31. Edward Harrison's date of birth is unknown, but Halley was forty-two in 1698. From his account of many years' sea experience before 1696, Edward Harrison cannot have been much younger than Halley in 1698.
32. IGI contains only some ten burials of males named Edward Harrison between 1698 and 1718. Most died in infancy and only one was adult in 1696, a man dying in 1703 aged twenty-six, who is too young to have accumulated that naval experience.

However, this primarily reflects the IGI's focus on weddings and baptisms—several score Edward Harrisons are recorded getting married, or being baptized (though many never reached adulthood).

33. Newly recognized phenomena included the spatially variable declination of compass needles, researched by Halley on the *Paramore* (Gurney, 2004). On 'sympathetic powder' and other quirky, or downright bizarre, personal perspectives on the workings of the universe, see Gingerich (1996).

34. Thacker's was the earliest use of 'chronometer' in print, though William Derham had used it in an unpublished paper dated 4 August 1714, read to the Royal Society on 4 November (Andrewes, 1996: 192 n. 13).

35. Thomas Savery had patented a pump driven by a steam engine in 1698, presenting his findings to the Royal Society, and publishing *The Miners Friend; or, an Engine to Raise Water by Fire...* in 1702, and was elected FRS in 1706.

36. Keeping a clock in a vacuum had been mentioned by an Italian clockmaker Matteo Campani (1620–87) in *c.*1665–70. Whether Thacker knew of this seems unclear—Campani is not one of the other workers he discussed. However, Thacker would very likely have known William Derham, who made the suggestion in a Royal Society paper in 1704 (Andrewes, 1996: 192, 204).

37. The 1714 pamphlet mentions many other suggestions for finding longitude, usually associated with particular individuals, though Thacker is coy about giving full names. Those mentioned, as we identify them, are: (i) 'Mr. W____n and Mr. D____n' (p. 2), William Whiston and Humphrey Ditton; (ii) 'Mr. H____bs' (p. 3), William Hobbs, *A new discovery for finding the longitude. Humbly submitted to the approbation of the Right Honourable the Lords spiritual and temporal*, London, 1714; (iii) 'Mr. Bill____y' (p. 3), Case Billingsley, *The Longitude at Sea*, London, 1714; (iv) 'Mr. Br____e', (p. 4), Robert Browne of Wapping, *Methods, propositions and problems, for finding the latitude; with the degree and minute of the equator upon the meridian. And the longitude at...*, London, 1714; (vi) 'Signior Al____ri', (pp. 4–5), Doroteo Alimari,... *to discover the longitude. To which are added, proper figures of some instruments...*, London, 1714; (vii) 'my friend (Mr) K____th' (p. 4 and p. 5), George Keith (MA) *Geography and Navigation Completed; Being a New Theory and Method whereby the True Longitude Of any Place in the World may be Found*, London, 1709; (viii) '*Mr. J.H.* "Writer of the Essay" ', (pp. 5–7); Isaac Hawkins, *An Essay for the Discovery of the Longitude at Sea, by Several New Methods Fully and Particularly Laid Before the Publick*, London, 1714; (ix) 'Mr Wa____n's Advertisement' (p. 7); which 'promises a Clock to make the longitude known to those of the meanest capacity', remains unidentified; (x) 'another Gentleman' (pp. 7–8), has a great Volume coming out—Quarto—successive dedications to King, Prince, Royal Society, the Reader', remains unidentified; (xi) 'the great Mr. L____'(p. 8), 'with an Industry equal to his Candour, should join the *Fluxions* and *Series, which he invented*, to his known Skill in the Laws of *Centripetal* and *Centrigugal* Forces...', Gottfried Leibniz; (xii) Mr. J C____r's *Marmeter*, in Quarto, (pp. 21–2), remains unidentified; (xiii) 'H____y, Esquire' (pp. 21–2), Francis Haldanby, *An attempt to discover the longitude at sea, pursuant to what is proposed in a late act of Parliament.* London, 1714.

38. The Sackler Archive Resource of the Royal Society; database of all Fellows since the foundation in 1660.

39. Loomes (1997) notes, in passing, numerous clockmakers without surviving work (despite his focus being on clocks, rather than makers), like the Barber family at Skipton. The two John Barbers discussed earlier, were son and grandson of another clockmaking John Barber, at Skipton, none of whose work survives, nor that of his brother, who worked in London until his death in 1698 (Loomes, 1998). Other examples include John Butterworth of Cawthorne, near Barnsley (Yorkshire), Thomas Carter of Bishop Auckland (Durham), James Cawson of Lancaster and (later) Liverpool, William Reader of Hull, and Timothy and Leonard Hepton, of Northallerton.
40. Loomes (1982, 1997, 2006).
41. East Yorkshire Record Office, PE1/51–90.
42. Surviving records from both Anglican and Congregational churches in Beverley consulted at East Riding of Yorkshire Archives and Records Service: Beverley St Mary PE1/; Beverley St John and St Martin PE129/; Beverley Lairgate Congregational Church, EUR1.
43. Printed in London, reprinted in the 1750s and again in 1770. The earlier items ranged from Arbuthnott's own writings, his collaborations; and his translation from Latin of Huygens's pioneering work on probability.
44. Arbuthnott's Doncaster connection is recorded in the list of all Royal Society Fellows since the 1660 foundation, at the Sackler Archive Resource. Available online at: <www.royalsoc.ac.uk/>.
45. Ward and Waller (1912), superseded by Loewenstein and Mueller (2002).
46. Unless, of course, the response was also by Arbuthnott. However, Ward is much more securely identifiable than Thacker, with other publications on longitude and related topics. See, *New Dictionary of National Biography*.
47. Made for him by John Rowley (fl.1698–1728) a leading London instrument-maker.
48. Ward's pamphlet resembles Thacker's, reviewing various longitude-determining methods—'all of them true in *Theory*, but either so abstruse, or so attended with such Difficulties, as have hitherto render'd them Impracticable at Sea' (Ward, 1715: 1), before advocating his timekeeper (also spring-driven), and suggesting solutions to problems caused by climatic variations, damp, corrosion, and an irregularly-moving ship. Ward claims to have worked on his machine since 1710, but, delayed by illness, thought Thacker had copied his idea in the meantime. Ward's discussion is more concerned with mathematical and astronomical difficulties than Thacker's, passing comparatively briefly over the clock, apart from its use of two opposed spiral springs to keep the balance in regular motion, and a dial plate marked in four-second intervals, for direct reading of time differences in minutes of longitude (Ward, 1715: 8–9). Ward cited Richard Street as clock and watchmaker with whom he had discussed his ideas, and several watches and a clock by Street survive showing Ward's influence (Turner, 1996: 127). The watches have minute hands rotating in 15 minutes, the minutes subdivided in 6 for 10-second readings (rather than 15 and 4, probably for visibility), with an equation of time table engraved on the cases. A long-case clock whose minute-hand revolves in 10 minutes is also known of (Turner, 1996: 123–7).

49. For help with translations, and locating their sources, we are grateful to Tom Holland.
50. Notwithstanding the emphasis on suppleness in the phrase, the overall context is a comparison of height. In Virgil's poem, two shepherds, Mopsus and Amyntas, are comparing their musical prowess. Mopsus compliments Amyntas by saying that he excels a third shepherd, Menalcas, to the same extent as the olive is superior to the willow. We owe this information to Tom Holland.
51. Venn and Venn (1922–7) volume 3; Foster (1891–2), though non-matriculation may only indicate non-Anglican beliefs, rather than either poverty or lack of education.
52. We are grateful to Tom Holland for answering questions on pupils' progression through the poets in Latin teaching of the late seventeenth and early eighteenth centuries.
53. International Genealogical Index ®, *FamilySearch*™, Internet Genealogy Service, The Church of Jesus Christ of Latter Day Saints, 50 East North Temple Street, Salt lake City, Utah 84150, USA. Our use of the IGI was confined to information collected by 'controlled extraction' from parish register originals, or Bishops' Transcripts of parish registers, ignoring the often-unreliable 'patron submissions' information. Even so, the IGI is problematic for various reasons. First, it is incomplete because not every English parish is covered: parish registers could not be microfilmed where they did not survive, and the CLDS could not always obtain permission to microfilm. Second, surviving registers could contain lost or illegible sections. Third, transcription errors when working with microfilms of variable quality are unavoidable. Fourth, far fewer non-conformist registers survive, which in some areas excludes significant parts of the population.
54. For example, in Porteus's study of the name Mell (Porteous, 1982).
55. IGI contains only one Jeremy/Jeremiah Thacker baptism in the eighteenth century—much later, but perhaps a non-conformist connection, Jeremiah Thacker, of Cannon Street Independent, Manchester, Lancashire, on 4 July 1762.
56. John Thaeker's recorded baptism date of 14 January 1692 pre-dates the move of New Year's Day from 25 March to 1 January in 1752, so is 1693 in current reckoning.
57. It is unclear where he had been born or educated. Sutton Coldfield had its own sixteenth-century grammar school: Bishop Vesey's Grammar School, founded in his home town in 1527 by the then Bishop of Exeter.
58. Anthony Thacker contributed regularly to *The Gentleman's Diary, or the Mathematical-Repository*, which appeared as an annual almanac with a substantial section of mathematical exercises and puzzles from 1741 through at least 1789. The mathematical content overlapped substantially each year with that of *The Ladies Diary, or Woman's Almanack*, but it is striking that the latter had published annually from 1704–5 through into the nineteenth century. Both are accessible to subscribing institutions through *Eighteenth Century Collections Online* at: <galenet.galegroup.com/servlet/ECCO>. See also *The diarian repository; or, mathematical register: containing a complete collection of all the mathematical questions which have been published in The Ladies diary, from the commencement of that work*

in 1704, to the year 1760; . . . By a society of mathematicians (London, 1774). Online at: <galenet.galegroup.com/servlet/ECCO>.

59. P.J. Wallis (1976) *An Index of British Mathematicians: A check list, part 2, 1701–1760*; R.V. Wallis and P.J. Wallis (1988) *Bio-bibliography of British Mathematics and its Applications. Part 2, 1701–1760*; R.V. Wallis and P.J. Wallis (1993) *Index of British Mathematicians: Part 3, 1701–1800*.

60. The only burials of 'Robert Thacker' in the period shown by the IGI were of infants and children.

61. For example, the Warwickshire Thackers were concentrated in and around Sutton Coldfield, and those from Derbyshire (also clear from hearth tax returns, Edwards [1982]) clustered around Wirksworth in Alderwasley, Alderhey, and Kirk Ireton.

62. Obvious research projects here include pursuit of exactly who were the key innovators in different clockmaking communities of practice, and how fast certain innovations worked through the various networks.

11
Some Concluding Remarks

At the outset (page 13), we emphasized that we do not not intend here to provide another overarching narrative of time, and we hope the book is not read as though such grand theorizing were its goal. We have attempted to trigger more research into a topic—the practices of clock time—which is often considered to be so opaque to systematic historical research at anything other than a heroic scale. Whatever the validity of our general conclusions, we hope it is now clear that such an inference no longer has to be drawn. Systematic historical research is possible, even in this difficult area.

Finding incidental evidence of non-institutional uses of clock time may always be fortuitous in archives generated by official administration, but we have attempted to demonstrate that it is recoverable from seemingly unpromising material. Much information about the shape of clock times as practices can be recovered from documents whose purposes had nothing to do with indicating uses of clock time. It is unlikely that full records of temporal practice could be available for any individual, but we explored vignettes of individual practice in Chapters 6 and 7 and of particular temporal communities in Chapters 8 and 9. Both individuals and specialized temporal communities were part of wider communities, and situated within national infrastructures of clock time and signalling, which we addressed in Chapters 4 and 5 respectively. Chapter 10 eschewed the 'lone hero' trope of recent works about John Harrison in trying to recover parts of the zeitgeist from which this key horological figure emerged.

Obviously, our thinking-together of various types of evidence on these topics, and the concepts outlined in Chapters 2 and 3, amounts to a partial and provisional history of clock-time practices in England. These closing thoughts therefore first highlight some key elements of this history of clock-time practices, and then point to four topics that now demand pressing attention.

For us, critical centuries-long trajectories in clock-time practices and everyday temporal environments are most effectively indexed by what might at first glance seem quite mundane developments (Table 11.1). This chronology differs sharply from chronologies of technology and of time-disciplining (Tables 2.1 and 2.3) in both the timing of critical changes or emergences, and in the interpretation of relationships among practices and disciplinary or technical changes. Whereas the key elements of 'horological' and 'disciplinary' chronologies

Table 11.1. Chronology of some major changes in English clock-time practices

1200		
	Longstanding temporal practices using other metrics	
1300	Clock times and equal hours reckoning pervade existing temporal practices	
		Clocks start to be used to measure out work processes
1400		
1500		'Telling the time' has become a normal procedure
	Almanacs widespread	Large numbers of public clocks
1600		
		Diary-keeping more common
1700	Use of seconds by specialized communities of practice	Clock time linked to rhythmic practices in general
		Commonplace in urban areas to ask the time of people in the streets
1800		
	Allowance for local times in transport and communications	

Source: Glennie and Thrift (2005).

embody explicitly directional dynamics of technological improvement and social control, practice-focused histories appear diffuse, centred as they are on what was able to be effected by particular practices, whether self-conscious, habitual, or unconsidered and unrecognized. Certainly, over the centuries, it is striking how many uses of clock time were incidental to carrying out practices and how comparatively rarely they were central. The incidental practices we emphasize are all but invisible to orthodox histories of timekeeping as a clock-bound metric, and they have remained under-theorized and under-researched in the historical record.

This complex of changing temporal practices was of great long-term importance, though not of a rapid 'revolutionary' pace. Rather, it was what we elsewhere characterized as a 'slow burn' (Glennie and Thrift, 1995: 192), giving everyday temporal environments 'a growing self-referential confirmatory-ness' arising from an increasingly dense population of material objects rich with temporal information. Each of these environments fed on the others to greater or lesser extent, and the formation of this circular dynamic lies at the heart of our history of clock-time practices. Long-run accumulations of the sorts of cognitive understandings found in instructional and other literature, and the developing embodied knowledges of bodily dispositions, co-produced new anticipations of time, and new kinds of 'common sense' with regard to time, that together interacted to re-weave textures of everyday life. Part and parcel of this reweaving consisted of changes in everyday temporal environments, as increasing densities of clock-time-related practices and of clock-time related frames such as timetables and diaries reciprocally produced growing needs for more clock-time measurement, in the process confirming needs for clock-time practices that these intermediaries helped to equip.

Of the many transformations occurring within the half-millennium span of this book we see three as of particular significance: the adoption of a single metric framework; its increasing subdivision; and interactions amongst more, and more specialized, communities of practice.

In the first of these three 'revolutions', clock times enter and pervade everyday life as mechanical clocks provide a critical impetus to standardized equal-hours timekeeping, first as a complement to, and then as replacement for, various earlier and looser frameworks of daily temporality. The increase in fifteenth- and sixteenth-century public clocks in towns is made more impressive by their substantial costs and the restricted distributions of specialized smithing skills. Our stress on clock-times as integral to novel ideas of 'urban living', contrasts with Dohrn-van Rossum's (1996) characterization of clocks as 'urban accessories', though large towns were too thinly distributed for clock times to yet become pervasive of everyday life. Nevertheless, the use of clock time to locate events within narrative was commonplace in narratives emanating from all parts of English society in the sixteenth century. People appear to have coped quite comfortably with several coexisting temporal frames whether metric like clock time or much looser and more flexible. As public signalling of hours spread, people commonly made opportunistic use of public time-signals for diverse organizational purposes of their own.

The clock times 'broadcast' as bells struck signals across urban environments were understood as a general provision serving multiple users, rather than as specific to a particular interest or authority. In consequence they seemed a 'natural' resource, and opportunistic uses of hour-signalling by a wide range of people can be shown in several kinds of sources. Particularly revealing are indications of people's ability to improvise temporal information in constructing hypothetical or fictitious accounts of behaviour, which are revealing of their appreciation of a need to meet others' expectations of adequation.

By the mid-sixteenth century, widely (but unevenly) signalled clock times were constitutive of many social practices, being both defined by and contributing to various specialist and everyday skills. Publicly struck hours were a routine descriptive resource, but the extensive 'foraging' required to tell the time really accurately ensured that, for everyday purposes, fast and frugal algorithms based on small amounts of robust temporal information were preferred, and adequation was relatively loose for most of the time. Exactitude always involved a trade-off between effort and accuracy in finding the time to an appropriate level, and whether accuracy or precision was more important in a particular context. Even though many uses, including conventions of politeness, became more temporally demanding, the availability of signals and improvements in accuracy meant a reduction in the necessary effort. Many people took improvements for granted, continuing to apprehend the required effort as at the threshold of 'worthwhile'. Even though, over time, less 'foraging behaviour' was required, people still weighed up a trade-off between the (now greater) accuracy of information obtained, and the (diminishing) effort involved in its acquisition. New heuristics of rough estimation made headway wherever they proved useful, sometimes providing adequate alternatives to the purposeful use of dedicated devices. Increasingly, there were diverse workable combinations of specialist knowing and robust heuristics among people holding divergent views of adequacy depending on their situation and skills.

The second 'revolution' is an increasing subdivision of hours, and then the use of minutes (and, more rarely, seconds) for all sorts of purposes, not least by people of low status and limited training. This pre-dated the late seventeenth-century horological revolution's decisive leap in timekeeping accuracy, and the substantially falling costs of more accurate timing, which transformed the trade-off between effort and adequacy in accurate timing. Astrology was at least as important as 'science' in this revolution, as part of a broader enthusiasm for technologies that measure. We reverse the usual horological picture, in which minute-indication appears after the invention of the pendulum regulator and the anchor escapement, in that there seems to have been substantial latent demand for precision prior to the technological advances. However, we emphasize that people's everyday practice usually fell well short of their capability for precision—the adequation-effort payoff was still a powerful consideration. How readily temporal practices, aesthetics, and sensibilities became general varied, depending on whether they 'worked' on an everyday basis. For example, the aesthetics of seconds migrated to several areas of everyday life more readily than did practices of seconds timekeeping, which for most people did not 'ground' in everyday life. The rapid but selective take-up of seconds attests to a pent-up demand for accuracy, but more as an aesthetic than as a utilitarian issue.

As greater numbers of better instruments circulated a richer array of information, they cumulatively produced environments that functioned more and more effectively as timekeeping devices, becoming increasingly self-referential and confirmatory. So whereas in 1550 complicated foraging behaviour was required to find the time to high levels of adequation, by 1750 the need for complicated foraging had largely disappeared into better timekeeping devices, and more self-referential environments. Long-run changes became increasingly centred on positive feedback relationships among the density and performance of timing devices, the information they provided, and the uses to which that information was put.

More self-referential environments and growth in mutually confirming temporal practices became woven into everyday life, becoming 'second nature' to people in extended distributions of various embodied temporal skills. Of these skills, we see capacities to handle routines, to improvise, to build a repertoire of

Some Concluding Remarks   411

experiences and skills that 'domesticated' or 'routinized' improvisation, and to estimate effectively, as of the greatest long-run importance. Taken for granted now, these were skills becoming much more widespread and highly developed in early modern England, reinforcing clock times as integral elements of everyday lives, social interactions, and localities. Critically for our argument, this occurred whether or not people were subject to explicit disciplinary impulses to explicit clock timekeeping.

The third revolution is the emergence of more specialized temporal communities, centred on practices involving small units of time, and ideas of precision and accuracy. We have sought to trace how and when precision in clock times mattered, and to whom, and to explore the dynamics of relations among specialized communities of practice, and their interactions with everyday practices. Growing numbers of communities of practice, varying in composition, size, and stability, were defined by particular timing practices, priorities, concern for accuracy, and adequacy. Some were of long-standing, some were completely new, and some older communities of practice faded away, especially those that were bypassed by changing technologies, skills, or needs. The nature of times in communities of practice also varied greatly, from the formal, numerical, and highly textual senses of time among professional astronomers, while others were chiefly based on informal circuits of talk.

Many specialized communities of practices, mainly of a secular nature, emerged in the seventeenth and (especially) eighteenth centuries, often demanding yet more exact timekeeping, thereby contributing to positive feedbacks with more and better clocks, and more and more detailed temporal information. Consequently, a more demanding approach to adequacy became a general feature of society, in part because technologists' success in building better adequacy into better-performing clocks and watches removed it as an everyday concern requiring action.

Specific relations with particular communities and places produced overlapping ebbs and flows of spatial contrast, with differentiation among people and places as new practices, dispositions, and devices appeared and spread, and convergence where these novelties were sufficiently mobile and stable to approach general knowledge or common practice. Long-run access to more detailed clock times became less and less problematic in an ever-widening range of places. For much of the period covered in this book, however, access to clock time had a substantial and durable geography of necessary effort. This geography entered into everyday anticipations as a practical grasp of feasibility, and into local aspirations for provision.

Clocks and other timekeeping devices populated everyday life more densely than most accounts recognize, partly because public clocks were more numerous than has been realized, but mainly because many more objects and practices could be interpreted temporally. Examples include the movements of Bristol's post- and coach-services (Chapter 4), the sense of pace accruing to Samuel Pepys as he registered time elapsed at landmarks on his walk to Woolwich (Chapter 6), and various embodied timings among seafarers (Chapter 8), and other specialized communities of practice. Such takings of clock time into human bodies help explain the emergence of new anxieties about punctuality and delays, and greater impatience about being kept waiting,

Besides—or rather intertwined with—the growth and spread of increasingly detailed clock-time metrics and the proliferation of specialized communities of practice, were other long-run changes: the reducing cost and effort needed for time-finding and time-telling; the increasing ecological rationality of urban and rural environments; and more accuracy-demanding approaches to adequation. All the changes helped build a 'slow revolution' of co-evolving transformations of practices, dispositions, devices, and environments, and in the many relations through which these were linked, as they became denser, more durable, and ever more taken for granted.

Geographies of clock-time practices were a complex product of many interacting factors, the practices of particular, very unevenly documented temporal communities, and how readily their practices moved beyond them. Communities' locations shaped their interactions with others, with strong clusterings of those sharing key priorities and skills, most obviously in and around early modern London, simultaneously the pre-eminent location for many groups interested in precision timekeeping, including astronomers, navigators, clockmakers, precision engravers, naval bureaucrats, as well as the Royal Society, Gresham College, the Greenwich Observatory, the Clockmakers Company, and the printing and educational industries through whose products the relevant knowledges were circulated. It was in London, too, that we have our clearest picture of the many aesthetics through which emerging ideas about precision and accuracy found expression. Besides utility, these aesthetics included piety, ingenuity and artifice, novelty, taste and sensibility, politeness, and bodily control. These multiple aesthetics help explain the rapid diffusion of favourable dispositions towards precision timing well ahead of the concrete uses for such precise measurement.

Overall, therefore, we see horological technology, social discipline, and work-discipline as three among many originary spheres for changing times in early modern England. Everyday sociality, consumption, and various specialized communities of practice were important sites of spontaneous clock-time usage, including the generation of new timing practices as circulating notions of timing were negotiated in the practices of diverse activities and localities. The familiarity of socially marginal groups such as provincial children with hours and minutes is striking, and demonstrates the 'reach' of such 'thick' everyday temporal practices.

Thus, over several centuries, 'time' was reconstituted as event-ful by new practices that relied on clocks and related devices as doings, in a coevolution of technologies and the uses made of them, which wove clocks and other intermediaries into the everyday operation of many communities of practice and their relations with others. These temporal revolutions intertwined with accelerating growth in the density of material objects to produce artificial environments increasingly taken for granted, in which systematic calculative practices became familiar parts of everyday life (Thrift and French, 2002). Amplifying and adding to contemporary trends such as the rise of political arithmetic, times were both a particular instance of, and a potent contribution to, this laying down of metrological networks through society. In Nardi and O'Day's (1999) phrase, clocks were 'keystone species' as temporal information proliferated and precision became linked into everyday practices in 'ever more prodigally suggestive ways' (Glennie and Thrift, 2005: 194).

The book's exploratory intent also means that we are well aware that on several issues we have barely scraped the surface. To close the book we want to indicate some of the topics that have come to the forefront of our research. The first of these is the extent to which considerable amounts of innovation in clock design were the result of general tinkering and experimentation of various kinds. Until recently, the influence of these kinds of activities has been underestimated in work on innovation which has largely assumed that technological authority rests with a few key producers, whereas we have shown that mechanical knowledge could be widely distributed. Indeed, given the vagaries of clocks in the period we have studied, any other outcome is difficult to envisage (Franz, 2005). However, of late this situation has begun to change as a result of a number of developments. In history, the work of writers like Mokyr (2002) and Franz (2005) has pointed to the scale of tinkering and experimentation, the way in which it could be built in to a stock of useful, best-practice knowledge, and the way that it was able to be mobilized during the industrial revolution through new networks of communication. These networks improved access to this knowledge and reinforced its presence by increasing the overall likelihood of producers knowing something that is known and knowing how to find it; 'it is the hallmark of an innovative producer to know what he or she does not know but is known to someone else, and then try to find out' (Mokyr, 2002: 9). In other words, the industrial revolution, by this account, increased the number of the doors of knowledge an innovative producer could walk through. In business and economics, more and more attention is being paid to means of boosting small-scale experimentation, thus allowing more learning to be created more rapidly and so increasing the overall rate of innovation (Thomke, 2003; Prahalad and Ramaswamy, 2004). Particular attention has been paid to increasing the capacity for this experimentation by turning to so-called open innovation models (Chesbrough, 2003; von Hippel, 2005) which will pull in the skills and knowledge of all manner of agents—from consumers to enthusiasts to inventors—so allowing a large range of innovations to be created, some of which will survive. In some of the incarnations of these models, the aim seems to be drawing in all of the knowledge gleaned from experimentation and tinkering, from the small triumphs, blind alleyways and outright mistakes, which had been lost as innovation became purely corporate property, thus reinventing something like the days of Mokyr's industrial revolution.

Our view is that the scale and intensity of clockmaking were such that it is likely that technical development (of both clocks *and* the means of making them) was an all-but-continuous activity. This developmental activity was itself transmitted, as an orientation and a stock of practical knowledge, from generation to generation, not least through new institutions that included a series of communities of practice which we have only touched on in this book. In other words, so intense was the generation and interchange of knowledge amongst clockmakers that they constituted a mini-industrial revolution before the industrial revolution. It is not only necessary to look beyond science to what Mokyr terms 'engineering knowledge', but to recognize that the boundaries between engineering knowledge, craft knowledge, and folk knowledge are extremely ill-defined.

The strongly hierarchical and 'gentlemanly' practice of early modern science—the object of so much study—is not a good model for these early-modern temporal knowledges where different 'levels'—engineering, craft, and folk—were much less clearly separated. There were plural, sometimes overlapping, reasons for interest in timing among different 'non-gentlemanly' groups, and for the movement of technical knowledges not originally intended to circulate at that level. Histories of science provide some excellent examples of key personnel and interfaces between different communities of practice, and the cultivation or inadvertent emergence of networks across different communities of practices (see, for example, Hunter *et al.*, 2003, on Robert Hooke). But tensions between gentlemanly, commercial, and artisan motives capture only part of the building of knowledge networks, and weaker connections and circulations call for more attention (Chapman, 1985).

The importance of these 'weaker' networks is generally under-rated. They account for our finding, at several points in this book, that people know more than one might think from the evidence of their direct contacts. Indirect links, mutual support, and conversation are the best candidates as to where people encountered that 'extra' knowledge, and offer an alternative to the 'individual genius' thesis exemplified by Sobel's presentation of John Harrison. But such circulating knowledges, more than could be found in the one text or the single person, are particularly difficult to show in the record.

A further implication is that the 'take-off' of temporal practices into wider use often depends on the robustness of the networks into which technological devices are interlocked. Practices are more effectively sustained where supportive knowledge is more broadly distributed than within a single key individual, like Harrison (recognized by the Admiralty's concern for spreading Harrison's technology to other makers) or Richard of Wallingford, where the technology was effectively lost (North, 2005). This dependence on 'critical mass' in developing practices made the specific geography of communities of practices particularly important.

One more implication is not just that knowledges are circulating, but that so are more general senses of ambition and aspiration. The pursuit of specific technical objectives, such as the Longitude Prize, receives considerable attention, but more general aspirations to the 'right' feel of practices merits closer attention in the future, despite the archival difficulties involved. The need is the more pressing given the gradual specialization of some of the skilled practices we have mentioned, and because the historical evidence about them increases largely because knowledge transfer was becoming more formalized. We must leave space for usually undocumented capacity of skills and knowledges to move beyond their initially quite specific purposes.

The second related topic is the extent to which repair and maintenance is an integral part of 'clock'. We think it is (Glennie and Thrift, 2007). Through a large part of the period we are considering, clocks were unreliable and were in continual danger of breaking down. They therefore needed constant maintenance and, all too often, repair. Whilst we cannot be sure about the rates of breakdown of clocks and then watches, we can be sure that, during our period, clocks and watches slowly became more reliable. But within this general tendency, there were significant variations. For example, standard clocks broke down more often than, for example, eight-day clocks that ran longer on a single winding and were likely to have better quality wheelwork.

Whatever the exact case, it is clear that, through our period, clocks were continually being tinkered with. Though they were finished consumer objects, they were subject to a series of revisions that sometimes made them into something quite different over time. At least four possibilities present themselves. First, they might be upgraded: the switch from verge and foliot to pendulums comes to mind: many clocks were converted. Second, they might be repaired, but in the process improved components were inserted. Third, they might be subject to home repairs that involved a good deal of bodging—about right was good enough. Fourth, they might be cannibalized, with parts from one clock being used in another. All of these possibilities were qualified by a degree of urgency: it might be that a clock was working well enough to be lived with whilst a solution was sought—given that what counted as acceptable levels of performance no doubt varied over our period—or the lack of accurate clock time might have proved sufficiently inconvenient, annoying, or even downright dangerous that a repair was needed immediately. Different strategies of repair might follow. These were also qualified by accessibility. If a satisfactory clockmaker could be found nearby, then this would again have obvious consequences.

In any case, our inclination is to treat this work of maintenance and repair as part of the timekeeping system. Why? For two main reasons. First, because it seems likely that a significant part of the work of many clockmakers consisted of maintenance and repair. Whilst there are no sources that allow us to be sure about what exact proportion of the work of clockmakers consisted of these activities, there are some intriguing pointers. Thus, church clocks regularly came with formal maintenance contracts. It was a common practice for whoever built a church clock to enter into a contract for the upkeep of the clock (Table 10.3). Similarly, some great houses entered into contracts to keep all their clocks going. Then there are the sheer number of clocks that were converted to pendulum running at the end of the seventeenth century. Then, it is clear that certain repairs (for example, of fusees) were only able to be carried out by a select few clockmakers. In part, this was a function of expertise (some clockmakers had 'the knack', others did not) but, later in the period in particular, it was also because only some clockmakers had access to the requisite tools (for example, gear-cutting engines) or had the scale of operation to produce batches of components. In other words, there was a distribution of expertise in repair.

The third topic is the history of watches. In this book we have addressed watches tangentially on the whole, as a minor subset of clocks. We have done this on purpose. The presence of watches has perhaps been exaggerated in the historical record because they are able to be used so easily as emblems of broader cultural changes like the consumer revolution or new architectures of gender and sexuality. In our period, though watches had been invented in the sixteenth century, they start to become general only in the eighteenth century as a result of a trail of innovations including the spiral leaf mainspring, the fusee, the use of jewels, the use of enamel on watch dials so that they could be better seen and, right at the end of our period, the self-winding movement. It is interesting to reflect on the extent to which these watches were actually used for timekeeping. For many, especially those with nomadic lives like sailors, they seem to have functioned more as a portable store of wealth and as an object that could be traded than they did means of time-telling (Earle, 1998).

The final topic is formal learning about clock time. At various points in the book, we have signalled the importance we attach to learning, but actual evidence of it is hard to come by. For example, even the formal learning of clock time at school has proved a remarkably difficult area in which to prospect. We have found far less material than we expected and most of that has been fragmentary. But, as is evident from Chapter 6, telling the time from the dial of a clock or watch does not seem to have been a part of 'what school taught' until later, perhaps the nineteenth century. We are sure that somewhere more evidence of clock-time learning lurks but we have not been able to find it.

To recapitulate, in this book we have tried to open up the historical trail blazed by E.P. Thompson to systematic historical research. In the process of doing so, we have substantially changed the bones of his account, both theoretically and empirically. In particular, by concentrating on the supposed minutiae of practices, on the fragile techniques and anxious glances, we have shown how new temporal knowledges were gradually accumulated, knowledges that prompted new bodily sensations and movements, produced new 'cognitive' heuristics and constructed new kinds of temporal environment, all of these different but related phenomena acting in lockstep (Petitot *et al.*, 1999). Yet this history of changes in temporal phenomenality still largely remains to be written—all we have done is to make a start—for, as Rée (1999: 386) puts it, 'The fact that the structures in which we live our lives are obvious does not mean that their significance is clear to us—for nothing is harder to understand than what is most familiar.' Perhaps this is why that significance has so often been downgraded when it should have been made the subject of sustained enquiry.

Bibliography

Abbott, A. (2001) *Time Matters: On Theory and Method*. Chicago: University of Chicago Press.
Ackroyd, P. (1998) *The Life of Thomas More*. London: Chatto and Windus.
Adam, B. (1990) *Time and Social Theory*. Oxford: Polity Press.
—— (1993) 'Within and beyond the time economy of employment relations—conceptual issues pertinent to research on time and work' *Social Science Information Sur Les Sciences Sociales* 32, 163–84.
—— (1994) 'Beyond boundaries: reconceptualizing time in the face of global challenges' *Social Science Information* 33, 597–620.
—— (1995) *Timewatch: The Social Analysis of Time*. Cambridge: Polity Press.
—— (2004) *Time*, Cambridge: Polity Press.
Aitken, G.A. *The Life and works of John Arbuthnot*. Oxford: Clarendon.
Allan, S. (1994) '"When discourse is torn from reality": Bakhtin and the principle of chronotopicity' *Time & Society* 3, 193–218.
Alliez, E. (1996) *Capital Times*. Minneapolis: University of Minnesota Press.
Aminzade, R. (1992) 'Historical sociology and time' *Sociological Methods & Research* 20, 456–80.
Anderson, R.C. (ed.), (1931) *The Book of Examinations and Depositions, 1622–1644. Volume II 1627–1634*. Southampton: Southampton Record Society.
Andrewes, W.J.H. (1985) 'Time and the astronomer, 1484–1884', *Vistas in Astronomy*, 28, 69–86.
—— (1996) 'Even Newton could be wrong: the story of Harrison's first three sea clocks'. In W.J.H. Andrewes (ed.), *The Quest for Longitude*. Cambridge, Mass: Harvard Collection of Scientific Instruments, pp. 189–234.
Andrews, K.R. (1984) *Trade, Plunder and Settlement: Maritime Enterprise and the Genesis of the British Empire 1480–1630*. Cambridge: Cambridge University Press.
Anon (1537) *The enquirie and verdite of the quest panneld of the death of Richard Hune wich was founde hanged in Lolars towe*. Antwerp, Peetersen.
Appadurai, A. (1996) *Modernity at Large: Cultural Dimensions of Globalization*. Minneapolis: University of Minnesota Press.
Arbuthnot, J. (1770) *Miscellaneous works of the late Dr. Arbuthnott*, 2 volumes. London: printed for W. Richardson & L. Urquhart.
Arkell, T. (2000) 'Interpreting probate inventories', in T. Arkell, N. Evans and N. Goose (eds), *When Death Us Do Part: Understanding and Interpreting the Probate Records of Early Modern England*, Local Population Studies, pp. 72–102.
Armitage, G. (1997) *The Shadow of the Moon: British Solar Eclipse Mapping in the Eighteenth Century*. Tring: Map Collector Publications.
Ashmole, E. (1717) *Memoirs of the Life of That Learned Antiquary, Elias Ashmole: Drawn Up By Himself By Way of a Diary*. London: Charles Burman.
Armstrong, C.A.J. (1983) 'The piety of Cecily, duchess of York: a study of late-medieval culture' in *England, France and Burgundy in the Fifteenth Century*. London: Hambledon.

Assad, M.L. (1999) *Reading With Michel Serres: An Encounter with Time*. Albany: State University of New York Press.

Aughton, P. (2000) *Bristol: a People's History*. Lancaster: Carnegie Publishing.

—— (2004) *The Transit of Venus: the Brief, Brilliant Life of Jeremiah Horrocks, Father of British Astronomy*. London: Orion Publishing Group.

Ault, W.O. (1972) *Open-field Farming in Medieval England: A Study of Village By-Laws*. London: Allen & Unwin.

Aveni, A. (2002) *Empires of Time: Calendars, Clocks and Cultures*, revised edition. Boulder: University Press of Colorado.

Baillie, G.H. (1951) *Clocks and Watches: an Historical Bibliography, volume I*. London: N.A.G. Press.

Ball, H. (1856) *The Social History and Antiquities of Barton-upon-Humber*. Barton-upon-Humber: M. Ball.

Barbour, R. (1982) *John Aubrey's Brief Lives*. Woodbridge: Boydell Press.

Barnett, J.E. (1998) *Time's Pendulum: the Quest to Capture Time—From Sundials to Atomic Clocks*. New York: Plenum.

Barnum, P.H. (1976) *Dives and Pauper*. Vol. 1, part 1. London: Oxford University Press for the Early English Text Society, no. 275, original series.

Barry, A., Osborne, T. and Rose, N. (eds.) (1996) *Foucault and Political Reason: Liberalism, Neo-Liberalism, and Rationalities of Government*. London: UCL Press.

Barry, J. (1985) *The Cultural Life of Bristol, 1640–1775*. University of Oxford: D.Phil. thesis.

—— (1988) 'The parish in civic life: Bristol and its churches, 1640–1750', in S.J. Wright (ed.), *Parish, Church and People: Local Studies in Lay Religion, 1350–1750*. London: Hutchinson, pp. 152–78.

—— (1995) 'Literacy and literature in popular culture: reading and writing in historical perspective' in Harris, T. (ed.) *Popular Culture in England, c.1500–1850*. London: MacMillan, pp. 69–94.

Barthes, R. (1976) *Sade, Fourier, Loyola*. New York: Hill and Wang.

Bartky, I.R. (1989) 'The adoption of standard time' *Technology and Culture* 30, 25–56.

—— (2000) *Selling the True Time: Nineteenth-Century Timekeeping in America*. Stanford: Stanford University Press.

Bassier, P. (1999) *Navigation Astronomique*. Paris: Vuibert.

Battersby, C. (1998) *The Phenomenal Woman: Feminist Metaphysics and the Patterns of Identity*. Cambridge: Polity Press.

Baulant, M., Schurmann, A.J. and Servais, P. (eds) (1988) 'Inventaires Apres-Deces at Vente de Meubles: Apports a une histoire de la vie economique et quotidienne (Xive–XIXe siecle)' *Actes du seminaire tenu dans le cadre du 9eme Congres International d'Historire Economique de Berne* (1986). Academia, Louvain-la-Neuve.

Beachcroft, G. and Sabin, A. (1938) *Two Compotus Rolls of St. Augustine's Abbey, Bristol, 1491–1492 & 1511–1512*. Bristol: Bristol Record Society publications volume 9.

Beck, U. (1992) *Risk Society: Towards A New Modernity*. London: Sage.

Bedini, S.A. (1991). *The Pulse of Time: Galileo Galilei, the Determination of Longitude, and the Pendulum Clock*. Florence: Leo S. Olschki.

Bedoucha, G. (1993) 'The watch and the waterclock: technological choices/social choices' in Lemonnier, P. (ed.) *Technological Choices: Transformation in Material Cultures since the Neolithic*. London: Routledge, pp. 77–107.

Beeson, C.F.C. (1977) *English Church Clocks 1280–1850: Their History and Classification*, revised edition. Ashford: Brant Wright Associates.

—— (1989) *Clockmaking in Oxfordshire, 1400–1850*. Oxford: Museum of the History of Science.

Bellchambers, J.K. (1969) *Somerset Clockmakers*. Taunton: S. A. Kellow.

Bender, J. and Wellbery, D.E. (eds) (1991) *Chronotypes: The Construction of Time*. Stanford: Stanford University Press.

Bennett, J. (2003) 'Knowing and doing in the sixteenth century: what were instruments for?' *British Journal for the History of Science* 36, 129–50.

Bennett, T. (1998). *Culture: a Reformer's Science*. London: Sage.

Beresford, M. (1967) *New Towns of the Middle Ages: Town Plantation in England, Wales and Gascony*. London: Lutterworth.

Bernstein, J.M. (1995) *Recovering Ethical Life: Jürgen Habermas and the Future of Critical Theory*. London: Routledge.

Bertman, S. (2001) 'Killing time', *KronoScope: Journal for the Study of Time*, new series, 1, 15–28.

Bettey, J.H. (ed.), (1981) *The Casebook of Sir Frances Ashley JP, Recorder of Dorchester 1614–35*, Dorchester: Dorset Record Society publication no. 7.

—— (1986) *Bristol Observed: Visitors' Impressions of the City from Domesday to the Blitz*. Bristol: Redcliffe Press.

Betts, J. (1993) *John Harrison*, Greenwich: National Maritime Museum.

—— (2006) in A. Wolfendale (ed.) *Harrison in the Abbey: Published in Honour of John Harrison on the Occasion of the Unveiling of his Memorial in the Abbey on 24th March 2006*. Westminster: Roundtuit Publishing.

Bhabha, H.K. (1994) *The Location of Culture*. London: Routledge.

Biagoli, M. (1993) *Galileo, Courtier: the Practice of Science in the Culture of Absolutism*. Chicago: University of Chicago Press.

Bickley, F. (ed.), (1930) *Hastings Manuscripts volume II*. London: Historical Manuscripts Commission.

Bird, C., Byrd, Y. (1989) *Norfolk and Norwich Clockmakers*. Chichester: Phillimore.

Birth, K. (1999) *Any Time is Trinidad Time: Social Meanings and Temporal Consciousness*. Gainesville: University of Florida Press.

Blackbourn, D. (1998) 'The Madonna of Marpingen: a likely story' *Common Knowledge*, 7, 112–22.

Blackbourn, P. (1993) *Marpingen: Apparitions of the Virgin Mary in Bismarckian Germany*. Oxford: Clarendon Press.

Blackburn, B. and Holford-Strevens, L. (1999) *The Oxford Companion to the Year*. Oxford: Oxford University Press.

Blaise, C. (2000) *Time Lord: Sir Sandford Fleming and the Creation of Standard Time*. London: Weidenfeld and Nicolson.

Blickle, P. (1998) 'Communalism as an organizational principle between medieval and modern times', pp. 1–15 in his *From the Communal Reformation to the Revolution of the Common Man*. Leiden: Brill.

Bloch, E. (1935/1999) *Heritage of Our Time*. Cambridge: Polity Press.

Bloch, M. (1977) *Feudal Society*. London: Routledge.

Bluedorn, A.C. (2002) *The Human Organization of Time: Temporal Realities and Experience*. Stanford: Stanford University Press.

Borges, J. (1970) *Labyrinths*. Harmondsworth: Penguin.
Borsay (1989) *The English Urban Renaissance: Culture and Society in the Provincial Town, 1660–1770*. Oxford: Oxford University Press.
Borst, A. (1993) *The Ordering of Time: From the Ancient Computus to the Modern Computer*. Cambridge: Polity Press.
Boulton, J. (2000) 'London, 1540–1760' in Clark, P. (ed.) *The Cambridge Urban History of Britain, Volume II 1549–1840*. Cambridge: Cambridge University Press, 315–45.
Bourdieu, P. (1977) *Outline of a Theory of Practice*. Cambridge: Cambridge University Press.
—— (1987) *Distinction: a Social Critique of the Judgement of Taste*. London: Routledge and Kegan Paul.
—— (2000) *Pascalian Meditations*. Cambridge: Polity Press.
Bowker, G.C. and Star, S.J. (1999) *Sorting Things Out*. Cambridge, Mass: MIT Press.
Boyne, R. (1998). 'Angels in the archive: lines into the future in the work of Jacques Derrida and Michel Serres' in S. Lash, A. Quick, and R. Roberts (eds), *Time and Value*, Oxford: Blackwell, pp. 48–64.
Bracken, H. (1738) *Farriery Improv'd: or, a Compleat Treatise upon the Art of Farriery*. London.
Bradbury, B. (1981) *A History of Cockermouth*. Chichester: Phillimore.
Brand, P. (2001) 'Lawyers' time in England in the later Middle Ages' pp. 73–104 in C. Humphrey and W.M. Ormrod (eds), *Time in the Medieval World*. York: York Medieval Press.
Brand, S. (1999) *The Clock of the Long Now: Time and Responsibility*. London: Weidenfeld and Nicolson.
Brayshay, M. (1991) 'Royal post-horse routes in Engalnd and Wales: the evolution of the network in the late sixteenth and early seventeenth century' *Journal of Historical Geography* 17, 373–89.
Brayshay, M., Harrison, P. and Chalkley, B. (1998) 'Knowledge, nationhood, and governance: the spread of the Royal Post in early modern England' *Journal of Historical Geography* 24, 265–88.
Brewer, J. and Porter, R. (1993) *Consumption and the World of Goods*. London: Routledge.
Brown, R.A. (1959) 'King Edward's clocks' *Antiquaries Journal*, 39, 283–6.
Bruno, G. (1993) *Streetwalking on a Ruined Map: Cultural Theory and the City Films of Elvira Notari*. Princeton: Princeton University Press.
Bryant, G.F. (2003) *The Later History of Barton-on-Humber, Part One: the Church in Late Medieval Barton-on-Humber*, Barton-on-Humber: Bryant.
Burgess, C. (1987) 'A service for the dead: the form and function of the anniversary in late medieval Bristol' *Transactions of the Bristol and Gloucestershire Archaeological Society* 105, 183–211.
—— (1988) '"A fond thing vainly invented": an essay on Purgatory and pious motive in late medieval England', pp. 56–84 in S.J. Wright (ed.), *Parish, Church and People: Local Studies in Lay Religion, 1350–1750*. London: Hutchinson.
—— (1990) 'Late medieval wills and pious convention: testamentary evidence reconsidered' pp. 14–33 in M.A. Hicks (ed.), *Profit, Piety and the Professions in Late Medieval England*. Gloucester: Alan Sutton.

—— (1996) 'Shaping the parish: St Mary at Hill, London, in the fifteenth century', pp. 246–86 in J. Blair and B. Golding (eds), *The Cloister and the World: Essays in Honour of Barbara Harvey*. Oxford: Oxford University Press.

—— (1998) 'London parishes: development in context', pp. 151–74 in R. Britnell (ed.), *Daily Life in the Late Middle Ages*. Stroud: Alan Sutton.

—— (ed.) (1999) *The Church Records of St Andrew Hubbard, Eastcheap c.1450–1570*. London: London Record Society, p. xxxiv.

—— (2002a) 'Pre-Reformation churchwardens' accounts and parish government: lessons from London and Bristol' *English Historical Review* 117, 306–332.

Burgess, C. (2002b) 'London parishioners in times of change: St Andrew Hubbard, Eastcheap, 1450–1750', *Journal of Ecclesiastical History*, 53, 38–63.

—— (2002c) 'Educated parishioners in London and Bristol', in C. Barron and J. Stratford (eds), *The Church and Learning in Late Medieval Society: Essays in Honour of Barrie Dobson*. Donington: Shaun Tyas, *Harlaxton Medieval Studies, XI*.

—— (2004) 'The broader Church? A rejoinder to "Looking beyond"' *English Historical Review* 119, 100–116.

—— and Kumin, B. (1993) 'Penitential bequests and parish regimes in late-medieval England' *Journal of Ecclesiastical History*, 44, 610–30.

—— and Wathey, A. (2000) 'Mapping the soundscape: Church music in English towns, 1450–1550' *Early Music History* 19, 1–46.

Burnett, D.G. (2003) 'Mapping time: chronometry on top of the world' *Daedalus* Spring, 1–19.

Butler, J. (1997) *Excitable Speech. Politics of the Performative*. London: Routledge.

Calhoun, C. (1993) 'Habitus, field, and capital: The question of historical specificity' in Calhoun, C., LiPuma, E., Postom, M. (eds) *Bourdieu: Critical Perspectives*. Cambridge: Polity Press.

Callon, M. (1986) *The Sociology of an Actor-Network: The Case of the Electric Vehicle*. Basingstoke: MacMillan.

Campbell, B.M.S., Keene, D., Galloway, J. and Murphy, M. (1993) *A Medieval Capital and its Grain Supply: Agrarian Production and Distribution in the London region, c.1300*, Cheltenham: Historical Geography Research Group, research series, 30.

Campbell, L. (1990) *Renaissance Portraits: European Painting in the Fourteenth, Fifteenth and Sixteenth Centuries*. New Haven: Yale University Press.

—— (1999) 'Time and the portrait'. In K. Lippincott, *et al.*, *The Story of Time*. London: Merrell Holberton, pp. 190–3.

Capp, B. (1979) *Astrology and the Popular Press: English Almanacs 1500–1800*, London: Faber and Faber.

—— (1989) *Cromwell's Navy: The Fleet and the English Revolution 1646–1680*. Oxford: Clarendon Press.

Carpenter C. (ed.), (1996) *Kingsford's Stonor Letters and Papers, 1290–1483* Cambridge: Cambridge University Press.

Castoriadis, C. (1991) 'Time and creation' in J. Bender and D.E. Wellbery (eds), *Chronotypes: the Construction of Time*. Stanford: Stanford University Press, pp. 38–64.

Cave-Brown-Cave, B.W. (1979) *Jonas Barber, clockmaker of Winster : a study of the lives and work of the Barber family who made clocks at Bryan Houses in Winster, Westmorland throughout most of the 18th century*. Ulverston: Reminder Press.

Chapman, A. (1985) 'Jeremy Shakerley (1626–?1655); astronomy, astrology and patronage in civil-war Lancashire' *Transactions of the Historical Society of Lancashire and Cheshire* 135, 1–14.

—— (1990) 'Jeremy Horrocks, the transit of Venus, and the "New Astronomy" in early seventeenth-century England' *Quarterly Journal of the Royal Astronomical Society* 31, 333–57.

—— (1995a) *Dividing the Circle: the Development of Critical Angular Measurement in Astronomy*, 2nd edition. Chichester and New York: Wiley-Praxis.

—— (1995b) *William Crabtree 1610–1644: Manchester's First Mathematician*. Manchester: Manchester Statistical Society.

—— (2005) *England's Leonardo: Robert Hooke and the Seventeenth-Century Scientific Revolution*, Bristol: Institute of Physics Publishing.

Chartier, R. (1994/97) *The Order of Books: Readers, Authors, and Libraries in Europe between the Fourteenth and Eighteenth Centuries*. Cambridge: Polity Press.

Chesbrough, H.W. (2003) *Open Innovation: The New Imperative For Creating and Profiting From Technology*. Boston: Harvard Business School Press.

Cipolla, C.H. (ed.) (1972–1978) *The Fontana Economic History of Europe*. 6 vols. London: Collins/Fontana.

Clanchy, M.T. (1993) *From Memory to Written Record: England 1066–1307*, 2nd edition. Oxford: Blackwell.

Clark, A. (2001) *Mindware*. New York: Oxford University Press.

Clark, G. (1991) 'Labour productivity in English agriculture, 1300–1860' in B.M.S. Campbell and M. Overton (eds), *Land, Labour and Livestock: Historical Studies in European Agricultural Productivity*. Manchester: Manchester University Press, pp. 211–35.

—— (1998) 'Work in progress? The Industrious Revolution' *Journal of Economic History* 58, 830–43.

Clark, P., Hosking, J. (1993) 'Population estimates of English small towns, 1550–1851.' Working paper no. 5. Leicester: Centre for Urban History, University of Leicester.

Claxton, G. (1999) *Wise Up: the Challenge of Lifelong Learning*. London: Bloomsbury.

Clayton, M. (2000) *Time in Indian Music: Rhythm, Metre, and Form in North Indian Rāg Performance*. Oxford: Oxford University Press.

Clément, C. (1995) *Syncope: The Philosophy of Rapture*. Minneapolis: University of Minnesota Press.

Clifford, D.J.H. (1990) *The Diaries of Lady Anne Clifford*. Stroud: Alan Sutton.

Clifton, G. (1995) *Directory of British Scientific Instrument Makers*. London: National Maritime Museum.

Cockayne, E. (2007) *Hubbub: Filth, Noise, and Stench in England 1600–1770*. New Haven: Yale University Press.

Clive Cohen, P. (1982) *A Calculating People: The Spread of Numeracy in Early America*. Chicago: University of Chicago Press.

Cohen, P.C. (1982) *Calculating People: the Spread of Numeracy in Early America*. Chicago: University of Chicago Press.

—— (1993) 'Reckoning with commerce: numeracy in eighteenth-century America' in J. Brewer and R. Porter (eds), *Consumption and the World of Goods*. London: Routledge, pp. 320–34.

Collins, G.E. (1902) *History of the Brocklesby Hounds, 1700–1901*. London: Sampson Low, Marston & Co.
Collinson, P. (1988) *The Birthpangs of Protestant England*. Basingstoke: MacMillan.
Cook, A. (1998) *Edmond Halley: Charting the Heavens and the Seas*. Oxford: Oxford University Press.
Cook, J. (2002) *Dr Simon Forman: a Most Notorius Physician*. London: Chatto & Windus.
Cooper, M. (2005) *Robert Hooke and the Rebuilding of London*. Stroud: Sutton Publishing Limited.
Cooper, W.R., (1996) 'Richard Hunne' *Reformation* 1, 221–51.
Corbin, A. (1998) *Village Bells: Sound and Meaning in the French Countryside*. New York: Columbia Press.
Coss, P.R. (1986) *The Early Records of Medieval Coventry*. Oxford: British Academy; Oxford University Press
Cotter, C.H. (1983) *A History of the Navigator's Sextant*. Glasgow: Brown, Son and Ferguson.
Couch, J. (1853) 'Description of a specimen of a pocket dial used in ancient times and referred to by Shakespeare in As You Like It, Act 2, Scene 7' *Transactions of the Natural History and Antiquarian Society of Penzance* VII, 163–4.
Cox, J.C. (1913) *Churchwardens' Accounts from the Fourteenth Century to the Close of the Seventeenth Century*. London: Methuen.
Craig, J.S. (1993) 'Co-operation and initiatives: Elizabethan churchwardens and the parish accounts of Mildenhall' *Social History* 18, 357–80.
Crane, N. (2002) *Mercator: the Man who Mapped the Planet*. London: Weidenfeld and Nicholson.
Cressy, D. (1989) *Bonfires and Bells: National Memory and the Protestant Calendar in Elizabethan and Stuart England*. London: Weidenfeld and Nicholson.
—— (1992) 'The Fifth of November remembered' in R. Porter (ed), *The Myths of the English*. Oxford: Polity Press.
—— (1993) 'Literacy in context: meaning and measurement in early modern England' in R. Brewer and J. Porter (eds), *Consumption and the World of Goods*. London: *Routledge*, pp. 305–19.
Crosby, A.W. (1998) *The Measure of Reality: Quantification and Western Society, 1250–1600*. Cambridge: Cambridge Univeristy Press.
Crossley, J. (ed.), (1847) *The Diary and Correspondence of Dr. John Worthington, part 1*. Manchester: Chetham Society vol. XIII.
Crump, T. (1999) *Solar Eclipse*. London: Constable.
Cubitt, S. (1991) *Timeshift: On Video Culture*. London: Routledge.
Davis, N.Z. (1983). *The Return of Martin Guerre*. Cambridge, Mass: Harvard University Press.
Davis, R. (1960) *The Rise of the English Shipping Industry in the Seventeenth and Eighteenth Centuries*. London: Macmillan.
Davies, K. (1989) *Women and Time: Weaving the Strands of Everyday Life*. PhD Thesis, University of Lund Department of Sociology, Lund.
—— (1994) 'The tensions between process time and clock time in care-work: the example of day nurseries' *Time & Society* 3, 277–304.

Davies, J.D. (1991) *Gentlemen and Tarpaulins: the Officers and Men of the Restoration Navy.* Oxford: Clarendon Press.

Davison, G. (1992) 'Punctuality and progress: the foundations of Australian standard time' *Australian Historical Studies* 25, 169–91.

Daybell, J. (2006) *Women Letter Writers in Tudor England.* Oxford: Oxford University Press.

Dear, P. (2001) *Revolutionizing the Sciences: European Knowledge and its Ambitions, 1500–1700.* Basingstoke: Palgrave.

Debord, G. (1995) *The Society of Spectacle.* New York: Zone Books.

Delaney, P. (1969) *British Autobiography in the Seventeenth Century.* New York: Columbia University Press.

Deleuze, G. (1994) *Difference and Repetition.* New York: Columbia University Press.

Derham, W. (1694) *The Artificial Clockmaker.* London: James Knapton.

Derrida, J. (1981) *Dissemination.* Chicago: Chicago University Press.

——(1992) *Given Time: 1 Counterfeit Money.* Chicago: Chicago University Press.

De Certeau, M. (1984) *The Practice of Everyday Life.* Berkeley: University of California Press.

De Souza, P. (2001) *Seafaring and Civilization: Maritime Perspectives on World History.* London: Profile Books.

De Vries, J. (1981) 'Peasant demand patterns and economic development: Friesland 1550–1750' in W.N. Parker and E.L. Jones (eds), *European Peasants and Their Markets: Essays in Agrarian Economic History.* Princeton, NJ: Princeton University Press, pp. 205–66.

——(1993) 'Between consumption and the world of goods' in J. Brewer and R. Porter (eds), *Consumption and the World of Goods.* London: Routledge, pp. 85–132.

——(1994) 'The industrial-revolution and the industrious revolution' *Journal of Economic History* 54(2), 249–70.

Deedes, R., Walters, J. (1907) *The Church Bells of Essex.* Essex.

Dietz, B. (1972) *The Port and Trade of Early Elizabethan London.* London: London Record Society, Vol. 8.

Dixon, R.M.W. (1997) *The Rise and Fall of Languages.* Cambridge: Cambridge University Press.

Doane, M.A. (2002) *The Emergence of Cinematic Time: Modernity, Contingency, The Archive.* Cambridge, Mass.: Harvard University Press.

Dohrn-van Rossum, G. (1996) *History of the Hour: Clocks and Modern Temporal Orders.* Chicago: Chicago University Press.

Dosse, F. (1999) *Empire of Meaning: the Humanisation of the Social Sciences.* Minneapolis: University of Minnesota Press.

Drake, S. (1990) *Galileo: Pioneer Scientist.* Toronto: University of Toronto Press.

Driver, F. (1993) *Power and Pauperism: the Workhouse System, 1834–1884.* Cambridge: Cambridge University Press.

——(1996) 'History at large: old hat, I presume? history of a fetish' *History Workshop Journal* Spring: 230–4.

——and Martins, L. (2002) 'John Septimus Roe and the Art of Navigation, c.1815–1830' *History Workshop Journal* 54, 144–61.

—— and Schwarz, B. (1998) 'Editorial' *History Workshop Journal*, Autumn, iii.
Drewe, F. (1991) *Ticehurst, Stonegate and Flimwell*, Chichester: Phillimore.
Druett, J. (2001) *Rough Medicine: Surgeons at Sea in the Age of Sail*. New York: Routledge.
Dudding, R.C. (ed.) (1941) *The First Churchwardens' Book of Louth, 1500–24*. Oxford: Oxford University Press.
Duffy, E. (1992) *The Stripping of the Altars: Traditional Religion in England, c.1400–c.1580*. New Haven, Conn: Yale University Press.
—— (2006) *Marking the Hours: English People and their Prayers, 1240–1570*. New Haven and London: Yale University Press.
Duley, A.J. (1977) *The Medieval Clock of Salisbury Cathedral*. Salisbury: Salisbury Cathedral.
Duncan, D.E. (1998) *The Calendar*. London: Fourth Estate.
Dunning, R.C. (1941) *The First Churchwardens' Book of Louth*. Oxford: privately printed.
Earle, P. (1998) *Sailors. English Merchant Seamen 1650–1775*. London: Methuen.
Ecclestone, M. (1989) 'The diffusion of English surnames' in *Local Historian* 19, 2.
Edelen, G. (ed.), (1968) *William Harrison's 'The Description of England', 1577*. Ithaca, NY: Cornell University Press.
Ehrman, J. (1953) *The Navy in the War of William III, 1689–1697: its State and Direction*. Cambridge: Cambridge University Press.
Elias, N. (1984/1992) *Time: an Essay*. Oxford: Blackwell.
Emmison, F.G. (ed.), (1994) *Essex Wills, Volume 9: the Bishop of London's Commissary Court 1569–1578*. Chelmsford: Essex Record Office; Friends of Historic Essex.
Epstein, S.R. (1998) 'Craft guilds, apprenticeship, and technological change in preindustrial Europe' *Journal of Economic History* 58, 684–713.
Ermarth, E.D. (1992) *Sequel to History: Postmoderniam and the Crisis of Representational Time*. Princeton: Princeton University Press.
Estabrook. C.B. (1999) *Urbane and Rustic England: Cultural Ties and Social Spheres in the Provinces, 1660–1780*. Stanford: Stanford University Press.
Estacio dos Reis, A. (1997) *Medir Estralis* [Measuring Stars]. Lisbon: CTT Correios.
Eversley, D.E.C (1967) 'The Home Market and Economic Growth in England' in *Land, Labour and Population in the Industrial Revolution: Essays Presented to J.D. Chambers*. London: Edward Arnold.
Fabian, J. (1983) *Time and the Other*. New York: Columbia University Press.
—— (1991) *Time and the Work of Anthropology: Critical Essays, 1971–1991*. Amsterdam: Harwood Academic Publishers.
Felski, R. (2000) *Doing Time: Feminist Theory and Postmodern Culture*. New York: New York University Press.
Fischer, D.H. (1989) *Albion's Seed: Four British Folkways in North America*. New York: Oxford University Press.
Flaherty, M.G. (1999) *A Watched Pot: How We Experience Time*. New York: New York University Press.
Fenton, E. (ed.), (1998) *The Diaries of John Dee*. Oxfordshire: Day Books.
Fentress, J. and Wickham, C. (1992) *Social Memory*. Oxford: Blackwell.
Ferguson, N. (1997) *Virtual History: Alternatives and Counter Factuals*. London: Picador.
Findlay, J.R. (1926) 'Obsolete methods of reckoning time' *Scottish Geographical Magazine* XLIII, 129–47.

Flaherty, M.G. (1999) *A Watched Pot: How We Experience Time*. New York: New York University Press.
Flinn, M.W. and Smout, C. (eds), (1974) *Essays in Social History*. Oxford: Clarendon Press.
Foister, S., Roy, A., and Wyld, M. (1997) *Holbein's Ambassadors: Making and Meaning*. London: National Gallery.
Forman, F.J. (1989) 'Feminising time: an introduction' in F.J. Forman and C. Sowton (eds), *Taking Our Time: Feminist perspectives on Temporality*. Oxford: Pergamon, pp. 1–9.
Forte, A.D.M. (1998) 'Kenning be kenning and course be course: maritime jurimetrics in Scotland and Northern Europe 1400–1600.' *Edinburgh Law Review*, 2, 56–89.
Foster, A. 'Churchwardens' accounts of early modern England and Wales: some problems to note, but much to be gained', pp. 74–93 in K. French, G. Gibbs, and B. Kumin (eds), (1997) *The Parish in English Life, 1400–1600*. Manchester: Manchester University Press.
Foster, J. (1891–92) *Alumni Oxoniensis: the Members of the University of Oxford, 1500–1714*. Oxford: Parker.
Foster, J. and Woolfson, C. (1989) 'Corporate reconstruction and business unionism: the lessons of Caterpillar and Ford' *New Left Review* 174, 51–66.
Foucault, M. (1973) *The Birth of the Clinic*. London: Tavistock Publications.
—— (1979) 'The life of infamous men' in P. Foss and M. Morris (eds), *Power, Truth, Strategy*. Sydney: Feral.
Fox, A. (2000) *Oral and Literate Culture in England, 1500–1700*. Oxford: Oxford University Press.
Fox, A. and Woolf, D. (eds), (2002) *The Spoken Word: Oral Culture in Britain, 1500–1850*. Manchester: Manchester University Press.
Frake, C.O. (1985) 'Cognitive maps of time and tide among medieval seafarers' *Man*, NS20, 20, 254–70.
Frankenberg, R. (1994) 'The politics of time' [review] *Time & Society* 3, 117–28.
Franz, K. (2005) *Tinkering: Consumers Reinvent the Early Automobile*. Philadelphia: University of Pennsylvania Press.
Freake, D. (1995) 'The semiotics of wristwatches' *Time & Society* 4, 67–90.
Franko, M. (1993) *Dance As Text: Ideologies of the Baroque Body*. Cambridge: Cambridge University Press.
French, K. 'Parochial fundraising in late medieval Somerset', pp. 115–32 in K. French, G. Gibbs, and B. Kumin (eds), (1997) *The Parish in English Life, 1400–1600*. Manchester: Manchester University Press.
Friel, I. (1995) *The Good Ship: Ships: Shipbuilding and Technology in England, 1200–1520*. London: British Museum Press.
Frow, J. (1997) Time *and Commodity Culture: Essays in Cultural Theory and Postmodernity*. Oxford: Oxford University Press.
Fury, C. (1998) 'Elizabethan seaman: their lives ashore' *International Journal of Maritime History* 10(1), 1–40.
Fyfe, A. (2003) *Science for Children*. Bristol: Thoemmes.
Galison, P. (2003) *Einstein's Clocks, Poincaré's Maps: Empires of Time*. London: Sceptre.
Gallagher, C. and Greenblatt, S. (2000) *Practising New Historicism*. Chicago: University of Chicago Press.

Garfinkel, H. (2002) *Ethnomethodology's Program: Working Out Durkheim's Aphorism.* Lanham: Rowman and Littlefield.
Garry, F.N.A, Garry, A.G. (eds) (1893) *The Churchwardens' Accounts of the Parish of St. Mary's, Reading, Berks, 1550–1662.* Reading: E. J. Blackwell.
Gasparini, G. (1995) 'On waiting' *Time and Society* 4, 29–45.
Gay, H. (2003) 'Clock synchrony, time distribution and electrical timekeeping in Britain 1880–1925' *Past and Present* 181, 107–40.
Gell, A. (1992) *The Anthropology of Time: Cultural Construction of Temporal Maps and Images.* Oxford: Berg.
—— (1998) *Art and Agency.* Oxford: Oxford University Press.
Geneva, A. (1995) *Astrology and the Seventeenth Century Mind.* Manchester: St. Martin's Press.
Gershuny, J. (2001) *Changing Times.* Oxford: Oxford University Press.
Gertler, M. (1988) 'The limits to flexibility: comments on the post-Fordist vision of production and its geography' *Transactions of the Institute of British Geographers*, new series 13, 419–32.
Giddens, A. (1979) *Central Problems in Social Theory: Action, Structure and Contradiction in Social Analysis.* London: Macmillan.
—— (1984) *The Constitution of Society.* Cambridge: Polity Press.
Gigerenzer, G. (2000) *Adaptive Thinking: Rationality in the Real World.* New York: Oxford University Press.
Gigerenzer, G. and Selten, R. (eds), (2001) *Bounded Rationality: the Adaptive Toolbox.* Cambridge, Mass.: MIT Press.
Gigerenzer, G. and Todd, P.M. (1999) *Simple Heuristics That Make Us Smart.* Oxford: Oxford University Press.
Gil, J. (1998) *Metamorphoses of the Body.* Minneapolis: University of Minnesota Press.
Gill, C. (1961) *Merchants and Mariners of the Eighteenth Century.* London: Edward Arnold.
Gingerich, O. (1996) 'Cranks and opportunists: "Nutty" solutions to the longitude problem' pp. 133–48 in W.J.H. Andrewes (ed.), *The Quest for Longitude.* Cambridge, Mass: Harvard Collection of Scientific Instruments.
Ginzburg, C. (1980) *Cheese and the Worms: the Cosmos of a Sixteenth-Century Miller.* London: Routledge and Kegan Paul.
Ginzburg, C. (1999) *History, Rhetoric, and Proof.* Hanover: Brandeis University Press.
—— (2001) *Wooden Eyes: Nine Reflections on Distance.* New York: Columbia University Press.
Gleick, J. (1999) *Faster: the Acceleration of Just About Everything.* New York: Random House.
Glennie, P.D. (1990) 'Industry and towns 1500–1730' in R.A. Dodgshon and R.A. Butlin (eds), *Historical Geography of England and Wales*, 2nd edition. London: Academic Press, pp. 199–222.
—— (1995) 'Consumption in historical studies' in D. Miller (ed), *Acknowledging Consumption: a Review of New Studies.* London: Routledge, pp. 164–203.
—— (2001) 'Town and Country in England 1570–1750' in Epstein, S. R. (ed.) *Town and Country in Europe, 1300–1800.* Cambridge: Cambridge University Press, 132–55.
—— and Thrift, N.J. (1996a) 'Consumers, identities, and consumption spaces in early-modern England' *Environment and Planning A* 25, 25–45.

Glennie, P.D. and Thrift, N.J. (1996b). 'Reworking E.P. Thompson's "Time, work-discipline and industrial capitalism"' *Time and Society* 5, 275–300.

—— (2002) 'The spaces of clock times' pp. 151–74 in P. Joyce (ed.), *The Social in Question: New Bearings in History and the Social Sciences* London: Routledge.

—— (2005) 'Revolutions in the times: clocks and the temporal structures of everyday life' in D. Livingston and C.W.J. Withers (eds), *Geography and Revolutions*. Chicago: University of Chicago Press, pp. 160–98.

Goldberg, P.J.P. (1995) *Women in England, 1275–1525*. Manchester: Manchester University Press.

Goodall, H. (2000) *Big Bangs: the Story of Five Discoveries that Changed Musical History*. London: Chatto and Windus.

Goody, J. (1987) *The Interface Between the Written and the Oral*. Cambridge: Cambridge University Press.

—— (1991) 'The time of telling and the telling of time in written and oral cultures' in J. Bender and D.E. Wellbery (eds), *Chronotypes: the Construction of Time*. Stanford: Stanford University Press, pp. 77–96.

Gosden, C. (1994) *Social Being and Time*. Oxford: Blackwell.

Grafton, A. (2003) 'Dating history: the Renaissance and the reformation of chronology' *Daedalus* Spring, 74–85.

Graham, G., *et al.* (1737) 'Observations of the late total eclipse of the sun' *Philosophical Transactions of the Royal Society* 175–201.

Graham, S., Thrift, N.J. (2007) 'Out of order: Understanding repair and maintenance.' *Theory, Culture and Society*, 24, 1–25.

Griffiths, J. (1999) *Pip, Pip: a Sideways Look at Time*. London: Flamingo.

Grigson, G. (ed.) (1984) *The English Year: From Diaries and Letters*. Oxford: Oxford University Press.

Grint, K. and Woolgar, S. (1997) *The Machine at Work: Technology, Work and Organization*. Cambridge: Polity Press.

Grosz, E. (1994) *Volatile Bodies Toward a Corporeal Feminism*. Bloomington: University of Indiana Press.

—— (1995) *Space, Time, and Perversion: Essays on the Politics of Bodies*. New York: Routledge.

—— (2004) *The Nick of Time: Politics, Evolution, and the Untimely*. Ithaca: Cornell University Press.

Guest, A.H. (1989) *Choreo-Graphics: a Comparison of Dance Notation Systems From the Fifteenth Century to the Present*. Amsterdam: Gordon and Breach.

Guibbory, A. (1986) *The Map of Time: Seventeenth Century English Literature and Ideas of Pattern in History*. Urbana, Illinois: Illinois University Press.

Gurevitch, G. (1964) *The Spectrum of Social Time*. Dordrecht: D. Reidel.

Gurney, A. (1998) *Below the Convergence: Voyages Towards Antarctica 1699–1839*. New York: Penguin.

—— (2004) *Compass: a Story of Exploration and Innovation*. New York: Norton.

Guttmann, A. (1978) *From Ritual to Record: the Nature of Modern Sports*. New York: Columbia University Press.

Haber, F.C. (1975) 'The cathedral clock and the cosmological clock metaphor' in Fraser, J.T, Lawrence, N. (eds) *The Study of Time*, Proceedings of the second conference at the International Society for the Study of Time. Berlin: Springer-Verlag.

Hägerstrand, T. (1970) 'What about people in regional science?' *Papers of the Regional Science Association* 24: 7–21.

—— (1973) 'The domain of geography' in R.J. Chorley (ed.), *Directions in Geography* London: Methuen, pp. 67–87.

—— (1982) 'Diorama, path and project' *Tijdschrift voor Economische en Sociale Geografie* 73, 323–39.

Hagger, A. and Miller, L. (1974) *Suffolk Clocks and Clockmakers*, with supplement published 1979. Ticehurst: Antiquarian Horological Society.

Halley, E. (1715) 'Observations of the late total eclipse of the sun on the 2nd of April' *Philosophical Transactions of the Royal Society* 29, 245–62 (no. 343); and 314–316 (no. 345).

Harding, R. (1999) *Seapower and Naval Warfare, 1650–1830*. London: UCL Press.

Hareven, T.K. (1982) *Family Time and Industrial Time: the Relationship Between the Family and Work in a New England Industrial Community*. Cambridge: Cambridge University Press.

—— (1991) 'Synchronizing individual time, family time and historical time' in J. Bender and D.E. Wellbery (eds), *Chronotypes: the Construction of Time*. Stanford: Stanford University Press, pp. 167–82.

Harland, J. (1984) *Seamanship in the Age of Sail*. London: Conway Maritime Press.

Harman, G. (2002) *Tool-Being: Heidegger and the Metaphysics of Objects*. Chicago: Open Court.

Harootunian, H. (2000) *History's Disquiet: Modernity, Cultural Practice, and the Question of Everyday Life*. New York: Columbia University Press.

Harris, J. (1971) *A Catalogue of British Drawings for Architecture, Decoration, Sculpture and Landscape Gardening, 1500–1900, in American Collections*. Upper Saddle River, NJ: Gregg Press.

—— (1985) *The Artist and the Country House: a History of Country House and Garden View Painting in Britain 1540–1870*. London: Sotheby's Publications.

Harrison, E. (1696) *Idea Longitudinis: Being, a brief Description Of the best known Axioms For finding the Longitude . . .* London.

Harrison, M. (1986) 'The ordering of the environment: time, work and the occurrence of crowds 1790–1835' *Past and Present* 110, 134–68.

—— (2000) 'From medical astrology to medical astronomy: sol-lunar and planetary theories of disease in British medicine, c.1700–1850' *British Journal of the History of Science* 33, 25–48.

Harvey, B.F. (1993) *Living and Dying in England 1100–1540: the Monastic Experience*. Oxford: Clarendon Press.

Harvey, D. (1989) *The Condition of Postmodernity*. Oxford: Blackwell.

Hawkyard, A. (1988) *The Counties of Britain: A Tudor Atlas by John Speed*. London: British Library, Pavillion Books.

Hayter, A. (2002) *The Wreck of the Abergavenny*. London: Macmillan.

Heidegger, M. (1967) *Being and Time*. Oxford: Blackwell.

Herren, J. (ed.), (1999) *A Medieval Miscellany*. London: Weidenfeld & Nicolson, Facsimile Editions.

Hewitt, P. (1994) *Turret Clocks in Leicestershire and Rutland*. Leicester: Leicestershire Museums, Arts and Records Service.

Hewson, J.B. (1983) *A History of the Practice of Navigation*. Glasgow: Brown, Son and Ferguson.

Higton, H. (2002) *Sundials at Greenwich: a Catalogue of the Sundials, Horary Quadrants and Nocturnals in the National Maritime Museum, Greenwich*. Oxford: Oxford University Press and the National Maritime Museum.

Hobbs, M. (ed.), (1994) *Chichester Cathedral: an Historical Survey*. Chichester: Phillimore.

Hobden, H. (1988) 'Navigation and the Horological Works of the Harrison Brothers' *Antique Clocks* 11(5), 22–6.

Hobden, H. and Hobden, M. (1999) *John Harrison and the Problem of Longitude*, 7th edition. Lincoln: Cosmic Elk.

Hodgett, G.J. (1975) *Tudor Lincolnshire*. History of Lincolnshire, volume 5. Lincoln: History of Lincolnshire Committee.

Hobhouse, E. (ed.) (1935) *The Diary of a West Country Physician [Claver Morris], 1684–1726*. 2nd ed. London: Simpkin Marshall.

Holmes, K. (1994) 'Making time: representations of temporality in Australian women's diaries of the 1920s and 1930s' *Australian Historical Studies* 26(102), 1–18.

Holmes, R. (2001) *Redcoat: the British Soldier in the Age of Horse and Musket*. London: Harper Collins.

Hopkins, E. (1982) 'Working hours and conditions during the industrial-revolution: a reappraisal' *Economic History Review* 35(1), 52–66.

Hore, J.P. (1886) *History of Newmarket and Annals of the Turf*, 3 volumes. Bailey, London.

Howell, S. (1992) 'Time past time present time future: contrasting temporal values in two Southern Asian societies' in S. Wallman (ed.), *Contemporary Futures: Perspectives from Social Anthropology*. London: Routledge, pp. 124–37.

Howse, D. (1996) 'The lunar-distance method of measuring longitude' In W.J.H. Andrewes (ed.), *The Quest for Longitude*. Cambridge, Mass: Harvard Collection of Scientific Instruments, pp. 149–61.

—— (1997) *Greenwich Time and the Longitude*. London: Philip Wilson, National Maritime Museum.

Hughes, P. (2006) 'The revolution in tidal science' *Journal of Navigation* 59, 445–59.

Humphrey, C. (2001) 'Time and urban culture in late medieval England' [York] pp. 105–18 in C. Humphrey and W.M. Ormrod (eds), *Time in the Medieval World*. York: York Medieval Press.

Hunter, M. and Gregory, A. (eds), (1988) *The Diary of Samuel Jeake*. Oxford: Oxford University Press.

Hussey, S. (1997) '"The last survivor of an ancient race": the changing face of Essex gleaning. *Agricultural History Review* 45: 61–72.

Hutchins, E. (1995) *Cognition in the Wild*. Cambridge, Mass.: MIT Press.

Hutchinson, G. (1994) *Medieval Ships and Shipping*. London: Leicester University Press.

Hutton, R. (1994) *The Rise and Fall of Merry England: the Ritual Year 1400–1700*. Oxford: Oxford University Press,.

Iliffe, R. (2000) 'The masculine birth of time: temporal frameworks of early modern natural philosophy' *British Journal of the History of Science* 33, 427–53.

Ingold, T. (1993) 'The temporality of the landscape' *World Archaeology* 25(2), 152–74.

—— (1995) 'Work, time and industry' *Time & Society* 4(1), 5–28.

—— (2000) 'Tools for the hand, language for the free: an appreciation of Leroi-Gourhan's *Gesture and Speech*' *Studies in the History and Philosophy of Biology and Biomedical Science* 4, 411–53.
—— (2001) *The Perception of the Environment*. London: Routledge.
—— (2004) 'Culture on the ground: the world perceived through feet' *Journal of Material Culture* 9, 315–40.
Ingram, M. (1995) 'From Reformation to toleration: popular religious cultures in England, 1540–1690' in T. Harris (ed.), *Popular Culture in England, c.1500–1850*. London: MacMillan, pp. 95–123.
James, W. (1996) *A Pluralistic Universe*. Lincoln: University of Nebraska Press.
Jardine, L. (1997) *Wordly Goods: a New History of the Renaissance*. London: MacMillan.
—— (1999) *Ingenious Pursuits: Building the Scientific Revolution*. London: Little, Brown.
Jennings, T. (2000) *British and Irish Chime Barrel Mechanisms, Their Music and the Community Response 1550–1930*. Loughborough: Jennings.
Johns, A. (1998) *The Nature of the Book: Print and Knowledge in the Making*. Chicago: University of Chicago Press.
Johnson, C. (1993) *History and Writing in the Philosophy of Jacques Derrida*. Cambridge: Cambridge University Press.
Johnson, J. (2002) *Who Needs Classical Music? Cultural Choice and Musical Value*. Oxford: Oxford University Press.
Johnston, R., Allsopp, G., Baldwin, J. and Turner, H. (1990) *An Atlas of Bells*. Oxford: Blackwell.
Johnson, S. (1755) *An Account of an Attempt to Ascertain the Longitude at Sea, By an Exact Theory of the Variation of the Magnetical Needle*. London.
Jones, M.K. and Underwood, M.G. (1992) *The King's Mother: Lady Margaret Beaufort Countess of Richmond and Derby*. Cambridge: Cambridge University Press.
Joyce, P. (2003) *The Rule of Freedom: Liberalism and the Modern City*. London: Verso.
Kahn, R.P. (1989) 'Women and time in childbirth and during lactation' in F.J. Forman and C. Sowton (eds), *Taking Our Time: Feminist Perspectives on Temporality*. Oxford: Pergamon, pp. 20–36.
Kant, I. (1956/1781, 1787) *Immanuel Kant's Critique of Pure Reason*. New York: St. Martin's.
Kassell, L. (2005) *Medicine and Magic in Elizabethan England: Simon Forman—Astrologer, Alchemist, and Physician*. Oxford: Clarendon Press.
Kemp, P.K. (1976) *The Oxford Companion to Ships and the Sea*. Oxford: Oxford University Press.
Kern, S. (1983) *The Culture of Time and Space 1880–1918/*. Cambridge, Mass: Harvard University Press.
Kerry, C. (1883) *The Municipal Church of St Lawrence, Reading*. Reading: Charles Kerry.
King, A. (1993) *From a Peel of Bells: John Harrison, 1693–1776*. Lincoln: Lincolnshire County Council, Usher Gallery Exhibition Catalogue.
King, A.L. (1996) ' "John Harrison, clockmaker at Barrow; near Barton upon Humber; Lincolnshire": the wooden clocks, 1713–1730' in W.J.H. Andrewes (ed.), *The Quest for Longitude*. Cambridge, Mass: Harvard Collection of Scientific Instruments, pp. 167–87.

King, P. (1997) 'Pauper inventories and the material lives of the poor in the eighteenth and early nineteenth centuries' in T. Hitchcock, P. King and P. Sharpe (eds), *Chronicling Poverty: the Voices and Strategies of the English Poor, 1640–1840*. London: MacMillan, pp. 155–91.

Kisby, F. (2002) 'Books in London parish churches before 1603: Some preliminary observations' in C. Barron and J. Stratford (eds), *The Church and Learning in Later Medieval Society*. Donington: Shaun Tyas, pp. 305–26.

Kitch, M. (1985) 'Migrants to early modern London' in A. Beier and R. Finlay (eds), *London 1500–1700: the Making of the Metropolis*. London: Longman.

Knights, D. and Vurdukakis, T. (1993) 'Calculations of risk: towards an understanding of insurance as a moral and political technology' *Accounting, Organisation and Society* 18, 729–64.

Koselleck, R. (2002) *The Practice of Conceptual History*. Stanford: Stanford University Press.

Kovach, B. and Rosenstiel, T. (1999) *Warp Speed: America in the Age of Mixed Media*. New York: Century Foundation.

Krise, R. and Squires, B. (1982) *Fast Tracks: the History of Distance Running*. Lexington, MA: Stephen Greene Press.

Kristeva, J. (1981) 'Women's time' *Signs* 7, 13–35.

Kubler, G. (1962) *The Shape of Time: Remarks on the History of Things*. New Haven: Yale University Press.

Kümin, B. (1996) *The Shaping of a Community: the Rise and Reformation of the Engliah Parish c.1400–1560*. Aldershot: Scolar Press.

—— (1997) 'The English parish in European perspective' in K. French, G. Gibbs and B. Kumin (eds), *The Parish in English Life, 1400–1600*. Manchester: Manchester University Press, pp. 15–32.

—— (2001) 'Masses, morris and metrical psalms: Music in the English parish, c.1400–1600' in F. Kisby (ed.), *Music and Musicians in Renaissance Cities and Towns*. Cambridge: Cambridge University Press, pp. 70–81.

—— (2004) 'Late medieval churchwardens' accounts and parish government: looking beyond London and Bristol' *English Historical Review* 119, 87–99.

Kwinter, S. (2001) *Architectures of Speed: Toward a Theory of the Event in Materialist Culture*. Cambridge, Mass: MIT Press.

Kyriacou, C.P. (2002) 'The genetics of time' in K. Ridderbus (ed.), *Time*. Cambridge: Cambridge University Press, pp. 65–84.

Landes, D.S. (1983) *Revolution in Time: Clocks and the Making of the Modern World*. Cambridge, Mass: Belknap Press (Harvard University Press).

Landes, D. (2003) 'Clocks and the wealth of nations' *Daedalus* Spring 2003, 20–6.

Langbauer, L. (1999) *Novels of Everyday Life: The Series in English Fiction, 1850–1930*. Ithaca: Cornell University Press.

Langford, A. (1750) *A Catalogue of the Genuine Stock-in-Trade of Mr. Stephen Quillet, Lately Burned Out in the Fire at Charing Cross*. London.

Langton, J. (2000) 'Urban growth and economic change: from the late seventeenth century to 1841' in P. Clark (ed.), *The Cambridge Urban History of Britain, volume II, 1540–1840*. Cambridge: Cambridge University Press, pp. 453–90.

Lash, S. and Urry, J. (1993) *Economies of Signs and Spaces*. London: Sage.

Latham, R. and Matthews, W. (eds) (1970–83) *The Diary of Samuel Pepys: a New and Complete Transcription* (10 volumes). London: G. Bell.

Latimer, J. (1893–1900) *The Annals of Bristol, Volume I, Sixteenth and Seventeenth Centuries*, 1970 reprinting. Bath: Kingsmead Press.

Latour, B. (1993) *We Have Never Been Modern*. New York: Harvester Wheatsheaf.

—— (1997). 'Trains of thought: Piaget, formalism and the fifth dimension' *Common Knowledge* 6, 170–91.

—— (2000) 'Good and bad science', paper available at: <www.ensmp.fr/PagePerso/CSI/Bruno_Latour.html/Articles/77-GERG.html>

—— (2005) *Reassembling the Social: an Introduction to Actor-Network Theory*. Oxford: Oxford University Press.

Laurence, A. (1996) *Women in England, 1500–1760*. London: Phoenix.

Lavery, B. (1983) *The Ship of the Line: the Development of the Battlefleet, 1650–1850*. (Two volumes) London: Conway Maritime Press.

—— (1989) *Nelson's Navy: the Ships, Men and Organisation, 1793–1815*. London: Conway Maritime Press.

—— (ed.) (1992) *The Line of Battle: the Sailing Warship, 1650–1840*. London: Conway Maritime Press.

Law, J. (1986) 'On its methods a long-distance control: vessel, navigators, and the Portuguese route to India' in J. Law (ed.) *Power, Action and Belief*. London: Routledge and Kegan Paul, 234–63.

Laycock, W.S. (1976) *The Lost Science of John 'Longitude' Harrison*. Ashford: Brant Wright Associates.

Leccardi, C. and Rampazi, M. (1993) 'Past and future in young women's experience of time' *Time & Society* 2, 353–79.

Leduc, J. (1999) *Les Historiens et le Temps: Conceptions, Problematiques, Ecritures*. Paris: Editions du Seuil.

Lefebvre, H. (2003) 'Time and history' in S. Elden, E. Lebas, and E. Kofman (eds), *Henri Lefebvre: Key Writings*. London: Continuum, pp. 177–86.

—— (2004) *Rhythmanalysis: Space, Time and Everyday Life*. London: Continuum.

Legg, E. (1976) *The Clock and Watchmakers of Buckinghamshire*. Milton Keynes: Bradwell Abbey Field Centre.

Le Goff, J. (1980) *Time, Work and Culture in the Middle Ages*. Chicago: University of Chicago Press.

—— (1984) *The Birth of Purgatory*. London: Scolar Press.

—— (1988) *The Medieval Imagination*. Chicago: University of Chicago Press.

—— (1988) *Your Money or Your Life: Economy and Religion in the Middle Ages*. New York: Zone Books.

Lemire, B. (1988) 'Consumerism in pre-industrial and early industrial Britain: the trade in second-hand clothes' *Journal of British Studies* 27, 1–24.

Le Roy Ladurie, E. (1978). *Montaillou: Cathars and Catholics in a French Village, 1294–1329*. London: Scolar Press.

Liddington, J. (ed.) (1998) *Female Fortune: Land, Gender and Authority/The Anne Lister Diaries and Other Writings*. London: Rivers Oram Press.

Linebaugh, P. and Rediker, M. (2000) *The Many-Headed Hydra: Sailors, Slaves, Commoners, and the Hidden History of the Revolutionary Atlantic*. Boston: Beacon Press.

Lippincott, K. (ed.) (1999) *The Story of Time*. London: Merrell Holbertson with the National Maritime Museum.

Little, B. (1971) *Sketchley's Bristol Directory, 1775*. Bath: Kingsmead Press.

Liuzza, R.M. (2001) 'Anglo-Saxon prognostics in context: a survey and handlist of manuscripts' *Anglo-Saxon England* 30, 181–230.

Lloyd, C. (1970) *The British Seaman, 1200–1860: a Social Survey*. Rutherford, NJ: Fairleigh Dickinson University Press.

Lloyd, G. (1993) *Time and Being*. London: Routledge.

Loewenstein, D. and Mueller, J. (eds) (2002) *The Cambridge History of Early Modern English Literature*. Cambridge, Cambridge University Press.

Long, R. (1742) *Astronomy in Five Books*. Cambridge.

Loomes, B. (1982) *Directory of British Clockmakers Before 1750*. London: N.A.G. Press.

—— (1997) *The Clockmakers of Northern England*. Ashbourne: Mayfield Books.

—— (1998) *Brass Dial Clocks*. Woodbridge: Antiques Collectors Club.

—— (2006) *Clockmakers and Watchmakers of the World: including Makers of Scientific Instruments, Sundials and Barometers*. London, N.A.G. Press.

Lorraine, T. (2003) 'Living a time out of joint' in P. Patton and J. Protevi (eds), *Between Deleuze and Derrida*. London: Continuum, pp. 30–45.

Louis, C. (1980) *The Commonplace Book of Robert Reynes of Acle: an Edition of Tanner MS 407*. London and New York: Garland Medieval Texts, 1.

Lowe, P.M. (1982) *History of Bourgeois Perceptions*. Chicago: Chicago University Press.

Luckmann, T. (1991) 'The constitution of human life in time' in J. Bender and D.E. Wellbery (eds), *Chronotypes: the Construction of Time*. Stanford: Stanford University Press, pp. 151–66.

Lundy, D. (2002) *The Way of the Ship: a Square-Rigger Voyage in the Last Days of Sail*. London: Jonathan Cape.

Lutzhoft, M.H. and Nyce, J.M. (2006) 'Piloting by heart and by chart' *Journal of Navigation* 59, 221–37.

MacDonald, M. (1981) *Mystical Bedlam: Madness, Anxiety, and Healing in Seventeenth-Century England*. Cambridge: Cambridge University Press.

—— (1996) 'The career of astrological medicine in England' in A. Cunningham and O. Grell (eds), *Religio Medici: Medicine and Religion in Seventeenth-Century England*. London: Scolar Press, pp. 62–90.

Macey, S.L. (ed.) (1994) *Encyclopaedia of Time*. New York: Garland Reference Library of Social Science.

Mackenzie, A. (2002) *Transductions: Bodies and Machines at Speed*. London: Continuum.

MacLean, G.M. (1990) *Time's Witness: Historical Representation in English Poetry, 1603–1660*. Madison, WI: University of Wisconsin Press.

Maddison, F. (1992) 'Navigation: instruments of navigation'. In S.A. Bedini (ed.), *The Christopher Columbus Encyclopedia*. New York: Simon & Schuster, 512–15.

Malcolmson, R. (1973) *Popular Recreations in English Society 1700–1850*. Cambridge: Cambridge University Press.

Marx, K. (1954/1887) *Capital Vol. 1*. London: Lawrence & Wishart.

—— (1973/1858) *Grundrisse: Foundations of the Critique of Political Economy*: London: Allen Lane.

—— (1976/1874) *Poverty of Philosophy*. Moscow: Progress Publishers.

Matthews, W. (1794) *Bristol Directory*.

May, J. and Thrift, N.J. (eds) (2001) *TimeSpace: Geographies of Temporality.* London: Routledge.

Meyer, G.D. (1955) *The Scientific Lady in England.* University of California Press.

Mayr, O. (1986) *Authority, Liberty & Automatic Machinery in Early Modern Europe.* Baltimore and London: Johns Hopkins University Press.

McCann, J. (ed.) (1976) *The Rule of St Benedict.* London: Sheed and Ward.

McClary, S. (2000) *Conventional Wisdom: The Content of Musical Form.* Berkeley: University of California Press.

McClintock, A. (1995) *Imperial Leather: Race, Gender, and Sexuality in the Colonial Conquest.* New York: Routledge.

McGrail, S. (1998) *Ancient Boats in North-West Europe: the Archaeology of Water Transport to AD 1500.* London: Addison Wesley Longman.

—— (2003) *Boats of the World: from the Stone Age to Medieval Times.* Oxford: Oxford University Press.

McKay, E. (2005) 'English diarists: gender, geography and occupation, 1500–1700' *History* 90, 191–212.

McKenna, J. (2000) *Clock and Watchmakers of the West Midlands.* Ashbourne: Mayfield Books.

McKenna, R. (1997) *Real Time: Preparing for the Age of the Never Satisfied Customer.* Boston: Harvard Business School Press.

McNeill, W.H. *Keeping Together In Time: Dance and Drill in Human History.* Cambridge, Mass.: Harvard University Press.

McRae, A. (1995) *'God Speed the Plough': Representations of Agrarian England,* Cambridge: Cambridge University Press.

Menzies, G., Oliver, J. and Payn, M. (2002) 'The determination of longitude by the Chinese in the early fifteenth century' in G. Menzies, *1421: the Year China Discovered the World.* London: Bantam Press, pp. 447–57.

Meyer, G.D. (1955) *The Scientific Lady in England: an Account of Her Rise, with Emphasis on the Major Roles of the Telescope and Microscope.* Berkeley: University of California Press.

Michael, M. (2002) 'Between the mundane and the exotic: time for a different sociotechnical stuff' *Time and Society* 12, 127–43.

Milburn, J.R. and King, H.C. (1978) *Geared to the Stars: the Evolution of Planetariums, Orreries, and Astronomical Clocks.* Bristol: Hilger.

Michelet, J. (1855) *Histoire de France.* Paris : Chamerot.

Miller, E. and Hatcher, J. (1995) *Medieval England: Towns, Commerce and Crafts, 1086–1348.* London: Longman.

Mokyr, J. (2002) *The Gifts of Athena: Historical Origins of the Knowledge Economy.* Princeton: Princeton University Press.

Molotch, H. (2003) *Where Stuff Comes From.* New York: Routledge.

Money, J. (1995) 'Teaching in the market-place, of "Caesar adsum jam forte: Pompey aderat": the retailing of knowledge in provincial England during the eighteenth century' in J. Brewer and R. Porter (eds), *Consumption and the World of Goods.* London: Routledge, pp. 335–80.

Moody, J. (ed.) (2001) *The Private Life of an Elizabethan Lady: the Diary of Lady Margaret Hoby, 1599–1605.* Stroud: Sutton Publishing.

Moore, A.J. (1999) *The Clockmakers of Bristol, 1650–1900*. Bristol: Moore.
Moore, A.J., Rice, R.W., and Hucker, E. (1995) *Bilbie and the Chew Valley Clockmakers*. Bristol: The authors.
Moore, J. S. (ed.) (1976) *The Goods and Chattels of our Forefathers: Frampton Cotterill and District Probate Inventories*. Chichester: Phillimore.
Morris, C. ed. E. Hobhouse (1935) *The Diary of a West Country Physician, 168–1726*, 2nd edition. London: Simpkin Marshall.
Morson, G. (1994) *Narrative and Freedom: The Shadows at Time*. New Haven: Yale University Press.
Mortimer, I. (2006) *The Perfect King: the Life of Edward III, Father of the English Nation*. London: Jonathan Cape.
Mumford, L. (1926) *The Golden Day: a Study in American Experience and Culture*. Oxford; London: Oxford University Press; Humphrey Milford.
Munn, N.D. (1992) 'The cultural anthropology of time: a critical essay' *Annual Review of Anthropology* 21, 93–123.
Myers, A.R. (1959) *The Household of Edward IV*. Manchester: Manchester University Press.
Nardi, B.A. and O'Day, V.L. (1999) *Information Ecologies: Using Technology with Heart*. Cambridge, Mass: MIT Press.
Naworth, G. (1642) *A new Almanacke and Prognostication for the yeere of our Lord and Saviour Iesus Christ, 1642: being the second from the bissextile or leape-yeere, and from the creation of the world, 5591: referred most especially to the meridian and latitude of the ancient city of Durham*. London: John Dawson for the Company of Stationers.
Nelson, W. (ed.) (1956) *A Fifteenth Century School Book. From a manuscript in the British Museum, MS Arundel 249*. Oxford: Clarendon Press.
Netz, R. (2004) *Barbed Wire: an Ecology of Modernity*. Middletown, Conn: Wesleyan University Press.
North, J.D. (1975) 'Monasticism and the first mechanical clocks' pp. 381–98 in J.T. Fraser and N. Lawrence (eds), *The Study of Time II: Proceedings of the Second Conference of the International Society for the Study of Time, Lake Yamanaka-Japan*. New York: Springer-Verlag.
—— (2005) *God's Clockmaker: Richard of Wallingford and the Invention of Time*. London: Hambledon.
Nowotny, H. (1994) *Time: the Modern and Postmodern Experience*. Cambridge: Polity Press.
O'Malley, M. (1990) *Keeping Watch: a History of American Time*. New York: Viking Penguin.
—— (1992) 'Time, work and task orientation: a critique of American Historiography' *Time & Society* 1, 341–58.
Orlin, L.C. (2002) 'Fictions of the early modern English probate inventory' in H.S. Turner (ed.), *The Culture of Capital: Property, Cities and Knowledge in Early Modern England*. London: Routledge, pp. 51–83.
Orme, N. (1973) *English Schools in the Middle Ages*. London: Methuen.
—— (1978) 'A Bristol library for the clergy' *Transactions of the Bristol and Gloucestershire Archaeological Society* 96, 33–52.
—— (1989) *Education and Society in Medieval and Renaissance England*. London: Hambledon Press.

—— (2001) *Medieval Children*. New Haven, Conn. and London: Yale University Press.
—— (2006) *Medieval Schools*. New Haven, Conn. and London: Yale University Press.
Osborne, P. (1995) *The Politics of Time: Modernity and the Avant-Garde*. London: Verso.
Ovens, R. and Sleath, S. (2002) *Time in Rutland*. Oakham: Rutland Local History and Record Society, Rutland Record Series, 4.
Overton, M. (1984) 'Probate inventories and the reconstruction of agricultural landscapes' in Reed, M. (ed.) *Discovering Past Landscapes*. London: Croom Helm, 167–204.
Overton, M., Whittle, J., Hann, A., and Dean, D. (2004) *Production and Consumption in English Households, 1600–1750*. London: Routledge.
Padfield, P. (1999) *Maritime Supremacy and the Opening of the Western Mind: Naval Campaigns That Shaped the Modern World, 1588–1782*. London: John Murray.
—— (2003) *Maritime Power and the Struggle for Freedom: Naval Campaigns That Shaped the Modern World, 1788–1857*. London: John Murray.
Parkes, D.N. and Thrift, N.J. (1980) *Times, Spaces, Places: a Chronogeographic Perspective*. Chichester: John Wiley.
Parry, J. (2006) 'The idea of the record' *Sport in History* 26, 197–214.
Patterson, G. (1990) *History and Communications: Harold Innis Marshall McLuhan and the Interpretation of History*. Toronto: University of Toronto Press.
Pawson, E. (1992) 'Local-times and standard time in New Zealand' *Journal of Historical Geography* 18(3), 278–87.
Penn, S.A.C. (1989) *Social and Economic Aspects of Fourteenth-Century Bristol*. University of Birmingham Ph.D. thesis.
Perkins, M. (2001) *The Reform of Time: Magic and Modernity*. London: Pluto Press.
Petitot, J., Varela, F.J., Pachoud, B., and Roy, J. (eds) (1999) *Naturalizing Phenomenology: Issues in Contemporary Phenomenology and Cognitive Science*. Stanford: Stanford University Press.
Pickford, C. (1991) *Bedfordshire Clock and Watchmakers 1352–1880: a Biographical Dictionary with Selected Documents*. Bedford: Bedfordshire Historical Record Society.
Pickstone, J.V. (2000) *Ways of Knowing: A New History of Science, Technology and Medicine*. Manchester: Manchester University Press.
Poole, A.L. (1934) 'The beginning of the year in the Middle Ages', pp. 1–27 in *Studies in Chronology and History*, Oxford.
Poole, R. (1998) *Time's Alteration: Calendar Reform in Early Modern England*. London: UCL Press.
Pope, D. (1981) *Life in Nelson's Navy*. London: Chatham Publishing.
Porathe, T., Svensson, G. (2003) 'From portolan charts to virtual beacons: an historical overview of mediated communication at sea.' *Proceedings of the International Visual Literary Conference*, October.
Porteous, J.D. (1982) 'Surname geography: a study of the Mell Family Name c.1538–1980' *Transactions of the Institute of British Geographers*, new series 7(4) (1982), 395–418.
Porter, R. (1995) *Enlightenment, Britain and the Creation of the Modern World*. London: Allen Lane.
Prahalad, C.K. and Ramaswamy, V. (2004) *The Future of Competition*. Boston: Harvard Business School Press.
Pred, R. (2005) *Onflow: Dynamics of Consciousness and Experience*. Cambridge, Mass: MIT Press.

Press, J. (1986) *The Merchant Seamen of Bristol, 1747–1789*. Bristol: Historical Association.
Price, J. de S. (1959) *Contra-Copernicus: A Critical Re-estimation of the Mathematical Planetary Theory of Ptolemy, Copernicus, and Kepler*. S.l.: s.u.
Probyn, E. (1996) *Outside Belongings*. New York and London: Routledge.
Proust, M. (1981) *Remembrance of Things Past*. S.C.K. Moncrieff *et al*. (eds), London: Chatto & Windus.
Pryce, W.T.R. and Davies, T. Alun (1985) *Samuel Roberts Clock Maker: An Eighteenth-Century Craftsman in a Welsh Rural Community*. Cardiff: National Museum of Wales, Welsh Folk Museum.
Putter, A. (2001) 'In search of lost time: missing days in Sir Cleges and Sir Gawain and the Green Knight', pp. 119–36 in C. Humphrey and W.M. Ormrod (eds), *Time in the Medieval World*. York: York Medieval Press.
Quill, H. (1966) *John Harrison: The Man Who Found Longitude*. New York: Humanities Press.
Radford, P.E. and Ward-Smith, A.J. (2003) 'British running performances in the eighteenth century' *Journal of Sports Science* 21, 429–38.
Randles, W.G.L. (1985) 'Portuguese and Spanish attempts to measure longitude in the sixteenth century' *Vistas in Astronomy* 28, 235–41.
—— (1995) 'Portuguese and Spanish attempts to measure longitude in the sixteenth century' *The Mariner's Mirror* 81, 402–8.
—— (1998) 'The emergence of nautical astronomy in Portugal in the XVth century' *Journal of Navigation* 51, 46–57.
Radley, A. (1991) *The Body and Social Psychology*. New York: Springer-Verlag.
Rath, R.C. (2003) *How Early America Sounded*. Ithaca: Cornell University Press.
Ravenhill, W. (1976) 'As to its position in respect to the heavens' *Imago Mundi* 28, 79–93.
Rawlings, A.L. (1993) *The Science of Clocks and Watches*. Upton: British Horological Institute.
Raybeck, D. (1997) 'The coconut-shell clock: time and cultural identity' *Time & Society* 1, 323–40.
Raymond, J. (2005) *The Invention of the Newspaper: English newsbooks, 1641–1649*. Oxford: Clarendon Press.
Rediker, M. (1987) *Between the Devil and the Deep Blue Sea: Merchant Seamen, Pirates and the Anglo-American Maritime World, 1700–1750*. Cambridge: Cambridge University Press.
Rée, J. (1999) *I See A Voice: Deafness, Language and the Senses—A Philosophical History*. London: Flamingo.
Reed, M. (ed.) (1988) *Buckinghamshire Probate Inventories, 1660–1714*. Buckinghamshire Record Society, volume 24.
Reeves, M. and Morrison, J. (eds) (1989) *The Diaries of Jeffery Whitaker, Schoolmaster of Bratton, 1739–1741*. Trowbridge: Wiltshire Record Society.
Reid, D.A. (1976) 'The decline of Saint Monday, 1766–1876' *Past & Present* 71, 76–101.
—— (1982) 'Interpreting the festival calendar: wakes and fairs as carnivals' in R. Storch (ed.), *Popular Culture and Custom in Nineteenth-Century Britain*. London: Croom Helm, pp. 125–53.

Revel, J. and Hunt, L. (eds) (1995) *Histories: French Constructions of the Past*. New York: The New Press.

Rheinberger, H. (1997). *Towards a History of Epistemic Things: Synthesizing Proteins in the Test Tube*. Stanford: Stanford University Press.

Rigaud, S.J. (1841) *Correspondence of scientific men of the seventeenth century, including letters of Barrow [&c.] in the collection of the earl of Macclesfield, II*, Oxf.

Riley, H.T. (1872) The manuscripts of the Corporation of Totnes in the *Third Report of the Royal Commission on Historical Manuscripts*. London: HMSO.

—— (1874) The manuscripts of the Corporation of Hythe in the *Fourth Report of the Royal Commission on Historical Manuscripts*. London: HMSO.

Roach, J.R. (1996) *Cities of the Dead: Circum-Atlantic Performance*. New York: Columbia University Press.

Roberts, D. (ed.) (1982) *Derbyshire Hearth Taxes*. Derbyshire Record Society.

Robey, J. (2001) *The Longcase Clock Reference Book*. 2 vols. Ashbourne: Mayfield.

Robinson, H.W. and Adams, W. (eds) (1968) *The Diary of Robert Hooke, M.A., M.D., F.R.S., 1672–1680*. London: Wykeham Publications.

Robinson, T. (1981) *The Long-case Clock*. Woodbridge: Antique Collectors Club.

Roche, D. (1989) *The Culture of Clothing: Dress and Fashion in the Ancien Regime*. Cambridge: Cambridge University Press with Past and Present Publications.

—— (2000) *A History of Everyday Things: the Birth of Consumption in France*. Cambridge, Cambridge University Press.

Rodger, N.A.M. (1986) *The Wooden World: an Anatomy of the Georgian Navy*. London: Methuen.

—— (1997) *The Safeguard of the Sea: a Naval History of Britain. Volume 1, 660–1649*. London: Harper Collins.

—— (2004) *The Command of the Ocean: a Naval History of Britain, 1649–1815*. London: Allen Lane.

Rogers, J.E.T. (no date/1884) *Work and Wages*. London: Swan Sonnenschein.

Rule, J. (1994) *Saturday Night and Sunday Morning* (Inaugural Lecture). Southampton: University of Southampton.

Rutz, H.J. (ed.) (1992) *The Politics of Time*. American Anthropological Association/American Ethnological Society Monograph no. 4, Washington, DC.

Rybczynski, W. (1991) *Waiting for the Weekend*. Viking: New York.

Sachse, W.L. (1938) *The Diary of Roger Lowe of Ashton-in-Makerfield, Lancashire, 1663–1674*. New York: Yale University Press.

Sanderson, J. (ed.) (1994) *Change-Ringing: The History of an English Art*, 3 volumes. Guildford: Central Council of Church Bell Ringers.

Sato, M. (1991) 'Comparative ideas of chronology' *History and Theory* 30, 26–50.

Savage-Smith, E. (2003) 'Islam' in *The Cambridge History of Science. Vol. 4: Eighteenth-Century Science*, R. Porter (ed.), Cambridge: Cambridge University Press, pp. 649–68.

Savitt, S. (ed.) (1995) *Time's Arrows Today: Recent Physical and Philosophical Work on the Direction of Time*. Cambridge: Cambridge University Press.

Scammell, G.V. (1981) *The World Encompassed: the First European Maritime Empires c.800–1650*. London: Methuen.

Schaffer, S. (1988) 'Astronomers mark time: discipline and personal equation' *Science in Context* 2, 115–45.

Schatzki, T. (2001) *Social Practices: a Wittgensteinian Approach to Human Activity and the Social*. Cambridge: Cambridge University Press.

—— (2002) *The Site of the Social: a Philosophical Account of the Constitution of Social Life and Change*. University Park: Pennsylvania State University Press.

—— (2003) 'Nature and technology in history' *History and Theory* 42, 82–93.

Schatzki, T.R., Knorr Cetina, K. and von Savigny, E. (eds) (2001) *The Practice Turn in Contemporary Theory*. London: Routledge.

Schivelbusch, W. (1986) *The Railway Journey*. Berkeley: University of California Press.

Schlör, J. (1998) *Nights in the Big City: Paris, Berlin, London 1840–1930*. London: Reaktion.

Schor, J. (1994) *Labor and Léisure: A Look at Contemporary Values*. Boston: Massachusetts Council of Churches.

Scott, C.E. (2002) *The Lives of Things*. Bloomington: Indiana University Press.

Seaver, P.S. (1985) *Wallington's World: a Puritan Artisan in Seventeenth-Century London*. Stanford, CA: Stanford University Press.

Secord, J. (1985) 'Newton in the nursery: Tom Telescope and the philosophy of tops and balls, 1761–1838' *History of Science* 23, 127–51.

—— (2000) *Victorian Sensation: the Extraordinary Publication, Reception and Secret Authorship of Vestiges of the Natural History of Creation*. Chicago: University of Chicago Press.

Seigworth, G.J. (2000) 'Banality for Cultural Studies' *Cultural Studies* 14, 227–68.

Sellen, A.J. and Harper, R.H.R. (2002) *The Myth of the Paperless Office*. Cambridge, Mass: MIT Press.

Sennett, R. (1998) *The Corosion of Character: The Personal Consequences of Work in the News Capitalism*. New York: W. W. Norton.

Serres, M. (1982). *Hermes: Literature, Science, Philosophy*. Baltimore: Johns Hopkins University Press.

Serres, M. and Latour, B. (1995) *Conversations on Science, Culture and Time*. Ann Arbor: University of Michigan Press.

Shapin, S. (1994) *A Social History of Truth: Science and Civility in the Seventeenth Century*. Chicago: Chicago University Press.

—— (1998) 'The philosopher and the chicken: on the dietetics of disembodied knowledge' in C. Lawrence and S. Shapin (eds), *Science Incarnate: Historical Embodiments of Natural Science*. Chicago: University of Chicago Press, pp. 21–50.

Shaw, J. (1994) 'Punctuality and the everyday ethics of time: some evidence from the Mass Observation Archive' *Time & Society* 3, 79–98.

Shenk, D. (1999) *The End of Patience: Cautionary Notes on the Information Revolution*. Bloomington: Indiana University Press.

Sherman, S. (1996) *Telling Time: Clocks, Diaries and English Diurnal Form 1660–1785*. Chicago: University of Chicago Press.

Shimada, S. (1994) 'Working hour, free time, family time in Japan: Dealing with the Western conceptions of time in Japanese society' Paper presented to 1994 ASSET conference, Dartington.

Simondon, G. (1989/1958) *Du Mode d'Existence des Objets Techniques*. Paris: Editions Aubier.

Simpson, L.C. (1995) *Technology, Time and the Conversations of Modernity*. New York: Routledge.

Sketchley, J. (1775) *Bristol Directory*. Bristol: James Sketchley; reprinted in B. Little (ed.) (1971) *Sketchley's Bristol Directory, 1775*. Bath: Kingsmead Press.
Smith, B.G. (1991) *The 'Lower Sort': Philadelphia's Labouring People, 1750–1800*. London: Cornell University Press.
Smith, B.R. (1999) *The Acoustic World of Early Modern England: Attending to the O-Factor*. Chicago: University of Chicago Press.
Smith, J. (1675) *Horological Dialogues: . . . the Use and Right Managing of Clocks and Watches. . . .* London.
Smith, M.M. (1994) 'Counting clocks, owning time: detailing and interpreting clock and watch ownership in the American south, 1739–1865' *Time & Society* 3, 321–39.
—— (1996) 'Old South time in comparative perspective' *American Historical Review* 101, 1432–69.
—— (1997) *Mastered by the Clock: Time, Slavery, and Freedom in the American South*. Chapel Hill: University of North Carolina Press.
Smith, M.Q. (1994) *St Mary Redcliffe: An Architectural History*. Bristol: Redcliffe Press.
Smith, T. (1986) 'Peasant time and factory time in Japan' *Past and Present* 111, 165–97.
Smith, W.D. (2002) *Consumption and the Making of Respectability 1600–1800*, New York & London: Routledge.
Smyth, A.W.H. (1867/2005) *The Sailor's Word-Book*. London: Conway Maritime Press.
Sobel, D. (1995) *Longitude: the True Story of a Lone Genius Who Solved the Greatest Scientific Problem of His Time*. London: Fourth Estate.
Sobel, D. and Andrewes, W.J.H. (1998) *The Illustrated Longitude*. London: Fourth Estate.
Sokoll, T. (ed.) (2001) *Essex Pauper Letters, 1731–1837*. Oxford: British Academy Records Series.
Sommerville, C.J. (1996) *The News Revolution in England: Cultural Dynamics of Daily Information*. Oxford: Oxford University Press.
Sorrenson, R. (1995a) 'The state's demand for accurate astronomical and navigational instruments in eighteenth century Britain' in A. Bermingham and J. Brewer (eds), *The Consumption of Culture, 1600–1800: Image, Object, Text*. London: Routledge, pp. 263–71.
—— (1995b) 'The ship as a scientific instrument in the eighteenth century' *Osiris* 2nd series, 11, 221–36.
Spengler, O. (1926) *The Decline of the West*. 2 vols. London: Allen and Unwin.
Spufford, M. (1984) *The Great Reclothing of Rural England: Petty Chapman and Their Wares in the Seventeenth Century*. London: Hambledon.
—— (1995) 'Literacy, trade and religion in the commercial centres of Europe' in K. Davids and J. Lucassen (eds), *A Miracle Mirrored: the Dutch Republic in European Perspective*. Cambridge: Cambridge University Press, pp. 229–83.
Stanford, M. (ed.) (1990) *The Ordinances of Bristol 1506–1598*. Bristol: Bristol Record Society Publications, 41.
Stark, S.J. (1998) *Female Tars: Women Aboard Ship in the Age of Sail*. London: Pimlico.
Stears, P.C.D. (ed.) (1972) *Yorkshire Probate Inventories 1542–1689*. Yorkshire Archaeological Society Record Series, volume 134.
Steel, D. (1999) *Eclipse: The Celestial Phenomenon Which Has Changed the Course of History*. London: Headline.

Stein, J. (1995) 'Time, space and social discipline: factory life in Cornwall, Ontario, 1867–1893' *Journal of Historical Geography* 21, 278–99.

Stengers, I. (1997) *Power and Invention: Situating Science*. Minneapolis: University of Minnesota Press.

Stephens, C. (1989) 'The most reliable time: William Bond the New England railroads and time awareness in nineteenth-century America' *Technology and Culture* 30, 1–24.

Sterelny, K. (2003) *Thought in a Hostile World: the Evolution of Human Cognition*. Oxford: Blackwell.

Sterne, L. (1759–69) *The Life and Opinions of Tristram Shandy, Gentleman*, nine volumes. London.

Stiegler, B. (1998) *Technics and Time: the Fault of Epimetheus*. Stanford: Stanford University Press.

Stimson, A. (1988) *The Mariner's Astrolabe*. Utrecht: HES.

Stitt, B. (ed.) (1947) 'Diana Astry's recipe book *c*.1700' *Bedfordshire Historical Records Society* 37, 86–168.

Strathern, M. (1992) *After Nature: English Kinship in the Late twentieth Century*. Cambridge: Cambridge University Press.

Street, B.V. (ed.) (1993) *Cross-Cultural Approaches to Literacy*. Cambridge: Cambridge University Press,

Styles, J. (1994) 'Clothing the North of England, 1660–1800' *Textile History* 25, 139–66.

Swerdlow, N. (1996) 'Astronomy in the Renaissance' in C. Walker (ed.), *Astronomy Before the Telescope*. London: British Museum Press.

Symonds, E.M. (1927–29) 'The diary of John Greene, 1635–57' *English Historical Review* 43, 385–94, 598–604; 44:106–17.

Taylor, E.G.R. (1930) *Tudor Geography, 1485–1583*. London: Methuen.

—— (1954) *The Mathematical Practitioners of Tudor and Stuart England*. Cambridge: [Published] for the Institute of Navigation at the University Press.

—— (ed.) (1963) *A Regiment for the Sea, by William Bourne, and Other Writings on Navigation*. Cambridge: Hakluyt Society, 2nd series volume 121.

—— (1971) *The Haven-Finding Art*, second edition. London: Hollis and Carter.

Taylor, E.G.R., and Richey, M. (1962) *The Geometrical Seaman*. London: Hollis and Carter.

Tedlock, B. (1992) *Time and the Highland Maya*, revised edition, ed. Albuquerque: University of New Mexico Press.

Tester, S.J. (1990) *A History of Western Astrology*. Woodbridge: Boydell.

Thacker, J. (1714) *The Longitude Examin'd*. London.

Theaker, R. (1665) *A Light to the Longitude or the Use of an Instrument call'd the Seaman's Director speedily resolving all Astronomical Cases and Questions concerning the Sun, Moon, Stars with several propositions whereby sea-men may find at what meridian and Longitude they are at, in parts of the world*. London.

Thirsk, J. (1978) *Economic Policy and Projects*. Oxford: Clarendon Press.

Thomas, K.V. (1964) 'Work and leisure in pre-industrial societies' *Past and Present* 29, 50–62.

—— (1971) *Religion and the Decline of Magic*. London: Weidenfeld and Nicholson.

—— (1987) 'Numeracy in early-modern England' *Transactions of the Royal Historical Society* 5th series 37, 103–32.

Thomas, N. (1991) *Entangled Objects: Exchange, Material Culture and Colonialism in the Pacific*. Cambridge, Mass: Harvard University Press.

Thomke, S. (2003) *Experimentation Matters: Unlocking the Potential of New Technologies for Innovation*. Boston: Harvard Business School Press.

Thompson, E.P. (1963/68) *The Making of the English Working Class*. Harmondsworth: Penguin Books.

—— (1967) 'Time, work-discipline and industrial capitalism' *Past and Present* 38, 56–97.

—— (1991) *Customs in Common*. London: Merlin Press.

Thompson, P. (1856) *The History and Antiquities of Boston, and . . . the Hundred of Skirbeck*. Boston.

Thomson, M.M. (1978) *The Beginning of the Long Dash: a History of Timekeeping in Canada*. Buffalo: University of Toronto Press.

Thorpe, C. (2004) 'Against time: scheduling, momentum, and moral order at wartime Los Alamos' *Journal of Historical Sociology* 17, 31–55.

Thrift, N.J. (1977) 'Time and theory in human geography' *Progress in Human Geography* 1, 65–101.

—— (1981) 'Owners time and own time: the making of a capitalist time consciousness 1300–1800' in A.R. Pred (ed.), *Space and Time in Geography: Essays Dedicated to Torsten Hagerstrand*. Lund: Gleerup Studies in Geography, pp. 56–84.

—— (1988) 'Vivos voco: ringing the changes in the historical geography of time consciousness' in M. Young and T. Schuller (eds), *The Rhythms of Society*. London: Routledge, pp. 53–94.

—— (1990) 'Transport and communications 1730–1914' in R.A. Dodgshon and R. A. Butlin (eds), *Historical Geography of England and Wales*, 2nd edition. London: Academic Press, pp. 453–86.

—— (1995) 'A hyperactive world' in R.J. Johnston, P.J. Taylor, M. Watts (eds) *Geographies of Global Change*. Oxford: Blackwell, 18–35.

—— (1996) *Spatial Formations*. London: Sage.

—— (2000a). 'Afterwords' *Environment and Planning D. Society and Space* 18, 213–55.

—— (2000b) 'Still life in nearly present time: the object of nature' *Body and Society* 6, 34–57.

—— (2003a) 'Remembering the technological unconscious by foregrounding knowledges of position' *Environment and Planning D. Society and Space* 21, 175–90.

—— (2003b) 'Space' in S. Holloway and G. Valentine (eds), *Contemporary Concepts in Geography*. London: Sage.

—— (2005) 'Movement-space: the changing domain of thinking resulting from the development of new kinds of spatial awareness' *Economy and Society* 33, 582–604.

—— (2006) 'Space' *Theory Culture and Society*, 23, 131–155.

—— and French, S. (2002) 'The automatic production of space' *Transactions of the Institute of British Geographers* 27(3), 309–35.

Thrower, N.J.W. (1996) *Maps and Civilization: Cartography in Culture and Society*. Chicago: University of Chicago Press.

Tindall, G. (2002) *The Man Who Drew London: Wenceslaus Hollar in Reality and Imagination*. London: Pimlico.

Tittler, R. (1991) *Architecture and Power: the Town Hall and the English Urban Community, c.1500–1640*. Oxford: Clarendon Press.

Todes, S. (2001) *Body and World*. Cambridge, Mass: MIT Press.
Toulmin-Smith, L. (ed.) (1964) *Leland's Itinerary in England and Wales*, 5 volumes. London: Centaur Press.
Tripp, C.H. (1935) *A History of Queen Elizabeth's Grammar School, Barnet*. Cambridge: Heffer.
Turner, A.J. (1996) 'In the wake of the Act, but mainly before' in W.J.H. Andrewes (ed.), *The Quest for Longitude*. Cambridge, Mass: Harvard Collection of Scientific Instruments, pp. 115–31.
Turner, A. (2002) 'Essential complementarity: the sundial and the clock', chapter 2 in H. Higton *Sundials at Greenwich: a Catalogue of the Sundials, Horary Quadrants and Nocturnals in the National Maritime Museum, Greenwich*. Oxford: Oxford University Press and the National Maritime Museum, pp. 15–23.
Turner, G.L.E. (1998) *Scientific Instruments 1500–1900*. Berkeley: University of California Press.
—— (2000) *Elizabethan Instrument Makers: the Origins of the London Trade in Precision Instrument Making*. Oxford: Oxford University Press.
—— (2002) 'Eighteenth century scientific instruments and their makers' in R. Porter (ed.), *The Cambridge History of Science. Volume 4. Eighteenth Century Science*. Cambridge: Cambridge University Press, pp. 511–35.
Turner, S. (1994) *The Social Theory of Practices*. Cambridge: Polity Press.
Underdown, D. (ed.) (1991) *William of Whiteway of Dorchester: His Diary 1618–1635*. Dorchester: Wiltshire Record Society, volume 12.
Urry, J. (1991) 'Time and space in Giddens' social theory' in C.G.A. Bryant and D. Jary (eds), *Giddens' Theory of Structuration: a Critical Appreciation*. London: Routledge, pp. 160–75.
—— (1994) 'Time, leisure and social identity' *Time & Society* 3(2), 131–50.
Vaisey, D. (ed.) (1984) *The Diary of Thomas Torner, 1754–1765*. Oxford: Oxford University Press.
Van de Woude, A. and Schuurman, A. (eds) (1980) *Probate Inventories: A New Source for the Historical Study of Wealth, Material Culture and Agricultural Development*. A.A.G. Bijdragen volume 23.
van Helden, A. (1986) *Measuring the Universe Cosmic Dimensions from Aristardus to Halley*. Chicago: University of Chicago Press.
Venn, J. and Venn, J.A. (1922–27) *Alumni Cantabrigienses: a biographical list of all known students, graduates and holders of office at the University of Cambridge, from the earliest times to 1751*, volumes 1–4. Cambridge: Cambridge University Press.
Verran, H. (2001) *Science and an African Logic*. Chicago: University of Chicago Press.
Vickers, N. (ed.) (1986) *A Yorkshire Town of the 18th Century: the Probate Inventories of Whitby, North Yorkshire. 1700–1800*. Studley: Brewin Books.
Virilio, P. (2003) *The Vision Machine*. Bloomington: Indiana University Press.
von Hippel, E. (2005) *Democratizing Innovation*. Cambridge, Mass.: MIT Press.
Voth, H-J. (1998) 'Time and work in eighteenth-century London' *Journal of Economic History* 58, 29–58.
—— (2000) *Time and Work in England, 1750–1830*. Oxford: Clarendon Press.
Walcott, M.E.C. (1871) 'Inventories and valuations of religious houses at the time of the Dissolution' *Archaeologia* 43, 201–49.

Walker, J.E.S. (1982) *Hull and East Riding Clocks*. Hull: Hornsea Museum Publications no. 3.
Walker, R. (2001) 'Review of Corbin, village bells', *The Ringing World* 2, November, 1089.
Wallis, F. (ed.) (1999) *Bede: the Reckoning of Time*. Liverpool: Liverpool University Press.
Wallman, S. (ed.) (1992) *Contemporary Futures: Perspectives from Social Anthropology*. London: Routledge.
Walsh, L.S., Carr, L.S., Main, G., and Main, J.T. (1988) 'Towards a history of the standard of living in British North America' *William and Mary Quarterly*, 3rd series 45, 116–70.
Walters, A.N. (1999) 'Ephemeral events: English broadsides of early eighteenth-century solar eclipses' *History of Science* 37, 1–43.
Ward, A.W. *et al.* (eds) (1907–21) *The Cambridge History of English and American Literature: An Encyclopedia in Eighteen Volumes*. Cambridge: Cambridge University Press.
Ward, A.W. and Waller, A.R. (eds) (1912) *The Cambridge History of English Literature, volume nine*. Cambridge: Cambridge University Press.
Ward, J. (1715) *A Practical Method to Discover the Longitude at Sea, By a New Contrived Automaton*. London.
—— (1740 [1967]) *The Lives of the Professors of Gresham College: to which is Prefixed the Life of the Founder, Sir Thomas Gresham*. New York: Johnson Reprint Corporation.
—— (ed.) (1995) *Women of the English Nobility and Gentry 1066–1500*. Manchester: Manchester University Press.
Warnier, J. (2001) 'A praxeological approach to subjectivation in a material world' *Journal of Material Culture* 6, 5–24.
Waters, D.W. (1958) *The Art of Navigation in Elizabethan and Early Stuart Times*. London: Hollis and Carter.
—— (1967) *The Rutters of the Sea*. New Haven: Yale University Press.
Watkinson, L., Ambler, R.W. (1987) *Farmers and Fishermen: The Probate Inventories of the Ancient Parish of Clee, South Humberside 1536–1742*. Hull: School of Adult and Continuing Education, University of Hull.
Weatherill, L. (1988) *Consumer Behaviour and Material Culture in Britain 1660–1760*. London: Routledge.
Webb, C.C. (1997) *The Churchwardens' Accounts of St Michael Spurriergate, York, 1518–1548*. York: Borthwick Texts and Calendars, volume xx.
Weimann, R. (1984) 'Fabula and Historia: the crisis of the "Universall Consideration" in *The Unfortunate Traveller*' *Representations* 8, 14–29.
Wells Cathedral (1994) *Wells Cathedral Clock*. Wells: Wells Cathedral.
Wendorff, R. (1980) *Zeit und Kultur: Geschichte des Zeitbewusstseins in Europa*. Weisbaden: Westdeutscher Verlag.
Wenger, E. (1999) *Communities of Practice*. Cambridge: Cambridge University Press.
Wheeler, J.S. (1999) *The Making of a World Power: War and the Military Revolution in Seventeenth-Century England*. Stroud: Sutton.
Whipp, R. (1981) '"A Time to Every Purpose": An Essay on Time and Work' in P. Joyce (ed.) *The Historical Meanings of Work*. Cambridge: Cambridge University Press, pp. 210–36.

Whitbread, Helena (ed.) (1992) *I Know My Own Heart: The Diaries of Anne Lister, 1791–1840.* New York: New York University Press.
Whiteman, A. (ed.) (1986) *The Compton Census of 1676: a Critical Edition.* London: British Academy, Records of Social and Economic History, new series 10.
Whitrow, G.J. (1988) *Time in History: Views of Time from Prehistory to the Present Day.* Oxford: Oxford University Press.
Wilbourn, A.S.H. and Ellis, R. (2001) *Lincolnshire Clock, Watch and Barometer Makers.* Lincoln: Hansord.
Wilcox, D.J. (1987) *The Measure of Times Past: Pre-Newtonian Chronologies and the Rhetoric of Relative Time.* Chicago: University of Chicago Press.
Wilcox, L.A. (1966) *Mr Pepys' Navy.* New York: Barnes.
Wildenbeest, G. (1988) '"Keeping up with the Times": Time and its Rhythms in the Countryside' *Netherlands Journal of Sociology* 24, 134–45.
Wilkinson, D. (2003) *Early Horse Racing in Yorkshire and the Origins of the Thoroughbred.* York: Old Byland Press.
Willmoth, F. (1993) *Sir Jonas Moore: Practical Mathematics and Restoration Science.* Woodbridge: Boydell Press.
Wilmerding, J. (1999) *Compass and Clock: Defining Moments in American Culture.* New York: Abrams.
Wilson, C. (2002) 'Astronomy and cosmology' in R. Porter (ed.) *The Cambridge History of Science. Volume 4. Eighteenth Century Science.* Cambridge: Cambridge University Press, pp. 328–53.
Wood, A. (1999) *The Politics of Social Conflict: the Peak Country, 1520–1770.* Cambridge: Cambridge University Press.
——(2002) *Riot, Rebellion and Popular Politics in Early Modern England.* Basingstoke: Palgrave.
Wood, D. (1989) *The Deconstruction of Time.* Atlantic Highlands: Humanities Press.
——(1990) *Philosophy at the Limit.* London: Unwin Hyman.
Woodford, S. (1983) *Looking at Pictures.* Cambridge: Cambridge University Press.
Woodhouse, C.J. (1996) *A Moment's Notice: Time Politics Across Cultures.* Ithaca: Cornell University Press.
Woodward, D. (ed.) (1984) *The Farming and Memorandum Books of Henry Best of Elmswell.* London: British Academy, Records of Social and Economic History, new series 8.
——(1995) *Men at Work: Labourers and Building Craftsmen in the Towns of Northern England, 1450–1750.* Cambridge: Cambridge University Press.
Woolley, Hannah (1765) *The Accomplish'd lady's delight in preserving physick, beautifying, and cookery.* London: B. Harris.
Wrightson, K. (1996) 'The politics of the parish in early modern England' in P. Griffiths, A. Fox and S. Hindle (eds), *The Experience of Authority in Early Modern England.* Basingstoke: Palgrave, pp. 10–46.
Wrigley, E.A. (1967) 'A simple model of London's importance in changing English society and economy, 1650–1750' *Past and Present* 37, 44–70. Reprinted in E.A. Wrigley (1987) *People, Cities and Wealth: the Transformation of Traditional Society.* Oxford, pp. 133–56.

Yates, J-A. (2000) 'Business use of information technology during the Industrial Age' in A. Chandler and J. Cortada (eds), *A Nation Transformed by Information*. Oxford: Oxford University Press.

Yates, F.A. (1966) *Art of Memory*. London: Penguin.

Yoder, J.G. (1988) *Unrolling Time: Christian Huygens and the Mathematization of Nature*. Cambridge: Cambridge University Press.

Zerubavel, E. (1981) *Hidden Rhythms: Schedules and Calendars in Social Life*. Chicago: University of Chicago Press.

—— (1982) 'The standardization of time: a sociological perspective' *American Journal of Sociology* 88, 1–23.

—— (2003) *Time Maps: Collective Memory and the Social Shape of the Past*. Chicago: Chicago University Press.

Zimmer, C. (2004) *Soul Made Flesh: Thomas Willis, the English Civil War and the Mapping of the Mind*. London: Heinemann.

Index

Ackroyd, Peter 11
Acle, Norfolk 188
Adams, George 297
Adams, John Couch 337
Agar, Seth 366
Alberti, Leon Battista 266
Allen, Thomas 190
almanacs 207, 247–50, 258, 286, 293
Andrews, Henry 364
Annales School 14
Anne, queen of England 391
Arbuthnott, John 391
Ashby-de-la-Zouche, Leicestershire 229
Ashley, Anthony 290
Ashley, Sir Francis 214
Ashmole, Elias 202, 208, 210
Ashwell, Samuel 247, 249
astrology 207–9, 269–71, 330, 332–5
astronomy 332–35, 340–2, 349–50
Astry, Diana 223–4
Aubrey, John 182, 189–92, 212, 253, 265, 266, 271, 354, 364
Augustine, St 13

Bacon, Francis 269
Bakhtin, Mikhail 81–2
Ball, Henry 373
Bank of England 269
Barlinch Priory, Somerset 231
Barrow-upon-Humber, Lincolnshire 360, 364, 370, 373, 375–6, 386, 396
Barton-upon-Humber 364, 366, 368, 369–76, 378, 381, 384–5, 396
Barringer, Laurens 369
Barry, Jonathan 107, 117
Bartlett, Richard 214
Bede 330–1
Best, Henry 217–19, 246–7
Betson, Thomas 277
Beverley, Yorkshire 371
Birchenlee, John 377
Blackstone Edge, West Yorkshire 268
Bolton, Lancashire 252
Boole, George 364
Borough, Stephen 311
Borough, William 311
Bourne, William 258, 286, 287, 290, 294–5
Boyle, Robert 224–5
Brahe, Tycho 6, 331, 341
Bressingham, Norfolk 220

Brest, France 337
Brewer, John 377
Bristol 16–17, 100, 101, 102–3, 104–10, 132, 162–3, 183
  cathedral 131
  church clocks 116–17, 123, 126, 128, 130
  clock ownership 109–10, 131–2
  clocks, regional distinctiveness of 115–16
  Custom House 105
  demographics 120–2
  hackney cabs 107, 120
  inns 117
  leisure activities 107–8, 121–2
  ownership of clocks and watches 124–6
  public library 129
  post coaches 106, 117–9
  Post office 106–7, 119–20
  public order/disorder 104–5, 129
  public houses 117, 121–2
  regulation of city activities 124, 128–9
  St Augustine's Abbey 130–1
  schools 124, 129–30
  Saint Monday 104–5
  surviving examples of Bristol clockmakers 114
  urban space, usage of 120–2
Bristol Directory 111
Bushman, John 255
Buzino, Oraxio 85
Byrd, William 365

Cabot, Sebastian 311
Cambridge University 364, 365, 382, 385, 392, 393
Campbell, Alexander 318
Campbell, John 289, 318
cartography 333–4, 336–42
Cassini, Giovanni Domenico 342, 351
cathedrals
  Bristol 131
  Carlisle 228–9
  Lincoln 76
  Norwich 77
  Saint Paul's 138, 213
  Salisbury 77
  Wells 77
  Westminster Abbey 138
Cecil, William, Lord Burghley 222
Cerne, Dorset 214

Cervantes Miguel de 269
Chancellor, Richard 311
Charles I, king of England
Chaucer, Geoffrey 307
Chelmsford, Essex 220
churchwardens 139, 148, 161, 188, 220, 380
churchwardens' accounts 148–53, 188, 228, 371, 373
Civil War, English 157
Clare, John 231
Cliff, Thomas 371
Clifford, Lady Anne 201–2
Clitheroe, Walter 390
clocks
  accuracy 140–1, 335–58
  alarms 29, 37–38, 41, 77, 116–17, 136, 204, 205, 255
  as luxury goods 109–10
  bells 38–9, 57, 82–83, 85, 116–17, 130, 136, 137, 158–9, 183, 186–7, 190, 204, 213, 226, 229–30, 231–2, 252, 254, 262–3, 376
  Bristol clocks, regional distinctiveness of 115–16
  church/parish 40, 41, 56, 75, 116–17, 126, 128, 130, 140, 142–5, 147–53, 153–8, 161–2, 175–6, 229–30, 264, 330–1, 366, 370–1, 373–6
  costs 77–8, 110, 123–4, 148, 161–2, 171–2, 191, 232
  dials 41, 42, 116, 158–61, 204, 226, 231, 254–5, 335
  difficulties in dating from artwork 11–12
  domestic 162–75, 366
  faces 82, 85, 91, 95, 116, 140, 226
  gaming 191–2, 217–2
  gleaning bells 186–7
  hourglasses 61 n. 1, 125–6, 220–2, 305
  household 24
  incense 85
  long-case 174
  mechanical 26, 27, 29–42, 77–8, 91–2, 136, 137–8, 141, 188, 329, 331, 335, 396
  medieval 37–8, 40, 75–8, 83, 91–2, 138
  multi-handed 253–6, 264
  music 83–4
  oppressive 48
  ownership 108–10, 124–6, 131–2, 162–75, 258–9, 273–4, 367–8
  pendulum 1–4, 12, 36, 40, 140, 161–2, 196, 198, 250–1, 253, 255, 257, 260, 276, 335, 355, 363, 366, 389
  public 116–22, 126–7, 175–6, 226, 231–3, 366, 396
  ship-board 294, 304

  single-handed 250–3
  stopwatch 271
  sun dials 25, 26, 33, 39, 41, 125–6, 127, 136, 141, 190–1, 220–2, 246, 255–6, 257, 269, 276, 288, 342, 377
  ticking 198–9
  verge and foliot 31–35, 40, 139, 161–2, 250, 396
  visual representations 38–9, 82–3
  watches 24, 30, 124–6, 162–75, 191, 254, 255, 257, 279, 304, 335, 342, 371, 396, 416
  water clocks 26, 30, 75
  wooden 369, 398
clock time 39, 40, 47–53, 75, 79–80, 89–90, 96–7, 103–4, 117–18, 122, 123–4, 127–32, 136–45, 202–3, 216, 231–4, 244–7, 249–50, 274–7, 279–80, 329–32, 348–50, 357–58, 407–11
  apprehending of 39, 329–32
  Bristol and 104–8, 122, 186
  cooking and 223–5
  cultural connections 90–3
  effects on women 108, 122, 193
  embodied practices 80–93
  English royal household and 184–86
  historic differences in understanding of 236
  importance to Euro-American society 65, 78–9
  Kelantanese peasantry's attitude towards 89–90
  learning of 86–9, 174–5, 279
  legal 210, 213–5
  many ways of counting and marking 27–8
  Mayan attitudes toward 331–32
  natural events/disasters and 210
  nautical 280–2, 304–18, 332–33, 342–8
  postal system and 222
  private 135–6, 141, 145, 162–75, 396
  publicness 24–25, 56, 116–22, 135–6, 135, 141, 144, 145, 222–3, 225–33, 396
  regulatory use 182–9, 222–3
  relation between clocks and time 53, 140–1
  relation to natural time 24, 49, 140–1, 187
  religious understanding of 129–30, 184–5
  schools and 124, 129–30, 225–31
  signalling 38–39, 41, 158–62, 183, 203–4, 213, 254
  telling time 42, 141–2, 213–5, 226–7, 230, 244
  visual representations 38–9, 82–3
  working day and 220, 232–3
  written representations 189–92, 193–225
  see also Bristol
  see also equation of time

clockmakers 36–37, 56, 110–15, 124–5, 143, 160–2, 171–2, 188, 190–1, 252, 253–7, 273, 279, 335–36, 359–400, 408–9
  apprentices 111–12, 363, 398–9
Cockermouth, Cumberland 230
Cole, Benjamin 297
Cole, Humphrey 256
Collins, Greenville 291
Columbas, Christopher 288
compass, *see* navigational devices
Compton Census 153
Cook, James 342, 353, 366
Copernicus, Nicolaus 7
Cortes, Martin 289
Coventry 229
Crabtree, William 365
Craythorn, Jonathan 390
Cromwell, Thomas 130
Crosby, Ayrshire 260
Cuckfield, Sussex 261

da Vinci, Leonardo 253
dance 86–7
Darwin, Charles 3
Davenant, Revd Edward 189–90
Davis, John 283, 289, 294
de Peiresc, Claude Fabri 334
Dee, John 202–3, 207, 208, 211, 224, 265, 268, 282, 311, 331, 333
Desaguliers, John 335
Descartes, René 269
Deptford 311, 314
dialogism 68
diaries/diarists 194–212, 245, 246, 261–74, 275
Dives et Pauper 188–9
Donne, John 196, 201
Dover 284
Drescher family 369
Dürer, Albrecht (*The Triumph of Time*) 48
Durham 247, 249
Dyer, William 110

East Grinstead, Sussex 204
East Hoathly, Sussex 204
East India Company 294, 306
Eastbourne, Sussex 204
Eckebrecht, Phillipp 341
Eden, Richard 311
Edward V, king of England 184
Edward VI, king of England 55, 56, 185
Elizabeth I, queen of England 55, 202, 310
Elkington, John 228
Elmswell, East Yorkshire 217–19

equation of time 26, 256–7, 362, 390
Evelyn, John 194

Fairfax, Ferdinando, Baron Fairfax of Cameron 369
Fine, Oronce 338
Flamsteed, John 336, 342, 345, 351
Forman, Simon 207–9, 265, 268
Foster, Sir John 222
Foster, Samuel 190, 229
Foucault, Michel 60–1
Fox, Hudson 371
Franklin, Benjamin 252
Frisius, Gemma 348, 354
Frobisher, Martin 297

Gailson, Peter 12
Galilei, Galileo 1–8, 35, 84, 334, 341, 342, 344, 347, 355
Gascoigne, Joel 337
Gascoigne, William 365
Goad, John 270
Goddard, John 189
Graham, George 335, 355, 363, 366, 377
Great Horwood, Buckinghamshire 187
Greene, John 207, 209, 210, 211, 212
Greenwich Mean Time 8, 317
Greenwich Observatory 293, 335, 336, 341, 345, 352, 365, 393
Grimsby, Lincolnshire 370–1
Guido of Arrezzo 84
Gunter, Edmund 287, 293, 302, 345

hackney cabs, *see* transport
Hadley, John 289
Hakluyt, Richard 294, 305, 311
Hall, Francis 190
Hall, John 371
Halley, Edmund 191, 258–60, 353, 356, 366, 387, 400
Halton, Buckinghamshire 187
Harrison, Christopher 387
Harrison, Edward 386, 387, 388, 396
Harrison, Henry 360–1, 368, 370, 373, 376, 380, 381, 384
Harrison, James 361–2, 373, 377, 382
Harrison, John 18, 257, 335–6, 348, 355, 359–400, 408–9
Harrison, Thomas 380
Harrison, William 175, 236–7, 257–8, 264, 276
Haskyns, John 189
Harvey, William 191, 253
Hastings, George 229, 246
Haythornthwaite, John 377

heartbeat (pulse), as a form of time
    keeping 2–3
Heidegger, Martin 47, 48–49, 68
Hevelius, Johannes 258
Henry VIII, king of England 213, 310
Herbert, Philip, 4th Earl of Pembroke 271
Hetzner, Paul 85
Hill, Richard 208
Hindley, Henry 256
Hoby, Lady Margaret 205, 206
Holbein, Hans, the Younger (*Thomas More and His Family*) 11
Holder, William 190
Hollar, Wenceslaus 268
Holme Lacy, Hertfordshire 190
Honeywell, William 210
Hooke, Robert 189, 191, 208, 269, 364
Hopetoun House, West Lothian 260
Horrocks, Jeremiah 365
hours
    defining of 25–7, 141, 192–3, 235, 249–50, 274–5, 329–32
    counting 27–8, 37, 40, 198, 203, 329–32
    equal 25–6, 26–7, 40, 141, 198
    in letter-writing 216–7
    minutes 207–8, 250–1, 330, 331
    Nuremberg hours 26, 27
    signalling 42, 140, 158–62, 183, 213
    unequal 25–6, 137
    see also clock time
    see also natural time
Howard, Richard 256
Hunne, Richard 213
Huygens, Christian 1, 35, 84, 161, 198, 257, 354–5

industry
    consumption 79
    factories 46
    Japanese 46–47
    proto-industrialization 46
    time regulation of 46, 50–1
    work habits 43–4
Isidore of Seville 333
Isle of Wight 296
Isles of Scilly 294

Jeake, Samuel 203, 205, 211–12, 246, 265, 269–71, 275, 276, 277, 331
Jesuit astronomers 5
Joan of Arc 186
Johnson, Dr Samuel 311–12
Jupiter (planet), moons of 5, 334, 337, 341, 342, 344, 351

Keswick, Cumbria 268
Kepler, Johannes 341
King, Gregory 364
Kirton, James 377
Knappe, Thomas 129
Knowles, Andrew 252

Lacy, John 365
Langworth, James 377
league (measure of nautical distance) 301–2
Langland, William 10
latitude 312
Layton, Richard 130–1
Leigh, Essex 295
Lewes, Sussex 204
Lincoln 369–70, 371
Lister, Anne 204–6, 211, 223, 261, 263–4, 266–8, 273, 397
Lizard Point, Cornwall 337
Lloyd, Humfrey 258
London 85, 103, 110, 138, 151–2, 161, 211, 213–14, 220, 247, 248, 255, 256, 257, 264, 297, 306, 361, 362, 363, 364, 365, 366, 370, 371, 378, 386, 388, 392
Longframlington, Northumberland 260
longitude 311–12, 335, 336–57, 359–400, 409
Louis XVI, king of France 337, 384

Machiavelli, Niccolò 269
Maskelyne, Neville 345, 346–7, 348, 352, 353
Magdalen College School, Oxford 230–1
Maldon, Essex 186
Malton, North Yorkshire 217
Margaret Beaufort, Countess of Richmond 185
Martin, Roger 55–7
Marvell, Andrew 365
Marx, Karl 48
Mary I, queen of England 55
Mary II, queen of England 223
Maurice of Orange 87
May, Robert 223
Maya/Mayan, *see* clock time
Mayer, Tobias 345
Mercator, Gerard 292, 338, 340
merchant navy 288, 305–9, 313, 314, 315
Millington, Edward 223
minutes, *see* hours
Milton, John 189, 190
military 87–8
Montrose, Angus 260
Moore, Jonas 365–6

Moray, Sir Robert 189
Morris, Claver 397
Morton, John Cardinal 365
Mowbray, John, Duke of Norfolk 184
Muscovy Company 311
music 83–9, 94

Napier, Richard 190, 208, 268, 293, 345
Narbrough, Sir John 292
natural time 24, 33, 80, 280–1, 283–4, 340–8
navigational devices 280–305
   astralobe 288, 289, 302, 307
   back staff 289
   cardes 285
   charts 290–1
   compass 283–4, 288, 297
   compass (pairs of) 293
   cross-staff 288–9, 302, 311
   globes 291–2
   Hadley's quadrant 289
   hourglass 305
   Jacob's staff 289
   landmarks 296
   lead-and-line/sounding line 282–3, 288, 297
   log-and-line 286–8
   logbook/journal 293–5, 308, 317
   manuals 311
   parallel rulers 293
   plotting 293
   quadrant (Davis quadrant) 287, 288, 289, 302, 311
   rotators 288
   rutter 282, 285, 289–90, 297
   sand glass 284–5, 287, 297
   seamarks 296
   sectors 293
   sextant 289, 302, 304, 318, 347, 349, 350
   slide-rule 293
   tables 292–3, 293, 304
   tide-computer 286
   traverse 286
   use of 300–5
Naworth, George 247, 249
Neale, George 257
Neckham, Alexander 283
Neville, Cecily, Duchess of York, mother of Edward IV 185
Newton, Sir Issac 3, 4, 364
Newton Longville, Buckinghamshire 187
Norwich, Norfolk 229
Norwood, Richard 302, 303
Nye, Nathaniel 247, 249

Ogilby, John 364
Oresme, Nicholas 92
Ortelius, Abraham 338, 339–40
Oughtred, William 190
Oxford University 227, 229, 364, 365, 385, 392, 393

Page, Thomas 112–4
Palmer, William 392
Paris 258
Partridge, John 189
Paulus Almanus 253
Pelham, Sir Charles 361, 383, 384
Pellet, Sir Amyce 222
pendulums, *see* clocks
Penn, William 189
Penrith, Cumbria 252
Pepys, Samuel 162, 192, 194, 196–8, 201, 203, 204, 205, 211, 212, 261, 264, 269, 273, 291, 397
Petavius, Domenicus 59
Petty, Sir William 189, 190, 255, 265–6, 276, 353, 364
Phillip Julius, Duke of Stettin-Pomerania 85
Place, Francis 393
planetariums 29
Plymouth 314
Poole, John 247, 248–9
Pope, Alexander 391
Porthouse family 252
Portsmouth 249, 314, 318
post *see* clock time, transportation
post coaches, *see* Bristol, transportation
Potter, Francis 190
Potter, Richard 303
probate inventories 108–9, 125, 163–8, 371–3
Prynne, William 190
Ptolemy 338
Pugh, Robert 189
pulse, *see* heartbeat
Puritans 44, 205, 229, 266, 269

Quare, Danial 257

Rashleighe, John 188
Record, Robert 311
Regiomontanus 344
Reynes, Robert 188, 194, 210
Riccioli, Giovanni Battista 3, 7
Richard of Wallingford 77, 409
Richardson, Richard 369
Richardson, William 369
Roberts, Samuel 171–2

Rooksby, John 371
Royal Navy 288, 291, 305, 309–18, 319, 352, 355
Royal Society 191, 258, 269, 364, 365, 366, 387, 390
Rule of Benedict 46
Rye, Sussex 246, 269

St Albans Abbey 77, 78
Saunderson, Nicholas 385–6, 396
Saxton, Christopher 337
Schortt, Thomas 129–30, 230
Seuse, Heinrich 92
Shakerley, Jeremiah 365
Shakespeare, William 365
Shawell, Leicestershire 228
Sheerness, Kent 314
Short, Thomas 214
Shovell, Sir Cloudesley 294
singing as a form of time keeping 2–3
Sketchley, James 111
  see also Bristol Directory
slave trade 306
Smith, John 366
Sobel, Dava 329, 359–60, 382, 384, 386, 395–7, 408
Solihull, Warwickshire 227, 246
Southampton 214–15, 249
Spengler, Oswald 48
Squire, Joseph 371, 377
Stamford, Lincolnshire 247
Steen, Jan (*The Dissolute Household*) 11
Stephen (Stefan Báthory), king of Poland 265
Stock, Lawrence 77
Stock, Roger 77
Stoke, Cheshire 230
Stonor, family 215–7
Stonor, Oxfordshire 215
Stroud, Gloucestershire 220
Sutton, Robert 369, 378, 383, 389
Swift, Jonathan 391
symbolic interactionism 68

Tamworth, Staffordshire 229, 246
Taylor, F.W. 50–1
Tele, Leonard 295
Tewkesbury, Gloucestershire 220
Thacker, Jeremy 348, 370, 386, 388–400
Tompion, Thomas 377
Thompson, E.P. 15, 24, 43–4, 46–7, 104, 131–2, 135, 147, 185–6, 193, 226–7, 245, 411

Thompson, Sanford 50–51
transport 106, 107, 117–20, 266–8
  postmasters 266
Treaty of Tordesillas 340
Trinity House 296, 311
time, *see also* clock time, natural time
  networks of interaction 71–8
  practices of 68–71
  sense 196
  theories of 65–8
  time-space paths 79–80
  women's relation to 69, 108
time-cues 116–22, 126–7
time discipline 44–7
Tompion, Thomas 257
Towneley, Richard 365
Turner, Thomas 203–4, 261, 262–4, 272
Tusser, Thomas 193
Tuttrell, Thomas 256

van Dyck, Sir Anthony 268
Venus (planet) 353
Villiers, Francis 189
Vivani, Vincenzio 1, 2
vulgaria 230–1

Wagenaer, Lucan Janszoon 290, 296
Walsingham, Sir Francis 222
Ward, John 391, 392, 395, 400
Ward, Seth 191
Washington International Meridian Conference (1884) 52
watches (pocket and otherwise); *see* clocks
Wentworth, Sir Thomas 369
Whitaker, Jeffery
William of Orange, 87
Williams, Raymond 24
Wing, Vincent 247
Winn, Rowland 384
Wolsey, Thomas Cardinal 365
Woolley, Hannah 224
Woolwich 314
women 70–1, 122, 306, 315
Wragby, Yorkshire 360, 385
Wren, Sir Christopher 134 n. 27, 161, 189, 190
Wynne, Henry 256

Yarmouth, Isle of Wight 220
York 256, 366, 371, 377

The manufacturer's authorised representative in the EU for product safety is Oxford University Press España S.A. of el Parque Empresarial San Fernando de Henares, Avenida de Castilla, 2 – 28830 Madrid (www.oup.es/en or product. safety@oup.com). OUP España S.A. also acts as importer into Spain of products made by the manufacturer.

www.ingramcontent.com/pod-product-compliance
Ingram Content Group UK Ltd.
Pitfield, Milton Keynes, MK11 3LW, UK
UKHW022229230426
12048UKWH00016BA/1158